Far-Infrared Properties of Solids

Optical Physics and Engineering

Series Editor: William L. Wolfe
Optical Sciences Center
University of Arizona
Tucson, Arizona

1968:

M. A. Bramson
Infrared Radiation: A Handbook for Applications

1969:

S. Nudelman and S. S. Mitra, Editors
Optical Properties of Solids

1970:

S. S. Mitra and S. Nudelman, Editors
Far-Infrared Properties of Solids

In preparation:

L. Biberman and S. Nudelman, Editors
Photoelectronic Imaging Devices

Vol. 1: Physical Processes and Methods of Analysis

Vol. 2: Devices and Their Evaluation

Far-Infrared Properties of Solids

Proceedings of a NATO Advanced Study Institute,
held in Delft, Netherlands, August 5-23, 1968

Edited by
S. S. Mitra and S. Nudelman
Department of Electrical Engineering
University of Rhode Island
Kingston, Rhode Island

PLENUM PRESS • NEW YORK – LONDON • 1970

Library of Congress Catalog Card Number 78-122627

ISBN-13: 978-1-4684-1865-1 e-ISBN-13: 978-1-4684-1863-7
DOI: 10.1007/978-1-4684-1863-7

Softcover reprint of the hardcover 1st edition 1970

A Division of Plenum Publishing Corporation
227 West 17th Street, New York, N.Y. 10011

United Kingdom edition published by Plenum Press, London
A Division of Plenum Publishing Company, Ltd.
Donington House, 30 Norfolk Street, London W.C.2, England

PREFACE

This book provides an account of modern aspects relating far infrared radiation to properties of solids; it encompasses both theoretical and experimental considerations. Written at the graduate level, it attempts a threefold purpose; an indication of the breadth of the subject, an in-depth examination of important areas, and reference material to complement a text for a course.

The treatment and organization of material here is compatible with a preceding volume of this series on "Optical Properties of Solids." Chapters 1-6 present material concerned principally with experimental considerations necessary to the carrying out of measurements in the far infrared spectral region. They also serve to provide considerable introductory material for the remaining chapters which deal with various areas that offer theoretical treatments utilizing and understanding far infrared properties of solids.

Several lectures presented at the Institute could not be included in this book for two reasons: (i) Final versions of the lecture notes suitable for publication never arrived from several lecturers; (ii) Some materials were deliberately left out from this book as they were also presented at an earlier NATO Institute and form part of a preceding volume edited by us in this series. In particular, it is recommended that Chapters 14 and 15, viz., infrared and Raman spectra due to lattice vibrations by S. S. Mitra and impurity induced lattice absorption by L. Genzel in "Optical Properties of Solids" be read concurrently with the present volume.

We are indebted to many people: the authors for their effort and patience; our host, the Governors and Senate of the Delft Technical University; Dr. H. A. Bakels, Secretary of the University who made necessary arrangements possible and provided the valued assistance of members of his staff, Mr. J. Goedbloed and Mr. C.H. Hoogstraaten; Mr. C. A. Chorus, Head of the Congress Bureau of the Ministry of Education and Sciences; and spontaneous secretarial services from our Mrs. M. Barney, J. Howard, A. Willis and F. Harvey.

Most of all, again as for the earlier Institute on Optical Properties of Solids, our heartfelt thanks go to NATO for their generous support of the Infrared Institute leading to this book. Without the help of Dr H. Arnth-Jensen, Head of the NATO Pure Science Bureau, the Institute could not have taken place. To F. Bentley for his interest, encouragement and solution of difficult administrative problems. Thanks are also due to the U.S. Air Force through the University of Cincinnati for additional support.

S. S. Mitra and S. Nudelman
University of Rhode Island

April 1970

CONTENTS

A COMPLETE SYSTEM FOR INTERFEROMETRIC FOURIER SPECTROSCOPY IN THE FAR-INFRARED

R. C. Milward+

Research & Industrial Instruments Company

Worsley Bridge Road, London SE26. England

ABSTRACT

The properties and theory of the Fourier transform technique as applied to far-infrared spectroscopy, have been briefly reviewed. The main features of a small modular Fourier spectrophotometer system comprised of a Michelson interferometer for the wavenumber region 10 - 500 cm^{-1} and a Lamellar Grating Interferometer for the 3 - 80 cm^{-1} region are described.

Attention has been paid to the operating techniques which are necessary to obtain optimum performance, and the typical applications for which these instruments are used. A number of absorption and reflection spectra (transformed by digital computation or by a special analogue computer) are displayed, to demonstrate the wide range and high resolution capability (0.1 cm^{-1}) of these instruments when used with Golay infrared detectors.

INTRODUCTION

Recent discoveries of a wide range of physical and chemical phenomena occuring in the low wavenumber or far-infrared region of the electromagnetic spectrum, 5 cm^{-1} to 500 cm^{-1} (wavelengths of 2,000 microns to 20 microns), have led to a growing demand for improved instrumentation for optical spectroscopy in this spectral region, which until the last decade had received little attention due to the serious experimental difficulties created by the lack of intense broad band radiant sources and sufficiently sensitive detectors for this characteristically low energy radiation. While

+ Present address: Société Coderg, 92 Clichy, France

moderately powerful "monochromatic" sources of radiation for the millimetre and sub-millimetre region have been recently developed[1], mainly using microwave techniques, they have not yet been fully utilized for continuous broad band spectroscopy.

The demand for improved instrumentation has been best realized in the form of interferometric Fourier spectrometers[2], which under these adverse conditions have yielded superior performances to those of conventional grating spectrometers[3], owing to their ability to record all parts of the spectrum of interest simultaneously and achieve high resolving powers using relatively large source apertures. These factors enable the otherwise low detector signal/noise ratios to be boosted to a level which allows good quality far-infrared spectra to be recorded in a fast and routine manner and permits higher limiting resolutions ($\sim 0.1\ cm^{-1}$) to be achieved in practice. Another useful property of Fourier spectrophotometers is that they allow wide ranges of the far-infrared spectrum to be recorded in a single scan, at constant resolution, and with none of the problems of "false energy" usually associated with conventional infrared grating spectrometers. The principle disadvantage of the interferometric method is, that as the spectrum of interest is not immediately displayed but has to be derived from Fourier analysis of an interference pattern ('interferogram'), considerable delays may be incurred in obtaining the spectrum unless adequate facilities are readily accessible in the form of high-speed digital computers or the special purpose analogue computers which are now available.

Two types of interferometric modulator have been employed for Fourier spectroscopy in the far-infrared, the Michelson interferometer[4], and the Strong and Vanasse-type lamellar grating interferometer[5]. The Michelson interferometer has been the most commonly used instrument as it is of relatively simple construction and has been found to give a good performance over the far-infrared spectral region 10 - 500 cm^{-1} when stretched "Melinex" (polyethylene terephthalate) films of suitable thickness are used as beam splitters[6]. It is not, however, particularly well suited for use at very low wavenumbers ($< 20\ cm^{-1}$), where radiant energy from conventional sources is very weak, as the Michelson configuration allows only half of the total energy available to be utilized and furthermore, multiple interference effects occurring within the thin beam splitter films themselves, bring about an unwanted total modulation of the interferometer transmission function. The variable groove depth lamellar grating interferometer, which acts as a divider of <u>wave front</u>, has received less attention due to its more difficult construction, but has been found to give a better performance[7] than the Michelson interferometer at wavenumbers $< 30\ cm^{-1}$, due to the fact that the lamellar grating facets act as perfect reflectors for the very far-infrared and that the design allows all of the radiant energy to be directed towards the detector. The lamellar grating interferometer suffers the restriction that it is a diffraction limited instrument, which must be made fairly large if it is to enjoy the same wide operating range and high 'throughput'[8] as a Michelson interferometer. If, however, its operating range is restricted to the lowest frequencies only, then it can be reduced to the same compact form as a Michelson interferometer.

The purpose of these lectures will be to describe the concepts which determine the design and operational performance of commercial interferometric spectrometers, of both the Michelson and lamellar grating type, and the applications to which they can be put. These factors will be discussed with specific reference to a small commercial Fourier spectrophotometer system comprised of a Michelson interferometer for the range 10 - 500 cm^{-1}, and a lamellar grating interferometer which has been designed to supplement the Michelson interferometer in the range 3 - 80 cm^{-1}. The performance of a small special purpose on-line analogue computer for performing Fourier transforms will also be illustrated.

THEORY

The theory and practice of Fourier transform spectroscopy has been extensively reviewed in the literature[2,9]; consequently, the discussion here will be briefly restricted to the basic factors which directly affect the operational performance of practical two beam interferometric spectrometers, and should be fully understood by all users of such instruments.

The technique of Fourier transform spectroscopy is based on the fact that in a two beam interferometer such as a Michelson interferometer, the intensity of the central light fringe measured at the output of the device is the Fourier transform of the incident optical power spectrum. The fundamental equation of the method relates the spectral input G (ν) at any frequency ν(cm^{-1}) with the measured interferogram intensity I(x) as,

$$G(\nu) = \int_{x=0}^{x=\infty} \left[I(x) - \tfrac{1}{2}\, I(o) \right] \cdot \mathrm{Cos}(2\pi\nu x) \cdot dx \qquad (1)$$

where x is the optical path difference introduced between the two beams, and I(o) is the interferogram intensity at zero path difference, measured in some arbitrary units. (See Figure 1.)

In practical interferometric spectrometers such as those described in this article, the interferogram intensities are measured at discrete values of optical path difference $n \cdot \Delta x$ up to some maximum value $N \cdot \Delta x$, where n is an integer equal to 0,1,2,3...N,

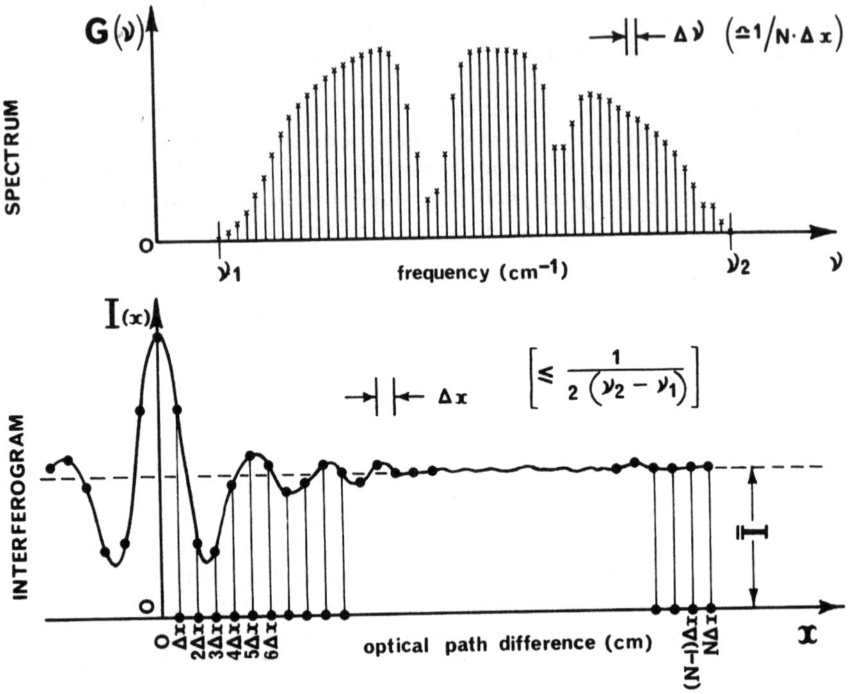

$$G(\nu)_{\nu=\nu_i} = \sum_{n=0}^{n=N} \left[I(n\cdot\Delta x) - \bar{I} \right] \cdot \cos\left(2\pi\, \nu_i \cdot n\,\Delta x\right) \cdot \Delta x$$

$$= \left[I(0) - \bar{I} \right] \cdot \Delta x$$

$$+ \left[I(\Delta x) - \bar{I} \right] \cdot \cos\left(2\pi\, \nu_i \cdot \Delta x\right) \cdot \Delta x$$

$$+ \left[I(2\Delta x) - \bar{I} \right] \cdot \cos\left(2\pi\, \nu_i \cdot 2\Delta x\right) \cdot \Delta x$$

$$+ \cdot \cdot \cdot \cdot \cdot \cdot \cdot \cdot \cdot \cdot \cdot \cdot \cdot \cdot$$

$$+ \left[I(N\cdot\Delta x) - \bar{I} \right] \cdot \cos\left(2\pi\, \nu_i \cdot N \cdot \Delta x\right) \cdot \Delta x$$

Figure 1. Computation of spectrum from measured interferogram. In practical instruments the average interferogram intensity $\bar{I}$ is $> \frac{1}{2}I(o)$; the ratio $\bar{I}(o)/2I$ provides a bulk measure of the interferometer's modulation efficiency.

and Δx is the basic interferogram sampling interval. To obtain unambiguous information, the value of Δx must be chosen so that it obeys the relation

$$\Delta x \leq \frac{1}{2\ (\nu_2 - \nu_1)} \tag{2}$$

where $(\nu_2 - \nu_1)$ defines the total band width of optical frequencies received by the detector. In far-infrared interferometers, low-pass filters are usually employed to restrict the frequency range of interest to $0 - \nu max.$ where $\nu max.$ is the upper cut-off frequency, so that

$$\Delta x \leq \frac{1}{2\ \nu max.} \tag{3}$$

(in physical terms this means that the amplitude of the shortest wavelength present must be measured at least <u>once</u> every half cycle).

For practical purposes, equation (1) then becomes approximated to a summation,

$$\underset{\nu=\nu_1}{G(\nu)} = \sum_{n=o}^{n=N} \left[I(n\cdot\Delta x) - \tfrac{1}{2}I(o) \right] \cdot Cos(2\pi\nu_1 n\Delta x)\cdot\Delta x \tag{4}$$

The interferogram intensities are converted to numbers, and a spectrum of frequencies is computed digitally, using equation (4). The value of $\frac{1}{2}I(o)$ may be obtained from the asymptotic value of the interferogram trace at large path differences, and must be subtracted from all the values of $I(x)$ before Fourier transform computations are allowed to commence. <u>NOTE</u>: that the computed spectral intensities $G(\nu)$ are directly proportional to the measured amplitudes of the oscillatory interferogram function $I(n\cdot\Delta x) - \frac{1}{2}I(o)$, and that the first (n=o) value of the summation ($\simeq\frac{1}{2}I(o)$) makes a constant contribution to the whole spectrum of frequencies and is independent of ν. Errors in measuring the true value of $I(o)$ thus result in baseline shifts in the computed spectra, which can reduce the accuracy of absolute transmittance or reflectance measurements. Since in practice interferograms rarely have a sampling point exactly at the zero path position, then care must be taken to correct measured interferograms for these slight displacements or "phase errors", which if uncorrected would cause distortions in the computed spectra, which become particularly severe at higher wavenumbers.

The optical resolution $\Delta\nu$ of an interferometric spectrometer is related to the maximum path difference $D(=N\cdot\Delta x)$ of the interferogram trace by the instrumental scanning function or "slit" function, which is of the form $2D\left[\sin 2\pi(\nu-\nu^1)D/2\pi(\nu-\nu^1)D\right]$, and has a peak width at half its height at centre frequency ν^1, of

$(\nu - \nu^1) \simeq 0.7/D$, so that the smallest interval resolved is approximately

$$\Delta\nu \simeq 0.7/D \tag{5}$$

In practice, however, the subsidiary side lobes of this scanning function can cause complications in the interpretation of spectra and it is desirable to suppress these subsidiary maxima and minima. This can be achieved by mathematically weighting or "apodising" the recorded interferogram values $I(n\cdot\Delta x)-\frac{1}{2}I(o)$ so that the amplitude modulation of the interferogram goes to zero at maximum path difference $N\cdot\Delta x$. Simple apodising functions of the form $(1 - n/N)$ or $\frac{1}{2}$ $(1 + \mathrm{Cos}\ (\pi n/N)$ have been commonly used for this purpose. Apodisation has the effect of smoothing the instrumental scanning function at the expense of increasing its half width by ~50 percent. The optical resolution now becomes

$$\Delta\nu = 1/D \tag{6}$$

The resolving power $R = (\nu/\Delta\nu)$ of an interferometric spectrometer is also limited by the angular beam spread of off-axis rays brought about by imperfect collimation of the finite source aperture. If the source subtends a solid angle Ω at the collimating mirror, then the maximum[10] value of R is given by

$$R = \nu/\Delta\nu = 2\pi/\Omega \simeq 8\ (f/d)^2 \tag{7}$$

where d is the source diameter, and f the collimator focal length. Thus either equations (5), (6) or (7) can limit the optimum resolution obtainable in practice. The spread of off-axis rays also leads to an average separation of wavefronts through the interferometer which is slightly shorter than that measured for the case of an axial ray, and causes all the frequencies in the computed spectrum to be reduced by a factor ~ $(1 - \Omega/4\pi)$. In order to obtain the true frequencies ν from the computed frequencies ν_{obs}, the simple correction

$$\nu = \nu_{obs}\ (1 + \frac{\Omega}{4\pi}) \tag{8}$$

can be incorporated into the spectrum computations. In practical instruments, however, this correction factor is extremely small, ~ $10^{-4}\ \nu_{obs}$, and only becomes significant when high resolution spectra are computed at higher wavenumbers. (Equation (8) is incorrectly formulated in references 1 and 7).

INSTRUMENTATION

The requirements of a laboratory interferometric spectrometer are that it should cover the widest possible spectral range with high resolution and/or high photometric accuracy, in convenient times of operation. In order for this to be achieved, the following standards should be met by the instrument. First, it should be mechanically and optically accurate to within a fraction of the shortest wavelength to be measured; second, it should perform with high efficiency as an interferometric modulator; and third, it should allow the maximum amount of optical flux to be transferred from the source to the detector, to boost the low signal/noise at the detector. Last of all, it should be of convenient size and construction and provide adequate facilities for investigating a wide variety of samples under varying environmental conditions of temperature, pressure, applied magnetic field etc.. Bearing these factors in mind, the Michelson and lamellar grating interferometers described in this article, were designed with a number of important common features, which may be summarized as follows: Both instruments make use of small (8 cm diameter optics) all-reflecting optical systems of f/1.5 - f/2 aperture, for high luminosity. The maximum throughputs of 0.20 and 0.12 cm^2 steradians for the Michelson and Lamellar grating interferometers are fairly well matched with the acceptance figures for commercial Golay detectors which vary between 0.06 and 0.36 cm^2 steradians, according to the size of detector window[11]. The sample optics of 12 mm image diameter and f/2 aperture beam, facilitate the study of very small solid or liquid samples. Each instrument uses a high pressure mercury lamp[12] as a broadband source of radiation for the far-infrared. This type of lamp behaves as a black body emitter of $T \simeq 3{,}500°K$ at frequencies $<25\ cm^{-1}$; at higher frequencies, the emission is characterized by the combined effect of the mercury plasma and the hot lamp envelope. The mercury lamp is fitted with a variable aperture stop of 10,5 and 3 mm diameter, which allows theoretical resolving powers (see equation 7) of approximately 10^3, 4×10^3, and 10^4 respectively. The largest value of Ω is $\sim 5 \times 10^{-3}$ steradians, which makes the frequency corrections due to finite aperture as given by equation (8) to be $\leq 4 \times 10^{-4}\nu$. Unicam Golay cells are used as detectors. Although these devices suffer from limited ultimate sensitivities (NEP $\sim 2 \times 10^{-10}$ watt), slow response times ($\sim$ 15 m sec) and small useable apertures (3-5 mm diameter, 60° - 90° acceptance angles), they are nevertheless very convenient to use in laboratory instruments. Their large outputs (responsivity $> 10^5$ volts/watt incident) and good linearity over a wide range of input signals (typically < 1% over 100:1 intensity range) make them particularly well suited to the requirements of commercial multiplex spectroscopy. 3 mm diameter diamond-windowed versions are normally used to cover the spectral range up to 500 cm^{-1} and 5 mm quartz-windowed versions to cover the range 3 - 200 cm^{-1}. It is necessary that these windows should be sufficiently bevelled to avoid unwanted effects of 'channel spectra' (the quartz jacket of the mercury lamp should be dimpled for the same reason).

Both instruments are of compact modular construction, which allows considerable flexibility of operation. External sources of radiation or other types of detector can be employed with only minor modifications to the system. A number of interchangeable accessory modules have been devised which allow absorption or reflection spectra to be obtained, at variable temperatures, or over long optical paths. The whole assembly of each instrument can easily be evacuated to <0.02 mm Hg to remove effects of atmospheric water vapour absorption. The construction of the Michelson interferometer makes provision for its use in complex refractive index studies, where the dispersive medium of interest is directly interposed in one arm of the interferometer.

The Michelson and lamellar grating interferometer are discussed in further detail below.

The Michelson Interferometer

This instrument is as shown in Figure 2. Radiation from a water cooled high pressure mercury discharge lamp of effective area ~0.8 cm^2 is collected and collimated by an f/1.5 off-axis paraboloid mirror. The collimated beam is divided by a stretched semi-reflecting semi-transmitting beam splitter film at 45° incidence, and is subsequently recombined after reflection from the two plane Michelson mirrors. The recombined beam is then brought to a focus by a condensing system made up of an f/2 off-axis paraboloid mirror and plane mirror combination. The image (largest size ~ 12 mm diameter) is formed at the centre of a sample chamber,

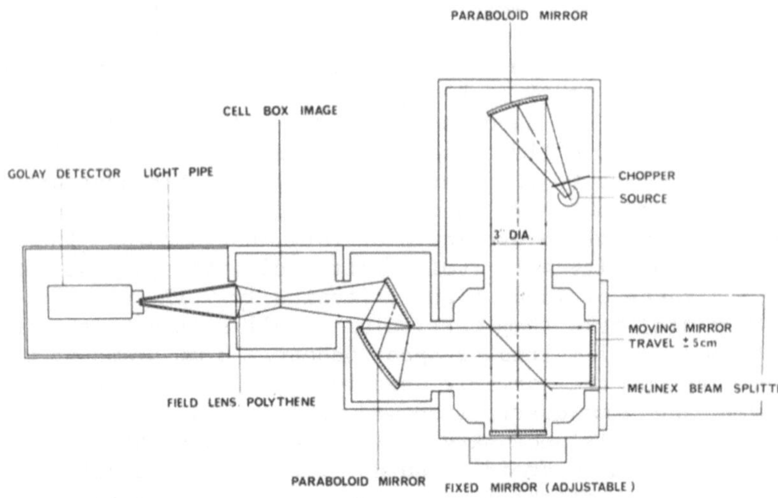

Figure 2. Ray diagram for the RIIC Michelson interferometer FS-720. (Details of the moving mirror drive unit are not shown). For low wavenumber applications (<200 cm^{-1}) a reduction in overall size may be gained by the use of a polypropylene lens in place of the off-axis paraboloid-plane mirror combination used as a beam condenser.

which is large enough to accommodate solid and liquid sample holders, and a 10 cm cell, for gaseous absorption measurements. The divergent beam from the sample chamber is then collected at the mouth of a tapered copper light cone, which condenses the beam by internal reflection[13] to a focus at the window of a Golay detector. A black polyethylene lens is used in conjunction with the tapered light pipe for the purpose of focussing, and also serving as a filter to attenuate visible and near-infrared radiation from the source to a negligible level.

Far-infrared interferograms are recorded in a continuous 'aperiodic' scanning mode by a uni-directional displacement of one of the Michelson mirrors. This is achieved by mounting the mirror on a close fitting piston and cylinder arrangement which is driven from a synchronous motor via a gear box and lead screw. A subsidiary Moire fringe system is used for measuring mirror displacement, in units of 4 microns and higher multiples, so that interferogram optical path differences (equals twice the actual mirror displacement) can be measured to a high degree of precision. This continuous-type of drive unit has a maximum displacement of 10 cm, which allows a maximum instrumental resolution of 0.05 cm^{-1}, or 0.10 cm^{-1}, according to whether single-sided or double sided interferograms have been scanned (see equation (6)). In order to achieve maximum modulation efficiency of the interferometer up to the highest frequencies of operation, it is necessary that the moving mirror does not vary in tilt or pitch by more than one or two seconds of arc[14] over most of its travel, which calls for a high degree of precision in the construction of the drive unit.

An alternative stepping drive unit, used with the lamellar grating interferometer and described in the following section can be used with the Michelson interferometer. This unit however, is only considered fully mechanically accurate for use at frequencies ≤ 300 cm^{-1}.

The fixed Michelson mirror has a simple system of adjustment, which remains stable under all instrumental conditions. The operational efficiency of the Michelson interferometer depends upon the choice of beam splitter film, which for maximum efficiency E, should have transmittance T and reflectance R both equal to 0.5, so that

$$E = T \cdot R = 0.25 \qquad (9)$$

In practice, however, T and R are rarely equal to 0.5, and furthermore, multiple reflections occurring within the thin beam splitter film itself produce total interference[6] in the reflected and transmitted beams, which cause the overall efficiency of the beam splitter to vary with wavenumber ν as

$$E \propto \sin^2 (2\pi\nu\, d') \qquad (10)$$

where d' is the apparent film thickness, which for a 45° beam splitter film of real thickness d and refractive index n is given by

$$d' = d \sqrt{(n^2 - \tfrac{1}{2})} \tag{11}$$

The instrument described here uses 'Melinex' films of refractive index $n \simeq 1.8$, and varying thickness as shown in Table 1, which allow either a wide range of frequencies to be covered in the first order pass-band of the beam splitter or much narrower ranges of the spectrum to be covered with greater efficiency at very low wave-numbers[6].

TABLE 1

USEFUL BEAM SPLITTER - FILTER COMBINATIONS

Frequency Range (cm^{-1})	Beam splitter Thickness (10^{-3}mm)	Low-pass Filter(s)
80 - 600	4	None
40 - 480	6	None
20 - 220	12	2 mm crystal quartz
10 - 100	25	3-5 mm crystal quartz and/ or 40 lines/cm grooved transmission filter.
5 - 50	50	5 mm crystal quartz + 24 lines/cm grooved trans-mission filter, or 1 mm lithium fluoride.

An improved type of beam splitter mount has been devised for the Michelson interferometer, which allows premounted films of varying thickness to be interchanged in a matter of seconds without causing any change in the optical alignment of the instrument. The interferometer is normally aligned using the thinnest beam splitter film, which produces the sharpest and most critical interferogram for alignment purposes. Since the beam splitters are used at 45° incidence, relatively close to the 'Brewster angle' for Melinex ($\simeq 60°$), the instrumental radiation is about 60% polarized in the plane of the beam splitter. One extra polarizer[15] usually suffices to increase the degree of beam polarization to ~99%, when absorption or reflection experiments with polarized light are required.

The Lamellar Grating Interferometer

The diffraction theory and principles of operation of the lamellar grating interferometer have been discussed in some detail by Strong and Vanasse[5], and Vanasse and Sakai[2]. These authors have shown that in order for the device to operate with maximum efficiency as an interferometric modulator, the lamellar grating must be illuminated with well collimated radiation at normal incidence, and only the zero'th order of diffracted radiation from the grating must be allowed to reach the detector. This is arranged by incorporating the lamellar grating in a simple spectrometer system with collimating and condensing optics, and entrance and exit apertures which define the acceptable limits of angular beam spread. For an instrument, such as the one described in this article, which uses an auto-collimating mirror of focal length 'f', and has an entrance aperture size 'w' the highest wavenumber radiation ν_H which can be accepted in zero order only is dependent upon the lamellar grating 'constant' (twice the width of an individual grating facet) 'a' by the relation

$$\nu_H \simeq f/a.w \qquad (12)$$

Thus for a given slit width and grating constant, the high frequency limit of efficient operation is directly proportional to the collimator focal length. The lamellar grating interferometer described here has been specifically designed to work over a restricted region of the low wavenumber spectrum ($\nu_H < 20$ cm^{-1}), so that the collimator focal length and overall size of optics can be made small, and compatible with those of the Michelson interferometer described in the previous section.

The optical ray diagram of the lamellar grating interferometer is shown in Figure 3. 'Chopped' radiation (12.5 Hz) from the

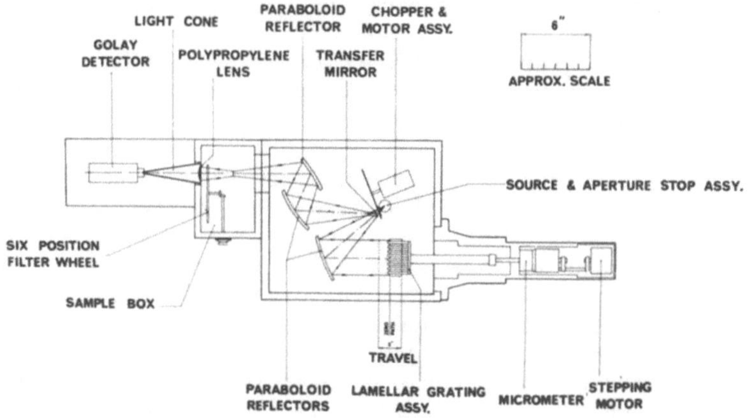

Figure 3. Ray diagram for the RIIC Lamellar grating interferometer LR-100 described in the text. The internal baffles required to prevent 'stray' radiation from reaching the detector are not shown, and the vertical separation of the source aperture and small transfer mirror is not indicated.

source is collimated by an f/2 off-axis paraboloid reflector and falls at normal incidence to the lamellar grating in the azimuth perpendicular to the grooves, but at a slight angle[16] from normal incidence in the azimuth defined by the groove length. This arrangement allows zero order radiation diffracted at normal incidence from the grating to be collected and brought to a focus, after reflection from a small transfer mirror positioned directly beneath and slightly forward of the source aperture. This mirror acts as an exit aperture and directs the radiation towards the sample chamber, via a condensing system of two-off-axis paraboloid reflectors. After the beam has been brought to an intermediate focus (largest image size $\sim$ 12 mm diameter for a 10 mm diameter source aperture) at the centre of the sample chamber, it diverges and is then condensed by internal reflection in an electroformed polished copper cone to the window of a Golay detector. The beam image size and intensity can be varied by means of a variable aperture stop (3 mm, 5 mm, and 10 mm, diameter) in front of the mercury lamp. Considerable filtering is essential with this instrument to remove all radiation of frequency > 100 cm^{-1}. This is achieved by a 3 - 5 mm wedged plate of crystal quartz, a plano-convex black polyethylene (high density polyethylene + 0.1% carbon) lens used in conjunction with the condensing cone, and grooved polypropylene grating transmission filters of the type first developed by Moller and McKnight[17]. Five different filters of rulings 40, 24, 16, 12, and 8 lines/cm, which have "cut-offs" at approximately 100, 60, 40, 30 and 20 cm^{-1} respectively are mounted on an externally actuated rotating wheel positioned between the sample focus and condensing cone. This facility allows filters (or solid samples) to be easily changed or the beam blanked off with a shutter, without releasing the instrumental vacuum.

The lamellar grating is comprised of 17 intermeshing aluminum plates (8 moving in 9 stationary) of thickness 0.47 cm whose front faces have been ground and lapped flat to within $\sim$ 2 microns over their whole length and are mounted so that these faces can be aligned and kept parallel to within a few seconds of arc. The individual plate thickness which controls the diffraction properties of the lamellar grating has been chosen so that it is large enough to allow the device to work accurately with unpolarized radiation[18] to wavenumbers as low as 3 cm^{-1}, the practical limit imposed by the size of the detector window and very poor signal/noise ratios, but not too large to unduly restrict the higher wavenumber limit to which the instrument is capable of performing with maximum efficiency. As shown in equation (12) this limit is determined by the 'grating constant', the size of source aperture and collimator focal length, and is estimated to be $\simeq$ 18 cm^{-1} for this instrument. At wavenumbers > 18 cm^{-1}, higher odd-order diffracted radiation becomes detected with the normally reflected zero order radiation and causes a loss of observed modulation of the zero order which worsens until the upper useable limit of the instrument is reached.

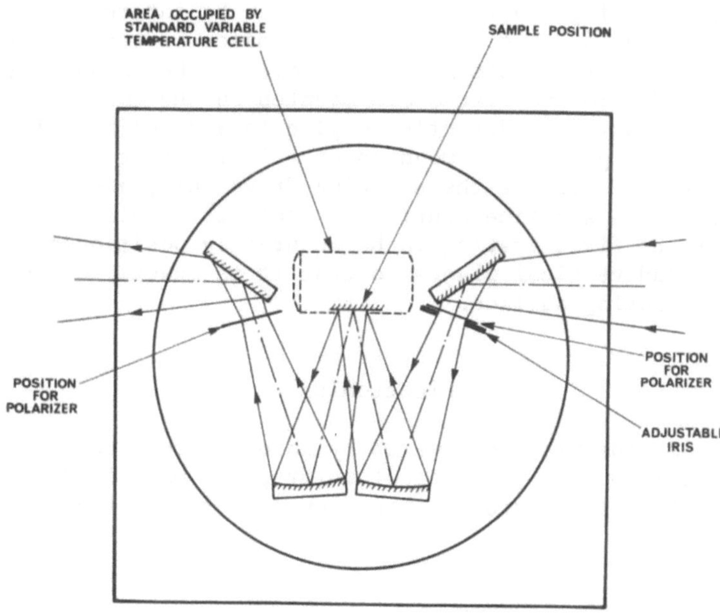

Figure 4. Ray diagram of the specular reflectance unit ($\simeq 12°$ angle of incidence) developed for use with the Michelson and lamellar grating interferometers. (This unit replaces the standard sample chambers shown in Figures 2 and 3).

This decrease in effeciency is to some extent compensated for by the sharp increase in source output at higher wavenumbers, so that the instrument can be used up to 70 - 80 cm^{-1} to provide a generous overlap with the Michelson interferometer's lowest range of operation. The instrumental performance suffers at these higher frequencies however, due to a high content of interferometrically unmodulated radiation falling upon the detector which reduces the dynamic range of the detection system. The radiation reflected from the lamellar grating is found to be unpolarized, even at the lowest wavenumbers. Two sets of interchangeable lamellar units are normally used with the instrument. One set comprises a moving grating of length 5 cm and a fixed grating of length 2.5 cm. This allows the groove depth to be varied ± 2.5 cm so that interferograms of ± 5 cm optical path difference can be scanned. The other grating comprises a moving and fixed set of plates both of length 5 cm, which allows the groove depth to be varied from zero to 5 cm in one direction, and allow single-sided interferograms of 10 cm optical path difference to be scanned. An alternative set of lamellar gratings of identical grating area and path length but a different grating constant, 0.69 cm, can be interchanged and used with the instrument when improved performance is required in the range 40 - 80 cm^{-1}. These gratings increase the value of ν_H to $\simeq$ 27 cm^{-1}. The moving plates are mounted on the end of a non-rotating piston which is directly coupled to a stepping drive mechanism. The fixed plates are mounted on a leaf spring framework which is bolted to the body of the stepping drive unit, and can be adjusted by coarse and fine cantilever controls. This allows the surface of the fixed grating to be accurately positioned in the horizontal and vertical planes so that it is exactly coplanar with the moving grating surface.

Electronics

Both instruments are powered by an electronic unit especially developed for Fourier spectrophotometers, the main features of which are shown in Figure 5. The sequence of operations is as follows.

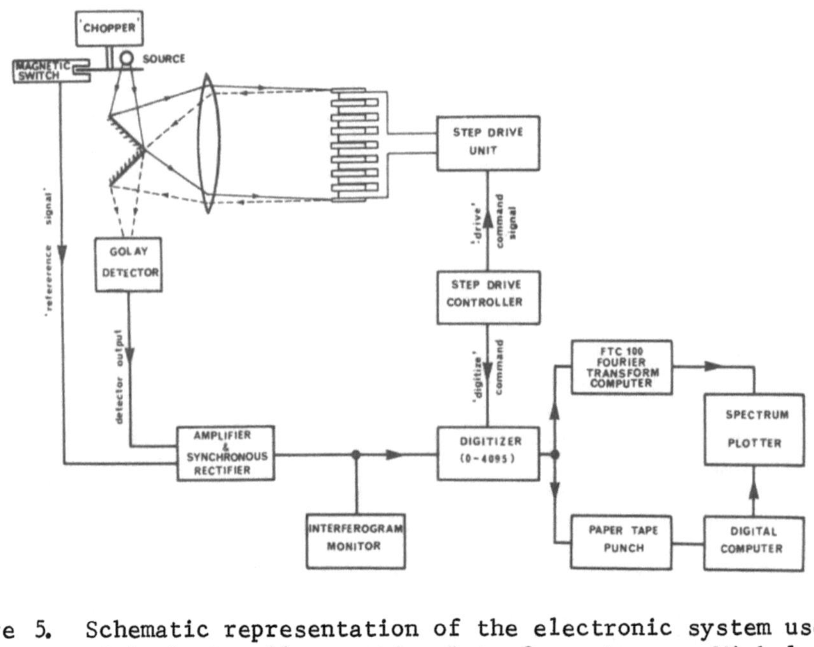

Figure 5. Schematic representation of the electronic system used with the lamellar grating interferometer, or Michelson interferometer with step drive. The system is identical to that used with the standard Michelson interferometer described in the text, except that the step drive control mechanism is replaced by a continuous drive unit, with a Moire fringe system for triggering the digitizer. (Power Supplies etc., are not shown.)

The modulated output signal from the Golay is fed directly to a low-noise narrow band amplifier, with calibrated gain continuously variable from 10 db to 90 db. The amplified detector signal is then synchronously rectified and smoothed to give a d.c. electrical voltage, which is proportional to the intensity of the optical interferogram signal received at the detector. This electrical output is then fed to a digitizer which converts the signal into 12-bit binary numbers (maximum resolution 1 part in 4096) for the purpose of Fourier transform computation. The digitizer is triggered at the appropriate intervals of optical path difference by command pulses sent either from the Moire fringe system for the continuous drive unit, or from the step drive control unit at the end of successive drive pause periods. The digitized interferogram values are then normally transferred to punched paper tape, to be read into a digital computer. Alternatively, when spectra of only moderate resolving power are required, it is more convenient to feed the digitized interferogram values into a small online hybrid Fourier transform computer[19], which has the facility for digitally storing two 1024 word interferograms and plotting out their spectra within 6-24 minutes, in either single beam or double beam ratio form with an amplitude accuracy ~ 1%. The maximum resolution of this device is restricted by the size of the computer memory and the resolution of the recording wave analyzer used to plot out spectra, to 1 part in 200 of the optical band width. The system can be used to plot-high resolution spectra (< 0.5 cm^{-1}) when the optical band widths are made small (< 100 cm^{-1}), so that large interferogram sampling intervals can be used, to effectively increase the path length of interferograms that can be stored by the computer

In order for optical interferograms to be accurately recorded, it is necessary that the output of the detector-amplifier-digitizer system should vary linearly with the strength of the optical signal, and that the amplitude resolution of the digitizer be as high as possible. The linearity of output of the detection system described here was ~ 0.5 percent over a 100:1 range of optical intensities, when care was taken to operate the Golay detector and amplifier at signal levels well below their limits of saturation[20]. The 12-bit binary accuracy (1 in 4096) of the digitizer provides a fair compromise between considerations of amplitude resolution, and the typical signal/noise ratios encountered in practice, which are usually $\leq 2{,}000$. An electrical backing-off facility for the rectified amplifier output allows much of the unimportant background level of interferogram signals to be subtracted and the remaining important oscillatory features to be expanded to occupy the entire range of the digitizer, so that interferogram values can be digitized with maximum accuracy.

PERFORMANCE

Equation (1) shows that the intensities in the computed wavenumber spectrum are directly proportional to the amplitudes of the decayed oscillatory interferogram function ($I(x) - \frac{1}{2} I(o)$) In order to record spectra at either very high resolution and/or photometric accuracy, it is therefore necessary that the amplitudes of oscillation of the interferogram function are amplified and digitized as accurately as possible, and the interferogram signal/noise values are high enough so that meaningful features are distinguishable above the noise level. When strongly absorbing samples are investigated, it is often necessary to employ long (>2 seconds) electronic time constants to maintain the signal/noise at the required level, at the expense of scanning interferograms at a slower rate. For accurate spectral intensity measurements[21], it is necessary that the interferogram values close to I(o) be recorded with the highest precision. This can only be achieved if interferograms are scanned slowly enough to avoid distortions due to the inherent time constant of the electronic system. The speed of scanning therefore depends upon the sharpness of the interferogram pattern, and the electronic time constant. In practice, it is found that interferograms can usually be scanned at speeds in the range 2.5 - 25 microns/sec. of optical path difference for the Michelson interferometer and 10 - 40 microns/sec. for the lamellar grating interferometer. The stepping drive unit offers the advantage that the pause period can readily be adjusted during the recording of an interferogram so that the central maximum fringe can be scanned at a sufficiently slow rate, while the portions away from zero path which exhibit much less amplitude modulation, can be scanned at a faster rate, to minimize the total scan period.

In order to achieve the highest degree of reproducibility in the computed spectra, optical filtering should be employed wherever possible so that the observed interferogram is constituted solely of the frequencies of interest, i.e. when the number of spectral elements are reduced to a minimum[22]. The normal practice with a Michelson interferometer is to use a beam splitter of the appropriate thickness to maximize the spectral energy in the wavenumber region of interest, and remove unwanted radiation of higher frequencies by means of a suitable low-pass filter (see Table 1). For the purpose of performing absolute transmittance or reflectance measurements with a high degree of accuracy, it is necessary to satisfy a number of conditions. First, a suitable shape of spectral background should be arranged (by choice of the appropriate beam splitters and filters, etc.), against which transmitted or reflected spectra can be ratio-ed, in order to get the most reproducible results. Secondly, since existing commercial far-infrared interferometers work only in the single beam mode, great care must be taken to ensure that all interferograms are recorded under

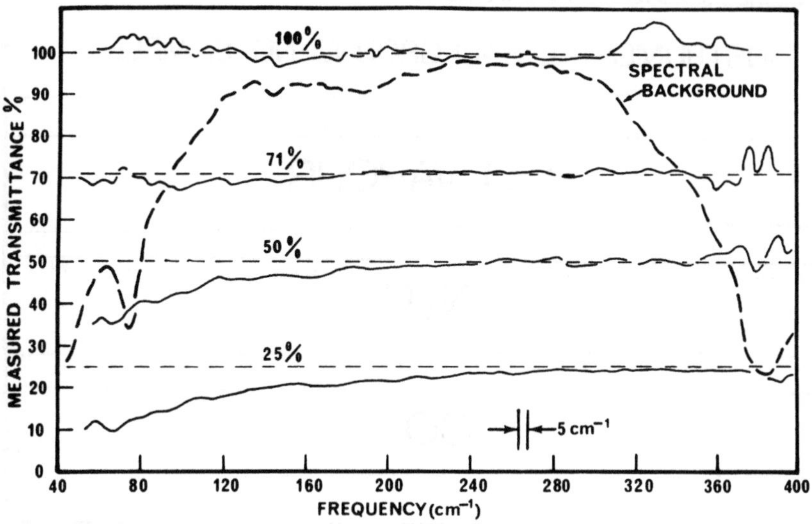

Figure 6. Typical results of a linearity disc check upon the outputs of the FS-720 Michelson interferometer - FTC-100 analogue computer system. The plotted transmission spectra obtained over a wide spectral range are compared with the theoretical values. The 100% curve was obtained by ratioing two consecutively recorded 'background' spectra against each other. The shape of the instrumental background (obtained with a 6 micron 'Melinex' beam splitter) is also shown. The results for 50% and 25% transmission show considerable deviation at the lowest wavenumbers, which is attributed to the effect which instrumental drift during the scanning of their interferograms has had upon the intensities in the low wavenumber region of their spectra.

exactly the same conditions of instrumental vacuum, source brightness and detector responsivity (commercial Golay detectors need $\sim$ 1 hour to attain reasonable stability after connection) if intensity errors are to be minimized. Thirdly, reproducible results will only be obtained for those regions of the spectrum where the signal/noise is high in the single-beam spectra. Under typical operating conditions where the above criteria are satisfied, photometric accuracies $\sim$ 2% can be obtained for absolute transmittance or reflectance measurements obtained from interferograms composed of a large number of spectral elements ($\sim$ 100). While it is necessary to resort to digital computation for the most exact results, the double-beam accuracy of the FTC-100 analogue computer is fully adequate for most laboratory purposes (see Figure 6). The highest

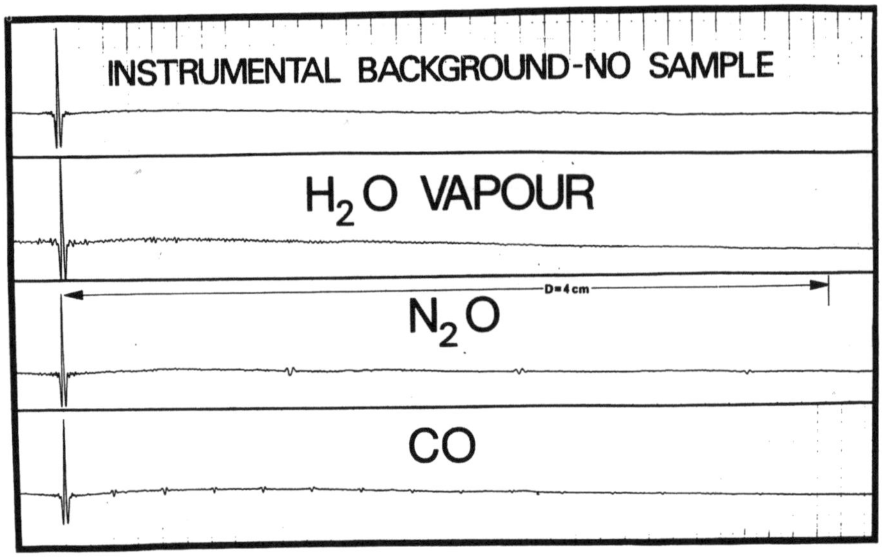

Figure 7. Interferogram traces recorded up to 4 cm optical path difference with the lamellar grating interferometer, with the instrument evacuated or filled to a low pressure (1 - 10 cm Hg) with H_2O, N_2O and CO. The interferogram trace for water vapour reflects the complicated nature of the H_2O pure rotational spectrum, but the interferograms for N_2O and CO exhibit clear 'signatures' spaced at intervals of optical path difference $(\Delta\nu ROT)^{-1}$, where $\Delta\nu$ ROT (cm^{-1}) is the average spacing of rotational energy levels of the 'symmetric-top' N_2O and CO rotators (0.84 cm^{-1} and 3.84 cm^{-1} for N_2O and CO respectively). The interferogram traces show slight effects of shadowing, which would be transformed into ultra-low frequencies (< 1 cm^{-1}) in the final spectra and are of no consequence.

practical resolution that can be attained with a far-infrared interferometric spectrometer is usually restricted by considerations of signal/noise, since no advantages are gained by scanning interferograms beyond a point where any meaningful features which contain spectral information are indistinguishable from the noise level. In general, the interferogram signal/noise values should be much higher than the total number of spectral elements to be resolved in a single scan, e.g. to achieve 0.1 cm^{-1} resolution over an optical band width of 200 cm^{-1}, the interferogram signal/noise should be $> 2,000$. The highest practical resolutions are thus usually achieved in fairly transparent media such as gases, where high signal/noise ratios can be maintained. (See for example Figure 7).

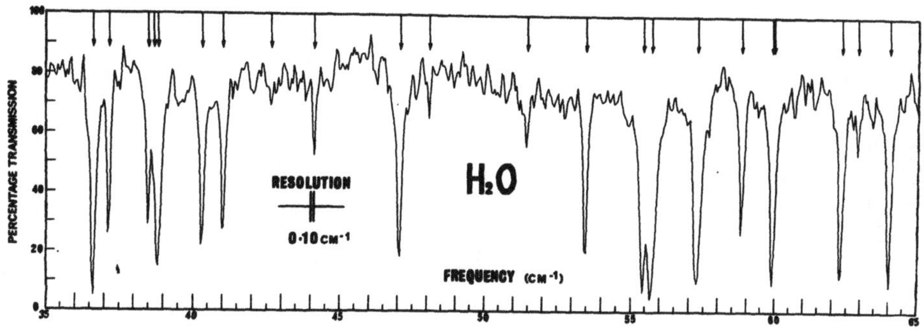

Figure 8. A digitally computed portion of the water vapour spectrum, recorded at high resolution with the FS-720 Michelson interferometer under conditions of 13 mm Hg pressure and 1 metre absorbing path. The measured frequencies (calculated at 0.05 cm^{-1} intervals and joined up by a digital plotter) should be increased by $4 \times 10^{-4} \nu$ to correct for effects of finite aperture. The vertical arrows mark the measured frequencies of H_2O pure rotational absorption lines as listed in reference 23. The splitting of the doublet at $\sim$55.5 cm^{-1} is 0.30 cm^{-1}. The 'sample' and 'background' interferograms of $\pm$ 10 cm optical path difference were each scanned in $\sim$168 minutes, using an electronic time constant of 0.5 second. The spectral bandwidth was restricted to the range 0 - 100 cm^{-1} by use of a 25 micron beam splitter and suitable low-pass filters. The spectrum computations were performed on an I.B.M. 1140 computer using a conventional Fourier transform program. The total computation time for transforming, ratioing and plotting two spectra of 600 frequency points from interferograms of $\pm$ 3,000 input points was $\sim$300 minutes.

Following common practice, the resolution capability of the Michelson interferometer has been demonstrated from measurements of the pure rotational spectrum of water vapour, and HCl vapour.

A low wavenumber portion of the water vapour transmission spectrum recorded with the Michelson interferometer is shown in Figure 8. The frequencies and relative strengths of all the

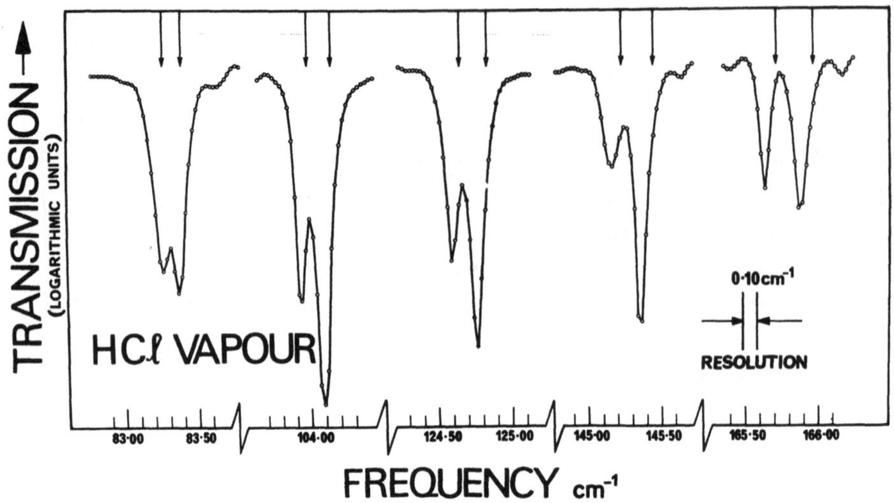

Figure 9. Details from the pure rotational spectrum of HCl vapour, recorded at high resolution with the FS-720 Michelson interferometer under conditions of 2 mm Hg vapour pressure - 1 meter absorbing path. The calculated frequencies (shown by the small circles) should be increased by $4 \times 10^{-4}\nu$ to correct for the effects of finite aperture. The vertical arrows mark the calculated frequencies of pure rotational transitions in HCl^{35} and HCl^{37} as listed in reference 1 (p. 105). The calculated splittings of the doublets at 83, 104, 125, 145 and 166 cm^{-1} are 0.13, 0.16, 0.19, 0.22 and 0.25 cm^{-1} respectively. The 'sample' and 'background' interferograms of $\pm$ 10 cm path difference were each scanned in $\sim$ 340 minutes, using an electronic time constant of 0.5 sec. The spectrum computations were performed on a CDC-3200 computer, using a "fast" Fourier transform program. The total computation time for transforming, ratioing and plotting two spectra of 6,250 frequency points from double sided interferograms of 12,000 input points was $\sim$ 12 minutes.

absorption lines are in very close agreement with those shown in the most recent water vapour spectra obtained by Richards[7], and Hall and Dowling[23], who used lamellar grating interferometers[7,24] in conjunction with liquid helium cooled doped-germanium bolometers. The weak $5_{41} - 6_{34}$ H_2O rotational transition just resolved in the spectra of Hall and Dowling[23], is only evident in Figure 8 as a slight shoulder on the line centered at 38.8 cm^{-1}. The symmetrical transitions at 55.41 and 55.70 cm^{-1} have been clearly resolved. From examination of the 'unoccupied' spectral regions between absorption lines, the average signal/noise in the double beam absorption spectrum may be estimated at ~ 15. The very narrow doublet centered at 59.9 cm^{-1} shown in the results of Hall & Dowling[23], and Hard & Lord[3] to be split $\simeq$0.07 cm^{-1} has not been resolved. While this small splitting is less than the nominal resolution limit for the measurements, it is also probable that the 13 mm Hg vapour pressure used in these measurements has caused the doublet to be pressure broadened beyond the limits of resolution.

The pure rotational spectrum of HCl vapour is characterized by a series of regularly spaced absorption lines at intervals of approximately 20.8 cm^{-1}. Due to the natural abundance of both the Cl^{35} and Cl^{37} isotopes, each rotational energy level should be split, due to the difference in rotational moments of inertia of HCl^{35} and HCl^{37}. Figure 9 shows details of the HCl pure rotational absorption lines between 83 and 166 cm^{-1} measured with the Michelson interferometer, which exhibit splittings of magnitude 0.13 - 0.25 cm^{-1} which have been clearly resolved. The positions and separations of the HCl doublets shown in Figure 9 are in extremely close agreement with those calculated from infrared and microwave data. The relative absorption strengths of each doublet are in agreement with prediction, and the absorption 'base-line' has remained constant. The average signal/noise across the whole spectrum (not shown) is in the range 10-15.

The high luminosity of the Michelson interferometer allows the transmission spectra of strongly absorbing solids and liquids, and reflection spectra of solids to be routinely obtained with good resolutions ($\geq$ 0.5 cm^{-1}), in convenient times of observation, over most of its range of operation.

For spectroscopic studies below 50 cm^{-1}, the lamellar grating interferometer provides a more favorable shape of background than the Michelson interferometer, and is considerably more efficient at wavenumbers <20 cm^{-1}, as shown in Figure 10. This improved efficiency of the lamellar grating interferometer, which is in substantial agreement with the findings of Richards[7] from similar comparisons, allows high resolution (~0.25 cm^{-1}) absorption spectra to be recorded in the 3 - 20 cm^{-1} spectra region (see for example Figure 11) under favorable conditions of operation using Golay detectors, even though the mercury lamp radiative output decreases by almost two orders of magnitude over this region[12]. While 3 cm^{-1} appears to be the lowest practical limit for the lamellar grating interferometer when used with Golay detectors, better performances are attainable with the latest liquid helium cooled solid state detectors[25], such as the commercially available doped-germanium bolometers[26] or indium antimonide photo-detectors[27], which are capable of giving an order of magnitude improvement in signal/noise to allow wavenumbers as low as 2 cm^{-1} to be reached at high resolution.

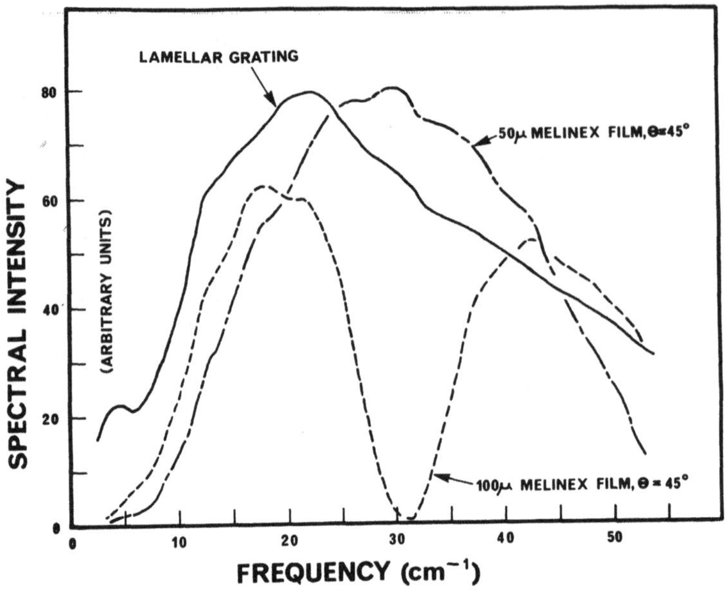

Figure 10. Low wavenumber energy comparison between the FS-720 Michelson interferometer (50μ and 100μ beam-splitters) and LR-100 lamellar grating interferometer (grating constant 1.00 cm). This was carried out using the same source, detector, optical filters and amplifier setting for each instrument. No correction has been made to take account of the higher 'throughput' of the Michelson interferometer on the theoretical assumption that the detector used for these comparisons (which had a 3 mm diameter window) was completely filled when used with either instrument. The enhanced efficiency of the lamellar grating interferometer at wavenumbers below 25 cm^{-1} and the more favorable shape of instrumental background for absorption studies up to 50 cm^{-1} are clearly demonstrated. In order to achieve good photometric accuracy at low wavenumbers, the optical band width of each instrument was restricted to the range 0 - 70 cm^{-1}.

While low detector signal/noise values are the main factor limiting the precision of very low wavenumber measurements, other spurious effects are sometimes caused by unwanted 'channel' spectra from localized parallel-sided areas of the mercury lamp quartz, jacket, grooved polypropylene filters and Golay window, which do not always ratio out correctly for double beam spectra.

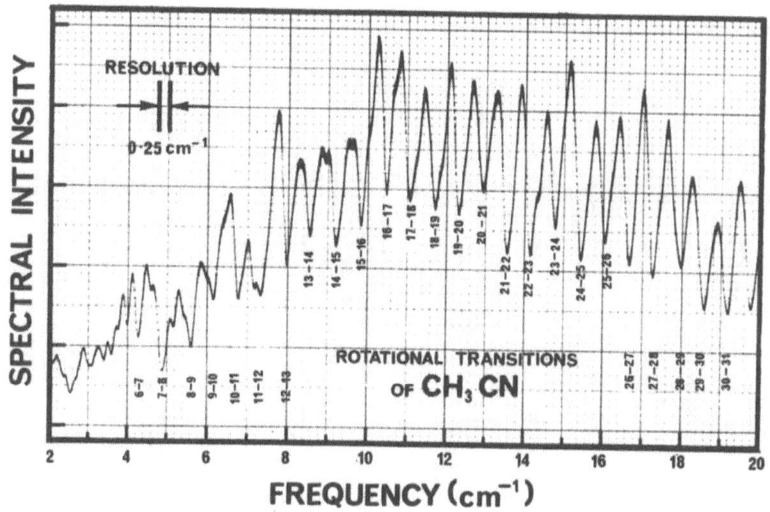

Figure 11. Portion of the low wavenumber pure rotational spectrum of CH_3CN vapor, recorded with the lamellar grating interferometer and transformed by the FTC-100 electronics. The peak absorption frequencies of average spacing 0.61 cm^{-1} are clearly observable against the instrumental background profile, and are in close agreement with recent microwave and optical measurements (see reference 35 and 36). The vapor pressure of CH_3CN was $\sim$ 1 mm Hg, and the absorbing path 1 meter. The interferogram of $\pm$ 4 cm optical path difference was scanned in 80 minutes, the electronic time constant was 4 seconds and the optical resolution was 0.25 cm^{-1}. A 20 lines/inch grooved transmission filter was employed, to restrict the optical band width to the range 0-30 cm^{-1}.

The upper useful frequency limit of the lamellar-grating interferometer described here is $\sim$80 cm^{-1}, which provides a favorable overlap with the Michelson interferometer.

APPLICATIONS

Double Beam Absorption Spectroscopy

<u>Gases</u>. Since the far-infrared gaseous absorption features of most interest to physical and chemical spectroscopists are those arising from molecular pure rotational or vibrational transitions, which usually give rise to very sharp absorption lines, then a high

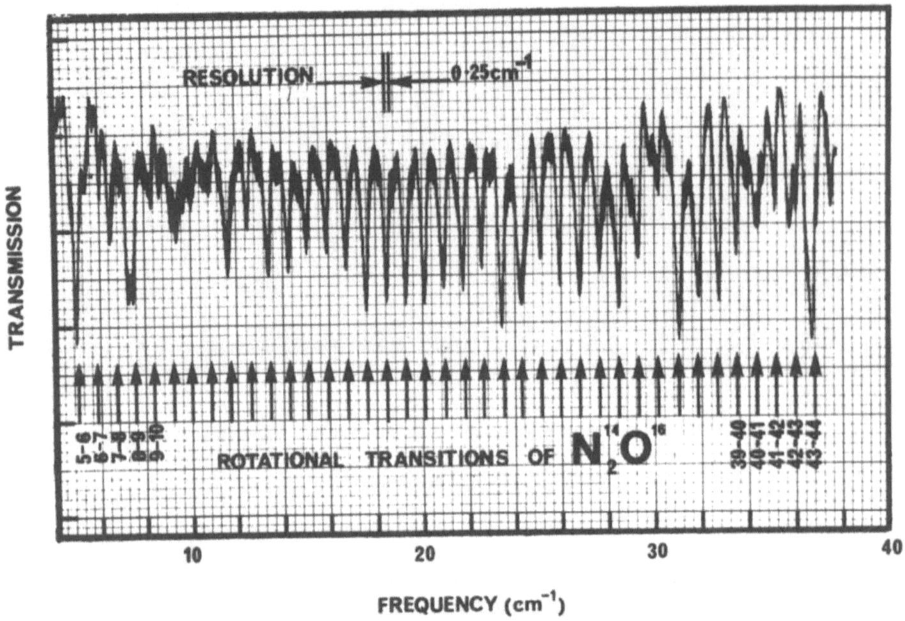

Figure 12. A portion of the pure rotational spectrum of nitrous oxide recorded at high resolution with the LR-100 lamellar grating interferometer under conditions of 10 cm Hg pressure - 1 meter absorbing path, and transformed by the FTC-100 electronics. The arrowed lines designate the calculated frequencies for pure rotational transitions of $N^{14}{}_2O^{16}$ as listed in reference 37, which have an average separation of ~ 0.84 cm^{-1}. The 'background' and 'sample' interferograms of $\pm$ 4 cm optical path difference took 80 minutes to scan. An electronic time constant of 2 seconds was employed, the interferogram sampling interval was 80 microns, and an average pause period of 5 seconds was used. Filters were employed to restrict the optical band width to the range 0-40 cm^{-1}.

instrumental resolution (0.1 - 0.5 cm^{-1}) is often demanded for these measurements. In order to avoid unwanted effects of pressure broadening and be able to work with materials with low ($<$ 10 Torr) vapor pressures, it is necessary to record spectra under conditions of low pressures and long optical paths. This can be arranged most conveniently with the current instruments by interposing a 1 meter (or longer) single-pass gas cell with polyethylene windows at a point where the optical beam is parallel (e.g. between the beam-splitter and condenser modules of the Michelson interferometer) or by transmitting the focussed beam through a long light pipe and condensing it upon the detector. Examples of high resolution gaseous absorption spectra are shown in Figures 8, 9, 11 and 12.

Solids. While many solid specimens are large enough (5 mm dia.) to be placed in the beam at the intermediate focus position, other specimens (particularly inorganic or organic crystals) are either too small or too fragile for this to be easily arranged. Such materials are normally examined in powder form, either as a pressed disc or as an embedded additive in pressed polyethylene discs or as a suspension in a transparent wax or liquid, such as 'Nujol' mull. It should be noted, however, that only a limited amount of information can be gained from specimens in powdered form by comparison to measurements on well oriented single-crystals. Figures 13 and 14 show examples of absorption spectra of solids in the form of powdered discs.

Low temperature transmission experiments may be performed by mounting the solid samples in the tail of a cryostat (see for example Figure 14 or, if a liquid helium cooled solid-state detector is used, by mounting the sample and detector in a low-temperature

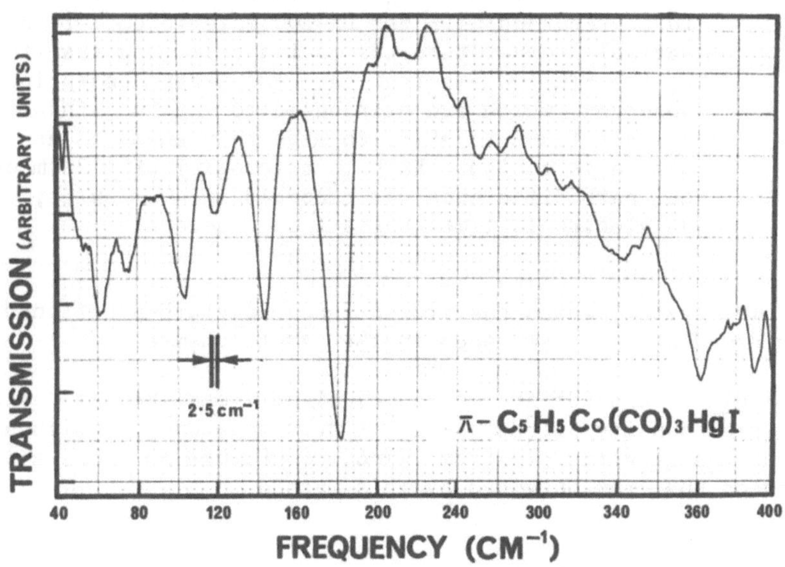

Figure 13. Far-infrared transmission spectrum of a powdered disc of C_5H_5 Co $(CO)_3$ HgI recorded with the FS-720 interferometer (6 micron beam splitter) and transformed with the FTC-200 electronics. Each interferogram of $\pm$ 4 cm optical path difference took 27 minutes to record, using an electronic time constant of 0.5 seconds. The strong absorption band at $\sim$ 180 cm^{-1} is associated with the HgI "stretching" vibration.

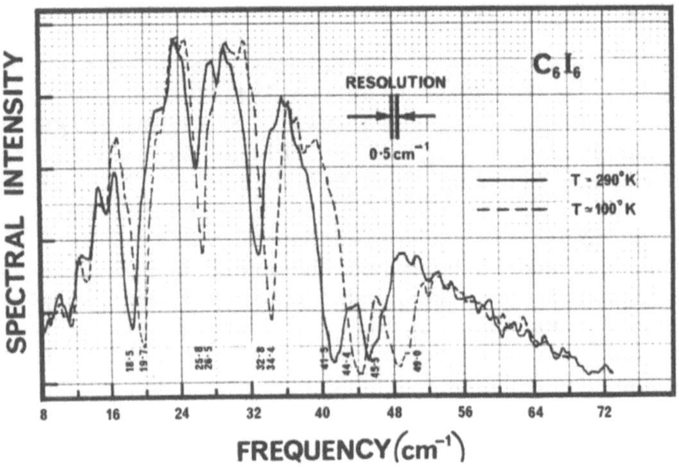

Figure 14. Low wavenumber (single beam) absorption spectrum of a powdered disc of $C_6 I_6$ at temperatures of 290°K and 100°K recorded with the LR-100 interferometer and VLT-2 variable temperature absorption cell, and transformed by the FTC-100 electronics. The sharp absorption peaks observed, which are attributed to lattice vibrations, are all found to shift to higher wavenumbers as the sample is cooled. Each interferogram of $\pm$ 2 cm optical path difference was recorded in 40 minutes, using an electronic time constant of 1 second.

light pipe system[7] which has its entrance aperture positioned at the intermediate optical focus in the sample chamber.

Liquids. Liquids are most conveniently examined in cells having high density polyethylene (or polypropylene) or high purity silicon (resistivity $> 10\,\Omega$ cm) windows. Figure 15 shows a liquid absorption spectrum obtained by this method.

Reflection Spectroscopy

In order to obtain meaningful information from specular reflectance measurements it is normally necessary to take into account the angle of beam incidence, the degree of beam polarization and also the relative angles between the beam electric vector and various crystallographic axes. The emergent radiation from the Michelson and lamellar grating interferometers can be rendered plane polarized using one or two 'wire grid' type polarizers[15]. (NB the

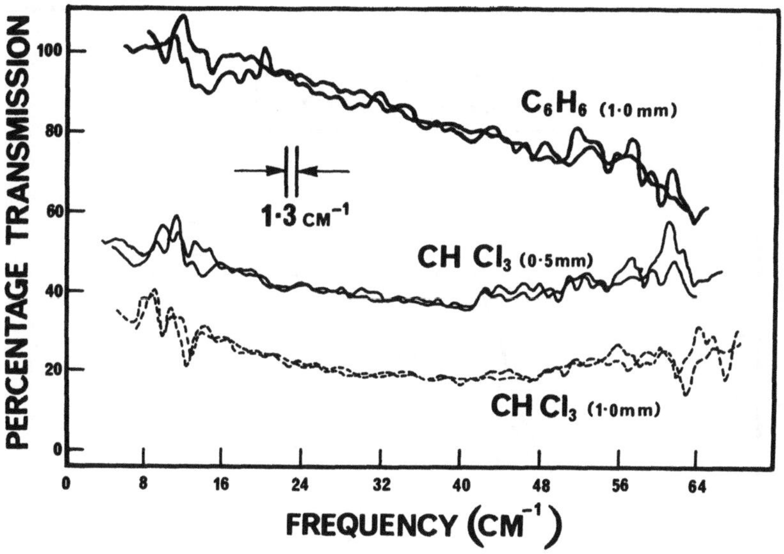

Figure 15. Transmission spectra of liquid benzene and chloroform recorded with the lamellar grating interferometer and transformed by the FTC-100 electronics. A liquid cell with wedged polyethylene windows and 1.0 mm or 0.5 mm spacers was used for these measurements. Each interferogram of $\pm$ 0.8 cm path difference took 8 minutes to record, using an electronic time constant of 0.5 seconds. Liquid benzene shows little absorption, but chloroform exhibits a strong (maximum absorption strength $\simeq$ 17 cm^{-1}) broad absorption peak centered at $\sim$40 cm^{-1}. This strong absorption, which has been observed in other polar liquids has been attributed to excitation of a 'quasi-lattice vibrational mode'.

Michelson interferometer is inherently polarized $\sim$60% in the plane of the beam splitter). The solid specimens are then orientated with their crystallographic axes at pre-determined angles to the direction of beam polarization. Figure 16 shows a polarized reflection spectrum obtained by these methods. The accessible construction of the reflectance unit shown in Figure 4 allows sufficient space for the samples to be mounted in a variable temperature cryostat so that reflection measurements can be made over a wide range of temperatures.

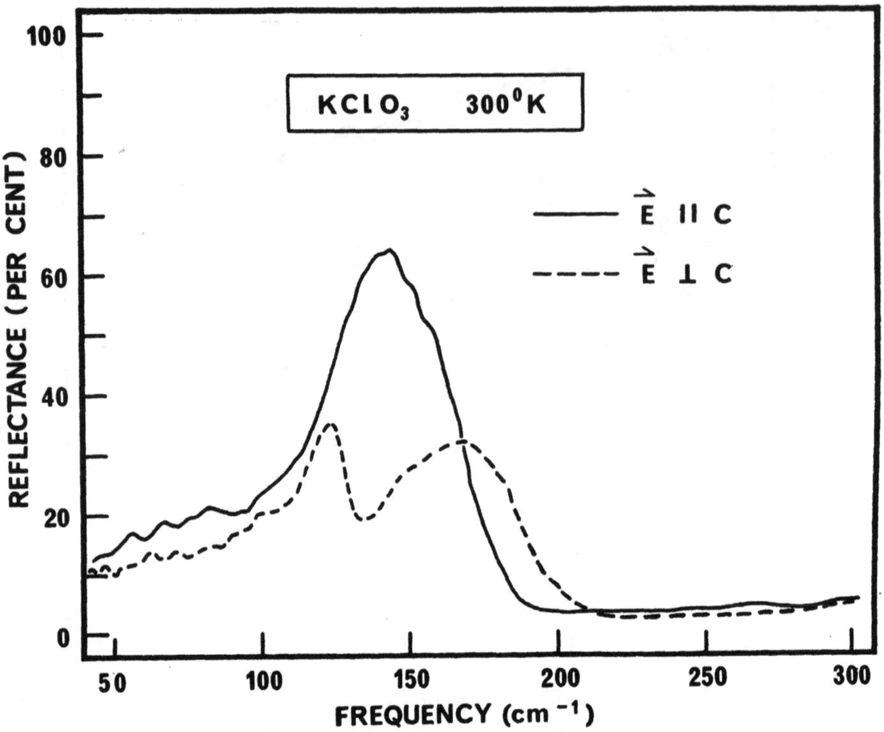

Figure 16. Polarized reflection spectrum of a $KClO_3$ single crystal ⊥ and ∥ to an axis of crystal symmetry, recorded with the FS-720 interferometer and reflectance attachment shown in Figure 3, and transformed by the FTC-100 electronics. Each interferogram of ± 0.2 cm optical path difference took 13 minutes to record, using an electronic time constant of 0.5 seconds.

Refractive Index and Dispersion Measurements

Complex refractive index measurements may be carried out using either of two basic techniques. The first method, which is convenient only for fairly transparent media, involves measuring the power transmission spectra of accurately parallel-sided layers of the material (either in the form of a solid plate, or a layer of gas or liquid between parallel windows), which should exhibit sinusoidal ripples (or 'channel spectra') caused by multiple interference effects occuring between the parallel boundaries of the medium. Since for normally incident radiation the periodicity $\delta\nu$ of such a channel spectrum is simply related to the effective optical thickness of the material, n.d where n is the refractive index and d the measured thickness of the medium, by

$$\delta\nu = \frac{1}{2\ n \cdot d} \tag{13}$$

then the values of n, the refractive index, can be calculated directly. If the absorption coefficient α of the material is also measured, the 'extinction co-efficient' $k(= \frac{\alpha}{4\pi\nu})$ can be calculated and hence the complex refractive index $n\ (=n+ik)$ can be determined[28].

The second method of obtaining refractive index measurements--the so-called 'asymmetric' Fourier transform method--is one where the dispersive medium of interest is interposed in one arm of the Michelson interferometer to introduce a shift and alterations of the interferogram fringes from which the refractive index of the material may be calculated[29,30,31]. Other variations of this method have been used to obtain the dielectric properties of solids and gases[32,33]. A particular advantage of this technique is that it allows the wave amplitude spectrum to be measured directly, often to an accuracy $\sim$ 1 in 10^2. Intensity measurements can thus be made to a precision $\sim$ 1 in 10^4, or two orders of magnitude greater than by conventional methods.

SUMMARY

Owing to their improved optical design and compact modular construction, the latest Michelson-type Fourier spectrophotometers, allow routine laboratory far-infrared spectra to be recorded in a faster and more convenient manner than was possible with earlier commercial instruments, for a wide variety of experiments. A practical resolution capability of 0.1 cm^{-1} for gaseous samples has been clearly demonstrated. While the advent of 'fast Fourier transform' computational techniques have greatly reduced computation times for high resolution spectra, as evidenced by a comparison of the computation times for the spectra shown in Figures 8 and 9, the full broadband resolution capability of the instruments described here may not be realized unless moderately large digital computers are available to cope with the large number of computational points involved. The high efficiency of the lamellar grating interferometer in the very far-infrared has made it possible to approach a wavenumber limit as low as 3 cm^{-1}, using commercial room temperature Golay detectors. Further advances in this direction are likely to be limited solely by problems of detector size and sensitivity.

While the recent proliferation of small general-purpose computers for the laboratory has alleviated some of the inconvenience suffered by users of Fourier spectrophotometers in having to wait for their spectra to be transformed on large often-over-loaded digital computers, considerable progress has been made with the small special-purpose analogue computers of the type mentioned here as they afford a reasonably accurate presentation of spectra within minutes of the termination of interferogram scans, and allow laboratory experiments to proceed in a much faster manner than was hitherto possible. While the interferometric systems described here are capable of yielding photometric accuracies of $\sim$ 2% for absolute transmittance and reflectance data in regions of the spectrum where the signal/noise is good, further improvements are likely to be achieved when a simultaneous double-beam method of recording interferograms is employed, to correct for effects of instrumental drift[34] and other errors inherent with a single-beam mode of operation.

ACKNOWLEDGEMENTS

The author is grateful for the assistance of his colleagues at R.I.I.C., particularly Dr. G. Hayward who recorded most of the spectra demonstrated in these lectures, and Dr. D. J. Neale for writing Fourier transform programs. The author would also like to thank Dr. J. Turner, Cambridge University, and Dr. G. R. Wilkinson, Kings College, London, for providing the samples whose spectra are shown in Figures 13, 14 and 16 respectively, and Mr. M. E. van Kreveld and Mr. W. van Leeuwen of the University of Amsterdam who wrote the "fast" Fourier transform programs and arranged for the digital computation of the spectrum shown in Figure 9.

REFERENCES

1. For a full discussion of these developments, see, Spectroscopic Techniques for Far Infra-red, Submillimetre and Millimetre Waves, Ed. D. H. Martin, Chapters 1, 5 and 6 (North Holland Publishing Co., 1967).
   See also Professor Genzel's lectures on "Far-infrared Sources".

2. For recent comprehensive reviews of the technique of Fourier Transform Spectroscopy for the Far-infrared, see:

   G. A. Vanasse and H. Sakai, Progress in Optics, 6, 261,

(North Holland Publishing Co., 1967).

P. L. Richards, Fourier Transform Spectroscopy, Ref. 1, Chapt. 2.

L. Mertz, Transformations in Optics, (John Wiley & Sons, 1967).

Dr. Gebbie's lectures on "Far Infrared Spectroscopy Using Interferometers".

3. For details of the most advanced instrumentation of this type, see for example:

F. K. Kneubuhl, J. F. Moser, and H. Steffen, J. Opt. Soc. Am., 56, 760, (1966).

T. M. Hard and R. C. Lord, Appl. Opt., 7, 589, (1968).

G. R. Wilkinson and D. H. Martin, "Grating Spectroscopy", Ref. 1, Chapt. 3.

4. H. A. Gebbie, Advances in Quantum Electronics, Ed. J. R. Singer, p. 155, (Columbia University Press, 1961).

Following the initial studies of Dr. H. A. Gebbie at the National Physical Laboratory, the first commercial far-infrared Michelson interferometers were marketed by Sir Howard Grubb Parsons & Co. (1962) and Research & Industrial Instruments Co. (1963).

5. John Strong and G. A. Vanasse, J. Opt. Soc. Am., 50, 113, (1960).

6. For descriptions of the far-infrared properties of "Melinex" films see: D. J. James and J. Ring, "Methodes Nouvelles de Spectroscopie Instrumentale" Supplement of J. de Physique, 1967, 28, No. 3-4, p. C2-150

and

J. E. Chamberlain, G. W. Chantry, F. D. Findlay, H. A. Gebbie, J. E. Gibbs, N. W. B. Stone, and A. J. Wright, Infrared Physics, 6, 195, (1966).

7. For example see:

P. L. Richards, J. Opt. Soc. Am., 54, 1474, (1964).

8. A measure of optical flux defined by the product of the source area, and the solid angle of the cone formed by the rays from the source to the extremities of the first collecting mirror.

(Units of cm^2 steradians.)

9. J. Connes, Rev. Opt., 40, 45, 116, 171, 231, (1961).

10. P. Jacquinot, Rept. Progr. Phys., 23, 267, (1960).

11. See technical literature of Unicam Limited, Cambridge, England, for full details.

12. Phillips HPK 125 watt. For details of the emissive properties of this type of lamp in the far-infrared, see R. Cano and M. Mattioli, Infrared Phys., 7, 25, (1967).

13. For a discussion of the imaging properties of light cones, see for example:

D. E. Williamson, J. Opt. Soc. Am., 42, 712, (1952).

R. C. Ohlman, P. L. Richards and M. Tinkham, J. Opt. Soc. Am., 48, 531, (1958).

J. C. Gill, Royal Radar Est. Memorandum 1612, (1959).

14. See for example:

Charles S. Williams, Appl. Opt., 5, 1984, (1966).

15. The latest type of 'wire grid' polarizers have proved to be extremely useful by virtue of their high efficiencies and small size. For a description of these see:

J. P. Auton, Appl. Opt., 6, 1023, (1967).

16. This small angle ($< 1^{\circ}$) causes all the computed frequencies ν to be underestimated by an amount $\simeq 2 \times 10^{-4}\nu$. An error of equal magnitude is also introduced by the optical beam spread caused by the finite angle subtended by the source (see equation (8)). However, the total error from these causes $\simeq 4 \times 10^{-4}\,\nu$ is negligible for the instrument described here, which operates in the range $3 < \nu < 80$ cm^{-1}, with a limiting resolution of 0.1 cm^{-1}.

17. K. D. Moller and R. V. McKnight, J. Opt. Soc. Am., 53, 760, (1963).

18. A 'blurring' of instrumental resolution occurs for unpolarized radiation at very low (< 5 cm^{-1}) wavenumbers where the corresponding wavelengths are becoming of comparable magnitude to the width of the cavities formed by the intermeshing lamellar plates, due to a difference in phase velocity of propagation

of the optical components polarized parallel and perpendicular to the cavity walls (see reference 7). This effect limits the resolution attainable with the lamellar grating described here to $\sim 0.1\ cm^{-1}$ at $3\ cm^{-1}$. The effect can of course be minimized by working with polarized radiation.

19. See technical literature of R.I.I.C. London S.E.26, England, for details of FTC-100 Fourier Transform Computer.

20. K. Hennerich, W. Lahmann and W. Witte, Infrared Phys., 6, 123 (1966), have found that Unicam Golay detectors are likely to operate with < 1% deviation from linearity, for radiant fluxes of total power $< 3 \times 10^{-6}$ watt, which typically produce output signals of ≤ 0.5 v from the detector.

21. For a thorough discussion of the photometric accuracies obtainable in the far-infrared with Fourier spectrometers, see:

D. R. Bosomworth and H. P. Gush, Canad. J. Phys, 43, 729, (1965).

22. It may be shown that the average signal/noise Ss for a computed spectrum of M spectral elements, is related to the average signal/noise of the interferogram measurements (including digitizing errors), Si, by

$$Ss \propto Si \frac{\sqrt{N}}{M} \propto \frac{Si}{\sqrt{M}}$$

where N is the number of interferogram data points used in the computation, and has a minimum value of 2M. Thus Ss can be improved by greatly oversampling interferograms to increase the value of N, at the expense of increasing computation time, or more readily by minimizing the value of M the number of spectral elements.

23. Richard T. Hall and Jerome M. Dowling, J. Chem. Phys., 47, 2454, (1967).

24. Richard T. Hall, Dale Vrabec and Jerome M. Dowling, Appl. Opt., 5, 1147, (1966).

25. See Professor Richards' lectures on "Far-infrared Detectors", and E. H. Putley and D. H. Martin, "Detectors", Ref. 1, Chapt. 4.

26. Texas Instruments Limited, Texas.

27. Mullard Limited, London.

28. This technique has recently been applied to a degree of precision in the far-infrared by C. M. Randall and R. D. Rawcliffe, Appl. Opt., 6, 1889, (1967).

29. J. E. Chamberlain, J. E. Gibbs and H. A. Gebbie, Nature, 198, 874, (1963).

30. J. E. Chamberlain, F. D. Findlay and H. A. Gebbie, Applied Optics, 4, 1382, (1965).

31. J. E. Chamberlain and H. A. Gebbie, Appl. Opt., 5, 393, (1966).

32. Edgar E. Russell and Ely E. Bell, J. Opt. Soc. Am., 57, 341, (1967).

33. R. B. Sanderson, Appl. Opt., 6, 1527, (1967).

34. Various workers have been able to correct for instrumental drift by servo-controlling the mercury lamp radiative output, by either using a second detector as a monitor or by feeding the 'unused' source radiation reflected from the back of the chopper blade directly to the usual detector, to derive a 'reference' signal which is modulated in exact anti-phase to the true signal and provides automatic compensation for small fluctuations in source brightness.

35. Kessler, Ring, Tramburato, and Gordy, Phys. Rev., 79, 54, (1960).

36. D. W. Robinson and D. A. McQuarrie, J. Chem. Phys., 32, 556, (1960).

37. K. Nahari Rao, C. J. Humphries and D. H. Rank, "Wavelength Standards in the Infrared", p. 148, (Academic Press 1966).

INTERFEROMETRIC SPECTROSCOPY

H. A. Gebbie

*National Bureau of Standards, Boulder, Colorado

<u>General</u>. The far infrared region has benefited considerably from the introduction of Fourier Spectroscopy, a topic of which the generalities have now been treated by a number of authors. Basically its advantages come from the ability of a two beam interferometer like Michelson's to pass a greater amount of information than any other "delay" instrument for a given size and spectral resolution.(1) This is of great importance in a spectral region where source brightness is low. It does, however, demand the additional process of Fourier Transformation to give spectra, but modern digital computers can do this very economically and rapidly to any required resolution.

The high information throughout can be exploited in a number of ways. Usually, the aim is to get improved spectral resolution, since in the far infrared, detector noise is always the fundamental limitation; but (though this is stressed less often) gains of another kind can be realized. The extra radiation intensity available with the interferometer allows measurements to be made on samples which would be too absorbing for conventional dispersion instruments. This is a particularly important advantage in solid state studies, since very often in these investigations the sample thickness cannot be reduced to compensate for an embarrassingly high absorption coefficient. The ordinary techniques of absorption spectroscopy by Fourier methods have now been so extensively used that little more need be said about them, but Michelson's form of the two beam interferometer can also be used to determine refractive index spectra. This is hardly any more

*Formerly with National Physical Laboratory, Teddington, Middlesex, England

difficult than getting absorption spectra. The sample is placed in one arm of the interferometer and, apart from the additional necessity of having specimens of adequate homogeneity and flatness to allow the spectra of phase retardation to be interpreted, the real part of the complex index can be determined over the same wide frequency ranges that we are used to dealing with in getting the imaginary part from absorption spectra. These index spectra, coupled with the Kramers-Kronig relations in appropriate form, are of very great value in determining oscillator strengths and promise to be applied quite widely in solid state physics. To realize this advantage of Fourier's Spectroscopy, however, it is essential to use Michelson's form of interferometer where the two beams are physically separated. An alternative form of interferometer which has been used for far infrared spectroscopy, the lamellar grating, does not have this valuable property.

The details of techniques for obtaining refractive index spectra have been treated elsewhere.[2]

Phase Modulation Techniques. Conventional spectroscopic systems invariably use amplitude modulation of the source to mitigate drift problems, but with interferometric techniques the alternative of phase modulation is often attractive. This has been exploited by the Connes in the near infrared planetary spectroscopy, but is particularly easily applied in the far infrared where the larger displacements and greater tolerances allow us to use simple transducers to produce oscillatory mirror displacements[3] which give the phase modulation. This relatively simple modification of a Michelson Interferometer gives a very effective way of discriminating against source variations and, at the expense of modifying the spectral background, can give some of the advantages of a double beam instrument. The modification of spectral background with phase modulation is well defined and will have to be allowed for in deriving absorption spectra, but again, this is easily done in a digital computer. Details of applications of this method will be available in a forthcoming publication.[4]

Molecular Maser Sources. The use of molecular masers has provided a valuable ancillary to wide-band spectroscopy in the far infrared. Though these sources are not tuneable, the relatively high power available from them, together with their monochromaticity, has proved to be a considerable asset. Masers using water vapor and HCN were the earliest to be used, but they have now been supplemented by transitions in DCN, ICN and SO_2, giving in total a considerable choice of wavelengths. However, the transition most used has been that of wavelength 337 microns in HCN which has found application in the spectroscopy of solids, liquids, gases and plasmas. Here, since we are concerned with the solid state, only two will be mentioned. The first is the use of monochromatic lines to measure the refractive index of solids. This has proved to be a very valuable supplement to the wide-band Fourier Method mentioned above. The Fourier method is particularly powerful in getting the shape of the dispersion curve, but the absolute value of the index which it gives is very sensitive to the determination of the exact position of a single grand maximum in the interferogram. The additional measurement of a good value at a single maser frequency is then a valuable verification point. The second application of masers, which is much more interesting, is their use in magnetic resonance spectroscopy.[5] The advantage of high frequency submillimeter carrier frequency replacing the usual microwave frequency is that resonances can be measured in systems with much shorter mean free times. This has been exploited in cyclotron resonance studies and will probably be extended to spin resonance measurements.

Recently an HCN maser has been used to study, for the first time, quantum effects in cyclotron resonance of a semiconductor.[6]

REFERENCES

1. H. A. Gebbie and R. Q. Twiss, Reports on Progress in Physics, 29 Part II. p. 729, 1966.

2. J. E. Chamberlain, F. D. Finlay and H. A. Gebbie, Applied Optics 4, 1302, 1965.

   J.E. Chamberlain, A. E. Cosltey and H. A. Gebbie, Spectrochemica Alta, 25A, 9, 1969.

3. 'New Scientist', 17th October, 1968, Radar for Engineers.

4. J. Chamberlain and H. A. Gebbie; will be submitted to Applied Optics forthcoming issue on Interferometric Spectroscopy.

5. K. J. Button, H. A. Gebbie and B. Lax, J. Quantum Electron, Q.E.2., 202, 1966.

6. K. J. Button, B. Lax and C. C. Bradley, Phys. Rev. Letters 21, 350, 1968.

FAR INFRARED PROPERTIES OF ATMOSPHERIC GASES

RALPH ZIRKIND

Polytechnic Institute of Brooklyn

Farmingdale, Long Island, New York

INTRODUCTION

This lecture deals primarily with the far infrared properties of atmospheric gases and appropriately is related to the spectroscopic instrumentation of the far infrared by virtue of it being the driving force for higher resolution. Further, the selection of the atmospheric gases is desirable for several cogent reasons: (1) variety of molecular types; (2) interest in the propagation properties at far infrared wavelengths and correction of observations; and (3) the possible application in studying the upper atmosphere.

Our atmosphere consists of

$$N_2,\ O_2,\ CO_2,\ H_2O,\ N_2O,\ CH_4,\ CO,\ O_3,\ \text{etc.}$$

With the exception of N_2, these molecules are electrically polar with extensive vibrational-rotational and pure rotational structure in the infrared region of the spectrum. O_2 possesses a rotational and microwave structure as a result of its magnetic dipole.

As the wavelength increases the different molecules play the predominant role. CO_2 is the significant one in the 14 to 17 micron region; however, N_2O is a contributor at the longer wavelength wing through the ground vibrational band. In addition, of course, a contributor throughout the region is the ν_2 - ground vibrational bands of the ozone isotopes;

$$O_3^{16} - 701.42\ \text{cm}^{-1},\ O^{16}O^{18}O^{16} - 697\ \text{cm}^{-1},\ O^{16}O^{16}O^{18} - 688\ \text{cm}^{-1}.$$

The H_2O molecule has extensive structure beyond 18 microns. Interestingly, atomic oxygen, O, can absorb at 44.3 and 62 microns through its magnetic dipole between the three 3P states. For the submillimeter and longer wavelength regions, the significance of the H_2O and O_2 are well known.

We will first consider the basic theoretical aspects as these are the parameters needed to determine the propagation properties.

MOLECULAR PROPERTIES

The import of far infrared spectroscopy is that with current spectroscopic resolution, using either optical or microwave techniques, the necessary capability is available to study line widths as a function of pressure and/or temperature, i.e., the collision interaction of molecules. Needless to say this parameter is basic to the absorption coefficient and, as stated earlier, the atmospheric molecules provide a good variety-linear, symmetric top and asymmetric rotor. Furthermore, the effect of self and foreign gas broadening of the line must be understood.

For an asymmetric rotor like H_2O, this spectral region provides a pure rotation spectrum instead of a complex vibrational-rotational one. Also, its low moment of inertia gives fewer lines.

A form of the absorption coefficient can be written as

$$\alpha(\text{cm}^{-1}) = \frac{P}{\Delta\nu}\left(\frac{8\pi^2 N}{3kTQ}\right)\mu^2\nu^2 g\, e^{-E/kT}\Sigma|\phi|^2 \qquad (1)$$

where P = partial pressure
$\Delta\nu$ = half width at half intensity
g = statistical weight
$\Sigma|\phi|^2$ = line strength (transition probability)
μ = dipole moment of rotor.

Now, the half width is related to the value at one atmosphere by the expression,

$$\Delta\nu = P\,\Delta\nu_o \qquad (2)$$

and

$$\Delta\nu_o = N\left(\frac{2kT}{\pi}\right)^{\frac{1}{2}}\rho^2\left(\frac{m_1 + M_2}{m_1 M_2}\right)^{\frac{1}{2}} \qquad (3)$$

where P = optical diameter
m_1 = absorber mass
M_2 = colliding mass

If we now rewrite (1), we obtain the result

$$\alpha\Delta\nu = P\left(\frac{8\pi^2 N}{3kTQ}\right)\mu^2\nu^2\, g\, e^{-E/kT}\,\Sigma\,|\phi|^2 \tag{4}$$

where the terms on the l.h.s can be measured independently and those on the right side are calculated. Thus we can verify either side of the equation.

The impact theory of pressure broadened lines was first developed by Anderson(1) and extended by Tsao-Curnutte(2). This quantum mechanical theory which is applicable to well-separated lines has been applied to the rotational lines of several linear molecules(3), asymmetric rotors(4) and recently to the symmetric top molecules(5). On the whole, the successes have been modest, e.g., several comparative values for self-broadening are given in Table 1.

TABLE 1. Comparative Values of Several Line Widths(5)

Theoretical (MHz/mmHg)	Experimental (MHz/mmHg)
27.5	20
44	38
36	18
164	16

Simply, Anderson's theory yields a quantity $S_p(b)$, subscript "p" refers to perturbing molecule, which defines the degree of effectiveness that a collision has at a distance of closest approach "b" in disturbing the radiation of a molecule in an initial state "i" or in final state "f".

Explicitly, the expression for $S_p(b)$ may be written in the following form(5):

$$S_p(b) = (C_\mu/b^n)\sum_{J_i'}\sum_{J_p'} C(J_i k_i J_i'; K_i 0 K_i)^2 C(J_p k_p J_p'; K_p 0 K_p)^2 f\mu(b)$$
$$+ (i\to f) + [(C_\mu/b^n)\sum_{J_p'} C(J_p k_p J_p'; K_p 0 K_p)^2 C(J_i k_i J_i; K_i 0 K_i)^2 \times$$
$$f_\mu(k) + (i\to f) + S_p(b)_m] \tag{5}$$

where

$$S_p(b_m) = (2C_\mu\ \phi_\mu/n^n)(2J_i+1)^{\frac{1}{2}}\ (2J_f+1)^{\frac{1}{2}} C(J_i k_i J_i, K_i 0K_i)\ \times$$

$$C(J_f k_f J_f; K_f 0K_f) W(J_i J_f J_i J_f; 1K_i) \sum_{\bar{p}'} C(J_p k_p J_p'; K_p 0K_p)^2 \times f_\mu(k)$$

and the notation is as follows:

$(i \rightarrow f)$ = a term obtained from the preceding by interchanging "i" by "f"

J_i' ≡ $J_i' = J_i$ is omitted in sum

C, W = Clebsch - Gordon, Racah coefficients respectively

C_μ, ϕ_μ = quantities defined by nature of intermolecular potential.

$$k = (2\pi b/v)\ B\ [J_i'\ (J_i' + 1) - J_i\ (J_i + 1) + J_p' \times (J_p' + 1) - J_p\ (J_p + 1)]$$

B = rotational constant.

For each of the molecular interactions with a potential function $V_p(b)$ we have the following values for n

Interaction, V_p	n	
Dipole - dipole	4	
Dipole - quadrupole	6	
Quadrupole - quadrupole	8	
Dipole - induced dipole	10	when the perturber is spherical symmetric
Quadrupole - induced dipole	12	when the perturber is spherical symmetric

Once the value of $S_p(b)$ is calculated then we can obtain the cross section,

$$\sigma = \int_0^\infty 2\pi b \; S_p(b) \; db \tag{6}$$

which for strong interactions, $S_p(b) \ll 1$, breaks down at large distances.

Anderson introduced the following artifact for the partial cross-section, σ,

$$\sigma = \pi \, b_o^2 \int_{bo}^\infty 2\pi b \; S_p(b) \; db \tag{7}$$

where

$$S\,(b_o) = 1$$

$$S\,(b) = 1 \qquad o \leq b \leq b_o$$

The total cross-section is then obtained by summing over all the rotational states

$$\sigma_{J_1} = \sum_{K_1 J_p K_p} \rho_{K_1} \, P_{J_p K_p} \, \sigma J_1 K_1 J_p K_p \qquad (J \rightarrow J+1) \tag{8}$$

Finally, the halfwidth is given by

$$\Delta\nu_J = Nv\sigma_J / 2\pi \tag{9}$$

where N = no. of perturbers

v = mean relative velocity.

The theory breaks down in the wings of the line: the observed absorption is greater than expected. The Lorentz description of line broadening is adequate near resonance, and at low pressures, and for narrow lines. The recent improvements by Ben-Reuven to this impact theory predicts a higher absorption in the low frequency wing; however, it involves unknown collisional coupling parameters. Recently Maryott and Birnbaum(6) have considered the line shape and collision effects in the microwave wing of far infrared rotational lines, and as one may expect relate the difficulties to the inadequate description during the "finite" collisional process. They suggest that "a reasonable explanation for the microwave absorption in gases with far-infrared rotational lines is the modification of the usual selection rules during a collision when the normal rotational motion is quenched by intermolecular forces." The application to water vapor is suggested. Thus we see that the halfwidth of the Lorentzian line is a function of the intermolecular collision.

The results of Benedict and Kaplan(7) in using Anderson's theory are noteworthy. They concerned themselves with the absorption of H_2O. For the self-broadening of H_2O, H_2O-H_2O, they considered only the dipole-dipole interaction (b^{-3}) for the pure rotation system. For broadening of H_2O by N_2 they took into account only the dipole-quadrupole interaction (b^{-6}) and omitted the quadrupole-quadrupole and dipole induced dipole interactions.

Their analysis showed that the dipole forces alone produced considerable broadening and the values of H_2O-H_2O is greater than H_2O-N_2 in the ratio of 6.8 to 1.9 depending upon the "J" value. This result is different than other earlier results. The variation of the halfwidth for H_2O with different gases is shown in Table 2.

TABLE 2. Dependence of Line Halfwidth with Perturber Molecule

Specie-Broadener	$\frac{\Delta\nu}{P}$ (MHz/mmHg)$_{exp.}$	$\frac{\Delta\nu}{P}$ (MHz/mmHg)$_{the.}$
H_2O - H_2O	19.1	19.5
- N_2	3.76	3.78
- O_2	2.72	1.75
-CO_2	6.1	
- Ar	2.0	
- Air	3.52	

For the 1.6 mm line (183,310.12 MC/sec) Rusk's(4) recent results are in excellent agreement with the theoretical predictions.

The wavelength dependence of the line halfwidth is shown for general interest in Table 3.

TABLE 3. Wavelength Dependence of Line Halfwidth

	Wavelength	$\frac{\Delta\nu}{P}$ (MHz/mmHg)
$H_2O - H_2O$	infrared	4.3
	1.35 cm	3.4 (water density low)
	1.35 cm	4.2 (water density high)
	1.6 mm	7.1 - 7.9
	1.6 mm	3.16 (Rusk)
$H_2O - N_2$	92 cm^{-1}	7.2
	170 cm^{-1}	4.8
	188 cm^{-1}	5.1

Another basic parameter, in addition to line strength, is line position. For many molecules the accuracy has not altered much; the 1.35 cm line of H_2O was observed in 1946 to be 0.7418 cm^{-1} and 0.78 cm^{-1} theoretically. Today theory and experiment agree; namely, 0.7416830 cm^{-1}.

TABLE 4. Summary of Atmosphere Molecule Properties

Specie	Type	Band	Halfwidth	$\frac{T}{T}^{-n}$	Source
CO_2	lin.	ν_3	0.064		Kaplan & Eggers (1956)
		$02^00 - 001$	0.084	.56	Kostkowsky (1955)
		$10^00 - 001$			
		$00^00 - 22^01$	0.07		Goldberg (1954)
H_2O	a.t.	$5_{-1} - 6_{-5}$	0.087		
			.11		
		$2_{20} - 3_{12}$	.116	1.22	Rusk (meas.)
$H_2 - H_2O$			0.357	.86	Benedict & Kaplan (calc.)
O_3	a.t.		0.089		Kaplan (1959)
			0.076		Walshaw (1955)
N_2O	lin.	$00^00 - 01^10$	0.16		Adel & Barker (1944)
		$00^00 - 02^00$	0.16		Goody & Wormell (1951)
		$00^00 - 10^00$	0.15		Goody & Wormell (1951)
CH_4	s.t.	0000 - 0020	0.08		Goldberg et al (1952)
CO	lin.	0 - 1	0.05 - 0.1		Benedict (1962)
		(3 - 100cm^{-1})	0.0965		
O_2	lin.	2cm^{-1}	0.02 - 0.07		Van Vleck (1947)
N_2	lin.		.03 - .08		

EXPERIMENTAL

During the past 35 years the far infrared and microwave spectral regions have been resolved with increasingly better resolution, namely, from 0.5 cm^{-1} to 0.06 cm^{-1}. Although this resolution, by short-wave standards, is not high, the identification of most rotational lines is possible with good positional accuracy.

In 1933[8] the rotational structure of water vapor for wave numbers less than 100 cm^{-1} was obtained and for the next twenty years no new significant work was done until Genzel and Eckhardt[9] reported the water spectrum $\gtrsim$ 50 cm^{-1}. In the ensuing years extensive research[10] - [14] has been conducted. One should emphasize that since 1942 the microwave spectrum[15], [16] has received considerable attention.

Atmospheric observations, i.e., solar spectra, have yielded much useful data: Gebbie[17] reported that the atmosphere is less opaque in the 350 microns region which was confirmed by Farmer & Key[18] by observation taken at an altitude of 3500 meters. At higher altitudes solar observations have been made by Bader et al[19] at 40,000 ft, Bader examined the region of 250 cm^{-1}, and at 30 Kms Gay[20] in the region of $\sim$ 200 cm^{-1}. From the latter two facts of general interest are (1) the temperature of the solar disc is approximately 3740°K and (2) the residual absorption was about 7.5% (5% instrumental and 2.5% atmospheric dust).

We will now consider the experimental results on the two molecular species, H_2O and O_2, in greater detail.

The pure rotational spectrum of water vapor in the spectral region of 5 to 125 cm^{-1} has been measured recently by Hall and Dowling[21] with a resolution between 0.06 cm^{-1}. An illustration of the detail is shown in Figure 1 where the doublet near 59 cm-1 has not previously been observed, i.e., when the spectral resolution is $\approx$ 0.1 cm^{-1}. The frequency assignments have been based upon the work in Refs. 22, 23, & 24. For comparative purposes a moderate resolution spectrum (18 to 82 cm^{-1}) is presented in Figure 2. The finer details are obviously missing but some lines appear. Hall's results have verified the many levels of the ground state.

The millimeter region was examined by Van Vleck[16] applying available data on line width; that is, the absorption at 6 cm^{-1} and $\sim$ 0.75 cm^{-1}. More recently Tolbert et al[25] measured the attenuation of water vapor in the region of 4 cm^{-1} which lies between the two aforementioned lines and found it to be about four times greater than predicted by Van Vleck. This may again be explained by the earlier discussion on absorption in the wings of lines.

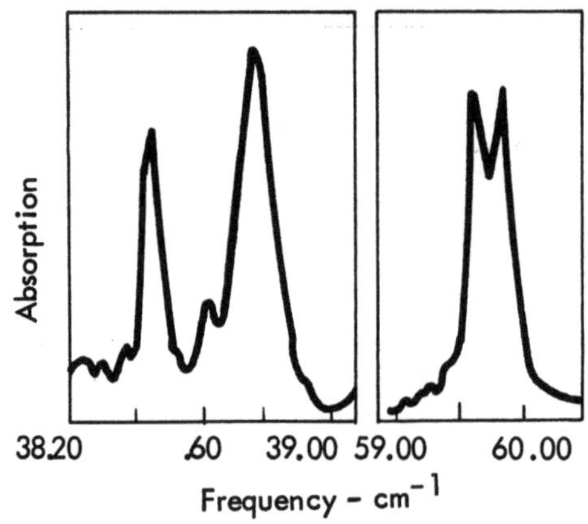

Fig. 1. Some features of water vapor rotational spectra.

The two significant microwave lines of oxygen are at 2 and ~ 4 cm^{-1}. The latter line has been examined in detail by Tolbert[25]. The peak values reported are slightly higher than other results, e.g. Ref. 16. The measured line shape is given in Figure 3 and the half-width is 2 MHz at 1 mm Hg pressure with a positive slope of 1.9 MHz mm Hg. The peak intensity is about 6.75 db/km with a contribution from the 60 GHz line at pressures greater than 100 mm Hg.

The observations made using the atmosphere as the source have contributed to our general knowledge. Here, of course, the lines are broadened by a combination of molecules and the path varies both in pressure and temperature.

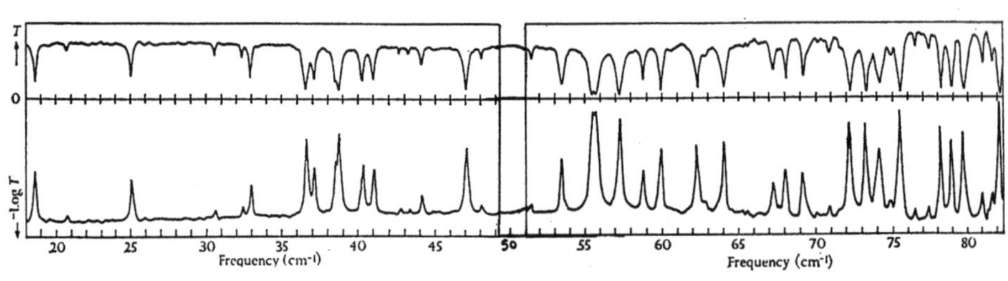

Fig. 2. Transmittance T and absorption coefficient -log T for water vapor in the far infrared taken with a Michelson interferometer and subsequent numerical transformation. (After Richards.)

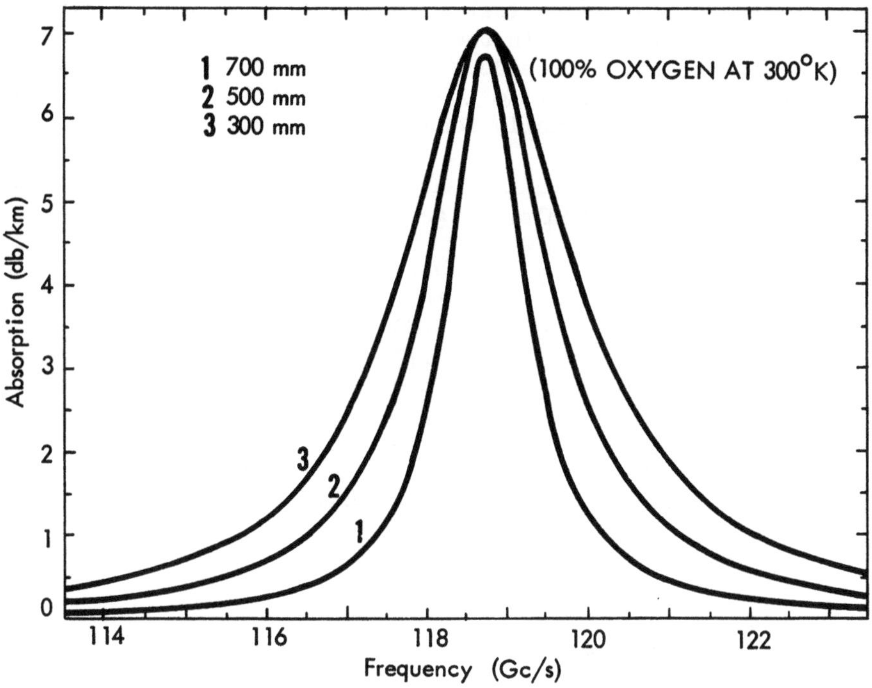

Fig. 3. Oxygen attenuation versus frequency for gas pressures of 700, 500, and 300 mm Hg. (After Talbert.)

In Figure 4, Gebbie's early observations in the 10 to 30 cm^{-1} region is shown (Jungfraujoch is at an altitude of about 3100 meters). The several H_2O lines are identified, in addition, there are several oxygen lines. An additional set of data[26] which extends down to 5 cm^{-1} is shown in Figure 5 under conditions of 1.5 air masses and 2 mm precipitable water. To enable the identification of the molecular oxygen lines Gebbie[27] obtained a laboratory spectrum (entrapped water was not completely removed). In addition to the observed lines given in Table 5, the 14.2 and 16.2 cm^{-1} features are attributable to molecular oxygen. Although some features are not identified, e.g. in 7.07 cm^{-1}, these may be attributed to $(H_2O)_2$ or other unknown species.

Finally, data were reported[19] that showed the spectrum of the sky in the region 15 to 65 cm^{-1} as seen from 40,000 ft. These are in agreement with the other data.

In summary, the theoretical status of line widths requires further study particularly in the wings and verification by experiment. On the experimental side, additional high resolution spectral data of the atmosphere in 1 to 100 cm^{-1} region is needed to identify molecular species not normally found in the laboratory.

TABLE 5. Molecular Oxygen Rotational Lines

K	K'	J	J'	Obs. ν (cm^{-1})	Theo. ν (cm^{-1})
3	5	4	4	25.80	25.812
3	5	4	5	27.79	27.824
5	7	6	6	37.79	37.384
5	7	6	7	39.24	39.358
7	9	8	8	48.93	48.929
7	9	8	9	50.88	50.874

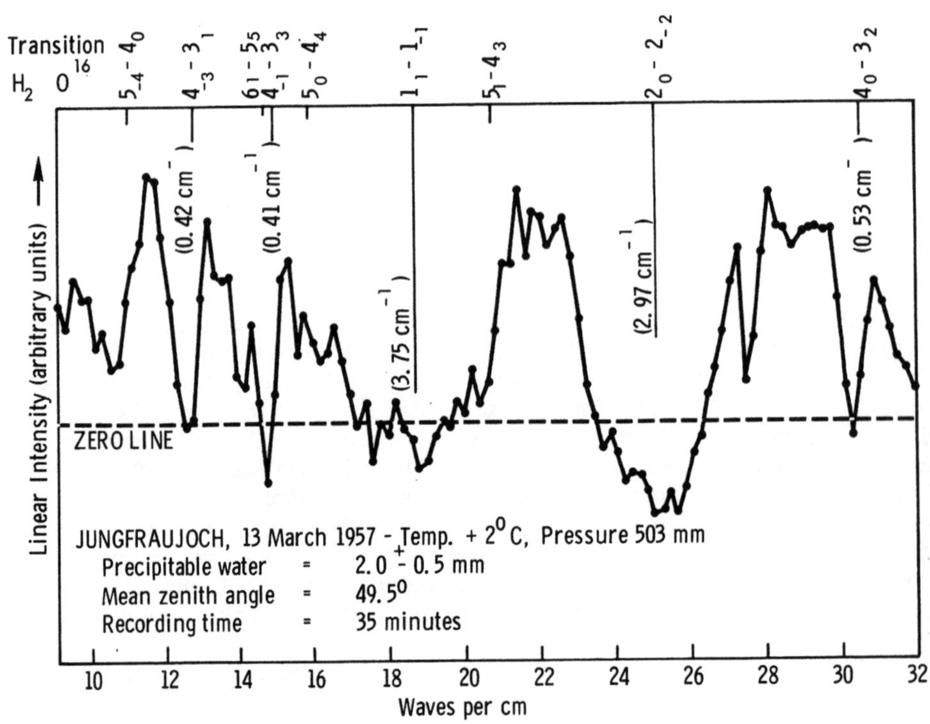

Fig. 4. Solar spectrum in submillimeter region. (After Gebbie.)

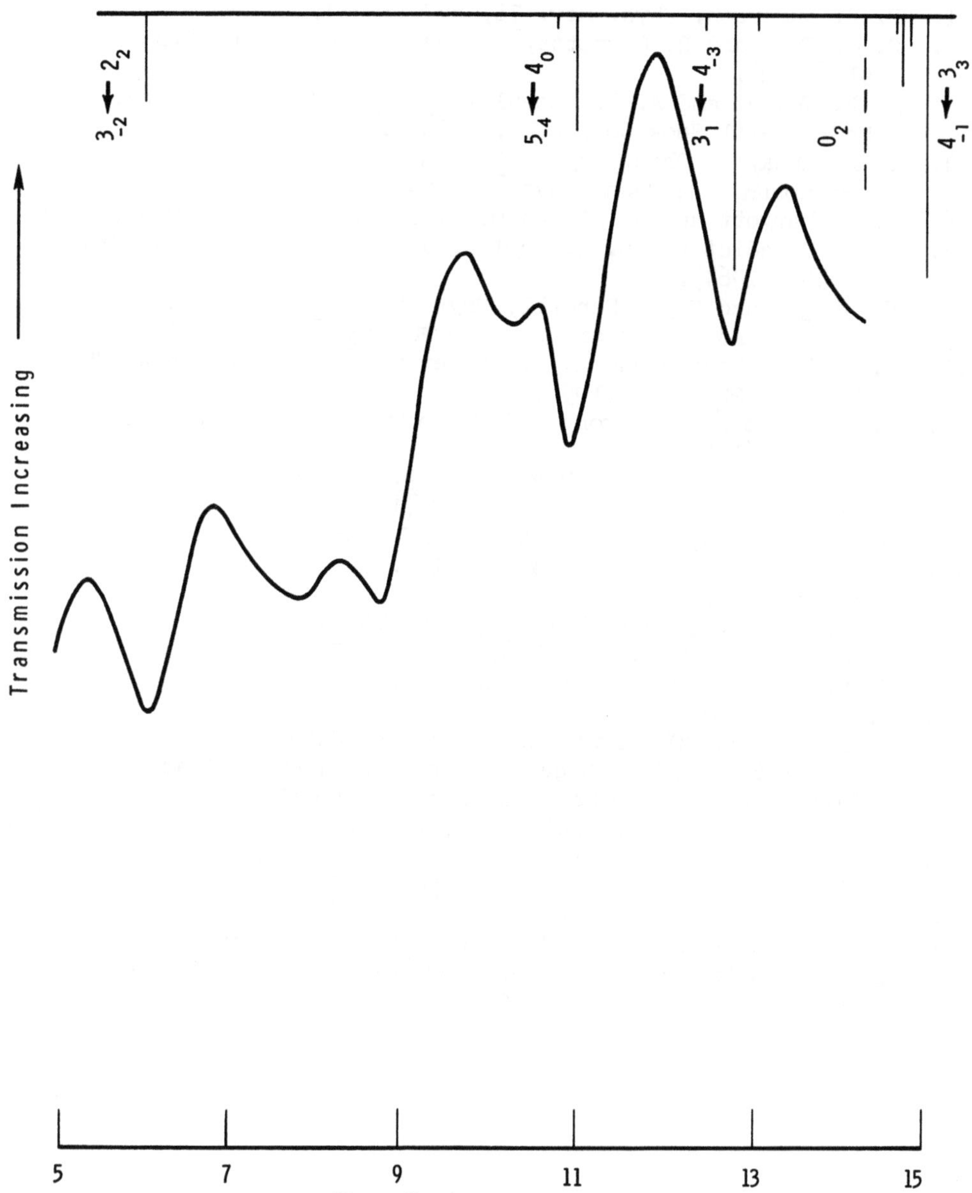

Fig. 5. Solar spectrum in submillimeter and millimeter region. (After Gebbie.)

REFERENCES

1. P. W. Anderson, Phys. Rev. 76, 647 (1949).
2. C. J. Tsao and B. Curnutte, J. Quant. Spect. Rad. Trans. 2, 41 (1962).
3. C. H. Townes and A. L. Schawlow, Microwave Spectroscopy, McGraw-Hill Book Co., N.Y., (1955).
4. J. R. Rusk, J. Chem. Phys. 42, 493 (1965).
5. G. Birnbaum, J. Chem. Phys. 46, 2455 (1967).
6. A. A. Maryott and G. Birnbaum, J. Chem. Phys. 47, 3200 (1967)
7. W. S. Benedict and L. K. Kaplan, J. Quant. Spect. Rad. Trans. 4, 453 (1964).
8. N. Wright and H. M. Randall, Phys. Rev. 44, 391, (1933).
9. L. Genzel and W. Eckhardt, Z. Physik 139, 578 (1954).
10. N. G. Yaroslavsky and A. E. Stanevich, Izv. Akad. Nauk SSSR, Ser. Fiz 22, 1145 (1958).
11. G. A. Vanasse, J. Strong and E. Loewenstein, J. Opt. Soc. Am. 49, 309 (1959).
12. A. Hadni and C. Janot, Rev. Opt. 39, 451 (1960).
13. K. N. Rao et al, J.O.S.A. 52, 862 (1962).
14. N. I. Furashov, Opt. i Spek 20, 427 (1966)
15. S. A. Zvevakin and A.P. Naumov, Izv. Vysskikh Uchebn. Zav., Radiofiz 6,674 (1963).
16. J. H. VanVleck, Phys. Rev. 71, 425 (1947).
17. H. A. Gebbie, Phys. Rev. 109, 1174 (1957).
18. C. B. Farmer and P. J. Key App. Opt. 4, 1051 (1965).
19. M. Bader et al, Nature 214, 377 (1967).
20. J. Gay et al, "Observations en Balloon dans le Domaine Submillimetre," Colloque International enUtilisation des Ballons pour La Recherches Scientifique CNES, Paris (1967).
21. R. T. Hall and J. M. Dowling, Jour. Chem. Phys. 47, 2454 (1967).
22. D. M. Gates et al, Natl Bur. Std (U.S.) Monograph 71 (1964).
23. M. Lichtenstein et al, J. Mol. Spect. 20, 391, (1966).
24. L R. Blaine et al, J. Res. Natl Bur. Std. (U.S.) 66A, 223 (1962).
25. C. W. Tolbert et al, J. Geo. Res. 69, 1349 (1964).
26. H. A. Gebbie and W. J. Burroughs Nature 217, 1241 (1968).
27. H. A. Gebbie et al, Nature 212, 66 (1966).

FAR-INFRARED SOURCES

Ludwig Genzel

Physikalisches Institut der Universität

Freiburg, Freiburg i.Br., Germany

A. INTRODUCTION

Although the far-infrared region of the spectrum ($3\ cm^{-1}$ to $300\ cm^{-1}$) can be considered today as instrumentally well developed it is still rather difficult to measure absorptivity or reflectivity with high resolution or high precision. This is mainly so because of the lack of powerful sources even though the techniques of spectrometry as well as the development of detectors has made considerable progress during the last 15 years.

The purpose of this article is to show the present status of available sources and also to describe briefly the trends for the future as well as some of the attempts which have been made to find new operating principles for sources.

One might distinguish generally the following types of sources: "Thermal sources, coherent microwave-type sources, quantum-oscillators, high-energy electron sources."

Some other special types of sources not discussed here are described in the references.

B. THERMAL SOURCES

1. Blackbody Radiation

From Planck's law, the intensity emitted from a blackbody per unit frequency interval is given by

$$I_b(\omega) \propto \omega^3 \left[e^{(\hbar\omega/kT)} - 1\right]^{-1} .$$

Since in the far-infrared kT >> $\hbar\omega$ for all sources, one has from the above

$$I_b(\omega) \propto \omega^2 T$$

showing the decrease of the available power on going to lower frequencies with ω^2 if the resolution $\Delta\omega$ is kept fixed throughout*. It shows furthermore the weak effect of increasing the blackbody temperature (compare Fig. 1 [4]). The normal thermal sources available in the far-infrared range have temperatures between about 1500°K and 6000°K. A worthwhile increase in T would be a factor of 10 requiring temperatures which are extremely difficult to attain.

In order to illustrate the spectral power in different wavelength regions, let us consider a blackbody of 5000°K temperature and 1 cm^2 area. Its total emission is about 10^3 Watts. From this, 99.99% fall into the region from 445 cm^{-1} to ∞ and 99.999% in the region from 200 cm^{-1} to ∞. If the far-infrared is defined as being the infrared below 200 cm^{-1}, only 10^{-3}% of the total power of this 5000°-blackbody, which is about 10^{-2} Watt/cm^2, is available for spectroscopy for the whole region. For a 0.1 cm^{-1} wavenumber interval, a spectral power of 9.4×10^{-5} Watts/cm^2 is available at 200 cm^{-1} and only 3.8×10^{-8} Watts/cm^2 at 10 cm^{-1}.

For a non-black thermal source with a spectral absorptivity $A(\omega)$, its emission is given by Kirchhoff's law

$$I(\omega) = I_b(\omega) \cdot A(\omega) \quad .$$

One recognizes that in addition to a high temperature it is necessary to have a source of high absorptivity.

2. The Mercury Arc Emission

The mercury arc emission in the far-infrared was discovered by Rubens and v. Baeyer in 1911 [16]. Since then, no more convenient source with continuous emission has been found so that today the normal laboratory spectroscopy of the far-infrared below

*) Instead of keeping $\Delta\omega$ constant, it is occasionally more appropriate to keep $\Delta\omega/\omega$ fixed. The blackbody radiation falls off then proportional to λ^{-3} .

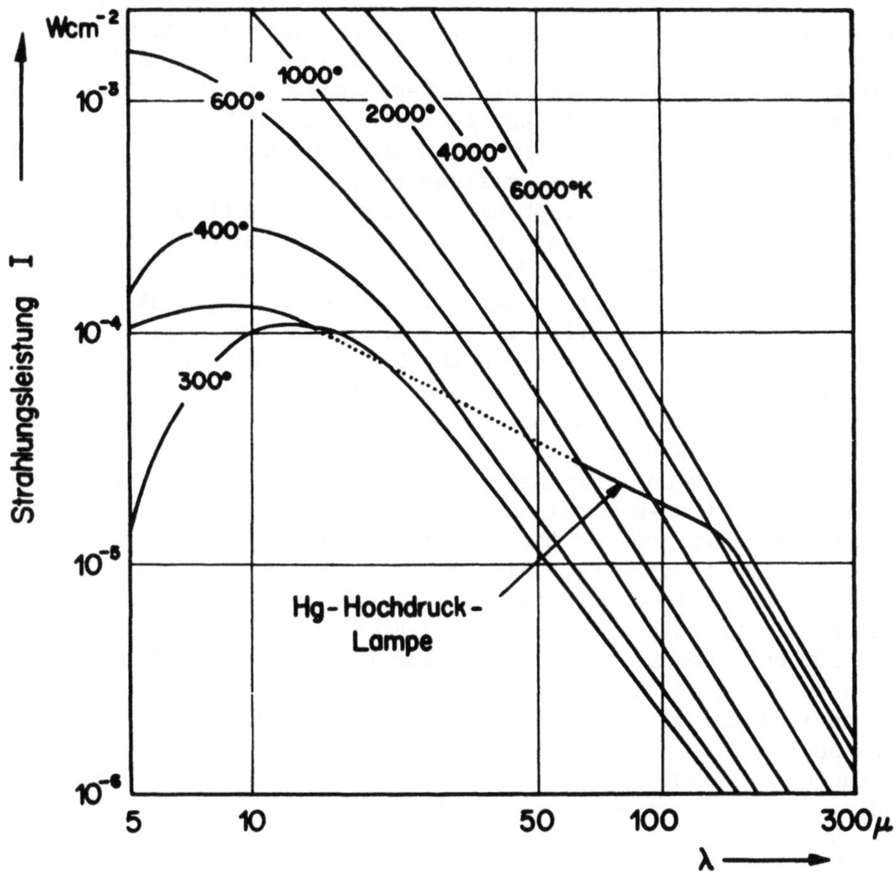

Fig. 1 Blackbody - radiation for different temperatures vs. wavelength (after [4]). The plot is for a constant $\lambda/\Delta\lambda$ of 100. The arrow shows mercury-arc emission after Bohdansky [24].

$100\ cm^{-1}$ is still done extensively with the high-pressure (1 to 10 atm) mercury lamp.

Due to the high pressure the ion-temperature equals the electron-temperature and is about $5000^\circ K$ to $6000^\circ K$ in the central part of the arc. From there it falls off parabolically to the fused silica envelope to about $1000^\circ K$. Because of the spatial dependence of temperature, and frequency dependence of the self-absorption of the plasma, the lamp appears in different spectral regions with different temperatures. At $10\ cm^{-1}$, the emission temperature is between $1000^\circ K$ and $200^\circ K$ because the radiation from the central arc is absorbed by the outer, cooler parts of the plasma which themselves emit at this lower temperature. Going to the $30\ cm^{-1}$ region, the self-absorption of the arc decreases so that the $5000^\circ K$ radiation from the central core can come out. This behaviour can be seen qualitatively in Fig. 2 showing an intensity measurement of the mercury arc with a non-dried grating spectrometer [55] of special construction to conserve the blaze condition.

Above $100\ cm^{-1}$ the arc emittance decreases again because the absorption factor $A(\omega)$ becomes smaller than 100 percent. This behaviour has been explained by Hettner [23] on the basis of free-free transitions (Bremsstrahlung) due to electron-ion impact and electron-neutral atom impact. Bohdansky [24] has measured the emittance of the mercury arc as a function of frequency and of pressure. Fig. 3 shows his experimental results in comparison with the theoretical calculations by Hettner. In practice, the mercury lamp is not very useful above $100\ cm^{-1}$ because the strong absorption of the fused silica envelope blocks the arc radiation.

An increase of the arc pressure above 1 atmosphere does not increase its far-infrared emission (Fig. 3) because of the increasing selfabsorption. One expects, however, on the other hand, a slightly better output in the very far-infrared from lamps with lower pressure, first because of the decreasing self-absorption of the outer parts of the plasma and second because then the electron temperature might exceed the ion temperature.

Attempts to push the hot central part of the arc nearer to the envelope by means of a magnetic field did not give a higher emission as one might expect from the above consideration. The reason for that failure is probably the fact that the arc discharge is wall stabilized.

3. The Neon Arc Emission

Recently, Hart [29] has developed a neon discharge lamp with an estimated electron temperature of about $15000^\circ K$ at a gas pressure of the order of 1 Torr. Because of the low pressure the light path through the lamp must be large in order to have a

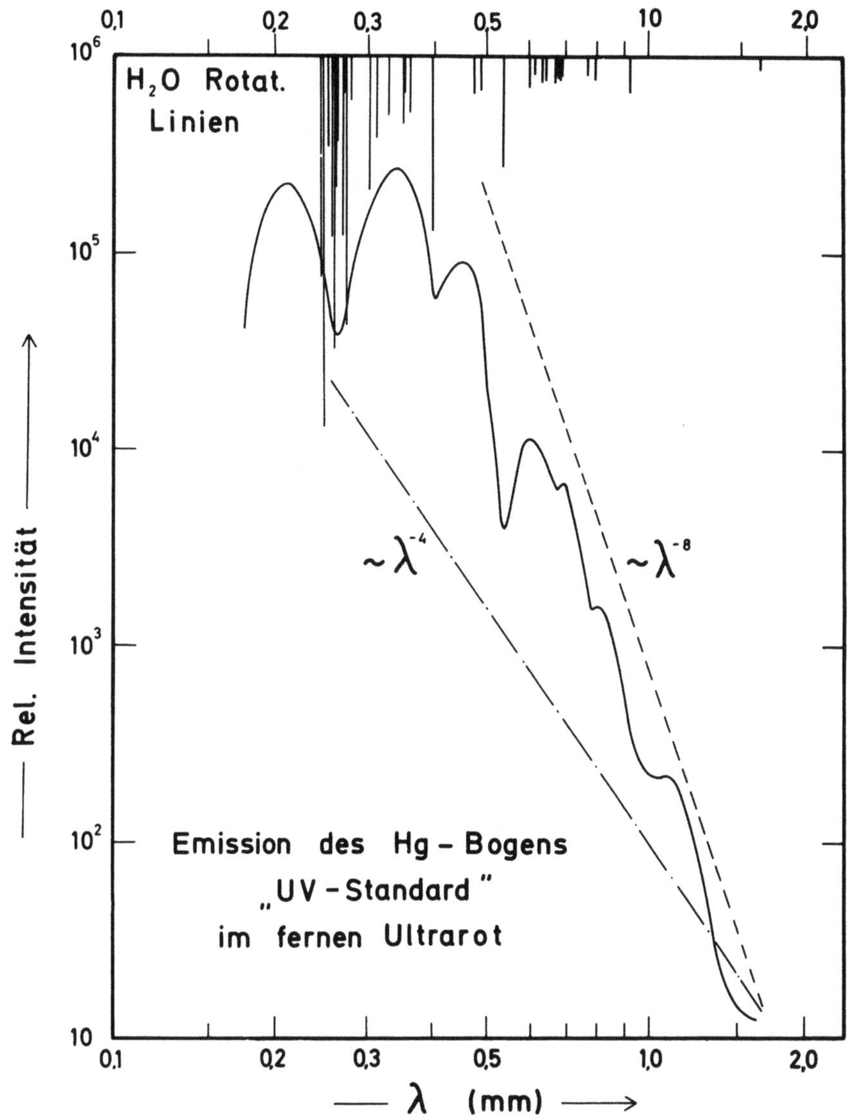

Fig. 2 Measured emission of the mercury - lamp in the submillimeter range showing a decrease steeper than λ^{-4}.

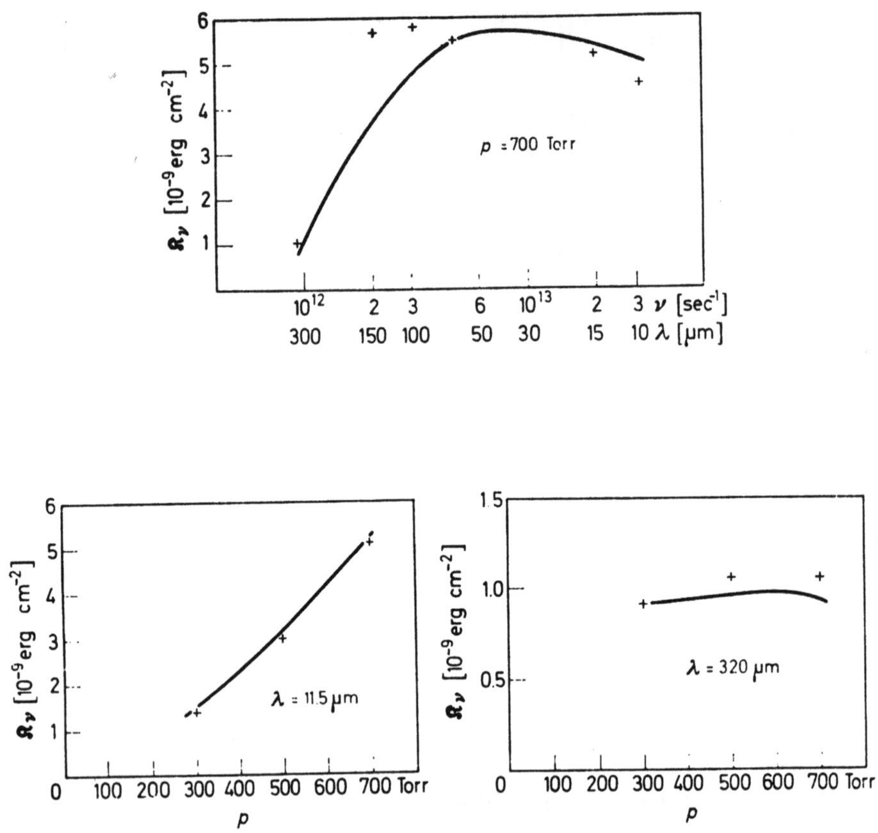

Fig. 3 Emissivity of the mercury arc.
Full lines: Theoretical curves after Hettner [23].
Crosses: Measurements after Bohdansky [24].
Upperhalf: Spectrum of emissivity.
Lower half: Emissivity vs. pressure in the lamp.

black source. This is achieved by an oversized waveguide system. Fig. 4 shows the construction of the neon lamp of Hart [29]. Fig. 5 shows the result of a comparison of the emission from this neon arc source with the emission from a mercury arc source as made by Nootz*. It was expected that the neon arc would be better than the mercury arc in the 5 to 25 cm^{-1} region because of its higher electron temperature and its lowered pressure. It turned out that only below 10 cm^{-1} is neon arc better than the mercury arc and even there much less than expected. This is probably due to an error in the estimation of the electron temperature in the neon discharge. Nevertheless, it has been shown by this development of Hart that one can improve the gas discharge type sources to some extent. Maybe one must make use of still lower gas pressures in order to increase the electron temperature further.

4. The Globar (SiC)

In the near and middle infrared the Globar is a well-known and convenient radiation source. Its emissivity at 1000 cm^{-1} is about 80 to 90 percent of a blackbody of 1250°K** which is about the temperature at which the Globar can be operated for reasonable length of time. In the far infrared, the absolute emissivity has not been measured yet but globar emission has been compared with the mercury arc emission by Plyler, Yakes and Gebbie [26, 41]. The result of this investigation (Fig. 6) is that below 140 cm^{-1} the mercury arc is increasingly better than the Globar. Above 140 cm^{-1} where the mercury arc emits only from its hot envelope, both sources are comparable and should be considered as rather weak sources.

The ThO_2-Source

Previously, Rubens and his coworkers [30] had found that the gas-heated Welsbach-mantle (thorium oxide with about 0.5 percent cerium oxide) is an excellent source for the far-infrared because its temperature can be as high as 1800°K and more, and also because of its low emissivity in the near infrared above 1250 cm^{-1} which diminishes the problem of prefiltering. Such a gas-heated source would be inconvenient today in far-infrared spectrometers which should be as free as possible from water vapor.

---

* D. Nootz, Diplomarbeit 1969, Physikalisches Institut der Univ. Freiburg.** Compare W. Brügel, Physik u. Technik der Ultrarotstrahlung p. 78, Vincentz Verlag, Hannover 1961.

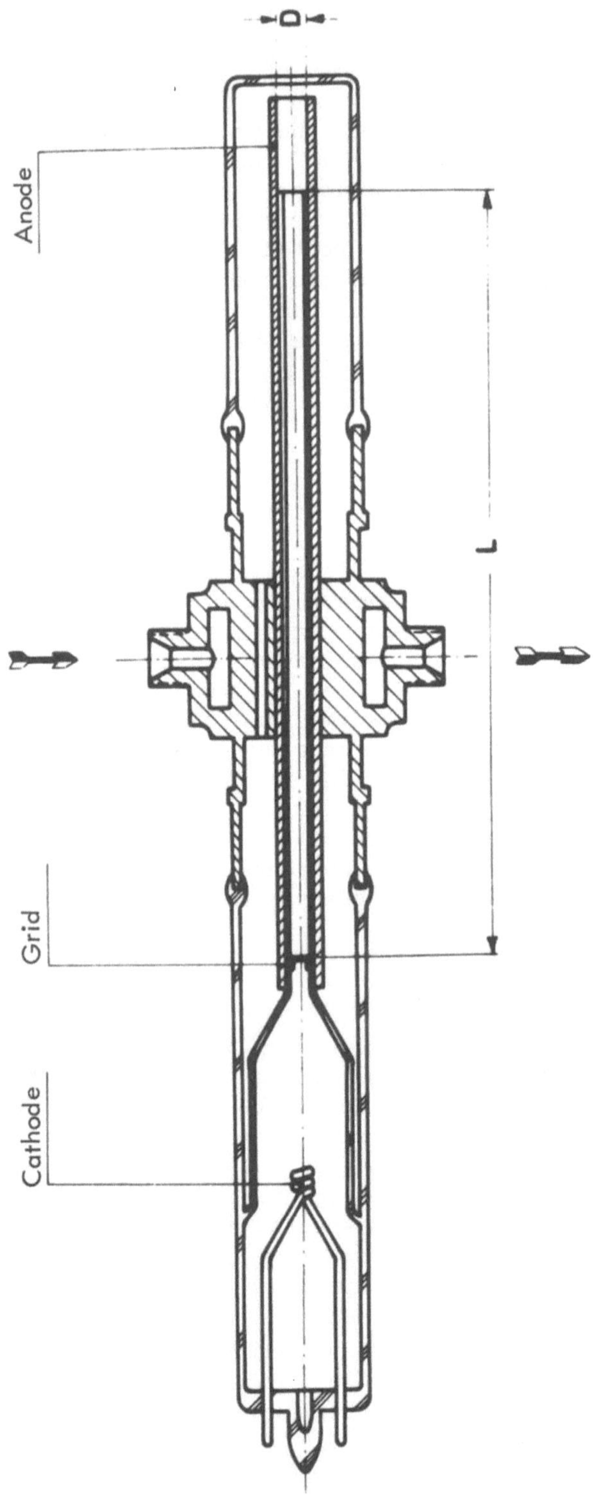

Fig. 4 The Neon - discharge lamp, after [29]. The lamp emits at the end of the waveguide near the anode.

J = 1A; U = 200 V; D = 6 mm; L = 150 mm.

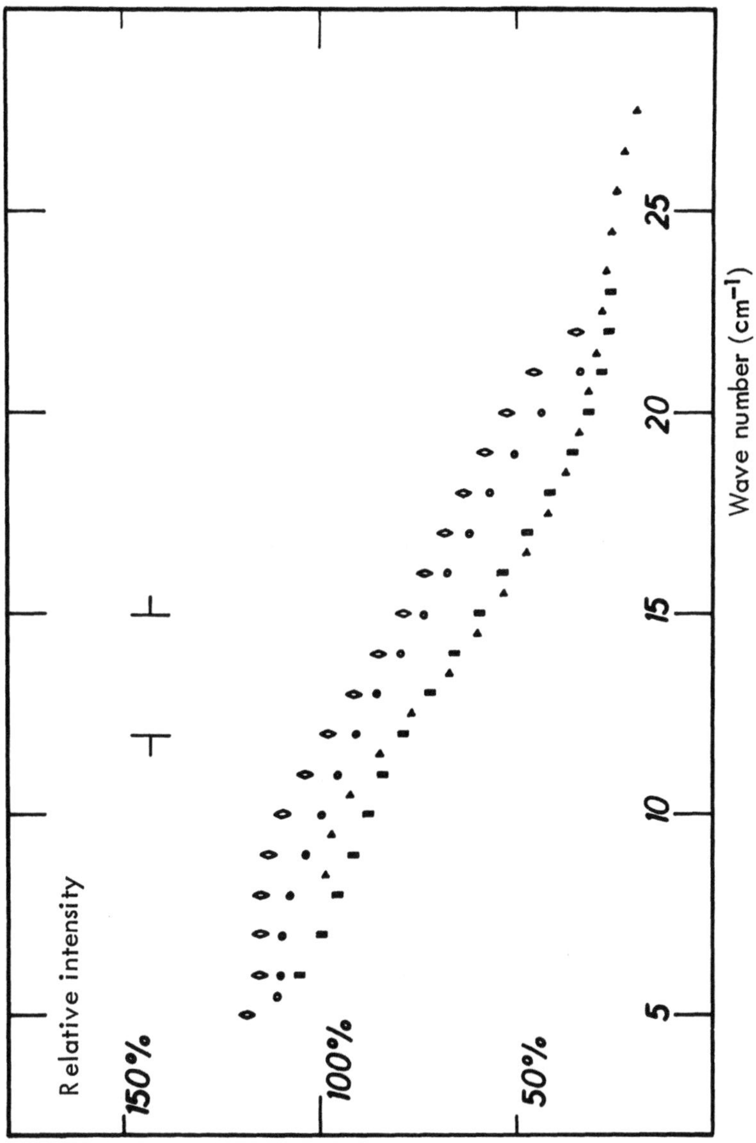

Fig. 5 The relative emission of the Neon - discharge lamp (Fig. 4) compared to the mercury - arc emission. After Nootz (l.c.).

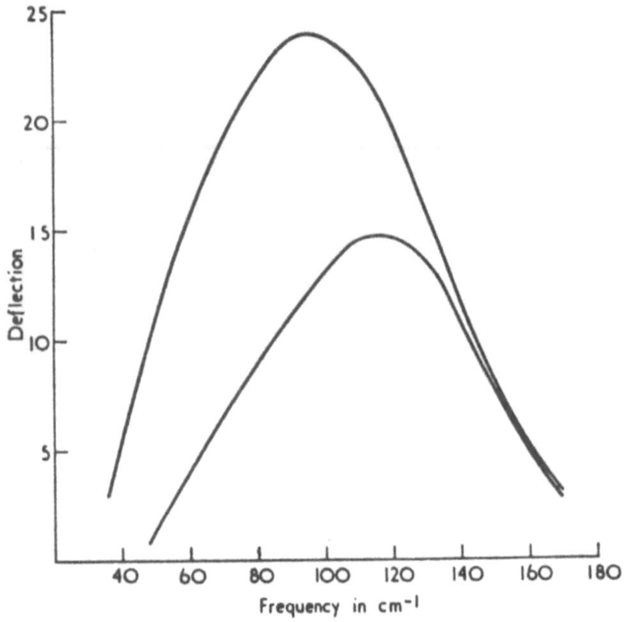

Fig. 6. The radiant energy of a mercury lamp, upper curve, and a globar source, lower curve, from 150 to 170 cm^{-1}. After [41].

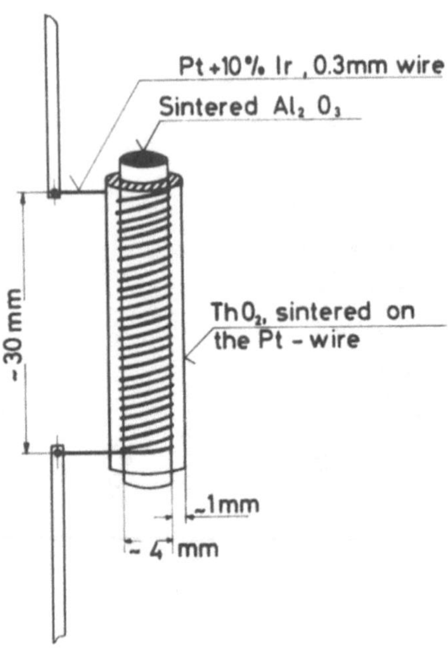

Fig. 7. The ThO_2 source. After [35].

A good source based on the Welsbach-mantle material should be heated electrically by a metal heater which can withstand temperatures up to about 1700 K and which does not emit too much near infrared radiation itself. A convenient glower of this sort, which can be easily made by the spectroscopist himself, has been described by Genzel and Neuroth [35] and later by Robinson[37]. The latter type uses a strip of platinum coated with ThO_2 while the former is made of a sintered Al_2O_3 cylinder wound with a layer of Pt + 10 % Ir wire on which the ThO_2 powder or Welsbach-mantle powder (Fig. 7 [35]) is sintered. This source needs a rather low current, less than 5 Amp., and can be made small so that no electrode cooling is necessary. From experience, this source is the most reasonable and convenient thermal source for use in the 100 to 300 cm^{-1} region of the far-infrared.

6. Thermonuclear Plasma Source

A rather unusual thermal source might become available for the far-infrared due to the attempts at producing extremly hot plasmas for thermonuclear reactions. Harding e.al. [39,40] have already shown that the 10^5 °K plasma of the ZETA instrument is an excellent far-infrared source, although rather inconvenient, as can be seen in Fig. 8.

C. COHERENT MICROWAVE TYPE SOURCES

1. General Considerations

An ideal source for the far-infrared would be a monochromatic and tunable oscillator as they are successfully used in the microwave region. In that region, the most convenient ones are the klystrons, magnetrons and the backward wave oscillators (BWO). They are partly tunable, the BWO best and the magnetrons nearly not.

It is extremely difficult to scale down the essential parts of these oscillators in order to make sources for the shorter millimeter or the submillimeter region because of problems in mechanical tolerance and in heat dissipation. Klystrons were constructed down to 2 mm wavelength (Van Iperen and Kuypers, 1965 [42])but it seems that this is already the short-wavelength limit. BWO's [43,44] are more favourable for the shorter millimeter region and are already available in the longer submillimeter range. Fig. 9 gives a survey of the available klystron and BWO-tubes with their power levels.

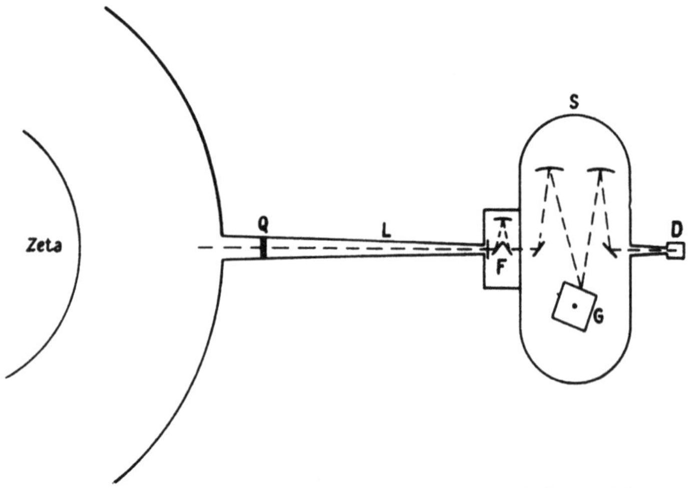

General experimental arrangement (schematic).

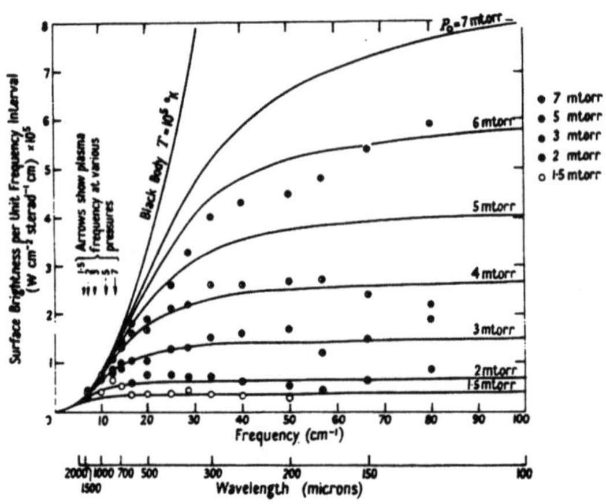

Fig. 8. General experimental arrangement of the ZETA discharge (upper half) as a far-infrared source and its measured surface brightness(lower half), after [39, 40].

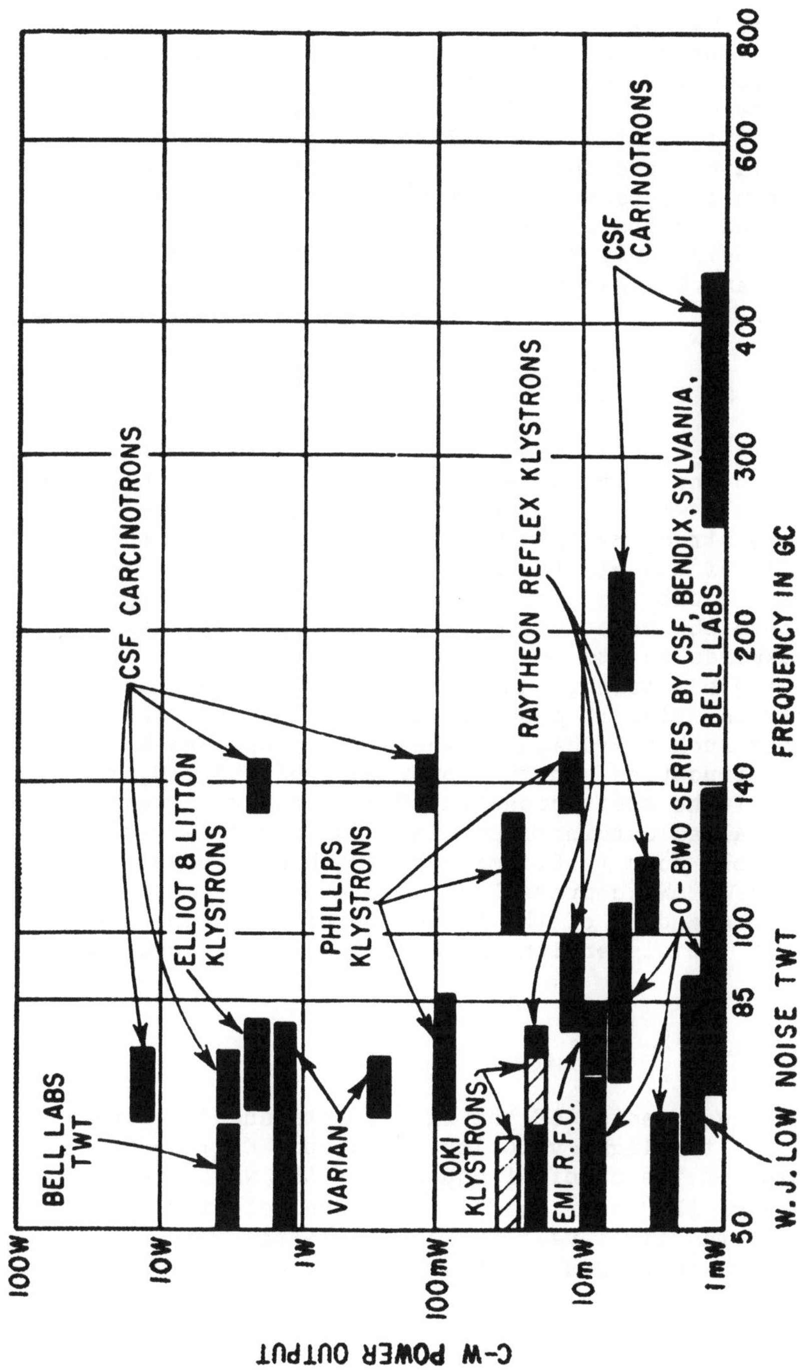

Fig. 9 Power levels for tubes either pulsed or operated CW above 50 Gc/s.

Very promising for the far-infrared is the possibility of generating harmonics with a nonlinear device using the above mentioned millimeter tubes as driving oscillators. The harmonics would be as monochromatic as the fundamental and can be tuned over the same fractional range. The problems arise here mainly in finding a nonlinear element which can transform efficiently the available high power of the driving oscillator into the harmonics. If these problems are solved then also parametric devices seem possible for the far-infrared (compare section D3).

In the following only the principle and performance of the BWO-tubes and the harmonic generation are considered in more detail.

2. Backward Wave Oscillator (BWO)*)

The BWO is a special kind of traveling wave oscillator and is based on the interaction of electrons with an electromagnetic field which is slowed down in its phase velocity by a slow wave structure to almost the same value as the group velocity of the electrons. In order to make the tube oscillate, a feedback is introduced so that the tube operates in a mode where the group velocity of the electromagnetic field is backwards to its own phase velocity and also to the velocity of the electrons, causing in this way a prebunching of the electrons. Fig. 10 shows in the upper part a schematic diagram of such a BWO and in the middle part two kinds of slow-wave structures. In the lower part of Fig. 10, the construction of the most successful C.S.F. backward wave oscillators, called carcinotrons, is shown. Such carcinotrons are now available down to 0.4 mm wavelength with power levels of the order of 1 mW (compare Fig. 12).

For an understanding of the principle of a BWO one should consider first the dispersion characteristic of a homogeneous waveguide

$$K^2 = \frac{\omega^2}{c^2} - K^2_{cutoff}$$

where K is the wavevector of the field, ω its angular frequency, and c the light velocity in vacuum. K_{cutoff} is the wavevector of the cutoff-frequency of the waveguide. As shown in Fig. 11 (upper half), the phase velocity ω/K always exceeds the light velocity c , while the group velocity $d\omega/dK$ is always smaller than c and has the same direction as the phase velocity.

*) This description of the BWO follows closely the survey article of Robson [45].

If now the waveguide has a periodic loading, one gets Bragg reflections at $K \cdot d = \pm n \cdot \pi$ (d = periodicity length), and stop-bands and pass-bands for successive frequency regions (Fig. 11 lower half). This is in complete analogy with the band structures appearing in the electronic or lattice-dynamical structure of crystalline solids. The dispersion is not restricted to the first Brillouin-Zone ($-\pi/d$ to $+\pi/d$), because contrary to lattice dynamics, the wave field is defined at any point of space and not just at the lattice points.

If one considers now only the lowest pass-band (Fig. 11), one has a dispersion law $\omega(K)$ which is periodic in K with the

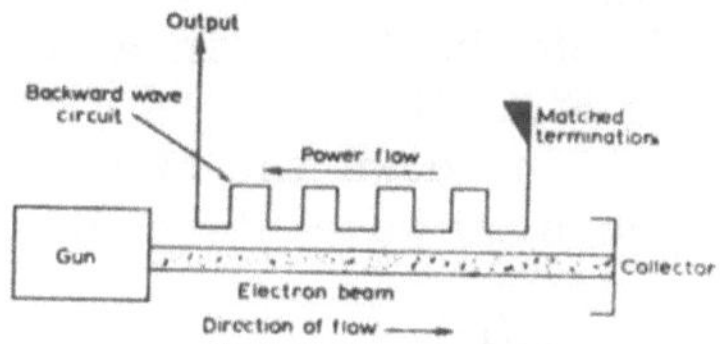

Schematic of backward wave oscillator.

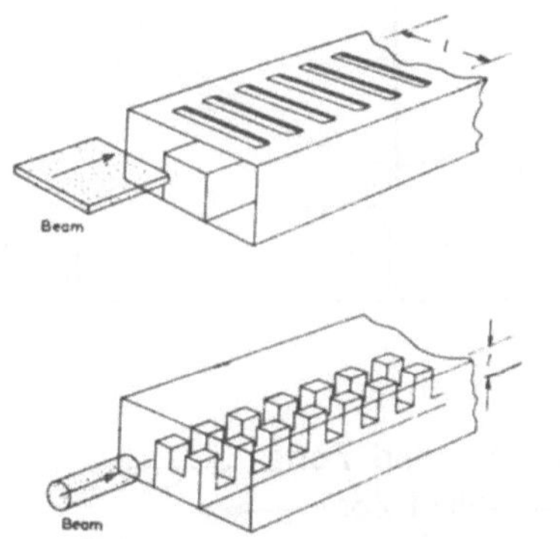

The Karp slow wave structure; (b) the Vane slow wave structure.

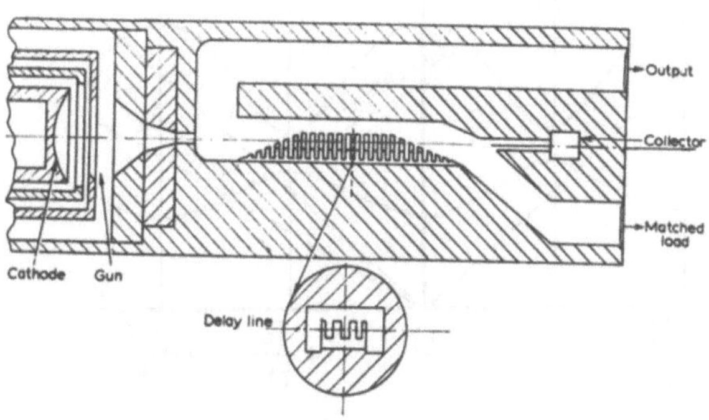

Sectional view (schematic) of C.S.F. backward wave oscillators.

Fig. 10. The backward wave oscillator (taken from [45]).

period $2\pi/d$. $n = 1$ defines the first zone, $n = 2$ the second zone and so on. According to the integer n one speaks about the n-th space harmonic. In the positive part of the second zone (+ π/d to $+2\pi/d$ one has just the conditions necessary for the BWO operation because there the phase velocity ω/K is positive and the group velocity $d\omega/dK$ is negative. This would be true also in the positive zone-parts between $3\pi/d$ to $4\pi/d$, between $5\pi/d$ to $6\pi/d$ etc., but the higher space harmonics have generally lower amplitudes. The phase velocity of the second space harmonic is almost the same as the electron velocity due to the interaction so that ω/K of the field can be tuned by the accelerating voltage for the electrons in the tube. In this way the BWO can be tuned in a considerable frequency range as seen in Fig. 12.

The limitation for scaling down the BWO's is given by problems of accuracy of the slow wave structure and of achieving

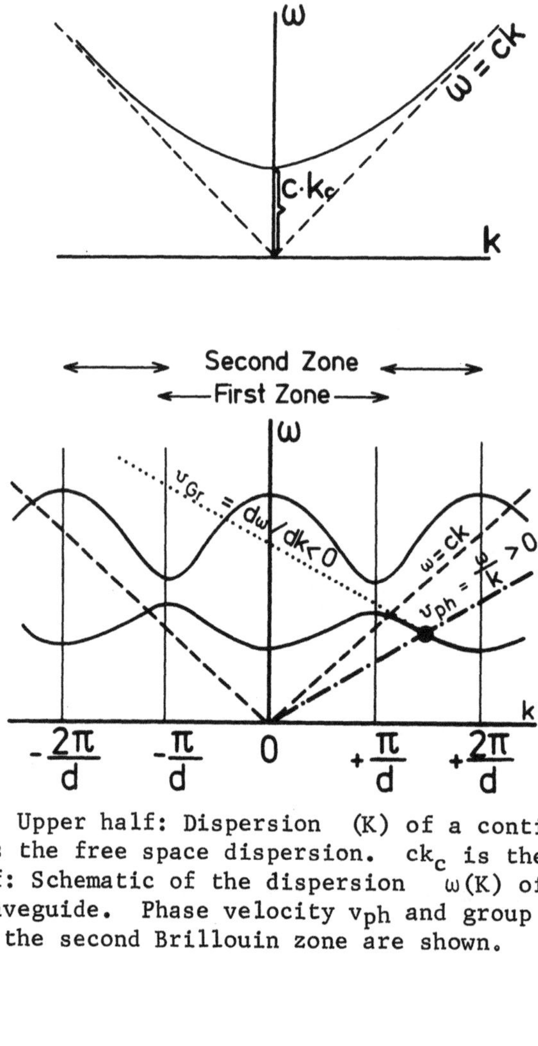

Fig. 11. Upper half: Dispersion (K) of a continuous waveguide. $\omega = cK$ is the free space dispersion. ck_c is the cutoff frequency. Lower half: Schematic of the dispersion $\omega(K)$ of a periodically loaded waveguide. Phase velocity v_{ph} and group velocity v_{gr} at a point of the second Brillouin zone are shown.

the necessary current to start oscillations from the noise. This starting current increases as ω^4 and is ultimately limited by the thermionic cathode emission. The total limit for BWO's seems to be about 100 cm^{-1} (0.1 mm wavelength).

3. Harmonic Generation with Nonlinear Elements

A nonlinear element driven by the fundamental power of an intense mm-oscillator or submillimeter-oscillator should produce harmonics which will be as monochromatic as the fundamental source. In principle, this would be the ideal tunable source because a fundamental oscillator which is tunable by only 10 percent of its frequency, as is possible with klystrons and BWO for example, could produce tenth and higher orders harmonics which are tunable over a range greater than their separation. Unfortunately, it is difficult in practice to produce harmonics higher than the tenth because of the lack of nonlinear elements with high conversion efficiency which

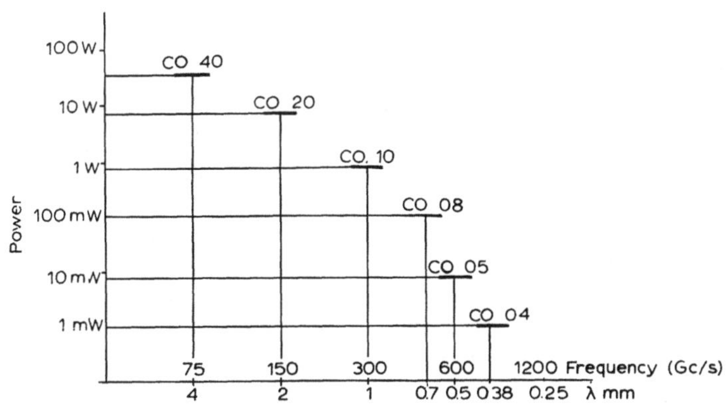

Performance of C.S.F. backward wave oscillators (after Yeou).

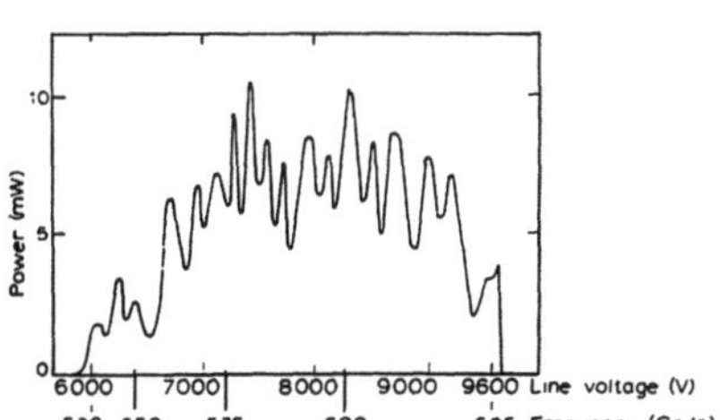

Output power as a function of line voltage for C.S.F. 0.5 mm B.W.O. (after Yeou).

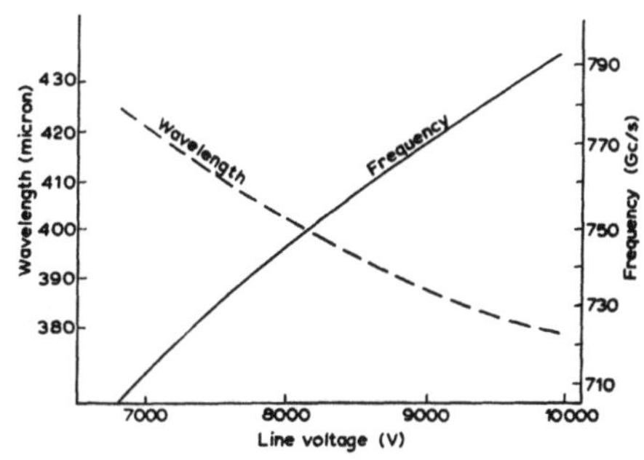

Fig. 12. Performance and characteristics of C.S.F. backward wave oscillators after [43,44], taken from [45].

can be driven with high fundamental power. This high power is necessary to obtain higher harmonics.

Consider the harmonic generator, and the field range of the essential non-linearity. The time for sweeping the field over this non-linearity determines which harmonics are strong. The faster one sweeps over this range the richer the spectrum will be in high harmonics. A high driving field now is just the way to sweep fast over the essential nonlinearity. This consideration also offers some understanding of the empirical fact that one achieves for a given harmonic a saturation with increasing driving power. It can be shown generally (Page [53]) that for resistive nonlinear elements the output power falls off with the harmonic number n at least as n^{-2}. For reactive nonlinear elements such a limitation does not exist.

Various kinds of nonlinear elements for harmonic generation or mixing in the far infrared have been tried experimentally or discussed theoretically: ferrites [54, 59, 60, 66], antiferromagnets [78], ferroelectrics [67], superconductors [62, 63], elements with multiphoton processes [61], phase-focused electron beams [52, 68, 79], plasma [58, 64, 65], semiconductor diodes (compare [74]). One can think also of crystals which are used for nonlinear optics in the visible or near - infrared (compare section D3). Here, the nonlinear behaviour can arise not only from electronic transitions, but also from lattice vibrations.

Nonlinear devices which already have proven successful for practical spectroscopic work in the far-infrared are those based on semiconductor diodes and on the plasma. These shall therefore be considered in the following. Detailed survey articles on these subjects have been published by Burrus [74] and by Backer [80].

The first detailed investigation of the harmonic generation with semiconductor diodes has been made by Nethercot [48] who used a multiplier with a silicon-tungsten contact driven in the K-band range. He found that the nth harmonic power P_n increases with the fundamental power P_1 approximately as $P_n \propto P_1^n$ and reaches a saturation value which is slightly greater for higher than for lower harmonics (Fig. 13). This saturation sets in for semiconductor-metal contacts at about 30 to 100 milliwatts incident power. Higher powers unfortunately can not be applied to these kinds of generators because of difficulties in heat dissipation.

A further severe limitation for the use of these multipliers down to shorter wavelengths in the far-infrared is the empirical fact that the conversion efficiency from the fundamental to a given harmonic - for example for a doubler or a tripler etc. - falls off as about $1/\nu^2$ due to increasing losses with increasing driving frequency ν. This means

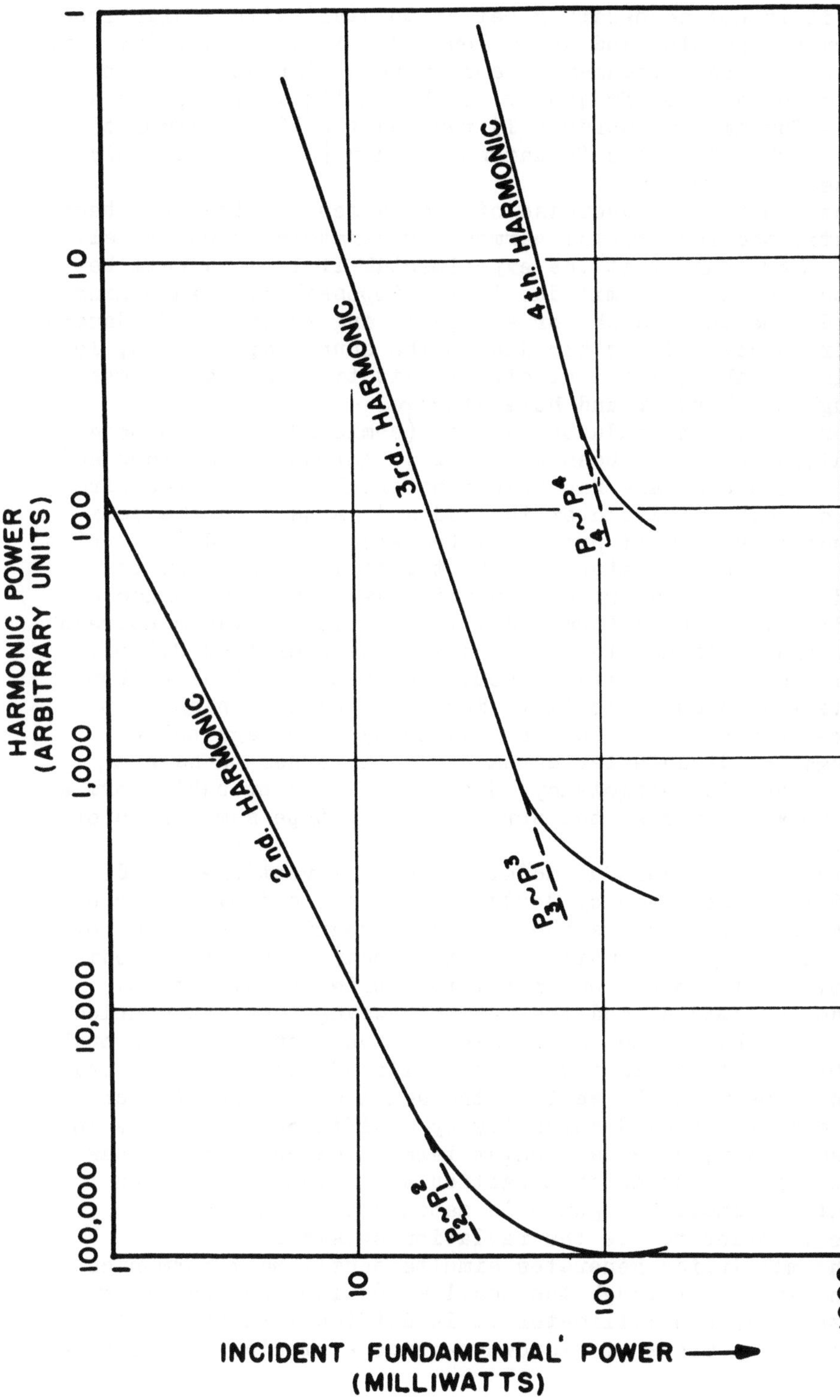

Fig. 13. Harmonic power from crystal frequency multiplier vs. incident fundamental power, after [48].

that it is not necessarily better to start with a high-frequency fundamental and use a low harmonic than starting with a low-frequency fundamental and using a high harmonic in order to get a given frequency of the multiplier. So, for a 32Gc/s fundamental Gordy [51] reached the 18 th harmonic while with a 280 Gc/s fundamental Bauer [77] reached only the second harmonic.

The general n^{-2} decrease of the harmonic output has been investigated for crystal harmonic generators by Ohl et.al. [56]. They found that the experimental limit lies well below the ideal n^{-2} limit due to the impossibility in making a good impedance match for all phase angles of the fundamental frequency and further due to the shunt capacitance, local heating at the point contact, transit time of the charges through the barrier and hole storage.

For metal-semiconductor diodes (semiconductor-semiconductor diodes have not been used here), germanium, silicon and gallium arsenide are favourable materials. A tungsten wire of about 0.04 mm diameter is used and is held on the semiconductor by a small pressure. The wire is etched to a point with radius of curvature of about 1 to 10 μ in order to avoid capacitive shortage. In this way, the capacitance of the diode, which is formal due to a depletion layer between metal and semiconductor, can be kept smaller than 0.1 pF. The semiconductor surface should be treated either by bombardment with 40 kV positive ions [56] or by a high-temperature phosphorus diffusion [74] or by a voltage pulse through the diode after contact [55] in order to have a good conversion efficiency. Such treatments probably form a thin n-type layer at the surface of a p-type semiconductor [80].

The first successful models of such a rectifier diode multiplier were made by Gordy et.al. [47] and by Nethercot [46,48]. Fig. 14 shows the crossed-waveguide multiplier of King and Gordy [47] which is still today, with minor modifications, the most convenient multiplier construction.

The fundamental and the harmonic waveguides are coupled by a small hole, the cat-whisker of tungsten is put through and brought to contact with the semiconductor which is nearly in plane with the wall of the small waveguide. The contact can be achieved carefully by a differential screw. In the small waveguide, a plunger terminates one end and has to be adjusted to match a particular harmonic. The open end of the small waveguide is matched by a horn to free space in order to use the radiation outside.

The multiplier generates simultaneously many harmonics which can pass through the small waveguide. For wavelengths shorter than one millimeter it is difficult to filter these harmonics by means of microwave-guide elements, so one uses far-infrared spectrometric devices to separate the harmonics. Often, simply a far-infrared grating spectrometer is used for the harmonics separation. The output horn of the multiplier is then placed at the entrance slit position of the spectrometer. Fig. 15 shows an example of such a measurement (Genzel et.al. [55]) where an especially constructed echelette grating for conserving the blaze condition has been used. A more convenient way to separate the different harmonics in free space is to use a metal-mesh Fabry-Perot filter (Ulrich et.al. [71]). As an example, Fig. 16 (taken from [71]) gives a record of different harmonics passed by such a tunable Fabry-Perot filter.

Another type of harmonic generator, the Froome plasma multiplier [57], can handle a much greater driving power than the metal-semiconductor harmonic generator. The construction of the Froome multipliers is shown in Fig. 17. A high pressure (≤ 600 atm) argon arc between platinum electrodes produces a very high cathodic electron density which responds rapidly to fast changes in the arc current, produced by the driving field. The nonlinearity is essentially the modulation of the cathodic field by means of the driving microwave field. The output of such a Froome multiplier is shown in Fig. 18 (taken from [80]). Although this principle seems to solve the problems for the lower submillimeter region, there are other severe limitations in practical use. One is that the high pressure necessary for stabilizing the operation of the multiplier requires a rather thick output window with its unavoidable absorption. Another difficulty is the need for careful adjustment of the arc. Any change in the driving frequency necessitates a new adjustment of the whole system, a fact which makes the continuous tuning impossible. Approximate calculations as well as experiments [75] show that the power P_n of the n-th harmonic falls off like $(n^2-1)^{-2}$ or as n^{-4} for the higher harmonics. Nevertheless, as long as no other principle of a bulk multiplier has proven successful, the Froome multiplier must be considered as best.

D. QUANTUM OSCILLATORS

1. General Considerations

Quantum oscillators use transitions of atomic-, molecular - or crystal - energy levels the population of which is inverted by means of a pumping action in order to make the stimulated emission the dominating process. Corresponding to the low photon energy in the far-infrared of 10^{-2} to 10^{-3} eV and the rather long lifetimes of excited states which are necessary to invert the population, only a few

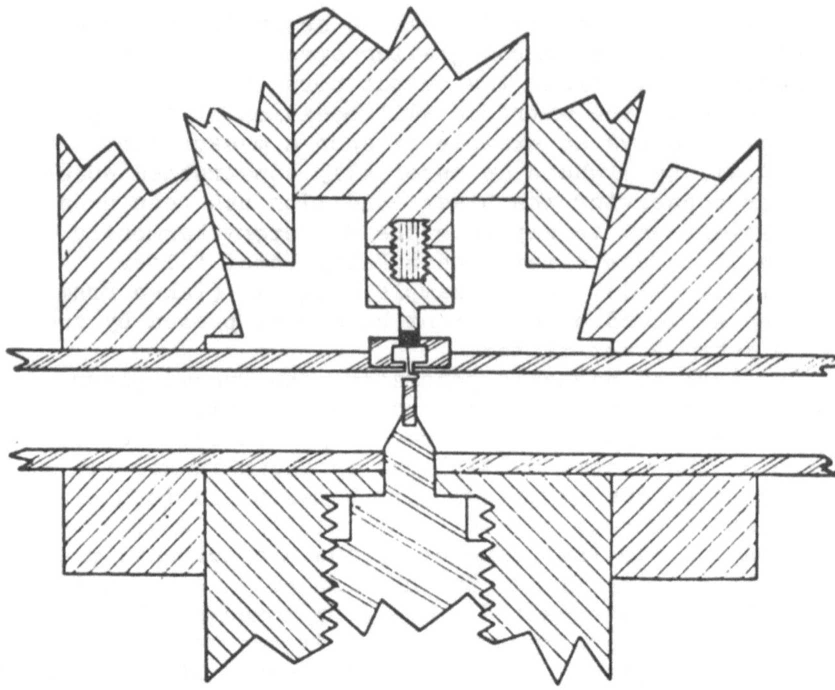

Fig. 14. The crossed waveguide harmonic generator, after King and Gordy (47).

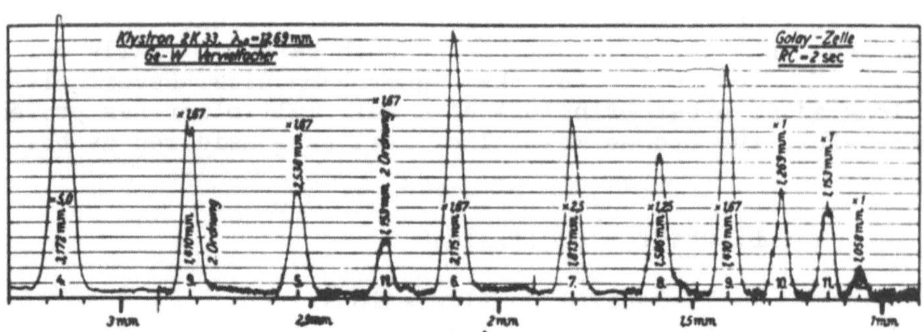

Fig. 15. Record of the spectrum of a harmonic generator with a germanium-tungsten contact and a 12.69 mm klystron as fundamental source. A grating spectrometer acted as harmonics selector. The numbers below the peaks characterize the harmonics, some of which appear in the second-order diffraction of the grating. After [45].

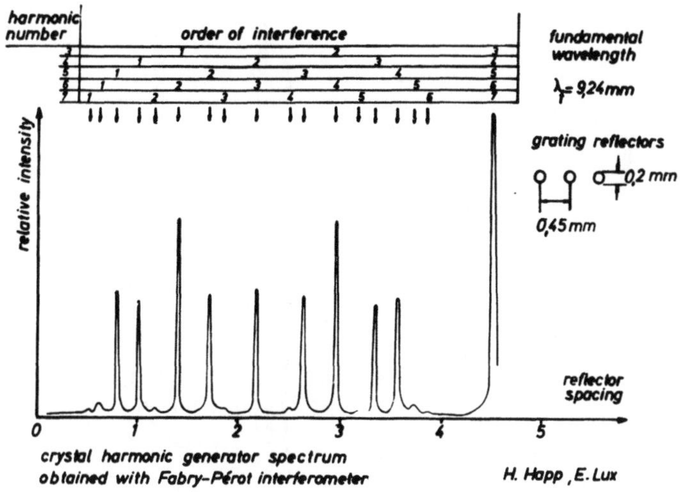

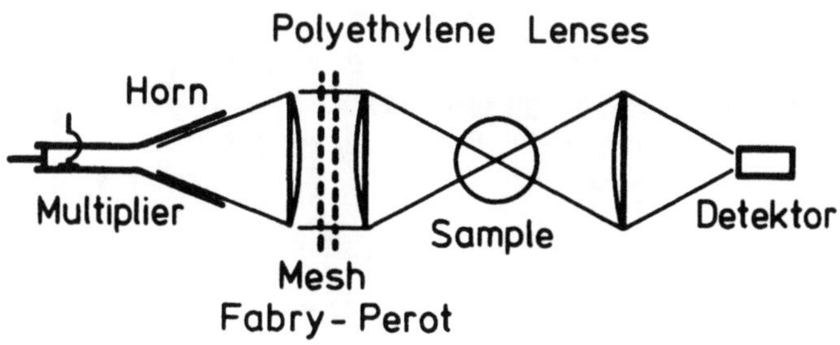

Fig. 16. The use of metal mesh Fabry-Perot for selecting the harmonics of a frequency multiplier.
Upper half: Record of a spectrum. The different peaks arise from the third to seventh harmonic orders of the Fabry-Perot. After [71].
Lower half: Quasioptical arrangement.

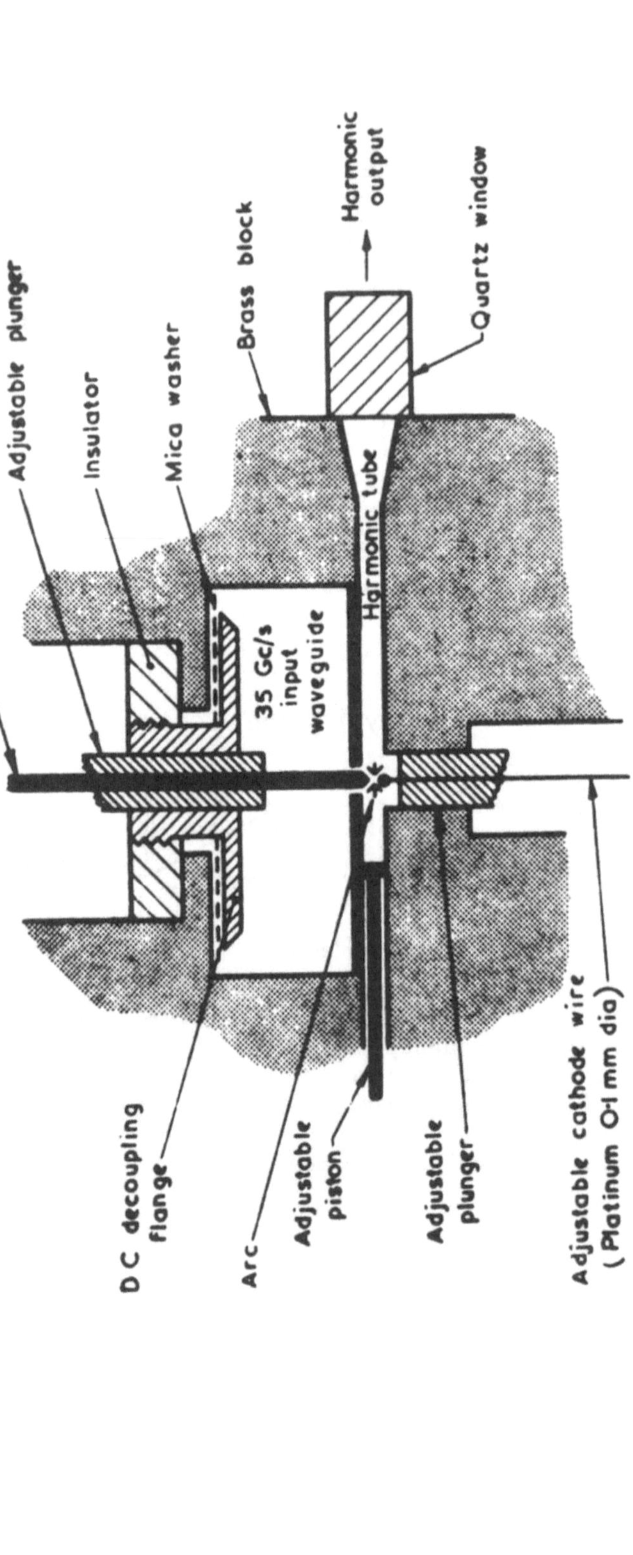

Fig. 17 Principle of the high-pressure arc harmonic generator, after Froome [57].

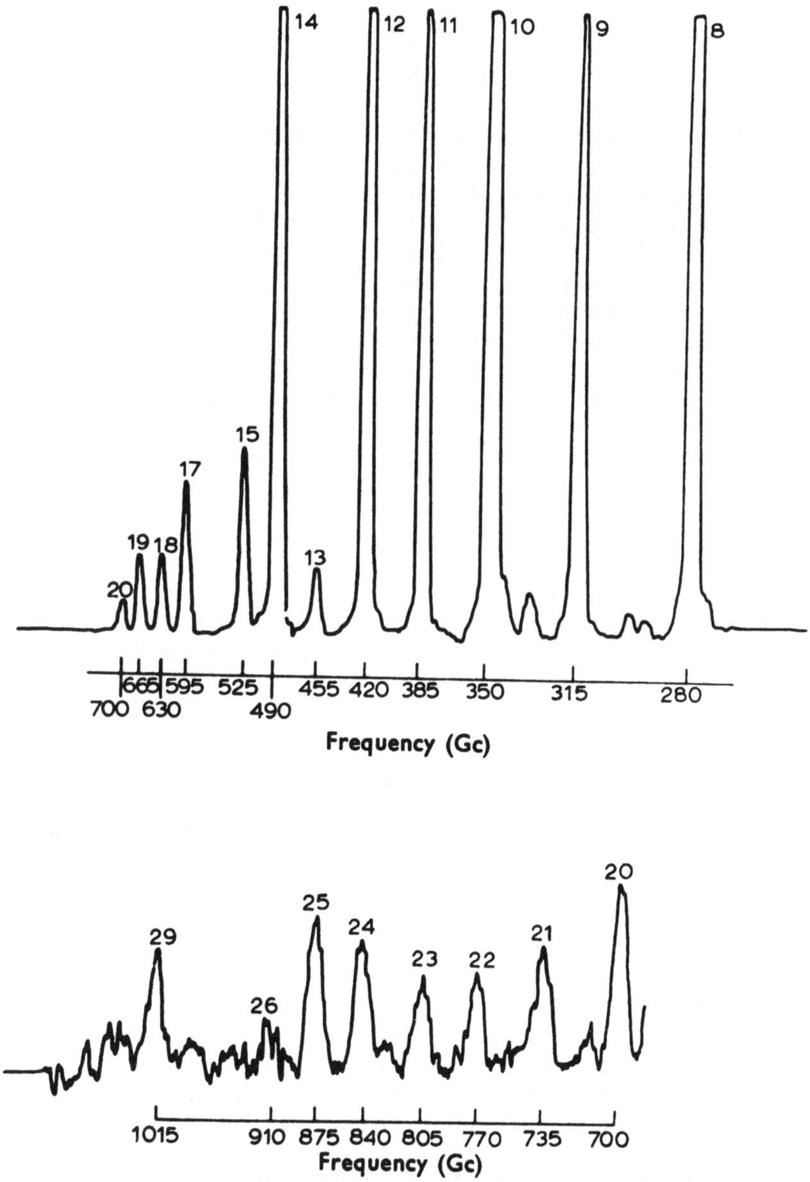

Fig. 18 The output of a plasma-metal junction microwave harmonic generator. The harmonics are numbered by multiples of the fundamental frequency, namely 35 Gc. Atmospheric water vapour absorbs the 16th harmonic at 0.54 mm. The output in the 13th appears low because of poor output coupling at this frequency. The deflexions between the 8th and 9th; 9th and 10th harmonics are grating 2nd order effects (from FROME [1963]).

systems have been found to oscillate in the far-infrared. These are at the moment the gas lasers with the noble gases Ne, He-Ne, Ar, Xe (Patel et.al. [87,88]), with H_2O and D_2O (Gebbie et.al. [89]) and with gases containing H,C,N or D,C,N (Gebbie et.al. [90,91], Mathias et.al. [92,93]). There is furthermore the Josephson-type superconducting tunnel junction as a possible generator of submillimeter wave radiation (Dayem et.al. and Langenberg et.al. [102, 103]). Some other types of quantum oscillators like molecular-beam masers (Barners [81], Grivet et.al. [86]), solid-state masers Hughes [83]) or the Gunn oscillator (Gunn [85] reach the millimeter region already and may be potential submillimeter oscillators in future,too. Very recently, first steps have been made [113, 120-122] to produce the difference frequency of two CO_2-laser lines by means of nonlinear magnetoplasma effects in InSb. This opens exciting possibilities for nonlinear optics and parametric oscillators for the far-infrared. In the following, only the submillimeter gas lasers and this nonlinear frequency mixing will be considered.

2. Gas Lasers

The noble gas laser emission in far-infrared is based on higher transitions of the noble gas atoms as those shown for example in Fig. 19 for the case of Ne [88]. The experimental set-up of a noble gas laser is shown in Fig. 20 (taken from [88]) The pressures used are for Ar:0.05 Torr, for Xe:0.03 Torr and for Ne:0.15 Torr. These lasers are generally used in CW-operation and yield output powers of the order of 10^{-8} to 10^{-9} Watt. The longest wavelength found is 132 μ with Ne as shown in the survey Fig. 21 and in table I. With the H_2O - or D_2O-lasers, powers of 10^{-1} to 10^{-2} Watts in pulsed operation and 10^{-3} Watts in CW-operation are reached (table I). The longest wavelength found here is 220 μ with H_2O. The mechanism of the H_2O lasers has been recently explained as vibration-rotational and pure rotational transitions of H_2O by Hartmann et.al. [116], Benedict [117], Pollack et.al. [118,119]. Extreme precise absolute frequency measurements by phase locking with klystron harmonics on the 118.6 μ H_2O laser line has been made by Frenkel et.al. [110].

Most important because of its high power and its many lines spread over the whole far-infrared (Fig. 21, table I) is the laser action with gases containing H,C,N and sometimes also I. The power levels reached are 10 Watts in pulsed operation and 10^{-1} Watts in CW-operation. The longest wavelength found is 774 μ with gas containing ICN (Steffen et.al. [98]). Some of the HCN laser lines have been assigned

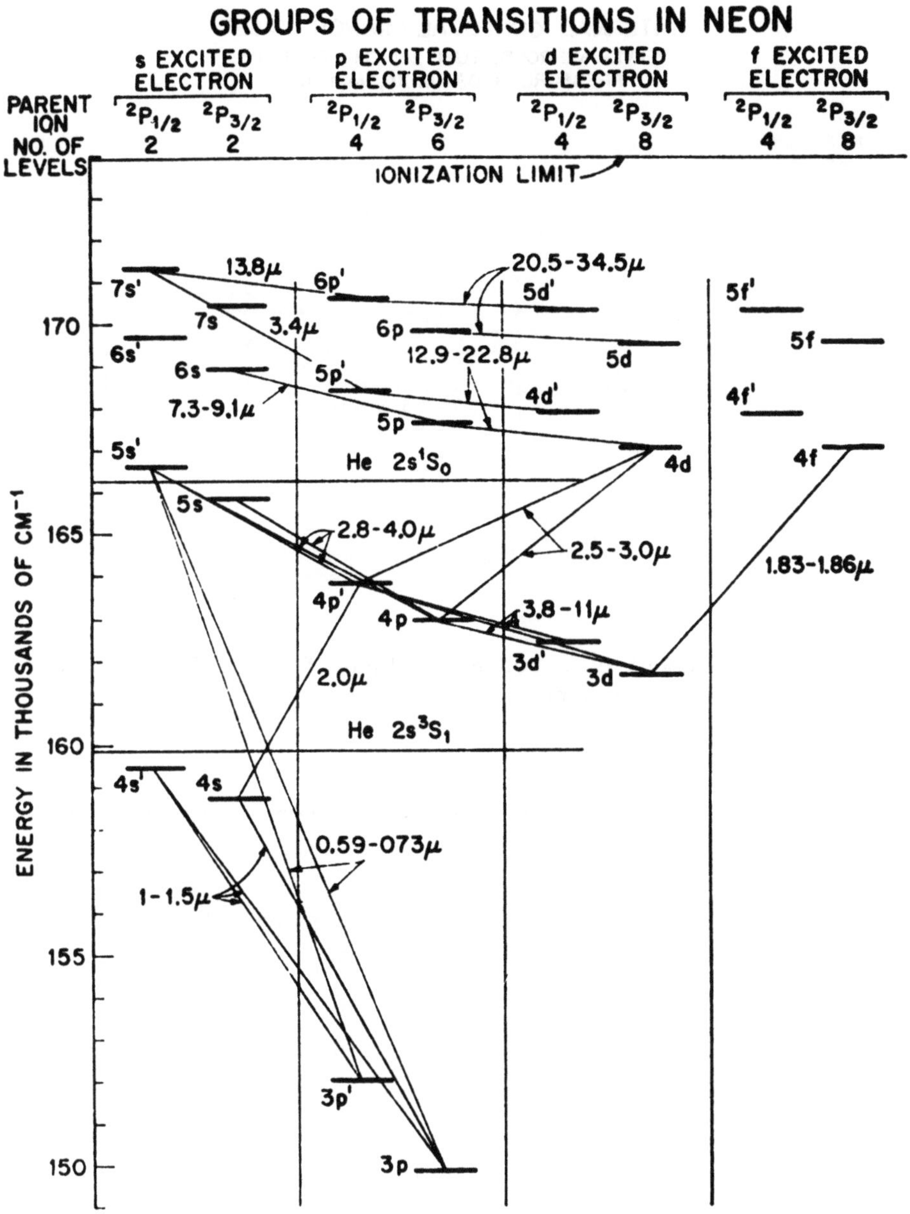

Fig. 19 Configurations of Ne above 150,000 cm^{-1} and wavelength ranges of laser lines connecting them. After [88].

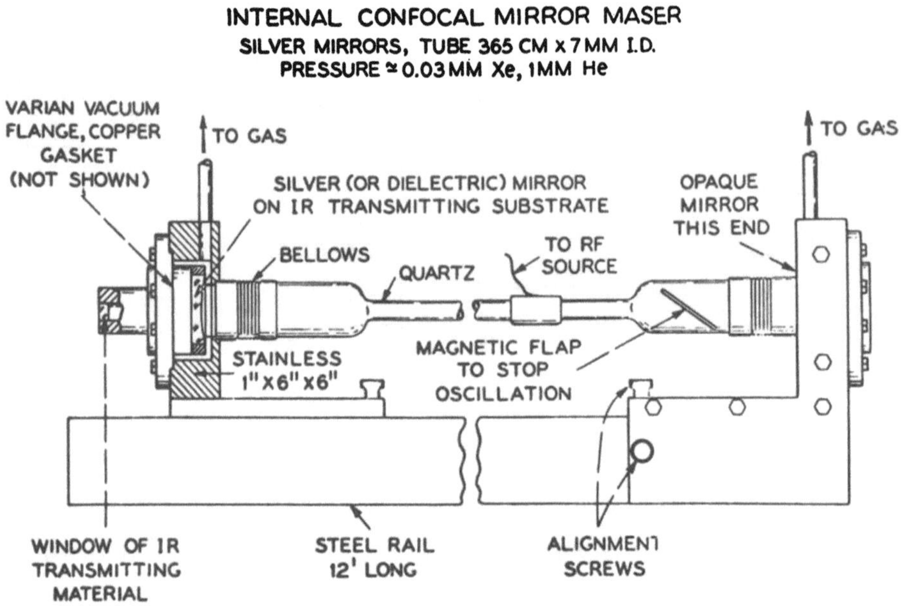

Fig. 20. Arrangement of a noble gas laser for the far infrared with internal confocal mirrors. After [88].

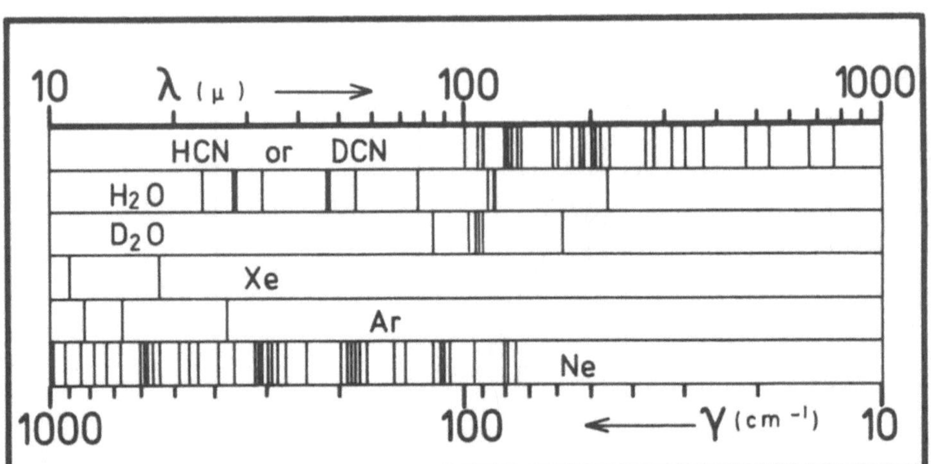

Fig. 21. Plot of gas laser lines in the 10 to 1000 cm^{-1} range. Some of the lines shown are split.

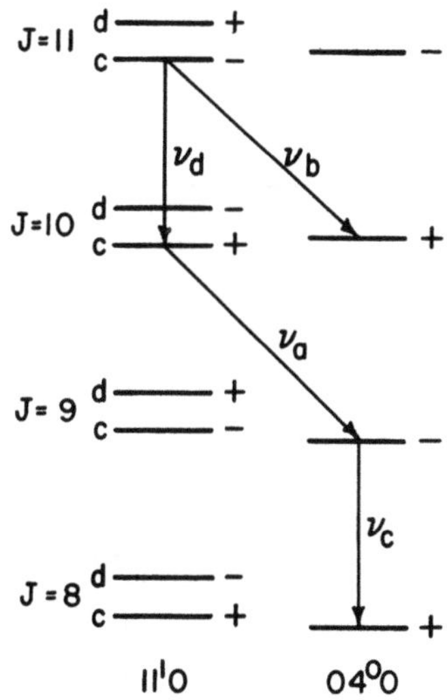

Fig. 22a. Energy level diagram for HCN. Transitions for the 30 cm^{-1} region. After (112).
$\nu_a = 29.71\ cm^{-1}$ $\nu_b = 32.17\ cm^{-1}$
$\nu_c = 26.84\ cm^{-1}$ $\nu_d = 32.29\ cm^{-1}$

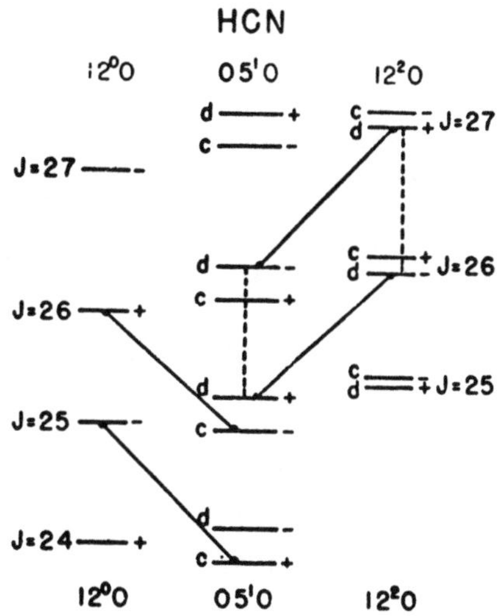

Fig. 22b. Energy level diagram for HCN. Transitions for the 78 cm^{-1} region. The solid lines indicate observed laser transitions. The dashed lines indicate predicted laser transitions. After [114].

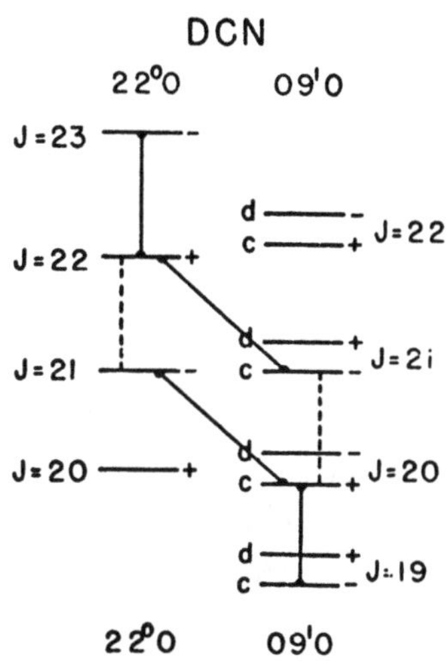

Fig. 22c. Energy level diagram for DCN. Transitions for the 51 cm^{-1} region. The solid lines indicate observed laser transitions. The dashed lines are predicted laser transitions. After [114].

Table I

Gaslaser lines in the far-infrared below 100 cm^{-1} *)

ν (cm^{-1})	λ (μ)	Gas contains	Power (Watts) Cw	pulsed	References
98.7586	101.257	H,C,N		0.2	[93]
96.777	103.33	D_2O			[100]
94.321	106.02	Ne	$5\cdot10^{-9}$		[88]
92.842	107.71	D_2O		0.1	[92,99]
91.844	108.88	D_2O			[100]
90.711	110.24	H,C,^{15}N			[93]
90.531	110.46	D_2O			[100]
89.493	111.74	D_2O			[100]
89.2809	112.006	H,C,N		0.2	[93]
88.2527	113.311	H,C,^{15}N			[93]
86.640	115.42	H_2O		$7\cdot10^{-4}$	[92]
86.1089	116.132	H,C,N		0.2	[93]
84.323416	118.591019	H_2O	10^{-3}	$2\cdot10^{-2}$	[89-91]
83.278	120.08	H_2O			[92]
80.39	124.4	Ne	$5\cdot10^{-9}$		[88]
79.30	126.1	Ne	$5\cdot10^{-9}$		[88]
79.2619	126.164	H,C,N		3	[93]
77.7430	128.629	H,C,N		9	[93,99]
76.4304	130.838	H,C,N		4	[93]
75.30	132.8	Ne	10^{-9}		[88]
74.1114	134.932	H,C,N		0.8	[93]
72.0627	138.768	H,C,^{15}N			[93]
60.5510	165.150	H,C,^{15}N			[93]
59.2	169	I,C,N,H			[98]
58.28	171.6	D_2O			[99]

*) Taken from J.F.Moser, H.Steffen and F.K.Kneubühl, Helv.Phys.Acta 41, 607 (1968)

Table I, continued

$\nu(cm^{-1})$	$\lambda(\mu)$	Gas contains	Power(Watts) Cw	pulsed	References
55.0088	181.789	D,C,N		0.1	[93]
52.6457	189.949	D,C,N		0.1	[93,114,115]
52.62936	190.0080	D,C,N			[111,114]
51.34409	194.7644	D,C,N			[114,115]
51.327	194.83	D,C,N		0.02	[93,114,115]
49.7366	201.059	H,C,N		0.05	[93]
48.92674	204.3872	D,C,N		0.04	[93,114,115]
47.3931	211.001	H,C,N		0.2	[93,99]
45.407492	220.22797	H_2O	10^{-6}	0.02	[116-119]
44.8533	222.949	H,C,N		0.08	[93]
37.3	268	I,C,N,H			[98]
35.2	284	H,C,N			[111]
32.28785	309,7140	H,C,N		0.4	[93,111]
32.16602	310.8870	H,C,N		1	[93,99,111]
29.83444	335,1831	H,C,N			[111]
29.75	336.1	H,C,N		1	[104,105]
29.71258	336.5578	H,C,N		10	[89-91,111]
26.84360	372.5283	H,C,N		0.6	[93,111]
21.5	465	I,C,N,H			[98]
18.58	538.2(split)	I,C,N,H		0.5	[98]
18.34	545.4	I,C,N,H			[98]
14.79	676	I,C,N,H		0.02	[106]
12.928	773.5	I,C,N,H		0.02	[98]

recently as rotational-vibrational transitions of the HCN molecule (Hocker et.al. [111], Lide et.al. [112,114]). These assignements were made possible by very precise frequency measurements of the 310μ, 337μ and 373μ laser lines by harmonic frequency mixing with V-band klystron radiation (Javan et.al. [111]). These measurements showed that the 373μ and 310μ transition of HCN run simultaneously with the 337μ transition and also that the 335μ and 284μ transition run simul-

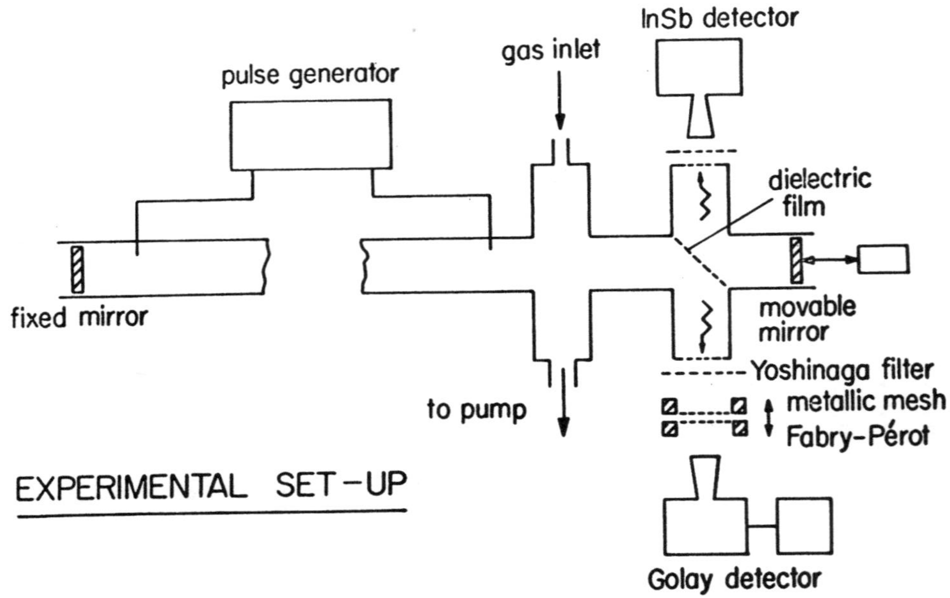

Fig. 23. Schematic of the arrangement of a HCN laser. After [98, 104-107].

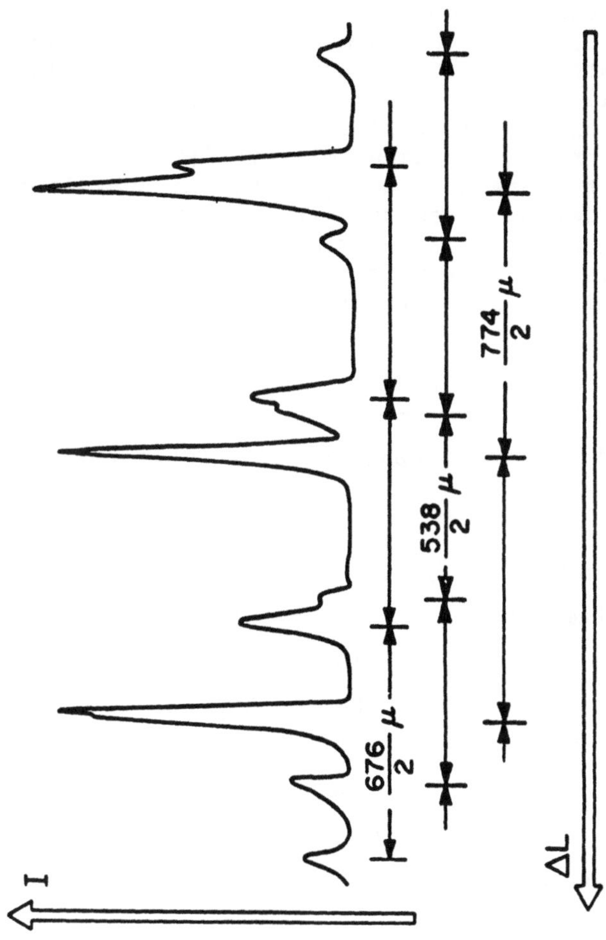

Fig. 24. Resonator interferogram of the longwave emissions of a HCN laser. After [98, 104-107].

taneously with the 311μ transition. The authors proposed therefore that these lines originate in pure rotational transitions of the vibrational ground state of the HCN molecule. A corresponding assignement could be made for DCN (Fig. 22). The origin of the many other lines of the HCN laser is not satisfactorily explained up to now although some other schemes have been proposed (Broida et.al. [97], Chantry et.al. [96]).

An experimental set-up of an HCN laser is schematically drawn in Fig. 23 (Kneubühl et.al.[94]) The radiation is coupled out here by a 45° dielectric film. This can be also done by a small hole of about 1 to 2 mm diameter in one of the laser mirrors. Still better seems to be the use of a metal mesh as one of the laser mirrors (Prettl et.al.[101]). Since the laser operates simultaneously in many lines and perhaps in different modes too, an outer metal-mesh Fabry-Perot interferometer (Prettl et.al. [101], Ulrich et.al. [71]) can serve to separate the lines. These can be measured then by recording a resonator interferogram which one gets

Table II

Typical Operational Data for HCN-Lasers

Pulsed Lasers:	
Resonator Length L	5 to 7 Meters
Resonator Diameter d	7 to 10 Centimeters
Peak Voltage	16 to 50 k Volts
Peak Current	400 to 500 Amps.
Repetition Rate	1 to 6 c/sec
Pulse Duration	~ 1.5 μsec.
Gas Pressure	0.1 to 0.5 Torr
Fresnel - Number $d^2/L\lambda$	0.2 to 2
Continuous Wave (CW) Lasers:	
Resonator Length L	1.5 to 2 Meters
Resonator Diameter d	~ 10 Centimeters
Current	~ 1.5 Amps.
Gas Pressure	~ 0.3 Torr
Fresnel-Number	~ 0.3

by moving continuously one of the laser mirrors. This method is possible because of the long wavelengths involved and due to the fact that the widths of the laser transitions are smaller than the mode separations. A resonator interferogram of the 538μ, 676μ and 774μ lines remaining after the other lines have been filtered out by the outer Fabry-Perot is shown in Fig. 24.

Submillimeter laser resonators are characterized, in contrast to lasers in the visible or near infrared, by their low Fresnel-numbers of about 0.2 to 2. Therefore special diffraction theories for this case must be used. Detailed experimental and theoretical studies on such resonators with plane and concave mirrors, taking also into account the influence of the laser tube, have been performed by Steffen et.al. [104], Schwaller et.al. [105,106] and Kneubühl et.al. [107]).

Typical operational data for far-infrared HCN lasers are compiled in Table II.

3. Difference Frequency Generation as Far-Infrared Sources

Nonlinear optics for frequency-multiplication and mixing are already a standard technique in the visible and near infrared, because of the variety of available nonlinear crystal-materials and the possibilities of phase-matching by means of birefringence.

The far-infrared now can be reached in this way both by frequency multiplication starting from microwave sources as described in section C 3 and also by the generation of difference-frequencies starting from powerful lasers such as the ruby laser or the CO_2 laser. First observations of difference-frequency generation using two slightly shifted ruby lasers have been reported by Zernike [113] and by Yajima and Inoue [120]. They used bulk nonlinear media which could not be phase matched for the far-infrared and the visible. Phase-matching in nonlinear mixing processes is especially important for the far-infrared generation first because it permits long crystals and second because the power output at the difference-frequency $\omega_3 = \omega_1 - \omega_2$ is proportional to the square of ω_3 itself and therefore is rather low in the far-infrared. This power-output is given by (compare [122])

$$P(\omega_3) = \frac{32\pi^2}{3c^3} \cdot \frac{\chi^2 \omega_3^2}{n_1 n_2 n_3} \cdot \frac{P(\omega_1) \cdot P(\omega_2) \cdot T^3}{W^2} \cdot \quad I$$

where χ is the second-order nonlinearity, n_1, n_2, n_3 are the refractive indices at the three frequencies ω_1, ω_2 and ω_3 respectively; W^2 is the cross-sectional area of the incident beams at ω_1 and ω_2; T is the power transmission coefficient for each surface of the medium and I is a coherence factor which contains the phase mismatch and the losses in the nonlinear material. Starting with ω_1 and ω_2 in the 1000 cm^{-1} region where the powerful Q-switched CO_2 laser is available with its many close-lying lines, one could cover easily by difference-frequency generation the whole far-infrared region below 100 cm^{-1}.

The problem arises with finding the proper medium which can be phase-matched at 1000 cm^{-1} and below 100 cm^{-1}. Heteropolar crystals have normally just in this region their optic active TO-modes (Reststrahlen bands), which cause a rather drastic change of the refractive index above and below the TO-modes. Cubic crystals have furthermore no birefringence. Therefore one would be limited to some special non-cubic crystals having TO-modes of not too high oscillator strength, like for instance tellurium or selenium.

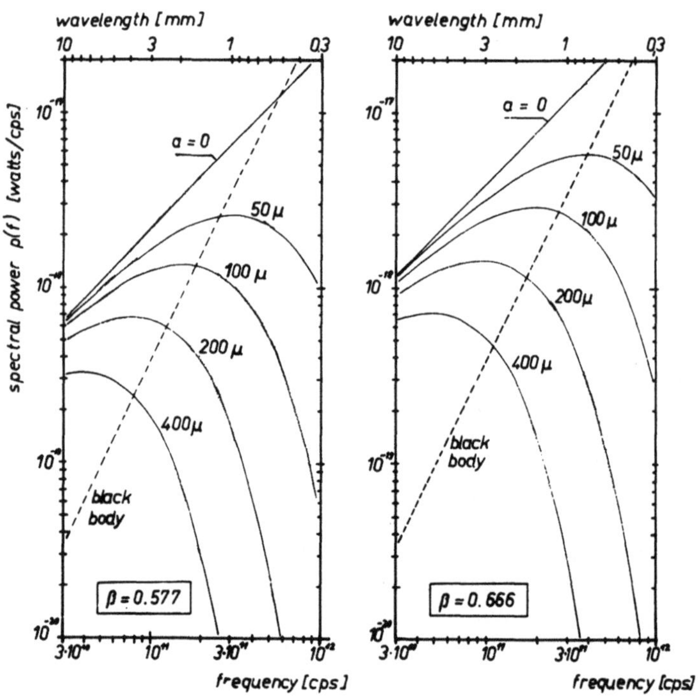

Fig. 25. Calculated spectral power of a Cerenkov coupler with a dc electron beam of J = 5mA for two electron velocities. The "effective" beam distance a is the parameter.

A very elegant way to overcome the phase-matching difficulties has been described very recently by Patel and coworkers [121,122]. They use the fact that the Drude free carrier absorption can compensate the lattice vibration contribution to the refractive index in the far-infrared. The perfect phase-matching is then performed by using a magnetic field which provides an adjustment of the dielectric constant at the difference-frequency be means of the Voigt effect so that also cubic semiconductors, which are normally not birefringent, can be phase-matched. Patel and coworkers used InSb with a carrier concentration of about 1 to 2 times 10^{15} cm^{-3} and applied a magnetic field of the order of 10 K Oe. Using two separate but synchronously Q-switched CO_2-lasers which had gratings within the optical cavities to select ν_1 = 1045 cm^{-1} (9,6μ) and ν_2 = 945 cm^{-1} (10,6μ) with about 1 kWatt peak power each, focused on to the InSb sample, they could detect ν_3 = 101 cm^{-1} with microwatts of coherent power.

Instead of using the Voigt configuration for the applied magnetic field, one can use also the Faraday configuration. It should be possible then to generate circularly polarized heliconlike waves in semiconductors in the submillimeter range. Also parametric oscillation in the far-infrared seems to be possible now by pumping a semiconductor in the 10.6μ range with CO_2 lasers.

E. HIGH - ENERGY ELECTRON SOURCES

A large number of attempts have been made to use the emission of highly accelerated electrons for a possible far-infrared source. None of these attempts have been really successful in the sense that they have produced useful and convenient sources comparable with those described before. Nevertheless, it might be interesting to discuss briefly some of them.

1. The Cerenkov - Source

In contrast to the normal way of getting Cerenkov radiation by shooting electrons through a material, an electron beam of velocity $v = \beta c$ is shot grazing the surface of a dielectric with refractive index $n > 1$. If the distance d between the surface and the beam is small compared to the wavelength which should be emitted, then polarization interaction of the electrons with the dielectric can take place. If furthermore the condition $v > c/n$ is fulfilled, then Cerenkov-radiation is excited in the dielectric travelling in the direction perpendicular to the Mach cone. This direction is given by cos $\theta = c/nv$, where θ is the angle to the electron beam. The spectral distribution of the Cerenkov-radiation depends on the spectral range in which the condition $v > c/n$ is fulfilled and also on the distance relation λ/d. Only electrons with a distance d to the surface which is smaller than $\lambda/10$ are sufficiently near for the interaction. For a real extended electron beam, an effective beam distance "a" to the surface must be considered in the theory. Fig. 25 shows the spectral power of such a dc Cerenkov source which has been calculated by Ulrich [130]. The result is that in the millimeter region a Cerenkov source can easily be better than a blackbody of 5000°K but hardly in the submillimeter range. This has been experimentally proved by Ulrich [130] with an arrangement shown in Fig. 26 which allowed the measurement of the Cerenkov power by comparison with a blackbody. The spectral distribution, measured with a lamellar Fourier spectrometer, is shown in Fig.27. In these experiments, unbunched electrons of 100 to 200 kV were used with a fused silica dielectric.

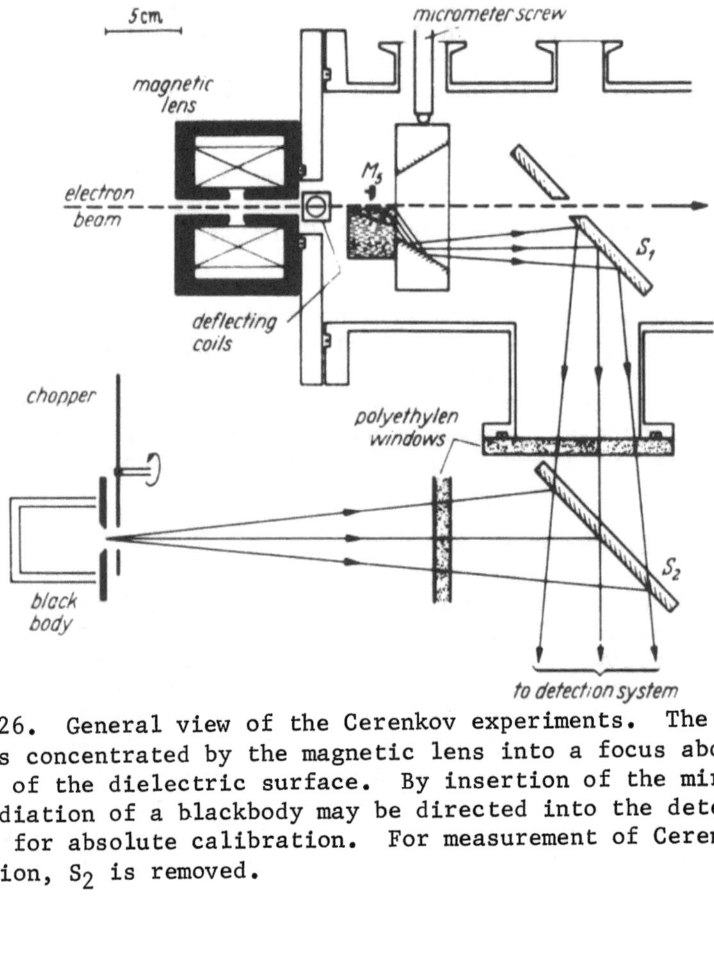

Fig. 26. General view of the Cerenkov experiments. The electron beam is concentrated by the magnetic lens into a focus above the center of the dielectric surface. By insertion of the mirror S_2 the radiation of a blackbody may be directed into the detection system for absolute calibration. For measurement of Cerenkov radiation, S_2 is removed.

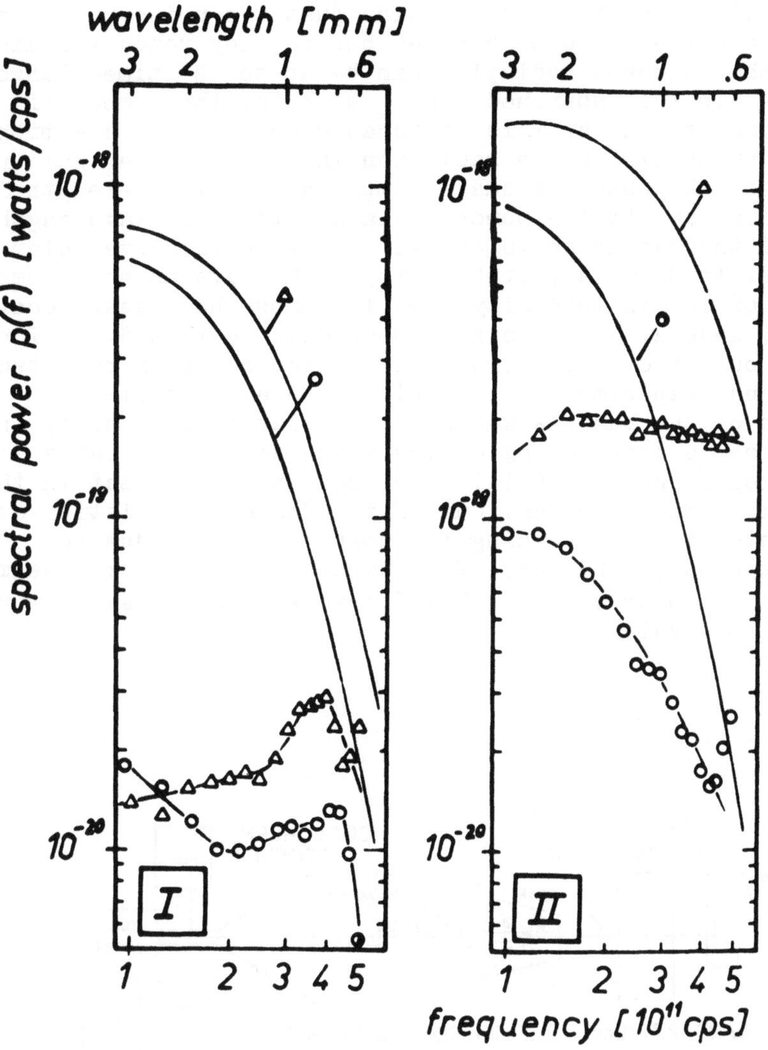

Fig. 27. The calculated (full line) and measured (points) Cerenkov spectra for two kinds of Cerenkov couplers (I and II) at two different effective beam distances. Coupler I: V = 114 kV, J = 5 mA, a = 218 and 254 . Coupler II: V = 173 kV, J = 5 mA, a = 197 and 310 . For details, compare [130].

The situation may drastically change if prebunched electrons are used instead of the continuous electron beam. In the latter case, the radiation is additive in intensity for the different electrons in the beam because of their random distribution in time. One obtains just the noise radiation. If the electrons are prebunched before they reach the dielectric, the Cerenkov radiation can be up to N times larger than in the unbunched case, N being the number of electrons per bunch. This is because now all the electrons of the bunch excite the radiation in phase if the spatial extent of the bunch is small compared to the wavelength in consideration. If the bunch is larger than the wavelength, part of the radiation is cancelled by destructive interference. In this way, prebunching might give an enhancement factor of 10^6 in intensity over the unbunched case. Accordingly, several attempts have been made in this direction but no output of far-infrared radiation has been reported from these experiments, probably because the spatial extension of the bunches was still too large. Of course, one would need a prebunching device which already operates at frequencies in the submillimeter region or at least in the lower millimeter region. One of the Cerenkov radiation experiments with prebunching is shown in Fig. 28 due to Coleman et.al. [127-129]. They used a 1 MeV electron beam bunched at 10 cm wavelength and reached 1 Watt peak power at 8.3 mm wavelength.

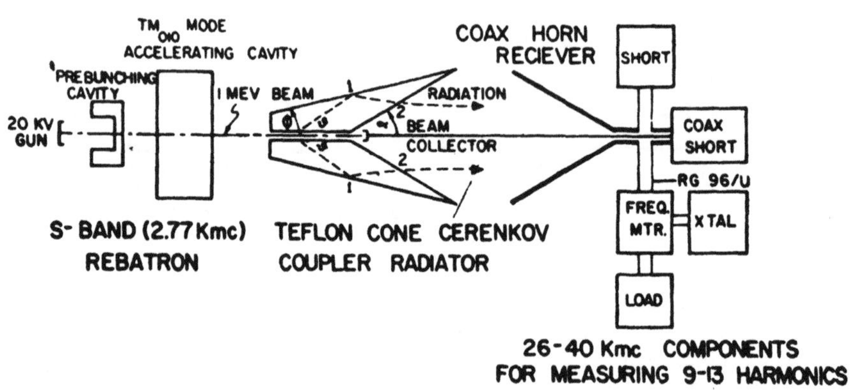

Fig. 28. Schematic diagram of Cerenkov coupler experiment using the 9th to 13th beam-current harmonic from S-band Rebatron. After Coleman et al. [125, 127 - 129].

2. The Smith - Purcell Source

In close connection to the Cerenkov effect is the Smith - Purcell effect [123,124,126,131]. Here, a bunched or unbunched electron beam grazes the surface of a periodic structure, for instance over a phase grating perpendicular to the grooves. If, for instance, this grating is made of metal then the origin of the radiation, emitted in free space, can be visualized as due to the positive mirror charges which are induced by the electrons acting like oscillating dipoles. To obtain appreciable radiation, the electrons must be relativistic which means that the outcoming radiation is concentrated strongly in the forward direction. If θ is the angle between the electron velocity and the direction of observation, the wavelength emitted in this direction is given by

$$\frac{\lambda}{d} = \frac{c/n}{v} - \cos\theta$$

where d is the periodicity of the structure and n is the refractive index of the medium in which the radiation propagates. This formula shows that the emitted wavelength spectrum is continuous with the shortest wavelength in forward direction and with longer wavelength at increasing angle θ. A 200 kV electron-beam over a 100μ grating for instance would emit 430μ for $\theta = 0^o$ and 560μ for $\theta = 30^o$. The emitted power is proportional to

$$n\ell^2 d^{-4} \cos^2\theta \cdot L \cdot V^{3/2} J$$

where $\ell/2$ is the effective distance of the electron beam to the surface of the periodic structure, L is the interaction length, V is the accelerating voltage and J is the current. This shows that a small constant d of the periodicity is preferable and that a 5 kV beam could give an emission in the 60 μ region with a 10 μ grating. Generally, it is not expected that the Smith - Purcell source would be better than the Cerenkov source.

3. The Undulator

Motz and coworkers [132,133] have proposed a device called "Undulator" in which an electron beam is affected array of alternatively poled magnets which periodically deviate the beam from the straight direction (Fig. 29). Looking towards the beam, each electron is seen to act like an oscillating dipole, radiating towards the observer. Due to the Doppler effect the frequency of the radiation can be shifted into the submillimeter region. Motz and Mallory [133] obtained in this way at 1 mm wavelength a power of 8 mWatts and at 400μ a power of 40 μWatts, using a 3,5 MeV electron beam from the Stanford linear accelerator, bunched at 3 cm wavelength.

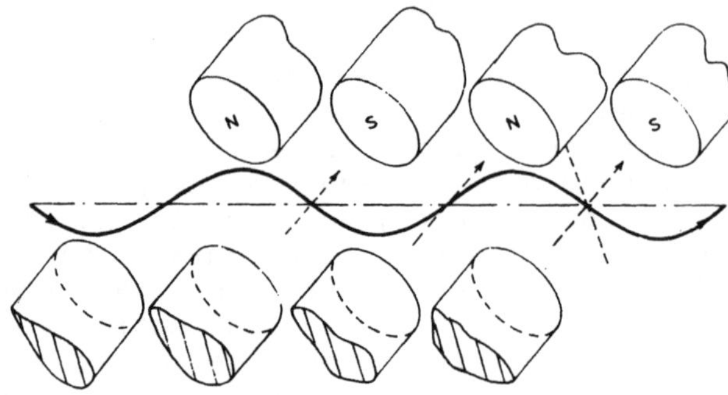

Fig. 29. Magnetic undulator. After Motz [132,133].

4. The Rebatron - Harmodotron

Coleman and coworkers [134,135,138] proposed a method of generating millimeter and submillimeter radiation by exciting a single higher order TM_{omn} mode in a cylindrical cavity. These modes were excited by a prebunched beam of 1 - 1.5 MeV coming from a "Rebatron" accelerator-buncher [136,137]. This bunched beam contains current frequency components at high harmonics of the bunch-frequency, to one of which the selective Harmodotron cavity is tuned in a high order mode. Using a TM_{018} mode, resonant at 8.31 mm wavelength, Coleman and Sirkis [134] could get one Watt peak power, using a 5 mA pulsed beam bunched at 2775 Mc/s. With the 34. harmonic of the Rebatron buncher, a wavelength of 3.18 mm could be detected. The submillimeter range apparently has not been reached up to now.

5. The Synchrotron Radiation Source

The radiation of an electron synchrotron is known as one of the best sources for the far ultraviolet region. The continuum, which is emitted, ranges down to the far-infrared, but with decreasing intensity. The power in the infrared is proportional to $\omega^{1/3}$ and increases with decreasing curvature radius as $R^{-2/3}$. The radiation is centered in a very narrow cone in the instantaneous direction of the electron motion.

Lax et.al. [141] have proposed such a synchrotron source for the far-infrared with a storage ring accelerator for 100 MeV electrons and magnetic fields of about 100 KOe, Fig. 30. They estimate a possible output of 2μWatts in a 1 % bandwidth at 100 cm^{-1} . It seems doubtful, however, that such a big machine will be built just for the far-infrared.

6. The Cyclotron - Resonance Source

If fast electrons are brought to small orbits by means of strong magnetic fields, the emitted radiation at the cyclotron-resonance frequency $\omega_c = eH/mc$ could become appreciable. In the "Tornadotron" [139,140], a sudden decrease of cyclotron orbits by means of strong magnetic field pulses of some 10 KOe is achieved. Radiation below 3 mm already has been detected. Another experiment [142] used in addition, a strong longitudinal magnetic field, which had the effect of producing an adiabatic compression of the electron beam. Here, an emitted power of the order of 10 mWatts has been reported in the 1 to 2 mm region.

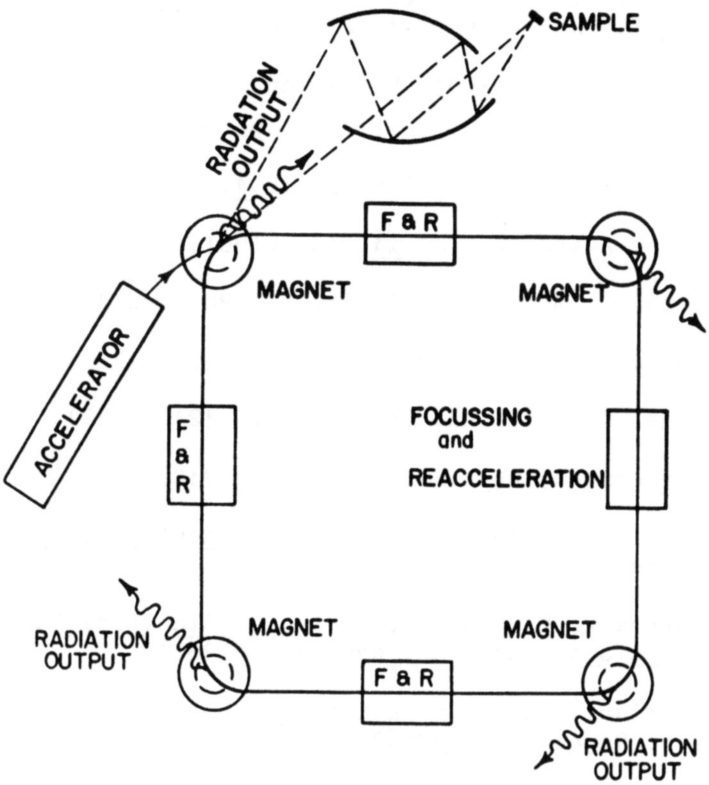

Fig. 30. Synchrotron radiation source (Radiatron). After Lax et al. [141].

Acknowledgment

The author is indebted to Professor E.E. Bell and Dr. S.D. Smith for the help in preparing this article.

REFERENCES

General References

1. R.A. Smith, F.E. Jones, The Detection and Measurement of Infrared Radiation. Oxford 1957, At the Clarendon Press.

2. J.T. Haughton, S.D. Smith, Infrared Physics. Oxford 1966, At the Clarendon Press.

3. D.H. Martin, Spectroscopic Techniques for Far Infrared, Submillimetre and Millimetre Waves. Amsterdam 1967, North-Holland Publishing Company.

4. J.F. Moser, H. Steffen, F.K. Kneubühl, Submillimeterwellen-Technik. Helv.Phys.Acta 41, 607 (1968).

Blackbody Radiation

5. A. Rubens, F. Kurlbaum, Berl.Ber. p. 929 (1900).

6. A. Rubens, F. Kurlbaum, Ann.Physik 4, 649 (1901).

7. A. Rubens, Astrophys. J. 14, 335 (1901).

8. H. Rubens, G. Michel, Berl.Ber. p. 590 (1921).

9. H. Rubens, G. Michel, Phys.Z. 22, 569 (1921).

10. H. Rubens, K. Hoffmann, Berl.Ber. p. 424 (1922).

11. C.H. Cartwright, Phys.Rev. 35, 415 (1930).

12. M. Czerny, A. Walther, Tables for Planck's Radiation Law, Springer (1961).

13. M. Pivovonsky, M.R. Nagel, Tables of Blackbody Radiation Functions, McMillan, New York (1961).

14. C.C. Ferriso, Blackbody Radiation Tables, General Dynamics, Report AE 62-0862 (1962).

15. R. Papoulas, Infrared Physics 4, 137 (1964).

Mercury Arc Emission

16. H. Rubens, O.V. Baeyer, Berl.Ber. p. 339 (1911), Berl.Ber. p. 666 (1911), PhilMag. 21, 689 (1911), Le Radium 8, 139 (1911), Berl.Ber. p. 802 (1913).

17. H. Rubens, Berl.Ber. p. 8 (1921).

18. G. Laski, Z.Physik 10, 353 (1922).

19. W. Kroebel, Z.Physik 56, 114 (1929).

20. B. Koch, Ann.Physik 33, 335 (1938).

21. W. Dahlke, Z.Physik 114, 205 (1939), Z.Physik 114, 672 (1939), Z.Physik 115, 1 (1940).

22. H. Säufferer, Z.Physik 131,376 (1952).

23. G. Hettner, Z.Physik 131, 385 (1952), Z.Physik 211, 483 (1968).

24. J. Bohdansky, Z.Physik 149, 383 (1957).

25. E.K. Plyler, D.J.C. Yates, H.A. Gebbie, J.Opt.Soc.Am. 52, 859 (1962).

26. A. Hadni, E. Decamps, J. Munier, Rev.Opt. 42, 584 (1963).

27. I. M. Arefyev, Opt.i.Spektr. 17, 300 (1964), Opt.i. Spektr. 17, 157 (1964).

28. O.K. Fillippov, V.M. Pivovaro, Opt.i.Spektr. 16, 522 (1964), Opt.i.Spektr. 16, 282 (1964).

29. P.A.H. Hart, Philips Res.Repts. 22, 77 (1967).

Other Thermal Sources

30. H. Rubens, Verh.Deut.Phys.Ges. 7, 346 (1905), Phys.Z. 6, 790 (1905), J.Phys.Theor.Appl. 4, 314 (1905), Ann. Physik 18, 725 (1905), Le Radium 2, 397 (1905), J.Phys. Theor.Appl. 5, 306 (1906), Phys.Z. 7, 186 (1906), Ann. Physik 20, 593 (1906), Verh.Deut.Phys.Ges. 8, 41 (1906), Verh.Ges.d.Naturf. Ärzte 2, 44 (1906), Berl. Ber. p. 8 (1921).

31. R.M. Badger, C.H. Cartwright, Phys.Rev. 33, 693 (1929).

32. J. Strong, Phys.Rev. 45, 887 (1934).

33. A.H. Pfund, J.Opt.Soc.Am. 26, 439 (1936).

34. T. Erb, H. Klumb, Z.Physik 114, 519 (1939).

35. L. Genzel, N. Neuroth, Z.Physik 134, 127 (1953).

36. J.A. Ketelaar, C. Haas, F.N. Hooge, R. Brockhuijsen, Physica 21, 695 (1955).

37. D.W. Robinson, J.Opt.Soc.Am. 49, 966 (1959).

38. W. Boldt, H. Reimann, Ann.Physik 6, 293 (1960).

39. G.N. Harding, M.F. Kimit, J.H. Ludlow, P. Portsous, A.C. Prior, V. Roberts, Proc.Phys.Soc.Lond. 77, 1069 (1961).

40. G.N. Harding, V. Roberts, Int.Conf.Ioniz.Phen.in Gases 2, 1977 (1962).

41. E.K. Plyler, D.J.C. Yates, H.A. Gebbie, J.Opt.Soc.Am. 52, 859 (1962).

Klystrons, Backward-Wave Oscillators

42. B.B. van Iparen, Philips.techn.Rdsch. 21, 217 (1959/60).

43. G. Convert, T. Yeou, B. Pasty, Proc.Symp.Millimeter Waves (Brooklyn Polyt.Press, New York 1959). p. 313.

44. G. Convert, T. Yeou, P.C. Mouton, Congrès Int.Tubes Hyperfrèquences (La Haye, 1962).

45. P.N. Robson, Coherent Sources Using Electron Beams. Chapter 6 in Ref. 3 (1967).

Harmonic Generation

46. J.A. Klein, A.H. Nethercot jr., Phys.Rev. 91, 1018 (1953).

47. W.C. King, W. Gordy, Phys.Rev. 93, 407 (1954).

48. A.H. Nethercot jr., Trans IRE MTT-2, 17 (1954).

49. C.M. Johnson, D.M. Slager, Rev.Sci.Instr. 25, 213 (1954).

50. J.M. Richardson, R.B. Riley, NBS Rept. 3580 (1956).

51. W. Gordy, M. Cowan, Phys.Rev. 104, 551 (1956), W. Gordy, Proc.Symp.on Millimeter-Waves (Polytechnic Press, Brooklyn, New York, 1960).

52. I. Kaufman, P.D. Coleman, J.Appl.Phys. 28, 936 (1957).

53. C.H. Page, Proc.IRE 46, 1738 (1958).

54. W.P. Ayres, IRE Trans. MTT-7, 62 (1959).

55. L. Genzel, H. Happ, R. Weber, Z.Physik 154, 1 (1959).

56. R.S. Ohl, P.P. Budenstein, C.A. Burrus, Rev.Sci.Inst. 30, 765 (1959).

57. K.D. Froome, Nature 184, 808 (1959), Nature 186, 959 (1960), Nature 188, 43 (1960), Nature 193, 1169 (1961).

58. C.B. Swan, Proc.Inst.Radio Eugrs. 49, 1941 (1961).

59. R.L. Jepsen, J.Appl.Phys. 32, 2627 (1961).

60. D.D. Douthett, J. Appl.Phys. 32, 3132 (1961).

61. J.R. Fontana, Adv. in Quantum Electronics. J.R.Singer, Editor, Columbia Univ.Press 1961, p. 612.

62. A.H. Nethercot jr., Phys.Rev.Lett. 7, 226 (1961).

63. W.C. Eppers, J.Appl.Phys. 33, 1608 (1962).

64. D.J. Knight, D. Walsh, 4 th Int.Congress on Microwave Tubes Scheveningen, Holland, 1962.

65. J.H. Krenz, G.S. Kino, Stanford Univ.,Calif.,Microwave Lab.Rept.N° 948, Sept. 1962.

66. I. Bady, IRE Trans. MTT-10, 55 (1962).

67. M.Di Domenico jr. et.al., J.Appl.Phys. 33, 1967 (1962).

68. J.R. Baird, P.D. Coleman, IEEE Trans. MTT-11, 312 (1963).

69. R.H. Bradsel, Rev.Sci.Instr. 40, 225 (1963).

70. P.D. Coleman, IEEE Trans. MTT-11, 271 (1963).

71. R. Ulrich, K.F. Renk, L. Genzel, IEEE Trans. MTT-11, 363 (1963).

72. M.R. Brown, Nature 200, 1270 (1963).

73. G. Jones, W. Gordy, Phys.Rev. 135 A, 295 (1964), Phys.Rev. 136 A, 1299 (1964).

74. C.A. Burrus jr., Proc. IEEE 53, 1256 (1965), Proc.IEEE 54, 575 (1966).

75. P.H. Knapp, D.H. Martin, Proc. IEEE 54, 528 (1966).

76. T. Gaddess, R. Temple, F. Barnes, Proc.IEEE 54, 672 (1966).

77. R.J. Bauer, M. Cohn, J.M. Cotton, R.F. Packard, Proc. IEEE 54, 595 (1966).

78. R.M. Hornreich, S. Shtrikman, IEEE Trans. MAG-2, 292 (1966).

79. V. Sells, F.C. Choo, E. Brannen, A.R. Froelich, Proc. IEEE 54, 552 (1966).

80. J.G. Baker, Harmonic Generators and Semiconductor Detectors, Chapter 5 in Ref. 3 (1967).

Quantum Oscillators

81. F.S. Barnes, Quantum Electronics. C.H. Townes, Editor, (Columbia Univ. Press, New York 1960) p. 57.

82. D.C. Blaine, Nature 191, 795 (1961).

83. W.E. Hughes, Proc. IRE 50, 1691 (1962).

84. N. Bloembergen, J.Ducuing, Phys.Lett. 6, 5 (1963).

85. J.B. Gunn, IBM J.Res.Dev. 8, 141 (1964)

86. P. Grivet, N. Bloembergen, Quantum Electronics. Proc. third Int.Congress, Paris (Columbia Univ.Press, New York 1964), Chapt. III.

87. C.K.N. Patel, W.L. Faust, R.A.Mc Farlane, C.G.B. Garrett, Appl.Phys.Lett. 4, 18 (1964).

88. R.A.Mc Farlane, W.L. Faust, C.K.N. Patel, C.G.B. Garett, Proc. IEEE 52, No 3 (1964), Phys.Rev. 133, A 1476 (1964).

89. A. Crocker, H.A. Gebbie, M.F. Kimmit, L.E.S.Mathias, Nature 201, 250 (1964).

90. H.A. Gebbie, F.D. Findlay, N.W.B. Stone, J.A. Robb, Nature 202, 169 (1964).

91. H.A. Gebbie, N.W.B. Stone, F.D. Findlay, Nature 202, 685 (1964).

92. L.E.S. Mathias, A. Crocker, Phys.Lett. 13, 35 (1964).

93. L.E.S. Mathias, A. Crocker, M.S. Wills, Electr.Lett. 1, 45 (1965).

94. F.K. Kneubühl, J.F. Moser, H. Steffen, W.Tandler, Z.Angew.Math.Phys. 16, 560 (1965).

95. M. Camani, F.K. Kneubühl, J.F. Moser, H. Steffen, Z.Angew.Math.Phys. 16, 562 (1965).

96. G.W. Chantry, H.A. Gebbie, J.E. Chamberlain, Nature 205, 377 (1965).

97. H.P. Broida, K.M. Evenson, T.T. Kikuchi, J.Appl.Phys. 36, 3355 (1965).

98. H. Steffen, J.Steffen, J.F. Moser, F.K. Kneubühl, Phys.Lett. 21, 425 (1966), Z.Angew.Math.Phys.17, 470 (1966), H. Steffen, F.K. Kneubühl, To be publ. in IEEE Trans.QE.

99. G.T. Flesher, W.M. Muller, Proc.IEEE 54, 543 (1966), W.M. Muller, G.T. Flesher, Appl.Phys.Lett. 8, 217 (1966). Appl.Phys.Lett. 10, 93 (1967).

100. D.P. Akitt, W.Q. Jeffers, P.D. Coleman, Proc.IEEE 54, 547 (1966), W.Q. Jeffers, P.D. Coleman, Proc.IEEE 55, 1222 (1967).

101. W. Prettl, L. Genzel, Phys.Lett. 23, 443 (1966).

102. A.H. Dayem, C.C. Grimes, Appl.Phys.Lett. 9, 47 (1966).

103. D.N. Langenberg, D.J. Scalapino, B.N. Taylor, Proc. IEEE 54, 560 (1966).

104. H. Steffen, J.F. Moser, F.K. Kneubühl, J.Appl.Phys. 38, 3410 (1967).

105. P. Schwaller, H. Steffen, F.K. Kneubühl, Z.Angew.Math. Phys. 18, 594 (1967).

106. P. Schwaller, H. Steffen, Phys.Lett. 23, 313 (1966), J.F. Moser, F.K. Kneubühl, Appl.Opt. 6, 827 (1967).

107. F.K. Kneubühl, H. Steffen, Phys.Lett. 25A , 639 (1967).

108. H. Steffen, B. Keller, F.K. Kneubühl, Electron.Lett. 3, 562 (1967).

109. M.A. Pollack, T.J. Bridges, W.J. Tomlinson, Appl.Phys. Lett. 10, 253 (1967).

110. L. Frenkel, T. Sullivan, M.A. Pollack, T.J. Bridges, Appl.Phys.Lett. 11, 344 (1967).

111. L.O. Hocker, A. Javan, Phys.Lett. 25A, 489 (1967), L.O. Hocker, D. Ramachandra, A. Javan, Phys.Lett. 24A, 690 (1967).

112. D.R. Lide, A.G. Maki, Appl.Phys.Lett. 11, 62 (1967).

113. F. Zernike, Bull.Am.Phys.Soc. 12, 687 (1967).

114. A.G. Maki, Appl.Phys.Lett. 12, 122 (1968).

115. L.O. Hocker, A. Javan, Appl.Phys.Lett. 12, 124 (1968).

116. B. Hartmann, B. Kleeman, Appl.Phys.Lett. 12, 168 (1968).

117. W.S. Benedict, Appl.Phys.Lett. 12, 170 (1968).

118. M.A. Pollack, W.J. Tomlinson, Appl.Phys.Lett. 12, 173 (1968).

119. M.A. Pollack, L. Frenkel, T. Sullivan, Phys.Lett. 26A, 381 (1968).

120. T. Yajima, K. Inoue, IEEE J. Quantum Electronics, QE-4 (1968).

121. T.Y. Chang, N.Van Tran, C.K.N. Patel, Appl.Phys.Lett. 13, 357 (1968).

122. N.Van Tran, C.K.N. Patel, Phys.Rev.Lett. 22, 463 (1969).

High - Energy Electron Sources

123. S.J. Smith, E.M. Purcel, Phys.Rev. 92, 1069 (1953).

124. J.A. Bradshaw, Proc.Symp. Millimeter Waves (Polytechnic Press, New York 1960), p. 223.

125. P.D. Coleman, C. Enderby, J.Appl.Phys. 31, 1699 (1960).

126. K. Ishiguro, T. Toko, Optica Acta 8, 25 (1961).

127. P.D. Coleman, Microwaves. Proc. 4 th. Int.Congress on Microwave Tubes (Eindhoven, Holland, Cetrex Publ.Co., Sept. 1962), p. 379.

128. F.J. Rosenbaum, P.D. Coleman, IEEE Trans. MTT-11, 3026 (1963).

129. P.D. Coleman, Proc.Symp. Quasi Optics (Polytechnic Press of the Polytechnical Inst. at Brooklyn, New York 1964) p. 199.

130. R. Ulrich, Diss.Univ.Freiburg i.Br. (1965) Z.Physik 194, 180 (1966), Z.Physik 199, 171 (1967).

131. F.S. Rusin, G.D. Bogomolov, JETP Lett. 4, 160 (1966).

132. H. Motz, J.Appl.Phys. 22 (1951).

133. H. Motz, K.B. Mallory, J.Appl.Phys. 26, 1384 (1955).

134. P.D. Coleman, M.D. Sirkis, J.Appl.Phys. 26, 1385 (1955).

135. M.D. Sirkis, P.D. Coleman, J.Appl.Phys. 28, 944 (1957).

136. P.D. Coleman, J.Appl.Phys. 28, 927 (1957).

137. I. Kaufman, P.D. Coleman, J.Appl.Phys. 28, 936 (1957).

138. R.H. Pantell, P.D. Coleman, R.C. Becker, IRE Trans. ED-5, 167 (1958).

139. G.E. Weibel, Symp. on Electronic Waveguides (New York 1958) p. 398.

140. H.O. Dressel, S.M. Stone, G.E. Weibel, 4 th Int.Congress on Microwave Tubes, Scheveningen, Holland, Sept. 1962.

141. B. Lax, D.T. Stevenson, Proc.Far Infrared Phys.Symp., Riverside, Calif., 1964.

142. I.B. Bott, Phys.Lett. 14, 293 (1965).

Miscalleneous Sources (Spark-Generator,Mass-Radiator)

143. A. Glagolewa-Arkadiewe, Z.Physik 24, 153 (1924). Natur. 113, 640 (1924).

144. M. Lewitzki, Phys.Z. 25, 107 (1924).

145. E.F. Nichols, J.D. Tear, Astrophys. J. 61, 17 (1925).

146. A. Kalugina, J.Exp.Theo.Phys. USSR 9, 362 (1939).

147. W. Eckhardt, Z.Physik 160, 121 (1960).

FAR INFRARED DETECTORS

P.L. Richards

Department of Physics, University of California, and
Inorganic Materials Research Division, Lawrence
Radiation Laboratory, Berkeley, California 94720

INTRODUCTION

There exist a large and growing number of useful far infrared detectors. A complete discussion of the properties of each of these detectors, including its practical and fundamental limitations, is far beyond the scope of this chapter. Fortunately, reviews do exist which present much of this material in a useful form. Of particular value are the chapter on detectors in Spectroscopic Techniques for Far Infrared, Submillimetre and Millimetre Waves[1] and large portions of The Detection and Measurement of Infra-red Radiation.[2] In this chapter some detailed information will be presented on the two far infrared detectors currently of greatest value for the spectroscopy of solids: the doped Ge bolometer and the InSb detector. In addition, heterodyne techniques, which are expected to be of increased importance in the future, will be briefly described and the superconductive Josephson effect detector will be mentioned as an example of a new development. The most commonly used far infrared detector, the Golay cell, will not be discussed since it is significantly less sensitive than the low temperature detectors and since it is sufficiently well engineered that near optimum performance can generally be achieved by simply plugging it in. Although an attempt has been made to keep this chapter self contained, supplementary reading in the above mentioned reviews is highly recommended.

THE BOLOMETER

Bolometer theory is useful not only for understanding the doped Ge bolometer, but also the InSb detector (in zero field) and other far infrared detectors such as the carbon resistance bolometer[1] and the superconductive bolometer[1]. We will first develop the general bolometer theory and then specialize it to cases of interest.

A bolometer consists of an active element characterized by a (preferably small) heat capacity C and (large) temperature coefficient of resistance $\alpha = 1/R\ dR/dT$. It is connected via a thermal link of conductance G (and small heat capacity) to a heat sink at a constant (low) temperature T_s. The resistance of the bolometer is usually[3] measured by connecting it in series with a load resistor R_L and a battery E. The voltage drop across the bolometer is then a c coupled via a capacitor to an amplifier. Since the most useful case is for $R_L \gg R$, we will assume a constant bolometer current I.

The bolometer absorbs a (large) fraction ε of the incident chopped radiation signal P_1 as well as the steady background radiation P from its surroundings.[4] The power input to the bolometer is thus $\mathcal{P} = \varepsilon P + \varepsilon P_1 e^{i\omega t} + I^2R$. The bolometer responds with a temperature $\mathfrak{T} = T + T_1 e^{i\omega t}$ and corresponding resistance $\mathcal{R} = R + R_1 e^{i\omega t}$. We simply require conservation of energy to obtain relations between these quantities.[5]

$$\varepsilon P + I^2R - G(T-T_s) + [\varepsilon P_1 + I^2R_1 - GT_1 - CT_1 \frac{d}{dt}]e^{i\omega t} = 0. \quad (1)$$

The static terms determine the operating temperature T while the dynamic part allows us to solve for the responsivity $S = IR_1/P_1$. We obtain[6]

$$S = \frac{\varepsilon IR\alpha}{G(1+i\omega\tau)-I^2R\alpha} \quad (2)$$

We have defined a time constant $\tau = C/G$ in the usual way.

If we neglect the second term in the denominator, (2) has a simple interpretation. The responsivity is the rate of change of bolometer voltage IR_1 with optical power P_1 focused on it. This is related by the absorptivity ε to the rate of change of voltage with temperature $IR\alpha$ divided by G the rate of change of power with temperature. The extra term in the denominator is the rate of change of bias heating power with temperature $d(I^2R)/dT = I^2R\alpha$. It is instructive to lump both terms in the denominator together so as to retain the intuitive form

$$S = \frac{\varepsilon IR\alpha}{G_e(1+i\omega\tau_e)} \quad (3)$$

where $G_e = G - I^2R\alpha$ and $\tau_e = C/G_e$ are called the effective thermal conductance and the effective time constant respectively. For a chopping frequency $\omega \ll 1/\tau_e$ the responsivity has its maximum value

$S = \varepsilon IR\alpha/G_e$. For higher frequencies S becomes complex (i.e., it lags the signal in phase) and its magnitude falls off as ω^{-1}.

The Germanium Bolometer

For a doped Ge bolometer operated at pumped He temperature we can assume the following properties: The resistance has the form $R \propto e^{\Delta/T}$ because of an effective semiconducting energy gap Δ, so $\alpha = -\Delta/T^2$. From the Debye law $C \propto T^3$, and for metal lead wires with impurity limited electrical conductance the Wiedemann-Franz law gives $G \propto T$. The operating temperature $T = T_s + (\varepsilon P + I^2R)/G$ is determined by the absorbed steady state radiant power εP and the resistive heating I^2R. The former is primarily due to room temperature radiation which should be minimized by the use of cooled filters. This is much easier to accomplish in the far infrared than it is at higher frequencies.[8] (Troublesome changes in bolometer responsivity sometimes occur when a cooled solid sample is investigated because the reststrahl band in the sample significantly reduces P.) The effect of this heating can be included in an effective sink temperature $T_o = T_s + \varepsilon P/G$ defined so that the operating temperature is simply $T = T_o + I^2R/G$. For a well designed far infrared bolometer T_o is very close to T_s. A practical bolometer is made large enough in area to accept the required radiation and thick enough to obtain a large value of ε. For this given bolometer volume the heat capacity C depends on T. The chopping frequency is selected (usually as slow as is convenient) to optimize the signal-to-noise ratio of the system and G_e is adjusted until $\omega\tau_e \simeq 1$.

A significant simplification in bolometer theory is obtained if the temperature dependence of the resistance and specific heat are expanded about the sink temperature T_o rather than the actual temperature T. Most simply we can assume that $\alpha = -\Delta/T_o^2$ and $C = AT_o^3$. This approximation is reasonable as long as $(T-T_o)/T$ is small. We may neglect the distinction between dynamic and static thermal conductance[5] for the same reason. Using $G_e = G[1-\alpha(T-T_o)]$ we find

$$S = \frac{-\varepsilon\Delta}{(1+i)}\sqrt{\frac{R\,(T-T_o)}{\omega A T_o^5[T_o^2 + \Delta(T-T_o)]}} \qquad (4)$$

Clearly, to maximize the responsivity the sink temperature T_o must be made as low as possible. The bolometer current I, which is the most easily accessible variable is adjusted for the optimum operating temperature. The important features of (4) to keep in mind are that S varies directly as the square root of the bolometer resistance R and inversely as the square root of the chopping frequency ω and the bolometer volume or the heat capacity constant A. The theory of the doped Ge bolometer has been developed in various approximations[9] by different authors, but these conclusions remain the same.

Bolometer Materials

Bolometers can be made from Ge doped with sufficient impurities that conduction occurs in the He temperature range by the hopping process. In this range, an effective gap Δ exists which is a small fraction of the donor or acceptor energy, and far infrared radiation is strongly absorbed due to photo-excitation of bound carriers. Gallium doped Ge with a room temperature resistivity of $0.11 < \rho < 0.13$ ohm-cm and In doped Ge with $\rho \simeq 0.06$ ohm-cm have been used successfully by the author at pumped He^4 temperatures. The properties of a bolometer made from this latter material are given in Table I. Germanium doped with $6x10^{16}/cm^3$ In and $2.4x10^{16}/cm^3$ Sb has been used very successfully by Drew and Sievers[11] for a bolometer operated at pumped He^3 temperatures as shown in Table I. The pioneering article by Low[10] gives the properties of a Ga doped bolometer.

Bolometer	Richards	Drew and Sievers
Impurity	Primarily In	In and Sb
Bolometer volume	$4.5 mm^3$	$8 mm^3$
Resistance R	$1x10^5$ ohm	$1.5x10^6$ ohm
Effective gap Δ	~ 14°K	3.9°K
Sink temp. T_s	1.1°K	0.37°K
Bolometer temp. T	1.5°K	0.43°K
Thermal conductance G	$5x10^{-5}$W/°K	$8x10^{-8}$W/°K
Time constant τ	$5x10^{-3}$sec	10^{-2} sec
Responsivity S	$4x10^4$V/W	$2x10^6$V/W
RMS noise V_n	$3.6x10^{-8}V/\sqrt{Hz}$	$5.3x10^{-8}V/\sqrt{Hz}$
Noise equivalent power NEP	$9x10^{-13}W/\sqrt{Hz}$	$3x10^{-14}W/\sqrt{Hz}$

Table I. Properties of two doped Ge bolometers. The bolometer materials were selected in each case for good performance at the sink temperature chosen. The advantage of low T_s is clearly shown.

Unfortunately no systematic study of the low temperature electrical properties of heavily doped Ge has been carried out[12] so the selection of bolometer materials remains an art. It does appear that R and Δ can be varied somewhat independently by using compensated material with two types of impurity. For reproducible results it is very important to work with pure starting material and take care that the doping is homogenous. Variations in impurity concentration cause much larger changes in He temperature than room temperature properties.

Very little has been published about the far infrared absorption of bolometer materials in the operating temperature range. It does appear that care must be taken to obtain adequate absorption at long wavelengths.[12,13] About 30 percent reflectance is expected in the far infrared from the low frequency lattice dielectric constant of Ge. Systematic measurements of absorptance as a function of doping (including compensation) would be of value for a fundamental understanding of hopping conductivity as well as for bolometer construction.

Bolometer Mounting

The bolometer must be mounted in a vacuum (or in a He exchange gas to provide thermal conductance to the sink) in such a way that the infrared signal can be efficiently absorbed. Two types of mounting are commonly used. Commercial detectors are usually sold in a metal optical Dewar with cooled windows designed to accept a focused beam. A somewhat cheaper and more flexible system used by the author and many others who make their own bolometers is to place the bolometer at the exit of a focusing cone which is fed from a metal light pipe[14] entering the top of a dewar. The bolometer is surrounded by a sealed integrating cavity as shown in Fig. 1.

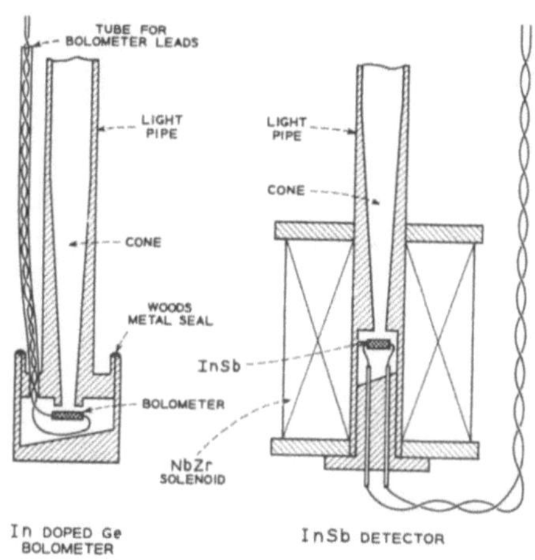

Fig. 1. A widely-used mounting for He temperature far infrared detectors. The assemblies are made from brass and immersed in liquid helium in a conventional glass cryostat. The Ge bolometer is operated in a vacuum while the InSb detector is immersed in liquid He. Each detector is backed by an integrating cavity[15] to enhance absorption.

The metal bolometer leads are generally attached to the carefully etched bolometer element by an In solder, and clamped to the bottom of the focusing cone to provide both mechanical support and a thermal link to the He bath. This system has about the same optical efficiency as the commercial one, but uses inexpensive conventional Dewars and can be readily combined with a variety of mountings for cooled samples as shown[16,17] in Figs. 2-3. A sealed He temperature window is often desirable in the design of such systems.

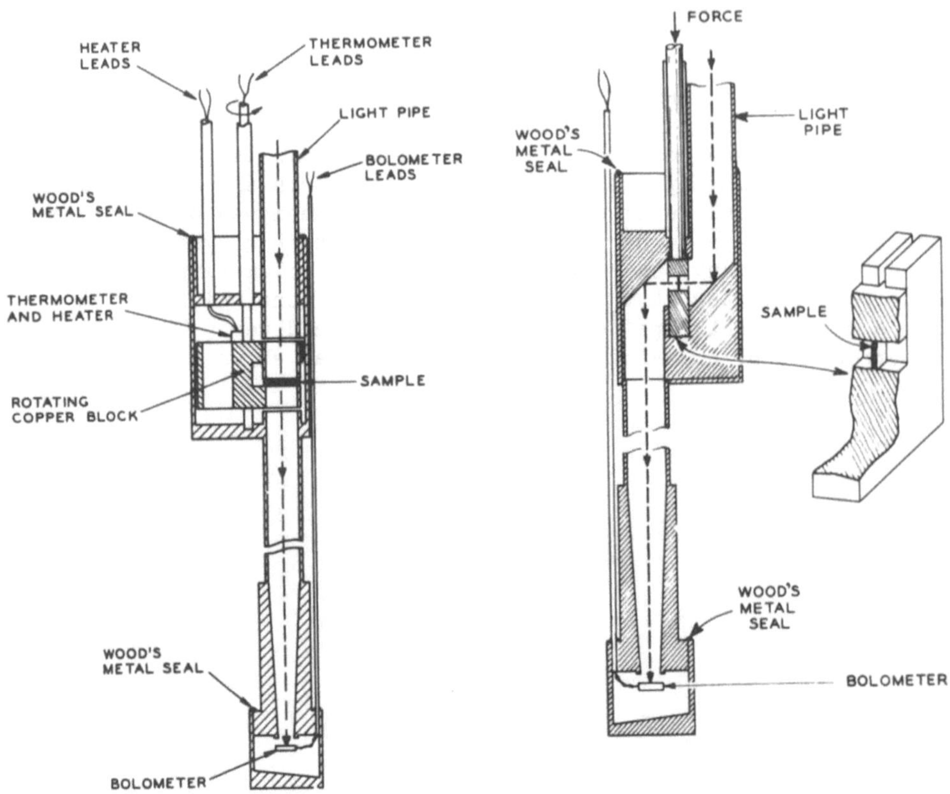

Fig. 2. Apparatus for measuring the transmittance of samples as a function of temperature[16] and uni-axial stress. The sample chamber in each case is evacuated. If sample temperature must be measured precisely it is preferable to mount sample in liquid He or in a He exchange gas.

It is used to isolate the detector vacuum space from a sample chamber filled with He. A sapphire disc, made wedge shaped, to avoid interference fringes and brazed to a metal ring[18] is reasonably transparent at low temperatures from 2-250 cm^{-1} and absorbs a considerable fraction of the higher frequency room temperature radiation.

Bolometer Testing

It is possible to measure the responsivity of a bolometer (aside from the absorptivity ε) by purely electrical means. All that is required is to introduce a known electrical heating power

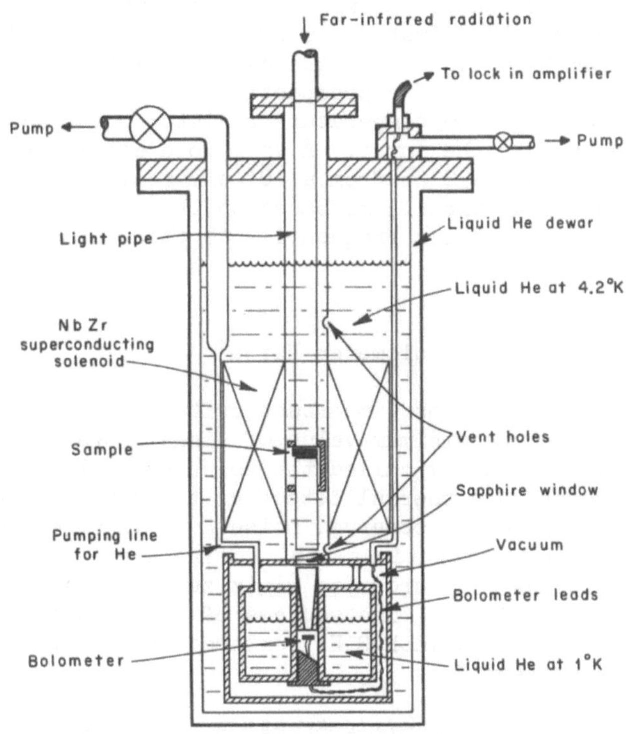

Fig. 3. Apparatus for measuring the transmittance of samples as a function of magnetic field.[17] Samples can be interchanged with minimal loss of He and, by closing the vent holes, the sample chamber can be evacuated for measurements at temperatures above 4.2°K.

at the proposed modulation frequency and to measure the ac voltage response. Practical methods for doing this are discussed in an article by Jones.[7] Anyone seriously interested in using bolometer detectors should be prepared to take these simple measurements. Accurate measurements of noise power require the use of an rms voltmeter and careful attention to bandwidths, but "eyeball" estimates are usually adequate for comparing detectors and for locating optimal operating conditions.

Amplifier

A lock-in phase sensitive amplifier is always used to provide adjustable integrating bandwidth around the chopping frequency. An ideal amplifier for use with the doped Ge bolometer should make a negligible contribution to the system noise. The amplifier noise usually arises from the first stage of amplification and often has a power spectrum $\propto \omega^{-1}$ in the frequency range most useful for Ge bolometers. The quality of a prospective preamplifier can best be judged by connecting a He temperature resistor equal to the bolometer resistance R to the input. A very good amplifier system can see Johnson noise at $\sim$30 Hz from a 10^6 ohm resistor at 4.2°K. The author has found that amplifiers based on the 6CW4 nuvistor tube are usually superior at low frequencies to those using field-effect transistors.

Bolometer Noise

In many experiments the most important criterion for the usefulness of the detector is the minimum value of signal power which can be detected. This is usually called the noise-equivalent power (NEP). In most low temperature thermal detectors there are several important contributions to the observed noise. It is often rather tedious to determine which contributions are dominant under a given set of operating conditions. In order to minimize the noise, however, its source must be identified. The sources of noise in a low temperature bolometer in a bandwidth B Hz centered about a frequency $\omega < 1/\tau$ can be written[19]

$$(\mathrm{NEP})^2 = \frac{4kTRB}{S^2} + 4kT^2GB + 8\varepsilon\sigma kAT_R^5B\sin^2\left(\frac{\phi}{2}\right) + \frac{CI^{\alpha}B}{\omega^{\beta}S^2} \qquad (5)$$

where each term is given in the form of the square of an optical noise power incident on the bolometer. The first term is the Johnson or Nyquist noise associated with the bolometer resistance R. Since $S \propto R^{1/2}$ this noise contribution is independent of R. The second term is the noise from statistical fluctuations in the bolometer temperature. At frequencies $\omega \gtrsim 1/\tau$ this term falls[2] off as $(1+i\omega\tau)^{-1}$. It is a minimum for slow (small G) bolometers. The third term arises from the statistical fluctuations in the black body radiation which strikes the bolometer of area A from a background temperature T_R viewed through a cone of angle ϕ. This term must of course be modified to include the effects of any cooled filters which limit the background radiation.[19] Typical values[1] are 10^{-13} to 10^{-14} W. The last term summarizes other possible contributions to the noise such as current noise in the bolometer, or current-independent inverse frequency flicker noise power from the amplifier. In a practical far infrared bolometer the last three terms in (5) may be of comparable magnitude so that careful study is required to determine the optimal operating conditions.[20] It appears possible to reduce bolometer current noise to a negligible value[10] by the use of single crystal material and careful attention to the method of making electrical contacts. It is usually the NEP and not the responsivity which determines the usefulness of a detector system. There may even be circumstances under which reducing S would improve system performance. In most spectroscopic applications the chopping frequency is selected to minimize the NEP. If the system noise is dominated by the last term in (5) with $\beta > 1$, then a high frequency bolometer is desirable. For all other cases ω should be chosen as small as practical. A commonly used value is 10Hz.

An ideal low temperature thermal detector for a given application is one which is limited by fluctuations in unavoidable background radiation. This condition can be met by broad band detectors in the near infrared,[10] but there is considerable room for improvement in the far infrared.

THE InSb DETECTOR

The InSb detector is sometimes called a photoconductive detector and sometimes an electronic bolometer. The reason for the variety in nomenclature is that under various operating conditons the parameters which describe the behavior of the charge carriers in the InSb, such as the plasma frequency, the cyclotron frequency, the impurity depth, and the photon energy can be the same order of magnitude. Therefore, few of the usual simplifying assumptions are made in the theoretical description of the detector. The InSb detector used by the author is a block of n-type material with a carrier density of $\sim 10^{14}/cm^3$. The leads are attached with In solder and it is mounted in a NbZr superconductive solenoid as shown in Fig. 1.

It is helpful to think of the indium antimonide detector as an electronic bolometer. Because of a relatively long electron-lattice relaxation time of 10^{-6} to 10^{-7} sec the electrons are the active element and the lattice, which is cooled by immersion in liquid helium, acts as a thermal sink. When the carriers absorb radiation, their temperature rises relative to the lattice and, because their mobility is limited by ionized impurity scattering, the resistance of the bolometer drops.

In zero magnetic field the carriers absorb photons in the neighborhood of the plasma and relaxation frequencies. In a magnetic field, absorption also occurs at the cyclotron resonance frequency. The magnetic field has a second important effect. In zero field the impurity centers in the n-type indium antimonide overlap so that the impurity energy levels merge into the conduction band. The conductivity thus lacks the exponential temperature dependence characteristic of most semiconductors. In a magnetic field the impurity centers contract and separate impurity levels develop. At low temperatures, therefore, a magnetic "freeze-out" of carriers occurs causing the large resistivity illustrated in Fig. 4. Measurements of the responsivity of this detector relative to an In doped Ge bolometer are shown in Figs. 4 and 5. In general, the responsivity is higher at pumped helium temperatures and in a magnetic field. The broad region of responsivity which falls off with frequency can be interpreted as absorption by the carriers in the vicinity of the plasma and relaxation frequencies.

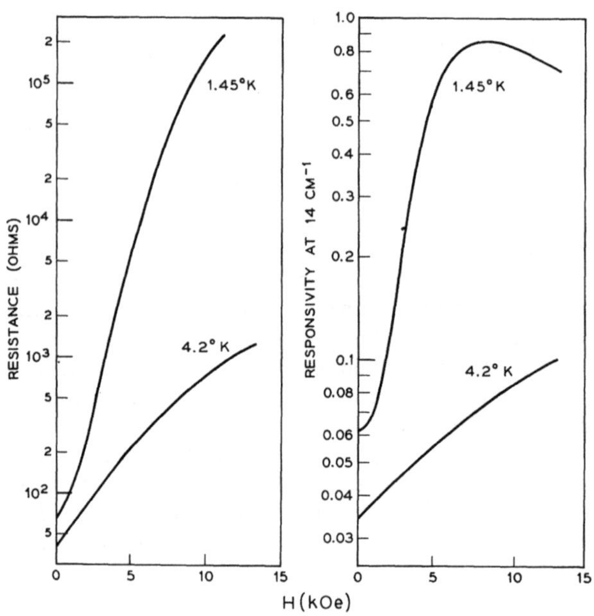

Fig. 4. Magnetic field dependence of resistance and responsivity (KV/W) of an InSb detector.[13] The bias current was set for minimum NEP.

Two peaks in responsivity are also seen near the cyclotron frequency, which is proportional to magnetic field. This observed splitting of the cyclotron resonance peak shows that our electronic bolometer model is oversimplified. The peaks are actually due to transitions between impurity states with different angular momenta associated with the first and second Landau levels. Thus the InSb detector in a magnetic field (Putley detector) does not show strictly cyclotron resonance response. Since it involves transitions between impurity levels with a consequent change in carrier mobility, the Putley detector is often referred to as a photoconductive detector.

The complicated spectral response of the InSb detector can be used to advantage in certain experiments. When experiments are done in the range $\omega \lesssim 50$ cm^{-1} using an Hg arc source the relative insensitivity of the InSb detector operated at low fields to higher frequency radiation greatly simplifies filtering problems. At higher frequencies the peak in detector sensitivity at approximately the cyclotron resonance frequency eB/m^*c shown in Figs. 5. and 6. can be used as an order sorter for a grating monochromator or even by itself as a low resolution monochromator. The width of the cyclotron resonance peak depends on the thickness of the detector and can be made somewhat narrower than shown in Fig. 6.

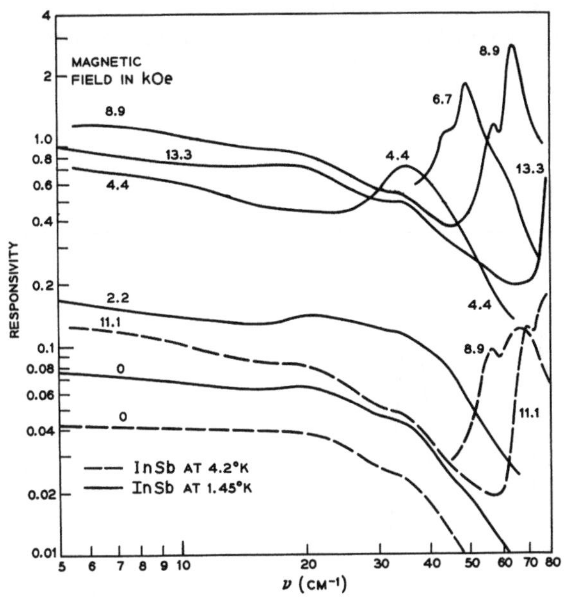

Fig. 5. Frequency dependence of the responsivity (inKV/W) of an InSb detector[13] for various values of temperature and magnetic field.

Sources of Noise

As is the case with the Ge bolometer, the InSb detector current is selected to minimize its NEP. The maximum useful current is often limited by noise arising from avalanche breakdown in the detector.

Because of its relatively low responsivity the InSb detector is usually amplifier noise limited. The strong variation of resistance with field makes it desirable to have several amplifier input impedances available. Fig. 4. shows that the square of the voltage responsivity of the InSb detector falls less rapidly than its resistance[21] as H is decreased. Therefore, if the detector is matched to the amplifier with an ideal noise-free transformer, optimum performance would occur at H = 0. Excellent results[1] have been obtained by using a transformer cooled to He temperatures. There is some evidence that the relation between responsivity and impedance, which determines the optimum operating point of the detector, varies somewhat between different indium antimonide samples.

The InSb detector has similar fundamental noise limitations from Johnson and background radiation noise to those given in (5) for the Ge bolometer. It is, however, expected to have temperature fluctuation noise only under those conditions for which it is primarily an electronic bolometer[22].

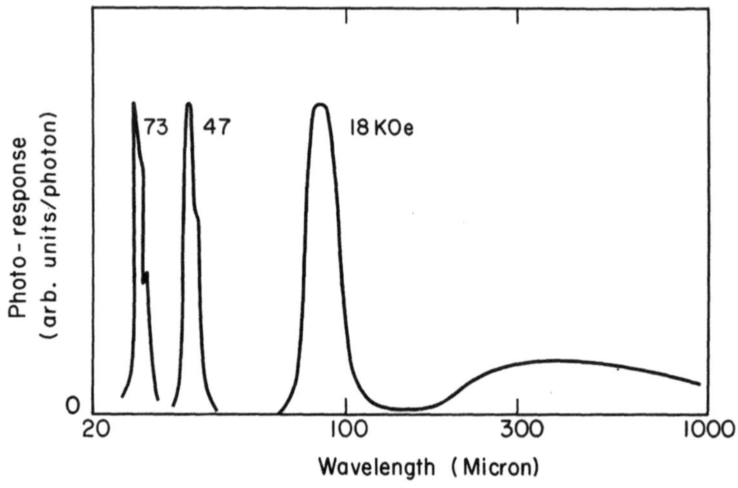

Fig. 6. Performance of the tunable InSb detector[1]

Speed of Response

The electron lattice relaxation time for low carrier density InSb is in the range from 10^{-6} to 10^{-7} sec under various conditions of temperature and magnetic field.[1] The electric bolometer is thus, of necessity, very fast. It is difficult to chop a continuous source of far infrared at frequencies much above 1 KHz but this is adequate to get above the low frequency flicker noise in most amplifiers. The InSb detector is particularly useful with pulsed radiation sources such as lasers or unstable plasmas. One difficulty encountered in actually using the potential fast response of the InSb detector is that cable and amplifier input capacitance reduces the input impedance below the detector resistance thus shorting out high frequency signals. One solution to this difficulty is to bias the detector with a constant voltage and use an amplifier in series with it connected as a low input impedance current detector. System response times of 10^{-6} sec have been obtained in this way.

HETERODYNE TECHNIQUES

The square law detectors described in the preceeding sections are often limited by noise associated with their wide spectral bandwidth. If all of the signal information is confined to a narrow infrared band as is the case with a laser spectrometer, the sensitivity can be increased by mixing the signal with a local oscillator (laser) in a square law detector and then observing the difference frequency in a finite bandwidth.[1] The total electric field vector at the detector is due to the signal plus the local oscillator. If the signals arrive in phase we have

$$E = E_s \cos \omega_s t + E_{LO} \cos \omega_{LO} t. \qquad (6)$$

The detector response r is proportional to E^2 so (7)

$$r \propto E_s^2 \cos^2 \omega_s t + E_{LO}^2 \cos^2 \omega_{LO} t + E_s E_{LO} [\cos (\omega_s - \omega_{LO})t + \cos (\omega_s + \omega_{LO})t]$$

Because of its finite bandwidth the detector cannot follow the infrared frequencies. It is assumed to be able to respond at the intermediate (IF) frequency $\omega_s - \omega_{LO}$ so the detector output is simply

$$r \propto E_s E_{LO} \cos (\omega_s - \omega_{LO})T + \text{constant}$$

Measurement of the signal is confined to a bandwidth B about the intermediate frequency so

$$r_{IF} = E_s E_{LO} \cos (\omega_s - \omega_{LO})t \qquad (9)$$

One advantage of this method is immediately apparent when we compare (9) with the response $r \propto E_s^2$ of the same detector operated in the conventional (video) manner. If the local oscillator is strong, $E_{LO} >> E_S$, then a much larger signal is obtained. The NEP for such a system may be correspondingly lower. In the heterodyne receiver using a phase sensititve detector only the noise in the post detection bandwidth B is important. For example, the noise power NEP $\simeq kTB$ when $h\nu << kT$ can be 4×10^{-21}W in a one Hz bandwidth for T = 300°K. In a square law video detector the effective noise bandwidth is the geometric mean of B and the (much larger) spectral sensitivity range.[23] Heterodyne detection promises to become more common in the far infrared, especially with laser spectrometers. Already many far infrared detectors (including the lowly Golay Cell) have been operated in heterodyne systems.[1]

JOSEPHSON EFFECT DETECTOR

As an example of a recently developed far infrared detector we now consider the superconducting Josephson effect detector. Although it is in a very early stage of development, it has several useful properties which make it appear promising. These include high sensitivity, extremely high speed, and potential as a tunable narrow band detector.

The Josephson effect, the tunneling of a lossless supercurrent through an oxide barrier separating two superconductors is one of the most complicated and interesting phenomena of modern solid state physics. An elementary review of these phenomena is available.[24] For the present discussion we need only observe that the current flow through such a junction has the form

$$I = I_o \sum_{n=-\infty}^{\infty} J_n \left(\frac{2eV_1}{\hbar\omega_1}\right) \cos\,(\omega_o + n\omega_1)t \qquad (11)$$

Where J_n is Bessel's function of order n, ω_o is the Josephson frequency, ω_1 the rf frequency and V_1 the rf voltage. When operating a Josephson junction as a detector we observe the dc (or near dc) component of this current flow which occurs when $\omega_o = n\omega_1 = 0$.

Broad Band Detector

When the detector is biased so as to observe the zero voltage current at $2eV_o/\hbar = \omega_o = 0$ then the dc component of (11) occurs for n = 0 and is thus independent of the rf frequency ω_1. A detector has been operated in this mode[25] using a point contact Josephson junction fabricated by simply touching two Nb wires together at liquid He temperatures. Broad band sensitivity was observed extending to $\simeq 40$ cm^{-1}. In favorable cases an NEP of 5×10^{-14} W/$\sqrt{\text{Hz}}$ could be achieved. Unlike a bolometer the sensitivity is expected to be independent of response frequency up to 10^{11} Hz.

Narrow Band Detector

If the detector is biased at a finite voltage then dc current flows for $\omega_o = n\omega_1$. Unfortunately, however, the amount of current flow is negligible for small V_1 since $J_n(\chi) \propto \chi^n$ for small χ. Thus, J_o is finite for small V_1, but not J_n for $n > o$. This difficulty can be avoided, however, if the detector is operated in a cavity which is resonant at ω_o. If the junction is biased at the constant voltage $V_o = h\omega_o/2e$ then the ac Josephson current will excite the cavity mode. Under these circumstances, there is a large V_1 and a large dc current flowing. An additional V_1 due to radiation coupled into the cavity will cause a large change in the dc current and will thus be detected. This situation is exactly analogous to a regenerative radio receiver. The receiver bandwidth is much narrower than the Q of the tuned circuit which controls the feedback.

A detector has been operated in this mode which has remarkable properties.[26] The detection bandwidth as shown in Fig. 7. is too small to be measured by the techniques of Fourier transform spectroscopy. An upper limit of $\delta\nu = 10^{-2}$ cm^{-1} has been established.

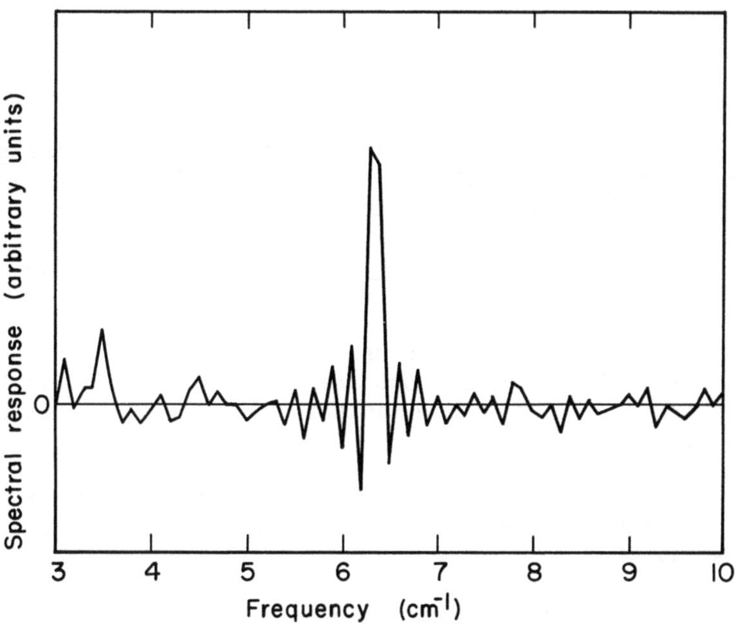

Fig. 7. Spectral response of regenerative Josephson effect detector[25] as measured with spectrometer with a $(\sin \nu)/\nu$ spectral window function.

Assuming this bandwidth the observed NEP is conservatively estimated to be 10^{-14} W/$\sqrt{\text{Hz}}$. If the bandwidth is actually narrower, the NEP is proportionally smaller. Since it is a square law detector, the effective noise bandwidth of this narrow band detector is $\sqrt{\delta\nu B}$.[23] The fundamental limit set by photon shot noise to the measurement of a monochromatic signal is NEP $\simeq 2h\nu\sqrt{\delta\nu B} \simeq 10^{-18}$W. The improvement over the fundamental limits for the Ge bolometer or the InSb detector is due entirely to the smallness of $\delta\nu$.

Josephson effect detectors are in a very primitive state of development at the present time, but show promise for the future. For example, in the proper type of cavity several resonances couple to the junction. The frequency of the detector can then be voltage tuned from one resonance to another. In principle the cavity dimensions can be changed so that the detector frequency could be swept continously. Such a detector, when used with a black body source, would eliminate the need for a monochromator or interferometer in the difficult 2 - 30 cm $^{-1}$ frequency range.

REFERENCES

1. Spectroscopic Techniques for Far Infrared, Submillimeter and Millimeter Waves, D. H. Martin, ed., (North Holland Publishing Company Amsterdam, John Wiley and Sons, Inc., New York, 1967).

2. The Detection and Measurement of Infrared Radiation, R.A. Smith, F.E. Jones, and R. P. Chasmar. (Oxford, 1968) 2nd ed.

3. A bridge circuit is sometimes used with a bolometer in each of two legs in order to cancel out the effect of a drift in the temperature T_S. In this case radiation is allowed to hit only one of the bolometers.

4. For He temperature bolometers we can neglect the power radiated by the bolometer, as well as the noise generated by fluctuations in this radiation.

5. More exactly, it is the dynamic thermal conductance dP/dT which appears in the dynamic terms of (1) and the static conductance G which determines the operating temperature T. See Ref. 2 for a detailed discussion.

6. If the load resistor R_L is not >>R then the responsivity will be multiplied[2] by the bridge factor $R_L/(R_L+R)$ and the bias heating power term $I^2R\alpha$ in the denominator of (2) by $(R_L-R)/(R_L+R)$. The responsivity may be singular if the two terms in the denominator of (2) cancel. This may occur when $\alpha > 0$ and $R_L > R$ because an increase in T increases R and thus increases the power dissipated in R from a high impedance source. It can also occur when $\alpha < 0$ and $R_L < R$ because an increase in T decreases R and increases the power dissipated in R from a low impedance source. These instabilities which rarely cause difficulty for low temperatures bolometers are discussed extensively by Jones.[7]

7. R. C. Jones, J. Opt. Soc. Am. 43, 1 (1953).

8. Useful filters for this purpose are described by Y. Yamada, A. Mitsuishi and H. Yoshinaga, J. Opt. Soc. Am. 52, 17 (1962).

9. A somewhat better approximation than that used here is desirable if an accurate value of the optimal operating temperature T is to be calculated. Rather than use the exponential dependence of R(T) assumed here, Low[10] has introduced the form $R = R_o(T_o/T)^4$ which has some calculational advantages.

10. F. J. Low, J. Opt. Soc. Am. 51, 1300 (1961).

11. H.D. Drew thesis, Cornell University, 1968.

12. A start in this direction has been made by S. Zwerdling, R.A. Smith, and J.P. Therrault (to be published).

13. P.L. Richards, Pure and Appl. Chem. 11, 535 (1965).

14. R. C. Ohlman, P. L. Richards and M. Tinkham, J. Opt. Soc. Am. 48, 531, (1958).

15. The integrating cavity can be analyzed from the theory of W.E. Lamb, Jr., Phys. Rev. 70, 308 (1946).

16. P.L. Richards, Phys. Rev. 138, A1769 (1965).

17. R. R. Joyce and P.L. Richards, Phys. Rev. (to be published 1969).

18. Availably commercially from Eimac Co., 301 Industrial Way, San Carlos, California.

19. F.J. Low and A. R. Hoffman, App. Optics 2, 649 (1963).

20. For this reason the concept of specific detectivity D* is often not useful for Ge bolometers.

21. More precisely, it is the differential detector resistance dV/dI, not R = V/I, which is important for matching to the amplifier. In the operating range dV/dI ≃ R/2 for the indium antimonide detector.

22. E. Putley, Infrared Phys. 4, 1, (1964).

23. R. A. Smith, Proc. Inst. Electrical Eng. 98, 43 (1951)

24. P.L. Richards, Sidney Shapiro, and C. C. Grimes, Am. J. Phys. 36, 690 (1968).

25. C. C. Grimes, P.L. Richards and S. Shapiro, Phys. Rev. Letters 17, 431 (1966), and J. Appl. Phys. (to be published).

26. P.L. Richards and S. A. Sterling (to be published).

DEVELOPMENTS IN DETECTION AND MICROSAMPLING TECHNIQUES FOR LASER-RAMAN SPECTROSCOPY

D. O. Landon

Spex Industries, Inc., Metuchen, New Jersey

INTRODUCTION

To be described in detail are 1: Recent developments in detection techniques, particularly the use of a unique system of synchronous detection using an interference filter chopper. Applications discussed will be the improved measurement of low frequency ($< 100\ cm^{-1}$) bands in solids, and the elimination of spurious Raman lines due to scattering of the exciting radiation. 2: Recent developments in microsampling techniques. The optical properties of laser radiation allow the excitation of Raman spectra from as little as 10^{-4} μL of sample. In principle, the physical size of sample need be no larger than the excited volume but many practical problems are encountered when the sample volume approaches this level. Various techniques will be discussed and the spectra obtained evaluated in terms of the present "state of the art".

DEVELOPMENTS IN DETECTION AND MICROSAMPLING TECHNIQUES FOR LASER-RAMAN SPECTROSCOPY

Several years ago, when the Model 1400 double spectrometer was perfected, it appeared that all of the instrumental problems in Raman spectroscopy had been solved simultaneously. Considering this early optimism, it's strange how many problems have since come to our attention. Of course the original thought, that low frequency Raman measurements ($< \sim 60\ cm^{-1}$) were made only in gases and transparent media, was more an expression of what spectroscopists could do in the past rather than of what they wanted to accomplish. We rapidly found out that there were ambitious souls who would gladly attempt to get a low frequency spectrum of a lump of coal or a mirror surface, if there were any hope of success.

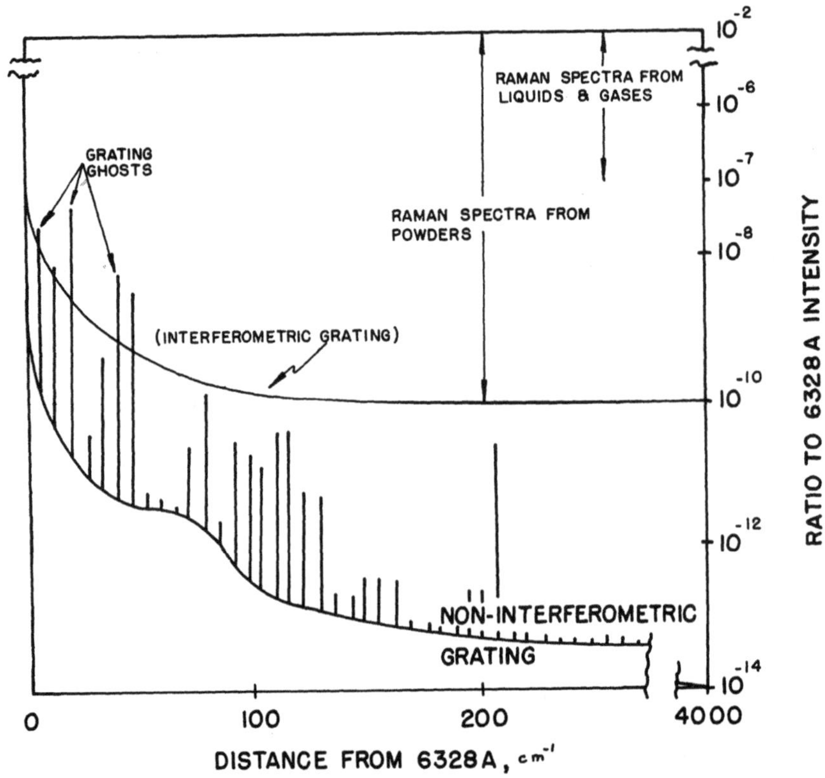

Figure 1. Bausch and Lomb gratings 1200 groove/mm, 5000A blaze.

Consider the instrumental problems that exist in respect to low frequency measurements. Figure 1 shows the scattering of light by the gratings in a typical double Raman spectrometer. (The gratings are the limiting feature.) As can be seen, in transparent media there are no limitations except instrumental resolution on the approach to zero frequency shift.

Opaque powders and translucent materials, those that scatter a considerable amount of exciting radiation into the spectrometer, do present a problem if Raman measurements are pushed below ~60 cm^{-1}. Figure 2 is a demonstration of the problem. The sample was a pressed pellet of the material and the spectrum was obtained by collecting the light scattered from one face. The Raman source was a 6328 A helium neon laser focussed on the sample face at 80 degrees incidence. There are plainly Raman features being obscured by the very visible grating ghosts--particularly noticeable here at 39 and 46 wavenumbers. Below about 20 wavenumbers, things appear nearly hopeless. Of course, this is in principle not true. If the grating ghosts can be eliminated, there are several existing devices to get rid of unwanted plasma emission lines such as neon 6331 or 6334. These devices are good diffraction gratings, narrow band filters,

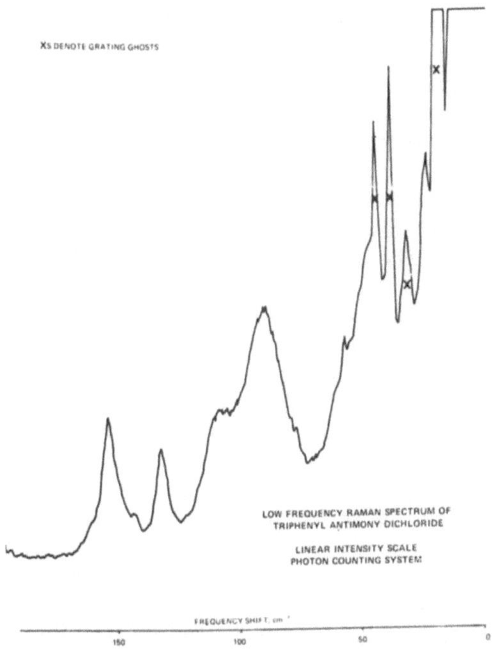

Figure 2. Low Frequency Raman Spectrum of Triphenyl Antimony Dichloride (Photon counting system).

Figure 3. Interference Filter Chopper Disc

prisms, or just plain distance. The problems would all be solved if there became available, an extremely narrow band device which behaved oppositely to our interference filter; that is to say, a device which would block a spectral region of a few Angstroms and pass all other wavelengths. In the days of mercury lamps and 2537 A excitation, cold mercury vapor performed this function beautifully. Unfortunately, we do not have an anti-laser.

One possible solution was suggested in the early days of considering the problem. Borrowing the idea from a device (the crystal chopper) used in infrared spectroscopy, led to its elaboration into an interference filter chopper. The principle here is to produce strong optical modulation of desired radiation and zero modulation of unwanted light; then detect the modulated signal synchronously at the modulation frequency. This idea was first suggested by me several years ago, but under such circumstances that I didn't really think the device was necessary. Enough contrary evidence accumulated that I was forced recently to undertake the experiment anyway. The first experiment was to assemble a very economical device; it used a stock one-inch-diameter 6328 A filter, a piece of exposed photographic plate, and a small velocity servo. The interference filter and exposed plate, adjusted to have the same transmission at 6328 A, were oscillated in the scattered light between the Raman sample and spectrometer. The 6328 A light is unmodulated, and consequently remains undetected. This includes the entourage of ghosts. A crystal phonograph cartridge was used to obtain the synchronizing signal for the lock-in system, which was a Brower Model 131 unit. This chopper could be run up to the astounding frequency of five Herz without setting the entire laboratory into motion.

However, its success in adversity led to bigger and better things.

Figure 3 shows the bigger aspect. This is a four-inch-diameter disc. One half is a narrow band 6328 A interference filter. The other half is an Inconel neutral filter having the same transmission at 6328 A.

Figure 4 concerns the better aspect. The filter has a transmittance of 73% at 6328 A, and an effective Raman half width of only 7.5 A - 19 wavenumbers.

Figure 5 shows how this was mounted in the experiment. This time, the sync. signal was generated by means of a magnetic pick-off, and the chopping frequency was 13 Herz. This frequency could easily be increased to 33 or 39 Herz without vibrational difficulties if it should prove desireable.

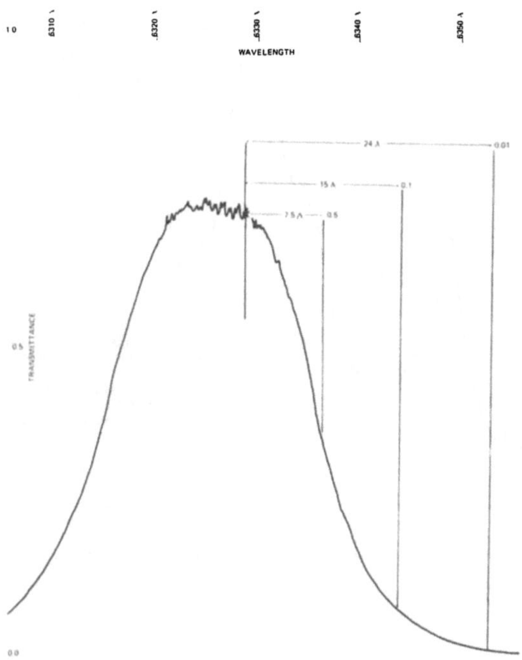

Figure 4. Interference filter test data

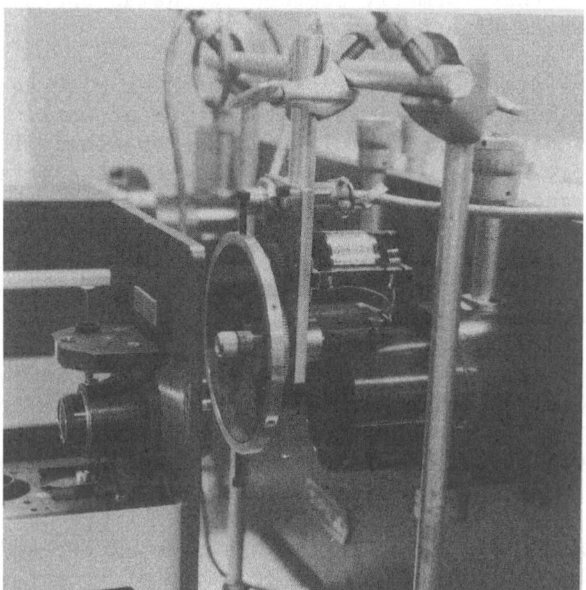

Figure 5. Interference Filter Chopper Mounted in Experiment

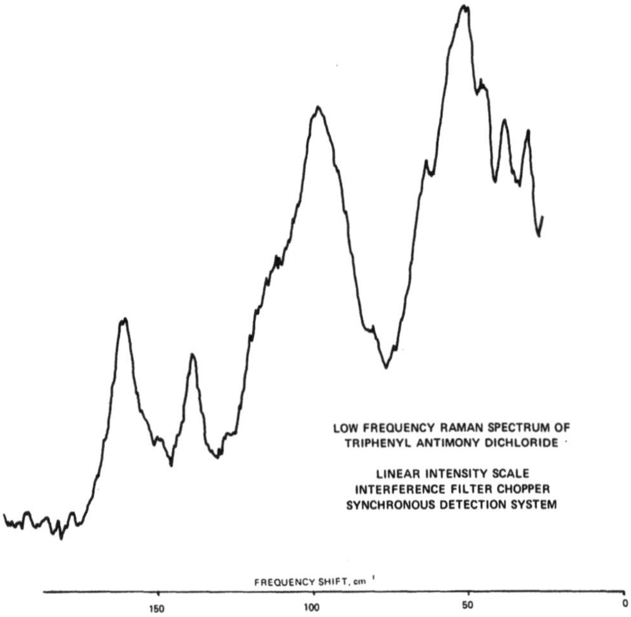

Figure 6. Low Frequency Raman Spectrum of Tripheny Antimony Dichloride (Interference Filter Chopper Synchronous Detector System)

Figure 6 shows the spectrum of our test sample as seen with the device. The Raman scan can now be started at 25 wavenumbers, and the ghosts that formerly adorned the Raman spectrum have disappeared (compare Figure 2).

There are distinct advantages in working with an exciting wavelength as long as possible in this type of work. The grating ghosts appear as per the formula $\frac{\lambda}{2400}$ for spacing. The longer the wavelength of exciting light, the lower the apparent frequency shift of the ghosts, and the fewer wavenumbers per wavelength interval. At 5000 A, there are 4 wavenumbers/Angstrom and a Rowland ghost spacing of approximately 8.3 wavenumbers, while at 6328 A or 6471 A there are 2.5 wavenumbers/Angstrom and approximately 6.5 wavenumbers/ghost.

With 4880 or 5145 A excitation, the seventh Rowland ghost appears at 58 cm^{-1}, while with 6328 or 6471 A light the same ghost appears at 45 cm^{-1}. In addition, the bandwidth of the interference filters is more favorable at the longer wavelength since a 5 A filter is 20 cm^{-1} at 5000 A, and is 12.5 cm^{-1} at 6328 A.

A perogative of the historian has been excercised in the above treatment, that of looking at the scramble of recent events and making them appear to be the result of organized and purposeful activity.

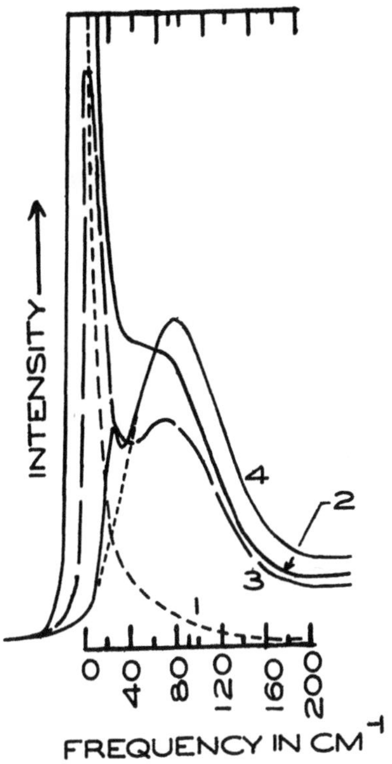

Fig. 7. The 4358-Å exciting line and the 75 cm^{-1} liquid-benzene Raman line at 26°C. *Curve 1 (short dashes)*, 0 mm (no) baffle. Amplification is nearly 100-fold less than that used for Curve 2. *Curve 2 (solid line)* baffle 9.6 mm, *Curve 3 (dashed line)* baffle 9.8 mm, and *Curve 4 (solid line)* baffle 10.0 mm with amplification double that of Curves 2 and 3. The frequency scale refers to the dispersion curve obtained with no baffle (cf. Table I). The dotted line in Curve 4 shows approximately how the unbaffled shoulder (residual) of the exciting line was separated from the benzene and other low-frequency lines.

The story of the interference filter chopper just presented is reasonably tidy, but neglects one of the factors that spurred us on to study the device.

Blatz[1] describes work on Low-Frequency Raman lines in organic liquids. He used a physical obstruction in the focal plane of a prism spectrograph to aid in the observation of spectra of liquids near zero frequency shift. This obstruction produced a reduction of intensity near the Raman exciting frequency and seemed to materially aid in identifying low frequency liquid lines. Early on, it appeared that the interference filter chopper might behave in a similar way in scanning Raman spectra with the double Spectrometer, and this coupled with the prior theoretical interest in the gadget led to initial work on the device.

Figure 7 is reproduced from Blatz's paper. Note that the amplification in the system changes by approximately 200X in going from curve 1 to curve 4. Figure 8 is also from Blatz' paper and shows the low frequency Raman spectrum of 1, 1, 2, 2, tetrachlorethane. Figure 9 shows the spectrum of this liquid as obtained

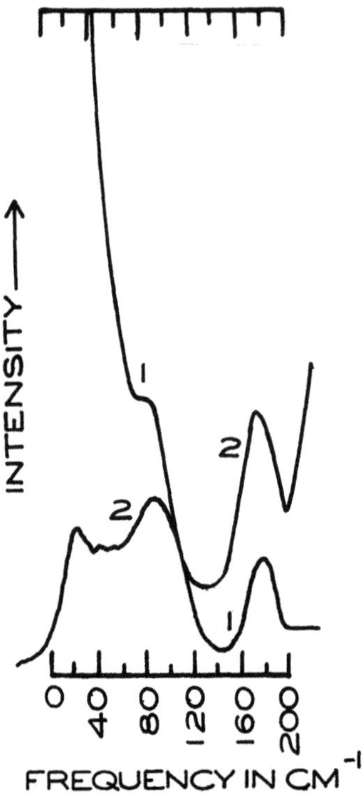

Figure 8. The low-frequency Raman spectrum of 1, 1, 2, 2, tetrachloroethane at 26°C. Curve 1, 0 mm (no baffle) and Curve 2, baffle setting 10.0 mm. There is a weak "liquid line" at 30 to 40 cm^{-1}.

under normal scanning conditions with the photon counting amplifier system. Figure 10 shows the same as Figure 9, but with our interference filter chopper operating in the scattered light.

A careful investigation of various spectra and analysis of the effect of this device finally led to the conclusion that the interference filter chopper is not appropriate for liquid line work. In fact, the narrower the pass-band of the interference filter, the less the observable effect on the liquid line and Rayleigh wing spectra. Stated in another way, since the Rayleigh wing phenomena is not caused by instrumental scattering, devices designed to eliminate instrumental scattering will not improve the observation.

Finally, the experiment in microsampling will be discussed briefly. Figure 11 shows some microcells that are covered. A is a small bore tube with an attached or integral plane parallel window. B is the same small bore tube with an integrally focused hemispherical lens of the type discussed by Bailey, Kint, and Scherer

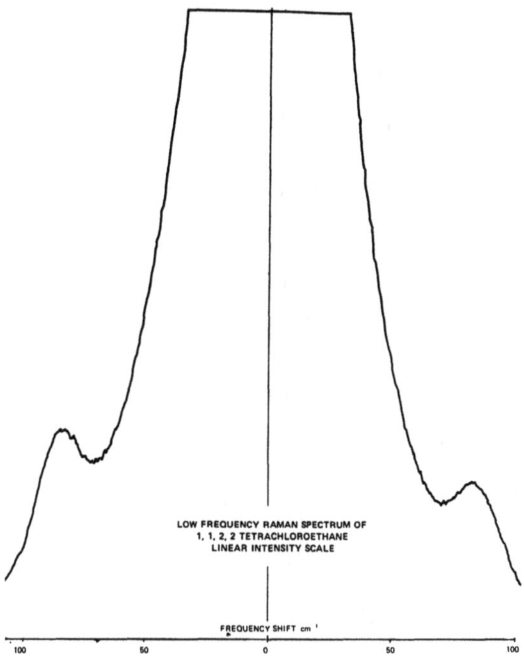

Figure 9. Low Frequency Raman Spectrum of 1,1,2,2 Tetrachloroethane (photon counting system)

in their paper in Analytical Chemistry last year.[2] In the A system, the laser objective lens focusses the laser beam in the cylinder. The length and bore of the capillary that can be used are limited by the divergence of the focussed laser beam. With a 90 mm f.l. lens, the divergence is $\sim\frac{1}{40}$, that is to say, the beam would be ~125µ diameter at each end of a 5 mm long tube.

In the B system the objective focusses the laser before the integral lens, and the distance between lenses is the sum of the focul lengths. The integral lens recollimates the laser beam through the bore. Bore size is limited by parallelism which in turn is limited by diffraction effects. The authors state that a 100 micron diameter beam (6328 A) expanded 40% in 5 mm lengths.

The C system, brought to my attention by Dr. Guido Pez of McMaster University probably cannot be made as small in sample volume as can A and B. However, unlike the B system and as in A,

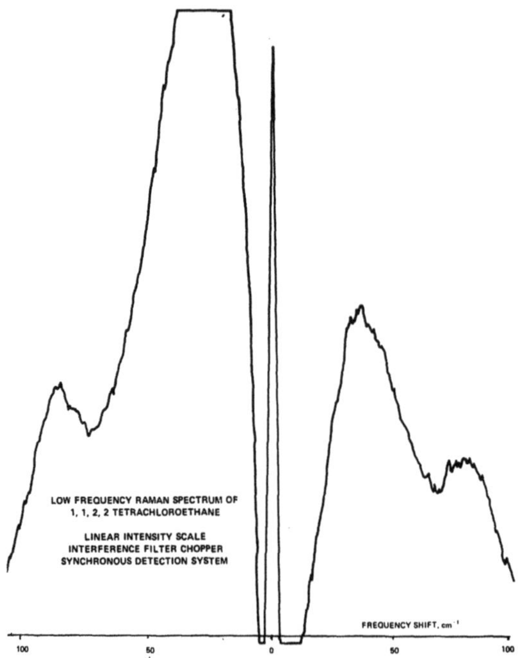

Figure 10. Low Frequency Raman Spectrum of 1,1,2,2 Tetrachloroethane (Interference filter chopper synchronous detector system)

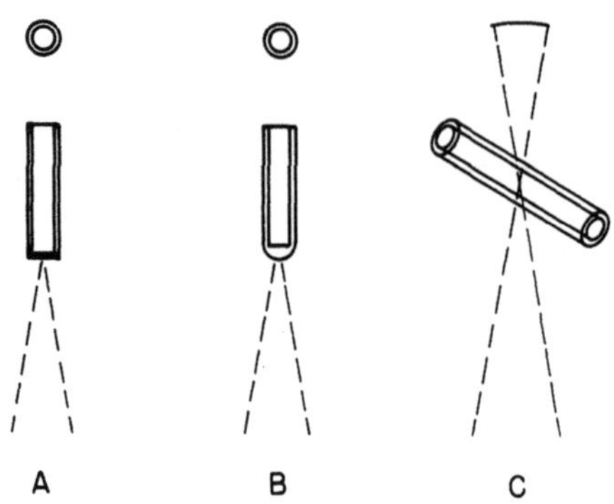

Figure 11. Microcell configurations

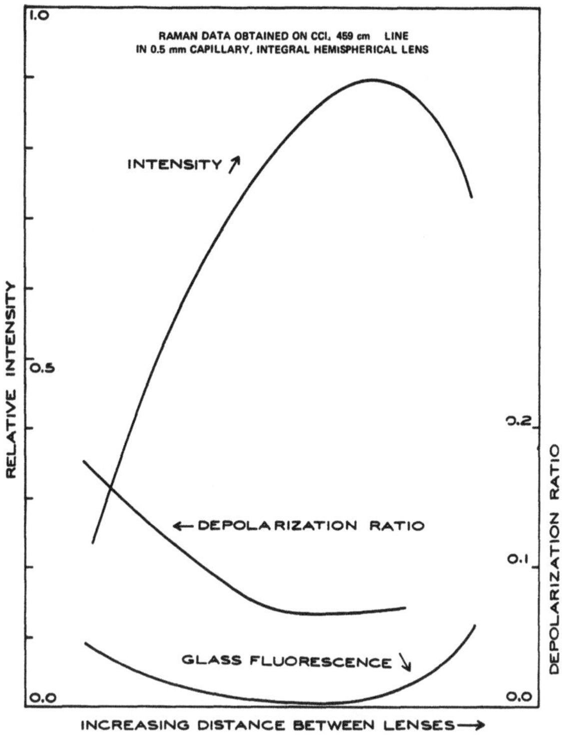

Figure 12. Raman data obtained on CCl_4 459 cm^{-1} line in 0.5 mm capillary, integral hemispherical lens.

multipassing of the laser beam proves possible. C is a standard melting point capillary. In our experiments, A and B gave identical results. Figure 12 gives the data for an 0.5 mm tube with integral lens. The measurements were made four ways, 1.0 mm and 0.5 mm bore, A and B systems. Identical results ensued. The persistant problem here is that the depolarization ratio for the line is significantly higher than the 0.006 value that we obtain in a high quality plane parallel window fused silica cell, such as the 12μL Raman cell. Bailey, Kint, and Scherer did much better, and one concludes that the optical finish on the cell window or hemispherical lens is the important thing. When Pez's mounting of the tube is used, the 0.006 value was again obtained. This system should prove excellent in handling G.C. fractions. The open tube can be chilled in the center while the carrier gas flows through. Then the tube can be sealed at both ends, analyzed, and stored permanently. The practical limit of sample size here appears to be approximately 1 μL. If smaller samples are to be handled, one of the other microcell configurations must be chosen (my own preference now being for that of Bailey, Kint and Scherer).

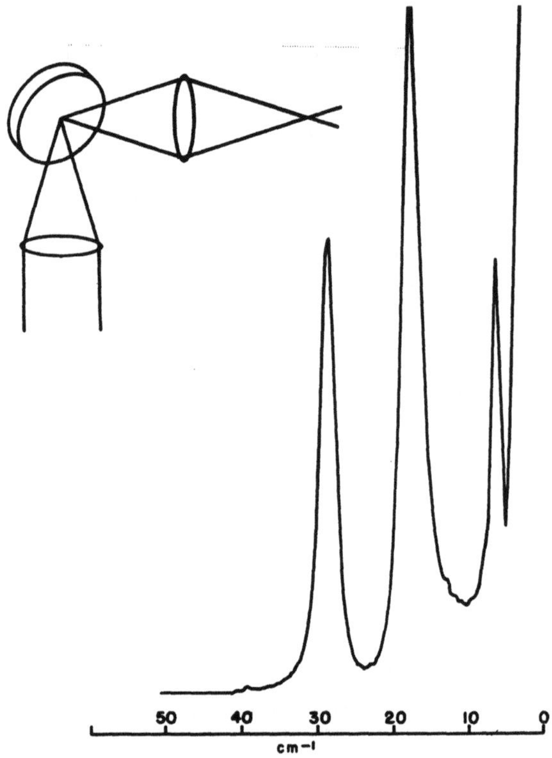

Figure 13. Close-in Raman spectrum of Mercuric Iodide by reflection technique.

Now, having made the above statement only three months ago (May, 1968), I find that a retraction is in order. Recently and with no great difficulty, we have obtained completely satisfactory spectra of liquids down to 0.05μL using the Pez technique; and this on a routine basis. Depolarization ratios of CCl_4 Raman lines are in excellent agreement with reported values on as little as 0.008μL of sample in a 0.1 mm bore tube. This information will be presented in full at the 1968 Eastern Analytical Symposium.

The truly surprising discovery made in exploring the transverse tube techniques came in the handling of powdered samples and condensed solids from gas chromatographs. With powdered samples placed in capillaries and sampled by the Pez technique, the ratio of Tyndall to Raman scattered light is reduced 100x as compared to previously studied reflection techniques. Furthermore, we have never yet failed to get a good spectrum by the "transillumination"

(Pez) technique. Figures 13 and 14 show a comparison of the previously used technique for sampling solids, and this new transillumination technique.

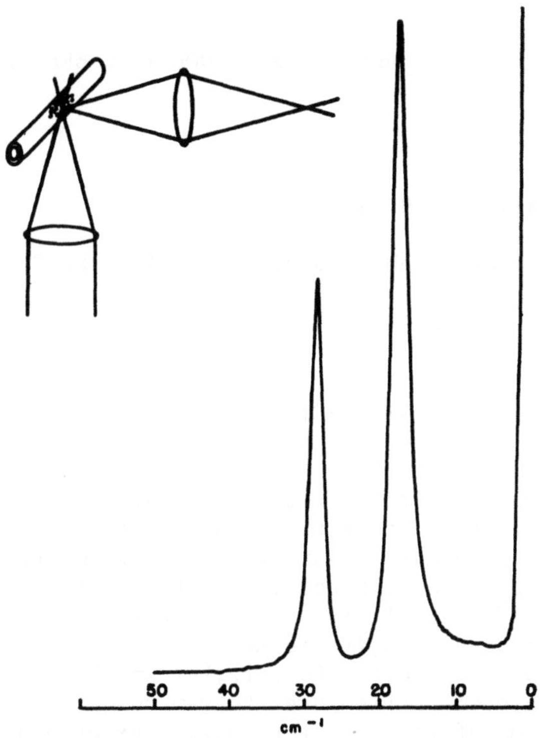

Fig. 14. Close-in Raman spectrum of Mercuric Iodide by trans-illumination technique.

REFERENCES

1. L. A. Blatz, J. Chem. Phys. 47, 842 (1967).

2. G. F. Bailey, S. Kint. J. R. Schever, Anal. Chem., 39, 1040, (1967).

A GRAPHICAL SURVEY OF FREE CARRIER MAGNETOOPTICAL EFFECTS IN SEMICONDUCTORS

E. D. Palik and B. W. Henvis

Naval Research Laboratory, Washington, D. C. 20390

INTRODUCTION

Since about 1955, numerous free carrier magnetooptical experiments have been devised to study the band structure of semiconductors. Among these are magnetoplasma reflection, rotation and ellipticity, cyclotron resonance, Faraday effect, Voigt effect and certain "helicon" effects. It is rather interesting that the classical model for bound and free electrons in a magnetic field which was developed by Drude,[1] Lorentz,[2] and Voigt[3] more than sixty years ago has been carried over to semiconductors essentially intact. Indeed, many of today's free carrier experiments are quantitatively explained on the basis of this early model. It is only necessary to introduce the effective mass (which may be anisotropic), the scattering time of the charged particle (which will vary with the scattering mechanism), and the dielectric constant of the host lattice. There are, of course, fine structure and new effects of an oscillatory nature which are explained only on the basis of quantum mechanics, but these will not be discussed here.

During the course of the magnetooptical work at NRL, we have found it instructive to calculate and plot, over a wide frequency range, the various optical constants and parameters which enter into the measurement of the effects. In this way we see at a glance where the phenomena will arise and how large the effects will be. This tends to unify our thinking in regard to all the free carrier effects, so that we do not consider each an isolated event.

In this paper we will present the pertinent formulas for the optical parameters, show plots of these parameters vs frequency for various values of magnetic field and scattering time, and look at where in frequency and magnetic field the various magnetooptical effects are usually measured. Rather than list the numerous papers on the various effects, we will refer only to several basic papers and review articles to which the reader may turn for detailed information about each effect. Review articles of particular interest have been written by Lax and Zwerdling,[4] Lax and Mavroides,[5] Moss,[6] Lax,[7] Palik and Wright,[8] Ginzburg,[9] and Smith[10].

DERIVATION: NON-POLAR SEMICONDUCTOR WITH FREE CARRIERS

The independent-particle model consists of N free charges/cm^3 in a host lattice of dielectric constant ϵ_∞. The rigid lattice contains N opposite charges to produce charge neutrality (ionized impurity atoms, for example). The free carriers have a constant isotropic effective mass m* and scattering time $\tau = 1/\nu$. When free carriers are added to a host lattice, the optical properties of the lattice change. Then, when an external magnetic field is applied, the random thermal motion of the free carriers is altered into cyclotron orbits in a plane perpendicular to the external magnetic field because of the Lorentz force. (Motion along the magnetic field is unaffected.) Consequently, the optical properties of the semiconductor change, bringing about all the magnetooptical effects. We omit lattice vibration effects for the moment, assuming that the background dielectric constant ϵ_∞ is, indeed, a constant as a function of frequency.

The two sample orientations generally used in experiments are shown in Fig. 1. In the longitudinal or Faraday orientation the radiation with which we study the effect propagates parallel to the external magnetic field. Generally we use circular polarization in such experiments. In the transverse or Voigt orientation, the radiation propagates perpendicular to the external magnetic field, and we use radiation linearly polarized parallel or perpendicular to the external magnetic field. The problem of arbitrary orientation of magnetic field to direction of propagation is discussed by Ginzburg[9] for a gaseous magnetoplasma.

The equation of motion of a charge e in a solid under the influence of an electric field $\vec{E}$ and an external magnetic induction $\vec{B}_o$ may be written as

$$m^*\dot{\vec{v}} = - m^*\nu\vec{v} + e\vec{E} + \frac{e(\vec{v} \times \vec{B}_o)}{c} \quad (1)$$

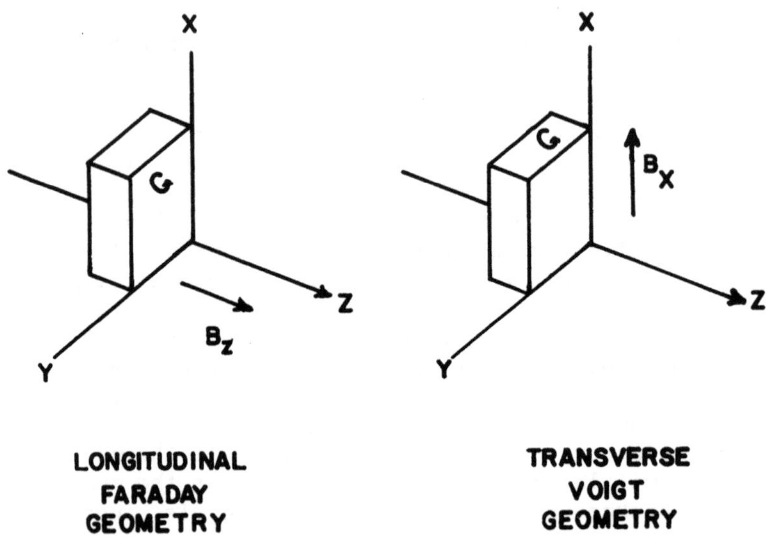

Figure 1. Faraday and Voigt geometries showing orientation of magnetic field, direction of propagation, and sense of electron cyclotron orbit.

where $\vec{v}$ is the drift velocity of the charge, and the magnetic induction of any applied EM wave is considered negligible. The sign of the charge is in e. Maxwell's equations may be written as

$$\nabla \cdot \vec{D} = 4\pi\rho \qquad \nabla \times \vec{E} = -\frac{1}{c}\dot{\vec{B}}$$
$$\nabla \cdot \vec{B} = 0 \qquad \nabla \times \vec{B} = \frac{4\pi}{c}\vec{J}'' + \frac{\overleftrightarrow{\epsilon}''}{c}\dot{\vec{E}} = \frac{4\pi}{c}\overleftrightarrow{\sigma}_{fc}\vec{E} + \frac{(\overleftrightarrow{\epsilon}_\infty + \overleftrightarrow{\epsilon}_{fc})\dot{\vec{E}}}{c} \tag{2}$$

where $4\pi\vec{J}''/c$ represents conduction current effects while $\overleftrightarrow{\epsilon}''\dot{\vec{E}}/c$ represents displacement current effects. The background lattice contributes a term $\overleftrightarrow{\epsilon}_\infty$ to $\overleftrightarrow{\epsilon}''$ while the free carriers contribute a term $\overleftrightarrow{\epsilon}_{fc}$. The conduction current term depends on $\overleftrightarrow{\sigma}_{fc}$. The constitutive equations $\vec{J}'' = \overleftrightarrow{\sigma}_{fc}\vec{E}$ and $\vec{D}'' = \overleftrightarrow{\epsilon}''\vec{E}$ supplement Eqs. 2. $\vec{B}$ is now the total magnetic field, but since the steady field effects are zero in Eqs. 2, the $\vec{B}$ can represent the magnetic field of the EM wave. With the assumption of exp $i\omega t$ dependences, the $\nabla\times\vec{B}$ equation can be rewritten as

$$\nabla \times \vec{B} = \frac{4\pi}{c}\left[\overleftrightarrow{\sigma}_{fc} + \frac{i\omega}{4\pi}(\overleftrightarrow{\epsilon}_\infty + \overleftrightarrow{\epsilon}_{fc})\right]\vec{E} = \frac{4\pi}{c}\overleftrightarrow{\sigma}'\vec{E} = \frac{4\pi}{c}\vec{J}' \tag{3}$$

where

$$\overleftrightarrow{\sigma}' = \overleftrightarrow{\sigma}_{fc} + \frac{i\omega}{4\pi}\overleftrightarrow{\epsilon}_{fc} + \frac{i\omega}{4\pi}\overleftrightarrow{\epsilon}_\infty = \overleftrightarrow{\sigma} + \frac{i\omega}{4\pi}\overleftrightarrow{\epsilon}_\infty$$

defines a complex conductivity. We can just as well rewrite the $\nabla\times\vec{B}$ equation as

$$\nabla \times \vec{B} = \frac{i\omega}{c}\left[\frac{4\pi\overleftrightarrow{\sigma}_{fc}}{i\omega} + \overleftrightarrow{\epsilon}_\infty + \overleftrightarrow{\epsilon}_{fc}\right]\vec{E} = \frac{i\omega}{c}\overleftrightarrow{\epsilon}'\vec{E} = \frac{i\omega}{c}\vec{D}' \tag{4}$$

where

$$\overleftrightarrow{\epsilon}' = \overleftrightarrow{\epsilon}_\infty + \overleftrightarrow{\epsilon}_{fc} - \frac{i4\pi}{\omega}\overleftrightarrow{\sigma}_{fc} .$$

We see that the free carriers contribute a real term to σ' of the form $\overleftrightarrow{\sigma}_{fc}$ and an imaginary term $i\omega\overleftrightarrow{\epsilon}_{fc}/4\pi$ while the background lattice contributes an imaginary term $i\omega\overleftrightarrow{\epsilon}_\infty/4\pi$. We will usually omit this last term when discussing conductivity (so that conductivity is due only to free carriers) and stress the dielectric tensor approach. The background lattice and free carriers contribute real terms to $\overleftrightarrow{\epsilon}'$ of the form $\overleftrightarrow{\epsilon}_\infty$ and $\overleftrightarrow{\epsilon}_{fc}$, respectively, while the free carriers contribute an imaginary term $- i4\pi\overleftrightarrow{\sigma}_{fc}/\omega$. In the present case $\overleftrightarrow{\epsilon}_\infty$ is considered a diagonal tensor with equal components. This would not be the case for a naturally birefringent crystal.

We forget about the background lattice contribution to the conductivity and solve for $\overleftrightarrow{\sigma}$ by use of Eq. 1. With the assumptions of exp $i\omega t$ dependences for $\vec{v}$ and $\vec{E}$, Eq. 1 can be solved for a complex velocity $\vec{v}$ without explicit knowledge of $\vec{E}$. Then the complex current density due to free carriers only is

$$\vec{J} = \overleftrightarrow{\sigma}\,\vec{E} = Ne\,\vec{v} \qquad (5)$$

from which $\overleftrightarrow{\sigma}$ can be obtained. With the successive reduction of primes, J now designates the complex current density due only to free carriers. Gaussian (unrationalized cgs) units are used.

The real and imaginary parts of the complex dielectric tensor, $\overleftrightarrow{\epsilon}_R$ and $\overleftrightarrow{\epsilon}_I$, are related to the imaginary and real parts of the conductivity $\overleftrightarrow{\sigma}_I$ and $\overleftrightarrow{\sigma}_R$ by the relations

$$\begin{aligned}
&\overleftrightarrow{\sigma} = \overleftrightarrow{\sigma}_R + i\overleftrightarrow{\sigma}_I && \overleftrightarrow{\epsilon}' = \overleftrightarrow{\epsilon}_R + i\overleftrightarrow{\epsilon}_I = \overleftrightarrow{\epsilon}_\infty - \frac{i4\pi}{\omega}\overleftrightarrow{\sigma} \\
&\overleftrightarrow{\sigma}_R = \overleftrightarrow{\sigma}_{fc} && \overleftrightarrow{\epsilon}_R = \overleftrightarrow{\epsilon}_\infty + \overleftrightarrow{\epsilon}_{fc} \qquad (6) \\
&\overleftrightarrow{\sigma}_I = \frac{\omega}{4\pi}\overleftrightarrow{\epsilon}_{fc} && \overleftrightarrow{\epsilon}_I = -\frac{4\pi}{\omega}\overleftrightarrow{\sigma}_{fc} \quad .
\end{aligned}$$

Hereafter, we drop the prime on $\overleftrightarrow{\epsilon}'$ and $\vec{D}'$.

Various solutions of Eq. 1 give the complex conductivity in terms of Cartesian coordinates, so that for the case of magnetic field along the z axis, the free carrier conductivity $\overleftrightarrow{\sigma}$ is

$$\overleftrightarrow{\sigma} = \begin{pmatrix} \sigma_{xx} & \sigma_{xy} & \sigma_{xz} \\ \sigma_{yx} & \sigma_{yy} & \sigma_{yz} \\ \sigma_{zx} & \sigma_{zy} & \sigma_{zz} \end{pmatrix} = \frac{\omega_p^2\epsilon_\infty}{4\pi} \begin{pmatrix} \dfrac{-i(\omega-i\nu)}{(\omega-i\nu)^2-\omega_c^2} & \dfrac{-\omega_c}{(\omega-i\nu)^2-\omega_c^2} & 0 \\ \dfrac{\omega_c}{(\omega-i\nu)^2-\omega_c^2} & \dfrac{-i(\omega-i\nu)}{(\omega-i\nu)^2-\omega_c^2} & 0 \\ 0 & 0 & \dfrac{-i}{(\omega-i\nu)} \end{pmatrix} \qquad (7)$$

The total $\overleftrightarrow{\sigma}'$ would have additional diagonal terms $i\omega/\omega_p^2$ due to the lattice. This tensor can be converted to the dielectric function tensor with the use of Eqs. 6. Then we find that

$$\overleftrightarrow{\epsilon} = \epsilon_\infty \begin{pmatrix} 1 - \dfrac{(\omega - i\nu)\omega_p^2}{[(\omega-i\nu)^2 - \omega_c^2]\omega} & \dfrac{i\omega_c\omega_p^2}{[(\omega-i\nu)^2-\omega_c^2]\omega} & 0 \\ \dfrac{-i\omega_c\omega_p^2}{[(\omega-i\nu)^2-\omega_c^2]\omega} & 1 - \dfrac{(\omega-i\nu)\omega_p^2}{[(\omega-i\nu)^2-\omega_c^2]\omega} & 0 \\ 0 & 0 & 1 - \dfrac{\omega_p^2}{(\omega-i\nu)\omega} \end{pmatrix} \qquad (8)$$

Note that $\overleftrightarrow{\epsilon}$ has diagonal terms ϵ_∞ (which give rise to the optical properties of the background lattice) and additional free carrier contributions. The quantity $\omega_p^2 = 4\pi Ne^2/m^*\epsilon_\infty$ defines the plasma frequency, $\omega_c = eB_o/m^*c$ is the cyclotron frequency, and $\nu = 1/\tau$ is the scattering frequency.

It is of interest to list the components of $\overleftrightarrow{\epsilon} = \overleftrightarrow{\epsilon}_R + i\overleftrightarrow{\epsilon}_I$:

$$\epsilon_{xx} = \epsilon_{yy} = \epsilon_\infty \left\{ 1 - \frac{\omega_p^2(\omega^2 - \omega_c^2 + \nu^2)}{[(\omega^2 - \omega_c^2 - \nu^2)^2 + 4\nu^2\omega^2]} + i\frac{\nu\omega_p^2(\omega^2+\omega_c^2+\nu^2)}{[(\omega^2-\omega_c^2-\nu^2)^2+4\nu^2\omega^2]\omega} \right\}$$

$$\epsilon_{xy} = -\epsilon_{yx} = \epsilon_\infty \left\{ \frac{-2\nu\omega_c\omega_p^2}{[(\omega^2-\omega_c^2-\nu^2)^2+4\nu^2\omega^2]} + i\frac{\omega_c\omega_p^2(\omega^2-\omega_c^2-\nu^2)}{[(\omega^2-\omega_c^2-\nu^2)^2+4\nu^2\omega^2]\omega} \right\}$$

$$\epsilon_{zz} = \epsilon_\infty \left\{ 1 - \frac{\omega_p^2}{(\omega^2+\nu^2)} - i\frac{\nu\omega_p^2}{\omega(\omega^2+\nu^2)} \right\} \qquad (9)$$

The corresponding σ_{ij} may be obtained from the equation

$$\epsilon_{ij} = \epsilon_\infty\delta_{ij} - i(4\pi/\omega)\,\sigma_{ij} \ . \qquad (10)$$

While $\overleftrightarrow{\sigma}$ and $\overleftrightarrow{\epsilon}$ tensors describe the medium, they must go back into Eqs. 2 if we are to determine what kinds of propagating waves can be maintained in the medium. The incident electromagnetic wave is assumed to be described by E_o' exp $i[\omega t-\omega z/c]$; inside the sample by E_o exp $i[\omega t-(n-ik)\omega z/c]$. The difference between E_o' and E_o involves the reflection loss at the incident surface and is ultimately found from the Fresnel equations. Usually $(\omega/c)(n-ik) = \kappa$ is called the wave vector of the radiation and as a vector has the direction of the allowed propagating wave. From Maxwell's equations one can obtain (with the use of $\nabla = -i\vec{\kappa}$) the wave equation

$$\nabla \times (\nabla \times \vec{E}) = -\nabla^2\vec{E} + \nabla(\nabla\cdot\vec{E}) = \frac{\omega^2}{c^2}\left(\vec{D}'' - \frac{i4\pi}{\omega}\vec{J}''\right) = \frac{\omega^2}{c^2}\overleftrightarrow{\epsilon}\,\vec{E}$$

or

$$\kappa^2\vec{E} - \vec{\kappa}(\vec{\kappa}\cdot\vec{E}) = \frac{\omega^2}{c^2}\overleftrightarrow{\epsilon}\,\vec{E} \tag{11}$$

the solution of which yields two "normal mode" values $\kappa^2_{1,2}$.

For the Faraday orientation in Fig. 1 we find

$$\kappa^2_{1,2} = \frac{\omega^2}{c^2}(\epsilon_{xx} \pm i\epsilon_{xy}) = \left(\frac{\omega}{c}\right)^2(n-ik)^2_{1,2} = \left(\frac{\omega}{c}\right)^2 \epsilon_{1,2}. \tag{12}$$

Insertion of these solutions into Eq. (11) gives $E_{1,2z} = 0$ and $E_{1,2x} = \mp iE_{1,2y}$. Thus the waves are circularly polarized. The z component of Eq. (11) dictates that $E_{1,2z} = 0$, except that $E_{1,2z}$ is not required to be zero at $\epsilon_{zz} = 0$. We will note this case shortly.

The model yields for the Faraday orientation with Poynting vector $\vec{S}\|\vec{B}_o$ and $\vec{E}\perp\vec{B}_o$

$$\epsilon_{1,2} = \epsilon_{\pm} = (n-ik)^2_{\pm} = \epsilon_{\infty}\left[1 - \frac{\omega_p^2}{\omega(\omega\mp\omega_c-i\nu)}\right]. \tag{13}$$

$\vec{E}_{\pm}$ are the right and left circularly polarized radiation electric vectors defined by $E_{1x} = -iE_{1y}$ for $\vec{E}_+$ and $E_{2x} = iE_{2y}$ for $\vec{E}_-$. Here the sign of the charge is included in e. Therefore, for an electron, resonance occurs for the lower sign in Eq. 13, which corresponds to left circular (cyclotron active) polarization when $\vec{S}\|\vec{B}_o$; $\vec{B}_o$ is positive when parallel to $\vec{S}$. Left circular polarization is a counterclockwise rotation of the electric vector when the observer looks back toward the source of radiation.

From Eqs. 12 and 13 the complex circular conductivity is $\sigma_{\pm} = \sigma_{xx} \pm i\sigma_{xy}$ in analogy to $\epsilon_{\pm}$. The real and imaginary parts of $\epsilon_{\pm}$ are given by $\epsilon_{\pm R} = n_{\pm}^2 - K_{\pm}^2$ and $\epsilon_{\pm I} = -2n_{\pm}k_{\pm}$. The conductivity and dielectric tensors are diagonalized in this configuration into the form

$$\begin{pmatrix} J_+ \\ J_- \\ J_z \end{pmatrix} = \begin{pmatrix} \sigma_+ & 0 & 0 \\ 0 & \sigma_- & 0 \\ 0 & 0 & \sigma_{zz} \end{pmatrix} \begin{pmatrix} E_+ \\ E_- \\ E_z \end{pmatrix} \qquad \begin{pmatrix} D_+ \\ D_- \\ D_z \end{pmatrix} = \begin{pmatrix} \epsilon_+ & 0 & 0 \\ 0 & \epsilon_- & 0 \\ 0 & 0 & \epsilon_{zz} \end{pmatrix} \begin{pmatrix} E_+ \\ E_- \\ E_z \end{pmatrix} \tag{14}$$

where the "normal mode" E fields inside the sample are E_+, E_-, and $E_z = 0$ (except when $\epsilon_{zz} = 0$).

The position vector $\vec{r}$ of a free carrier is obtained from the velocity $\vec{v}$, since $\vec{r} = \vec{v}/i\omega = \overleftrightarrow{\sigma}\vec{E}/Nei\omega$. Then the x,y,z coordinates for the Faraday configurations are

$$x_{1,2} = \frac{1}{Nei\omega}(\sigma_{xx}E_{1,2x}+\sigma_{xy}E_{1,2y}) = \frac{1}{4\pi Ne}\left[(\epsilon_{xx}-\epsilon_\infty)E_{1,2x}+\epsilon_{xy}E_{1,2y}\right]$$

$$y_{1,2} = \frac{1}{Nei\omega}(\sigma_{yx}E_{1,2x}+\sigma_{yy}E_{1,2y}) = \frac{1}{4\pi Ne}\left[\epsilon_{yx}E_{1,2x}+(\epsilon_{yy}-\epsilon_\infty)E_{1,2y}\right]$$

$$z_{1,2} = \frac{1}{Nei\omega}\sigma_{zz}E_{1,2z} = \frac{1}{4\pi Ne}(\epsilon_{zz}-\epsilon_\infty)\, E_{1,2z} \tag{15}$$

where the σ_{ij} and ϵ_{ij} are given by Eqs. (7) and (8), and the "normal mode" electric fields are $E_{1,2x} = \mp iE_{1,2y}$ and $E_{1,2z} = 0$. Further expansion gives

$$x_{1,2} = -\frac{1}{4\pi Ne}(\epsilon_\infty-\epsilon_\pm)E_{1,2x} = \frac{-e[(\omega\mp\omega_c)+i\nu]}{m^*\omega[(\omega\mp\omega_c)^2+\nu^2]}E_{1,2x}$$

$$y_{1,2} = -\frac{1}{4\pi Ne}(\epsilon_\infty-\epsilon_\pm)E_{1,2y} = \frac{-e[(\omega\mp\omega_c)+i\nu]}{m^*\omega[(\omega\mp\omega_c)^2+\nu^2]}E_{1,2y} \tag{16}$$

$$z_{1,2} = 0 \; .$$

The condition $z_{1,2} = 0$ holds except when $\epsilon_{zz} = 0$. Then $z_{1,2} = (-\epsilon_\infty/4\pi Ne)E_{1,2z}$ holds. Dropping the 1,2 notation, we may write $E_z = (-4\pi Nez/\epsilon_\infty) = -4\pi P_z$. The z electric field is related to charge displacement and thus to electric polarization P_z. An equation of motion in the z direction may be written as $\ddot{z} = (e/m^*)E_z = -\omega_p^2 z$. This represents a simple harmonic oscillator of resonant frequency ω_p. E_z is a "self field" which acts as a restoring force on the charges shifting their resonant frequency from $\omega = 0$ to $\omega = \omega_p$. Eq. (11) yields no relation between ω^2/c^2 and $\varkappa$ for the z direction, and no EM wave propagates along z although an electric field $E_z \sim \exp i\omega t$ does exist. This motion is a plasma oscillation (plasmon). When $\vec{B}_o = 0, \omega_p$ is located by a zero in the dielectric constant ($\epsilon_{zz} = 0$). However, when $\vec{B}_o \neq 0$ the plasma continues to oscillate at ω_p, although $\epsilon_\pm$ does not equal zero there.

For the Voigt orientation the direction of propagation is kept as the z axis, but $\vec{B}_o$ is turned to the x axis; the dielectric tensor, modified by relabeling and interchange of terms (z → x, x → y, y → z), is given by

$$\overleftrightarrow{\epsilon} = \begin{pmatrix} \epsilon_{xx} & 0 & 0 \\ 0 & \epsilon_{yy} & \epsilon_{yz} \\ 0 & \epsilon_{zy} & \epsilon_{zz} \end{pmatrix} .$$

With this change it can be shown that

$$\sigma_{||} = \sigma_{xx} \quad , \quad \epsilon_{||} = \epsilon_{xx}$$

$$\sigma_\perp = \frac{(\epsilon_\infty \omega/4\pi)\sigma_{yy} - i(\sigma_{yz}^2 + \sigma_{yy}^2)}{(\epsilon_\infty \omega/4\pi) - i\sigma_{yy}} \quad , \quad \epsilon_\perp = \epsilon_{yy} + \frac{\epsilon_{yz}^2}{\epsilon_{yy}} \quad . \tag{17}$$

The "normal mode" waves are $E_{1x} = E_{||}$, $E_{1y} = 0$ and $E_{1z} = 0$; $E_{2x} = 0$, $E_{2y} = E_\perp$ and $E_{2z} = (\epsilon_{yz}/\epsilon_{yy})E_{2y}$.

For $\vec{S}\perp\vec{B}_o$ and $\vec{E}_{||}\vec{B}_o$ the optical constants are

$$\epsilon_{||} = (n-ik)_{||}^2 = \epsilon_\infty \left[1 - \frac{\omega_p^2}{\omega(\omega - i\nu)}\right] . \tag{18}$$

For $\vec{S}\perp\vec{B}_o$ and $\vec{E}\perp\vec{B}_o$

$$\epsilon_\perp = (n-ik)_\perp^2 = \epsilon_\infty \left[1 - \frac{\omega_p^2[\omega_p^2 - \omega(\omega - i\nu)]}{\omega[\omega_p^2(\omega - i\nu) - \omega(\omega - i\nu)^2 + \omega\omega_c^2]}\right] . \tag{19}$$

The z equation of Eq. (11) gave the relation $D_z = D_z'' - i4\pi J_z''/\omega = 0$ from which the relation $E_{2z} = (\epsilon_{yz}/\epsilon_{yy})E_{2y}$ was obtained. The diagonalized conductivity and dielectric tensors are written in the form

$$\begin{pmatrix} J_{\|} \\ J_{\perp} \end{pmatrix} = \begin{pmatrix} \sigma_{\|} & 0 \\ 0 & \sigma_{\perp} \end{pmatrix} \begin{pmatrix} E_{\|} \\ E_{\perp} \end{pmatrix} \quad \begin{pmatrix} D_{\|} \\ D_{\perp} \end{pmatrix} = \begin{pmatrix} \epsilon_{\|} & 0 \\ 0 & \epsilon_{\perp} \end{pmatrix} \begin{pmatrix} E_{\|} \\ E_{\perp} \end{pmatrix} . \tag{20}$$

The "effective" ϵ_{zz} in the equation $D_z = \epsilon_{zz} E_z$ is zero, as displacement current effects cancel conduction current effects. The "effective" σ_{zz} is finite as J contains only free carriers contributions.

The x,y,z coordinates are

$$x_{1,2} = \frac{1}{Nei\omega} \sigma_{xx} E_{1,2x} = \frac{1}{4\pi Ne} (\epsilon_{xx} - \epsilon_{\infty}) E_{1,2x}$$

$$y_{1,2} = \frac{1}{Nei\omega} (\sigma_{yy} E_{1,2y} + \sigma_{yz} E_{1,2z}) = \frac{1}{4\pi Ne} [(\epsilon_{yy} - \epsilon_{\infty}) E_{1,2y} + \epsilon_{yz} E_{1,2z}]$$

$$z_{1,2} = \frac{1}{Nei\omega} (\sigma_{zy} E_{1,2y} + \sigma_{zz} E_{1,2z}) = \frac{1}{4\pi Ne} [\epsilon_{zy} E_{1,2y} + (\epsilon_{zz} - \epsilon_{\infty}) E_{1,2z}] . \tag{21}$$

Further expansion gives

$$x_1 = \frac{1}{4\pi Ne} (\epsilon_{xx} - \epsilon_{\infty}) E_{\|} = - \frac{e}{m^*} \left[\frac{1}{(\omega^2 + \nu^2)} + i \frac{\nu}{\omega(\omega^2 + \nu^2)} \right] E_{\|}$$

$$y_1 = z_1 = 0$$

$$x_2 = 0$$

$$y_2 = -\frac{1}{4\pi Ne} [\epsilon_{\infty} - \epsilon_{yy} - \frac{\epsilon_{yz}^2}{\epsilon_{yy}}] E_{\perp} = \frac{e[(\omega - i\nu)\omega - \omega_p^2]}{m^* \omega \{[(\omega - i\nu)^2 - \omega_c^2]\omega - (\omega - i\nu)\omega_p^2\}} E_{\perp}$$

$$z_2 = -\frac{1}{4\pi Ne} (\frac{\epsilon_{\infty} \epsilon_{yz}}{\epsilon_{yy}}) E_{\perp} = \frac{-ie\omega_c}{m^* \{[(\omega - i\nu)^2 - \omega_c^2]\omega - (\omega - i\nu)\omega_p^2\}} E_{\perp} . \tag{22}$$

A frequency-dependent Hall electric field E_{2z} arises as a result of charge motion in the z direction due to the y driving electric field of the radiation. This field is

$$E_{2z} = \frac{i\omega_p^2 \omega_c}{\{[(\omega - i\nu)^2 - \omega_c^2]\omega - (\omega - i\nu)\omega_p^2\}} E_{2y} . \tag{23}$$

The discussion of plasma frequencies is somewhat more complicated in the Voigt configuration(11). For $\vec{B}_o \neq 0$ a plasma oscillation at ω_p does not exist along the z direction, as it has been coupled into the cyclotron motion to form the resonance at $\omega^2 = \omega_p^2+\omega_c^2$ indicated in Eq. (19). We can think of the system as a coupled oscillator of both transverse (y) and longitudinal (z) character with respect to the direction of propagation. This coupled oscillator has the same resonant frequency for both its y and z motion.

The requirements that $\nabla\cdot\vec{B} = 0$ and $\nabla\times\vec{B} = (i\omega/c)\overleftrightarrow{\epsilon}\vec{E}$ are used to determine the $\vec{B}$ fields for each wave. Since $E_{1x} = E_{\parallel}$ and $E_{1y} = E_{1z} = 0$, it follows that $B_{1x} = B_{1z} = 0$ and $B_{1y} \neq 0$. Also, since $E_{2x} = 0$, $E_{2y} = E_{\perp}$ and $E_{2z} \propto E_{2y}$, it follows that $B_{2x} \neq 0$, $B_{2y} = B_{2z} = 0$. The $\nabla\times\vec{E} = -(1/c)\dot{\vec{B}}$ equation gives $\vec{\kappa}\times\vec{E} = (\omega/c)\vec{B}$ which also gives the direction of $\vec{B}$.

We have used Eqs. (13), (18), and (19) to obtain ϵ_R and ϵ_I and then n and k. A simplification occurs if $\nu \to 0$, which is called the lossless case. No absorption of radiation occurs, only reflection and transmission. We will also treat the practical case of a finite scattering frequency typical of the example we choose. Usually τ is obtained from the mobility $\mu = e\tau/m^*$.

The optical constants $2n^2 = \epsilon_R + (\epsilon_R^2+\epsilon_I^2)^{\frac{1}{2}}$ and $2k^2 = -\epsilon_R + (\epsilon_R^2+\epsilon_I^2)^{\frac{1}{2}}$ may be written explicitly in several ways one of which we include here. For the Faraday orientation(12)

$$2n_{\pm}^2 = \left(\epsilon_\infty - \frac{4\pi A_\pm}{\omega}\right) + \left\{\left(\epsilon_\infty - \frac{4\pi A_\pm}{\omega}\right)^2 + \left(\frac{4\pi C_\pm}{\omega}\right)^2\right\}^{\frac{1}{2}}$$

$$2k_{\pm}^2 = -\left(\epsilon_\infty - \frac{4\pi A_\pm}{\omega}\right) + \left\{\left(\epsilon_\infty - \frac{4\pi A_\pm}{\omega}\right)^2 + \left(\frac{4\pi C_\pm}{\omega}\right)^2\right\}^{\frac{1}{2}} \qquad (24)$$

where

$$A_\pm = \frac{\sigma_o(\omega\mp\omega_c)\tau}{1+(\omega\mp\omega_c)^2\tau^2} \qquad C_\pm = \frac{\sigma_o}{1+(\omega\mp\omega_c)^2\tau^2} .$$

The dc conductivity is $\sigma_o = Ne^2\tau/m^*$. For the Voigt orientation(13) the optical constants are

$$2n_\perp^2 = (\epsilon_\infty - \frac{4\pi A}{\omega}) + \left\{(\epsilon_\infty - \frac{4\pi A}{\omega})^2 + (\frac{4\pi C}{\omega})^2\right\}^{\frac{1}{2}}$$

$$2k_\perp^2 = -(\epsilon_\infty - \frac{4\pi A}{\omega}) + \left\{(\epsilon_\infty - \frac{4\pi A}{\omega})^2 + (\frac{4\pi C}{\omega})^2\right\}^{\frac{1}{2}} \qquad (25)$$

where

$$A = \frac{\sigma_o \tau \omega}{\beta\left[\left(\frac{\alpha}{\beta}\right)^2 + \omega^2\tau^2\right]} \qquad C = \frac{\sigma_o \alpha}{\left[\beta\left(\frac{\alpha}{\beta}\right)^2 + \omega^2\tau^2\right]}$$

$$\alpha = 1 + \left\{\frac{\omega^2 \omega_c^2 \tau^2}{\omega^2 + (\omega^2 - \omega_p^2)^2 \tau^2}\right\} \qquad \beta = 1 - \left\{\frac{\omega_c^2(\omega^2 - \omega_p^2)\tau^2}{\omega^2 + (\omega^2 - \omega_p^2)^2\tau^2}\right\}.$$

The parallel optical constants are obtained when $\alpha = \beta = 1$. Explicitly, we get

$$2n_{\|}^2 = \left(\varepsilon_\infty - \frac{4\pi\sigma_o\tau}{1+\omega^2\tau^2}\right) + \left\{\left(\varepsilon_\infty - \frac{4\pi\sigma_o\tau}{1+\omega^2\tau^2}\right)^2 + \left(\frac{4\pi\sigma_o}{\omega[1+\omega^2\tau^2]}\right)^2\right\}^{\frac{1}{2}}$$

$$2k_{\|}^2 = -\left(\varepsilon_\infty - \frac{4\pi\sigma_o\tau}{1+\omega^2\tau^2}\right) + \left\{\left(\varepsilon_\infty - \frac{4\pi\sigma_o\tau}{1+\omega^2\tau^2}\right)^2 + \left(\frac{4\pi\sigma_o}{\omega[1+\omega^2\tau^2]}\right)^2\right\}^{\frac{1}{2}}. \tag{26}$$

Note that the parallel optical constants are independent of magnetic field for the isotropic mass model and represent the zero field constants. This is not the case for anisotropic masses in cubic semiconductors(7,14).

Reflectance $\mathfrak{R}$ and transmittance $\mathcal{T}$ of a sample may be computed from

$$\mathcal{T} = e^{-2\gamma}\left[\frac{(1-R)^2 + 4R\sin^2\eta}{(1-Re^{-2\gamma})^2 + 4Re^{-2\gamma}\sin^2(\beta+\eta)}\right]$$

$$= \frac{(1-R)^2(1+k^2/n^2)}{(e^{\gamma}-Re^{-\gamma})^2 + 4R\sin^2(\beta+\eta)} \tag{27}$$

$$\mathfrak{R} = R\left[\frac{(1-e^{-2\gamma}) + 4e^{-2\gamma}\sin^2\beta}{(1-Re^{-2\gamma})^2 + 4Re^{-2\gamma}\sin^2(\beta+\eta)}\right] \tag{28}$$

where $\tan \eta = 2k/(n^2+k^2-1)$, $\beta = 2\pi nd/\lambda$, $\gamma = 2\pi kd/\lambda$, d is the sample thickness, and R is the reflectivity. Usually, interference fringes as such are not seen in fairly thick samples ($2\pi nd/\lambda \gg 1$) where low spectral resolution averages out the fringes. Then the transmittance T is

$$T = \frac{(1-R)^2(1+k^2/n^2)e^{-\alpha d}}{1-R^2e^{-2\alpha d}} \tag{29}$$

where $\alpha = 4\pi k/\lambda$ is the absorption constant and λ is the vacuum wavelength. For just one pass through the sample, the multiple reflection term in Eq. (29) is neglected.

The bulk reflectance (reflectivity) is given by the equation

$$R = \frac{(1-n)^2+k^2}{(1+n)^2+k^2} \;. \tag{30}$$

Both Eq. (29) and Eq. (30) represent changes of intensity of the radiation beam.

Along with the change in amplitude of the electric vector of the radiation upon reflection and transmission, there is a phase shift also. Upon normal reflection in air, the phase shift Δ is given by

$$\tan \Delta = \frac{2k}{1-n^2-k^2} \;. \tag{31}$$

For a dielectric, this phase shift is 180°. $\tan \Delta = \infty$ and $\Delta = 90°$ when $n^2+k^2 = 1$. $\tan \Delta$ is positive only when $n^2+k^2 < 1$ and Δ will range from 90° to 0°. Upon transmission the phase shift φ is given by

$$\varphi = \frac{\omega dn}{c} \;. \tag{32}$$

These phase shifts bring about changes in the state of polarization of the incident beam. We are now equipped to calculate all of the pertinent optical parameters which lead to the various magnetooptical effects.

In the next two sections we discuss the Faraday and Voigt configurations in terms of the pertinent dielectric constant for the lossless case $\nu = 0$. We will note in what spectral regions various experiments are done and pay particular attention to the reflectivity of the sample. Free carrier effects predominantly occur in the far infrared.

DISCUSSION OF DIELECTRIC CONSTANT IN THE FARADAY GEOMETRY

General Remarks

We first examine the lossless case for the Faraday geometry. When $\nu = 0$, $\epsilon_{\pm} = \epsilon_{\pm R} = n_{\pm}^2 - k_{\pm}^2$ holds. It is instructive to plot the positions of the zeroes, unities and infinities in ϵ to obtain an isoindex of refraction "contour" map[8] as shown in Fig. 2. Here we have plotted ω_c/ω_p vs ω/ω_p. Negative ordinate signifies the region of left or CCW or cyclotron-active circular polarization for $\vec{B}_o \| \vec{S}$, since the electron charge is negative, while positive ordinate signifies the region of right or CW or cyclotron-inactive polarization. In the shaded regions k is finite and n is zero. The dotted lines near $\epsilon = 0$ are lines of unit index of refraction. Of particular interest are the isoindex lines corresponding to n = 1, 0 and ∞. From Eq. 13 for $\nu = 0$ we get

$$\epsilon_{\pm} = n_{\pm}^2 - k_{\pm}^2 = \frac{\epsilon_{\infty}[\omega(\omega \mp \omega_c) - \omega_p^2]}{\omega(\omega \mp \omega_c)} . \tag{33}$$

The zeroes in $\epsilon_{\mp}$ occur at

$$\omega = \pm \frac{1}{2}\omega_c \pm \frac{1}{2}(\omega_c^2 + 4\omega_p^2)^{\frac{1}{2}} . \tag{34}$$

The infinities occur at

$$\omega = 0 \qquad\qquad \omega = \omega_c . \tag{35}$$

The unities occur at

$$2\omega = \pm\, \omega_c \pm \left(\omega_c^2 + \frac{4\epsilon_{\infty}\omega_p^2}{\epsilon_{\infty} - 1}\right)^{\frac{1}{2}} . \tag{36}$$

Marked features in the reflection spectrum occur at these values of $\epsilon_{\pm}$. For example, zeroes determine points of unit reflectivity; infinities determine resonant frequencies of the charge system and points of unit reflectivity; and unities determine zeroes in reflection.

Of some interest is the function $1/\epsilon_{\pm}$. It may be written in real and imaginary parts as

$$\frac{1}{\epsilon_{\pm}} = \frac{\omega}{\epsilon_{\infty}} \frac{\{(\omega \mp \omega_c)(\omega^2 \mp \omega\omega_c - \omega_p^2) + \nu^2\omega^2 + i\nu[\omega(\omega \mp \omega_c) - (\omega^2 \mp \omega\omega_c - \omega_p^2)]\}}{(\omega^2 \mp \omega\omega_c - \omega_p^2) + \nu^2\omega^2} . \tag{37}$$

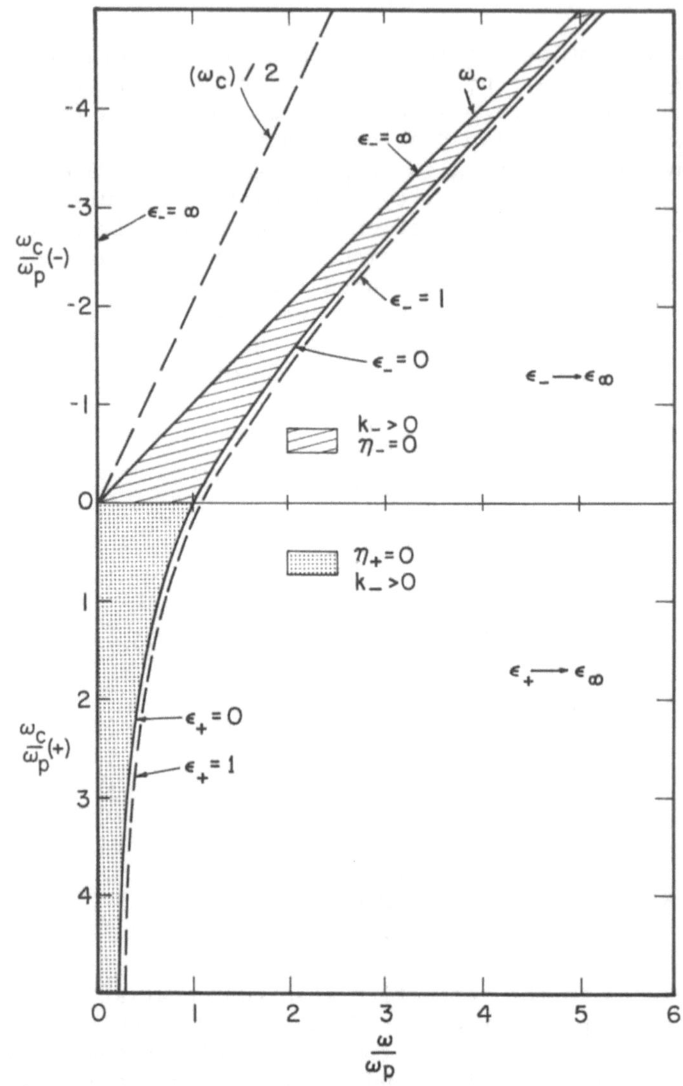

Figure 2. Index of refraction plot showing zeroes, unities and infinities in $n_\pm^2 = \epsilon_\pm$ for the lossless case $\nu = 0$, Faraday geometry. In shaded regions $k > 0$, $n = 0$ holds.

Horizontal lines across Fig. 2 can be plotted as ϵ vs ω/ω_p for fixed magnetic fields. Four cases of interest, $\omega_c=0$, $\omega_c \ll \omega_p$, $\omega_c \approx \omega_p$ and $\omega_c \gg \omega_p$ are illustrated in Fig. 3 for each circular polarization. Below each ϵ plot is a reflectivity plot. In real semiconductors where $\nu \neq 0$, the sharp corners are rounded off the curves and n and k are finite everywhere. For example, we show in Fig. 4 the effect as ν increases from zero for the particular case $\omega_c = (1/2)\omega_p$. Note that now ϵ_I is finite also.

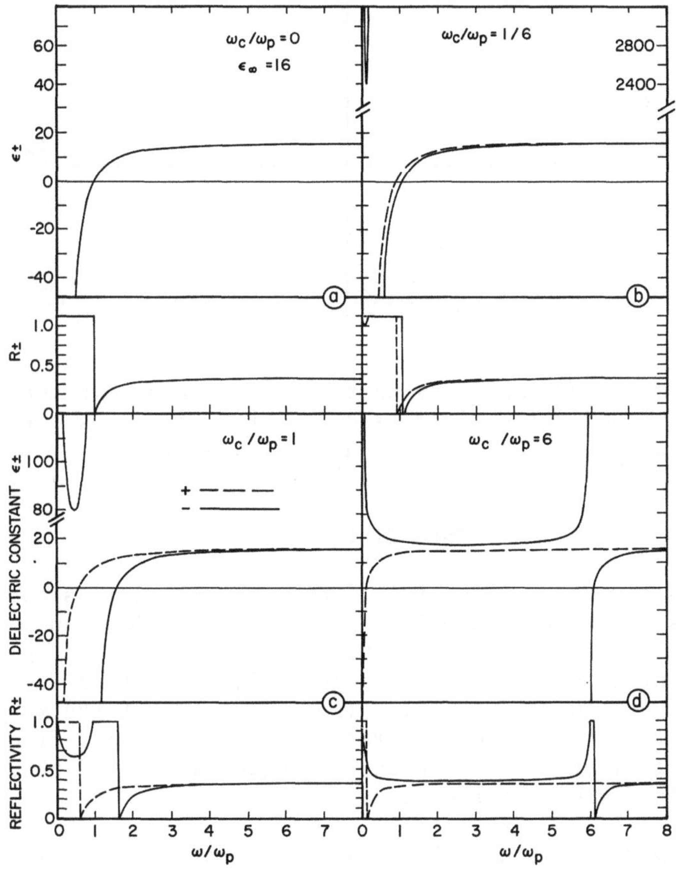

Figure 3. Dielectric constants $\epsilon_\pm$ and reflectivities $R_\pm$ for the cyclotron inactive (+) and cyclotron active (-) circular polarizations of the Faraday geometry, for the lossless case $\nu = 0$. Figures a, b, c and d represent the cases $\omega_c/\omega_p = 0$, 1/6, 1, and 6, respectively.

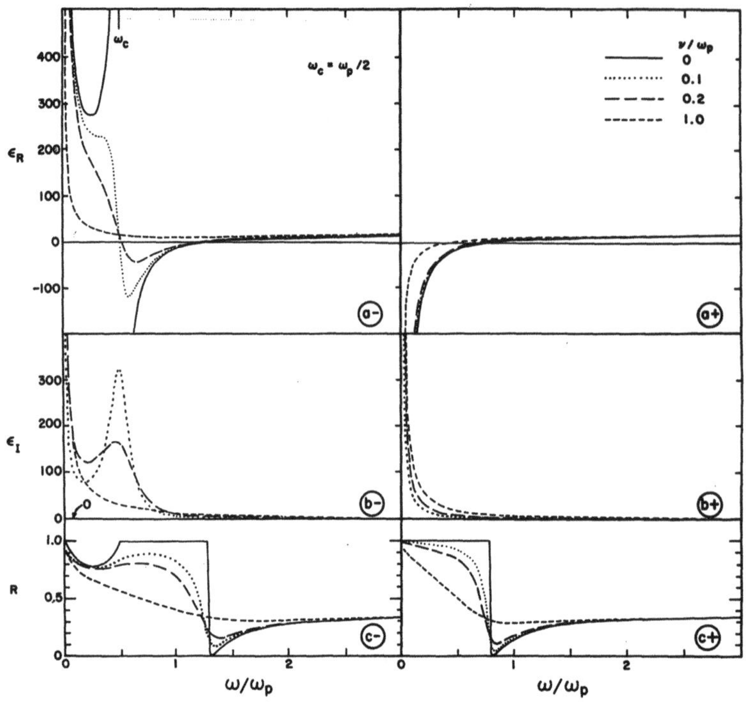

Figure 4. Real and imaginary parts of the dielectric constants $\epsilon_{\pm}$ and reflectivities $R_{\pm}$ for the Faraday geometry for ν/ω_p = 0, 0.1, 0.2, 1 and ω_c/ω_p = 0.5.

Zero-field Case

Such plots of ϵ are useful to discuss the various possible experiments. We start with $B_o = \nu = 0$. The medium can support a propagating EM wave for $\omega > \omega_p$, but is totally reflecting below ω_p, since ϵ is negative below ω_p. In general, a negative dielectric constant occurs just to the high frequency side of the resonant frequency of an oscillator (in this case the resonant frequency is $\omega = 0$) and indicates that the charge motion is out of phase with the electric field in the medium.

Since $\nu = 0$ obtains, there is no absorption, so all radiation not reflected is transmitted. For $\nu \neq 0$ and $B_o = 0$, it follows from Eq. 13 that

$$\epsilon_I = -2nk = -\frac{\epsilon_\infty \omega_p^2 \nu}{\omega(\omega^2+\nu^2)} = \frac{4\pi}{\omega}\sigma_R$$

$$\epsilon_R = n^2 - k^2 = \epsilon_\infty \left[1 - \frac{\omega_p^2}{(\omega^2 + \nu^2)}\right] = \epsilon_\infty - \frac{4\pi}{\omega}\sigma_I \ . \tag{38}$$

Power absorption per unit volume is given by

$$\wp = \tfrac{1}{2}\,\mathrm{Re}\,(\vec{J}\cdot\vec{E}^*) = \tfrac{1}{2}\,\mathrm{Re}\,\sigma\,\vec{E}\cdot\vec{E}^* = \tfrac{1}{2}\sigma_R\,|E|^2 \tag{39}$$

so σ_R or ϵ_I is a measure of power absorption.

For the conditions $\omega > \omega_p$, $\omega\tau \gg 1$ and $k^2 \ll 1$, it follows from Eq. 13 that the index of refraction is

$$n^2 \approx \epsilon_R = \epsilon_\infty \left(1 - \frac{\omega_p^2}{\omega^2} \right)$$

so that

$$\left(\frac{1 + \sqrt{R}}{1 - \sqrt{R}}\right)^2 = n^2 = \epsilon_\infty - \frac{\epsilon_\infty \omega_p^2 \lambda^2}{4\pi^2 c^2} \tag{40}$$

holds. Thus from a zero-field reflectivity measurement above ω_p both N/m^* and ϵ_∞ can be determined[15].

Free carrier absorption can also be measured in the same spectral region. For the conditions $\omega \gg \omega_p$, ν and $\epsilon_R \gg \epsilon_I$, the extinction coefficient may be written as

$$2k^2 = -\epsilon_R + (\epsilon_R^2 + \epsilon_I^2)^{\frac{1}{2}} \approx \frac{\epsilon_\infty \omega_p^4 \nu^2}{2\omega^6} \ . \tag{41}$$

The absorption constant $\alpha = 4\pi k/\lambda$ then becomes

$$\alpha = \frac{\sqrt{\epsilon_\infty}\,\omega_p^2 \nu \lambda^2}{4\pi^2 c^3} \tag{42}$$

which indicates that absorption measurements, in principle, will yield $N\lambda/m^*$. In practice, the λ^2 dependence is only approximate and analysis of free carrier absorption data requires more detailed treatment of ν[16]. Thus it is better to choose an experiment where effects are independent of ν.

There is also a reflection phase shift given by Eq. (31). For $\nu = 0$ for all frequencies above the frequency where $\epsilon = 1$, Δ is 180°. At $\epsilon = 1$, Δ abruptly falls to zero and remains there until $\epsilon = 0$; below ω_p it gradually rises back toward 180°. As absolute phase shifts are difficult to measure, this experiment has not been used very much(17).

While the reflectivity is unity in the lossless case for $\omega < \omega_p$, in a real semiconductor it is not quite unity. The property of skin depth in metals and semiconductors(18) can be examined with the use of the extinction coefficient k. For a good conductor with $\omega_p\tau > 1$ and $\omega_p^2\tau^2 \gg \omega\tau \ll 1$, we can write Eq. (26) as

$$2k^2 = -(\epsilon_\infty - 4\pi\sigma_o\tau) + \left\{(\epsilon_\infty - 4\pi\sigma_o\tau)^2 + \left(\frac{4\pi\sigma_o}{\omega}\right)^2\right\}^{\frac{1}{2}}$$

$$= -\epsilon_\infty(1-\omega_p^2\tau^2) + \epsilon_\infty\left\{(1-\omega_p^2\tau^2)^2 + \left(\frac{\omega_p^2\tau^2}{\omega\tau}\right)^2\right\}^{\frac{1}{2}} .$$

Then k becomes

$$k \approx \left(\frac{2\pi\sigma_o}{\omega}\right)^{\frac{1}{2}} = \omega_p\left(\frac{\epsilon_\infty\tau}{2\omega}\right)^{\frac{1}{2}} . \qquad (43)$$

Since the E field of the radiation is of the form $E \sim \exp i[\omega t-(n-ik)\omega z/c]$, the attenuation of the wave to 1/e of its initial value takes place in a skin depth

$$\delta = \frac{c}{k\omega} = \frac{c}{(2\pi\sigma_o\omega)^{\frac{1}{2}}} . \qquad (44)$$

The large value of $k (k \gg 1)$ and the small value of $n (n \ll 1)$ make it difficult to measure k by reflection techniques in the spectral range below ω_p. The wave is called evanescent as it does not propagate along the z axis in a sinusoidal manner, but is simply attenuated. A small portion of the EM energy is absorbed, but most is reflected.

General Magnetic Field Effects

With a magnetic field applied, we can expect circular birefringence and dichroism in both transmission and reflection, and these are the basis of all the magnetooptical experiments. The phase shifts $\Delta_\pm$ will produce a reflection Faraday rotation in a plane polarized beam given by

$$\theta_r = \frac{\Delta_- - \Delta_+}{2} . \qquad (45)$$

Also, since R_+ and R_- are different, there is an ellipticity to the beam given by

$$\epsilon_r = \frac{|r_+|-|r_-|}{|r_+|+|r_-|} \tag{46}$$

where the Fresnel coefficients are given by

$$r_\pm = \frac{1-n_\pm+ik_\pm}{1+n_\pm-ik_\pm} \tag{47}$$

and the reflectivity is given by

$$R_\pm = |r_\pm|^2 = |r_\pm^2| = r_\pm r_\pm^* . \tag{48}$$

Here, the absolute value signs signify the modulus, since r may be written as $r = |r|e^{i\varphi}$.

In transmission, there is the Faraday effect, magnetic birefringence in the circularly polarized components which make up a linearly polarized beam. The rotation is

$$\theta_t = \frac{\varphi_- - \varphi_+}{2} = \frac{\omega d}{2c}(n_- - n_+) . \tag{49}$$

An ellipticity to the beam can arise from dichroic effects and is given by

$$\epsilon_t = \frac{|1-r_+^2|\exp(-\alpha_+ d/2)-|1-r_-^2|\exp(-\alpha_- d/2)}{|1-r_+^2|\exp(-\alpha_+ d/2)+|1-r_-^2|\exp(-\alpha_- d/2)} . \tag{50}$$

In samples with $\nu \neq 0$, circular dichroism effects may be measured in absorption also. The prominent feature in this case is cyclotron resonance absorption of one of the circularly polarized components. If the dielectric constant of a system of oscillators can be written in the form of Eq. (13) with a resonant denominator containing a simple i term, then the coefficient of i (which is the damping constant) is a measure of the half width $\Delta\omega$ of the absorption line ($\Delta\omega = 2\nu$).

While the previous equations illustrate the general types of magnetooptical experiments, in practice, Faraday rotation, for example, is not measured over the entire spectral range but in a region where simple approximations hold. We now discuss several such experiments.

Low-field Approximations

In the vicinity of the plasma reflection edge, the application of an external magnetic field produces shifts in the reflection edges(14). The zeroes in $\epsilon_\pm$, in the approximation that $\omega \approx \omega_p \gg \omega_c$ and $\omega\tau \gg 1$, shift according to Eq. (34) so that

$$\omega_\pm \approx \omega_p \pm \frac{\omega_c}{2} + \frac{\omega_c^2}{8\omega_p} + \ldots \tag{51}$$

holds. The unities in $\epsilon_\pm$, which define the reflection minima, shift in a similar way. We remember that a negative sign is contained in ω_c when electrons are considered, so the left circularly polarized reflection edge shifts to higher frequency. The left and right circularly polarized reflection edges split by the cyclotron frequency ω_c.

There are also large effects in reflection rotation and ellipticity because of the dependences in Eqs. (45) and (46).

The Faraday effect, with the approximation $(\omega \mp \omega_c)\tau \gg 1$ is given by(19)

$$\theta_t \approx \frac{e^3 N d \lambda^2 B_o}{2\pi c^4 n m^{*2}} \tag{52}$$

where n is the index of refraction for $B_o = 0$, not $\sqrt{\epsilon_\infty}$. The advantages of this experiment are that it is independent of τ and it is performed at $\omega \gg \omega_p$ which is an easily accessible spectral region.

In the region $\omega < \omega_c \ll \omega_p$ Eq. (13) simplifies to

$$\epsilon_R = n_\pm^2 - k_\pm^2 \approx \epsilon_\infty \left(1 \pm \frac{\omega_p^2}{\omega\omega_c}\right). \tag{53}$$

Hence, the optical constants are

$$\begin{aligned} k_+^2 &\approx -\epsilon_\infty \left(1 + \frac{\omega_p^2}{\omega\omega_c}\right) \qquad n_+^2 = 0 \\ n_-^2 &\approx \epsilon_\infty \left(1 - \frac{\omega_p^2}{\omega\omega_c}\right) \qquad k_-^2 = 0 \end{aligned} \tag{54}$$

The helicon[20,21] or left circularly polarized wave described by the minus subscript can propagate through the sample, while the + wave is reflected. Again, we remember that ω_c carries a minus sign for electron charge. Thus ϵ_-, instead of being $\sim \epsilon_\infty$, may be orders of magnitude larger, as indicated in Fig. 3b. The wave is very slow moving and may have a velocity comparable to the velocity of sound in the medium. A plane parallel sample, because R_- is so high, acts as a Fabry-Perot cavity producing transmission fringe maxima of order m when $m\lambda_- = 2n_- d$. This equation leads to the relation

$$m^2 = \frac{8d^2 Nec}{B_o} \tag{55}$$

from which the carrier density is obtained.

This result is similar to the dc Hall effect results. If, in Eq. (7), we set $\omega = 0$, analysis of the equation $\vec{J} = \vec{\sigma}\,\vec{E}$ when $\omega_c \gg \nu$ yields the simple Hall effect results. Current density J_y along the y direction is given in terms of the Hall field E_x by

$$J_y = \frac{-Nec}{B_o} E_x = \sigma_H E_x = -\frac{E_x}{B_o R} \tag{56}$$

where $R = 1/Nec = V_x t/I_y B_o$ and $\sigma_H = -Nec/B_o = \sigma_{xy}$ is the Hall conductivity. V_x is the Hall voltage, t is the sample thickness and I_y is the current flowing through the sample. The Hall angle θ is defined a

$$\tan\theta = \frac{E_x}{E_y} = -\omega_c \tau \; . \tag{47}$$

Note that there is no simple analogy between the Hall angle and the Faraday rotation angle at low frequencies.

Cyclotron resonance is not usually measured by transmission techniques in this low field case, as the samples are opaque near ω_c as indicated in Fig. 3b. Azbel-Kaner cyclotron resonance can be measured in the Voigt geometry[22].

It is instructive to note that when reflectivity increases rapidly to unity toward lower frequency, this unity point locates a zero in ϵ_R. Also, when reflectivity increases rapidly to unity toward higher frequency, this point locates an infinity in ϵ_R. Also, when reflectivity increases rapidly to unity toward higher frequency, this point locates an infinity in ϵ_R (the resonant frequency of a transverse oscillator). This "rule of thumb" is applicable to the reflectivity of all materials and will be noted again when we introduce lattice vibrations into the sample.

High-field Approximations

As B_o increases ω_c and $\epsilon_- = 0$ move to higher frequency, ω_c almost catching up with $\epsilon_- = 0$, while $\epsilon_+ = 0$ moves to lower frequency. The various approximations just discussed begin to break down. Cyclotron resonance can now be measured as an absorption line in relatively thick samples and has a half width 2ν. In the helicon region below ω_c, there is now a considerable region where the sample is somewhat transparent to both circular polarizations, and ϵ_- drops toward ϵ_∞. In this region Faraday rotation may now be measured. For $\omega_p < \omega \ll \omega_c$ and $\omega\tau \gg 1$, rotation is given by[23,24]

$$\theta_t \approx \frac{2\pi e d N}{n B_o} \quad . \tag{58}$$

Rotation is independent of ω as can be inferred from the $\epsilon_{\pm R}$ curves in Fig. 3d.

Dispersion Relations

Dispersion relations between the components of ϵ_{ijR} and ϵ_{ijI} can be written as[25]

$$\epsilon_{ijR}(\omega)-\delta_{ij} = -\frac{2}{\pi} P \int_0^\infty \frac{\omega' \epsilon_{ijI}(\omega')d\omega'}{\omega'^2-\omega^2}$$

$$\epsilon_{ijI}(\omega) = \frac{2}{\pi} P \int_0^\infty \frac{\epsilon_{ijR}(\omega')d\omega'}{\omega'^2-\omega^2} \quad . \tag{59}$$

While it is not possible to write corresponding relations between $\epsilon_{\pm R}$ and $\epsilon_{\pm I}$, some useful relations can be written. For example, since

$$\epsilon_{\pm R} = n_\pm^2 - k_\pm^2 = \epsilon_{xxR} \mp \epsilon_{xyI} = 1 + \frac{4\pi}{\omega}(\sigma_{xxI} \pm \sigma_{xyR})$$

$$\epsilon_{\pm I} = -2n_\pm k_\pm = \epsilon_{xxI} \pm \epsilon_{xyR} = -\frac{4\pi}{\omega}(\sigma_{xxR} \mp \sigma_{xyI}) \tag{60}$$

hold, it follows that

$$(n_+^2-n_-^2)-(k_+^2-k_-^2) = -2\epsilon_{xyI} = \frac{8\pi}{\omega}\sigma_{xyR}$$

$$2(n_+k_+-n_-k_-) = -2\epsilon_{xyR} = -\frac{8\pi}{\omega}\sigma_{xyI} \quad . \tag{61}$$

In weakly absorbing regions where $k \ll n$, from the definition of rotation in Eq. (49) we obtain

$$n\theta \approx \frac{2\pi}{c}\sigma_{xyR} \quad . \tag{62}$$

Equations (61) used in Eqs. (59) yield dispersion relations between Faraday rotation $(n_+^2-n_-^2)$ and dichroism $(n_+k_+-n_-k_-)$. Such a treatment is useful since it relates the free carrier rotation at a given frequency ω to the dichroism occuring over all frequencies (but only strong near ω_c) or the interband rotation below the band gap to the dichroism above the energy gap[26].

For the high field approximation Eq. (62) gives Eq. (58), while for the low-field approximation Eq. (62) gives Eq. (52).

In weakly absorbing regions Eq. (50) becomes

$$\epsilon_t = \tanh\frac{\omega d}{2c}(k_--k_+) \tag{63}$$

so that the ellipticity in Faraday rotation can be used to determine dichroic effects. Dichroic effects can be directly measured also[27].

We have made some further calculations to illustrate the various experiments to be done in the Faraday geometry, using parameters typical of InSb but omitting the lattice vibration effects. We used $N = 6 \times 10^{16}$ cm^{-3}, $\epsilon_\infty = 16$, $m^* = 0.015\ m_o$, $e = -4.8 \times 10^{-10}$cm, and $d = 1$ cm where not specified otherwise. In Figs. 5-8 we plot the parameters n, k, φ, Δ, R, θ_λ, ϵ_t, θ_r, ϵ_r for $\nu = 0$ and 10^{12} sec^{-1} and for $B_o = 3$kG. This is very nearly the case $\omega_c \ll \omega_p$ described in Fig. 3b. The various reflection and transmission magnetooptical experiments discussed above can be seen in different spectral regions. It is sometimes useful to think in terms of n and k instead of ϵ.

Wave Vector Analysis

We can plot the wave vector κ vs ω by inserting the appropriate value of ϵ into $\kappa = (\omega/c)\sqrt{\epsilon}$. Some typical curves are shown in Fig. 9 for the lossless case $\nu = 0$. In this case κ is real and imaginary where ϵ_R is positive and negative, respectively. For $\nu \neq 0$, κ is complex. Such curves are useful in anticipating interactions of EM waves with various oscillators in the system, for example, the interaction of transverse acoustic phonons and

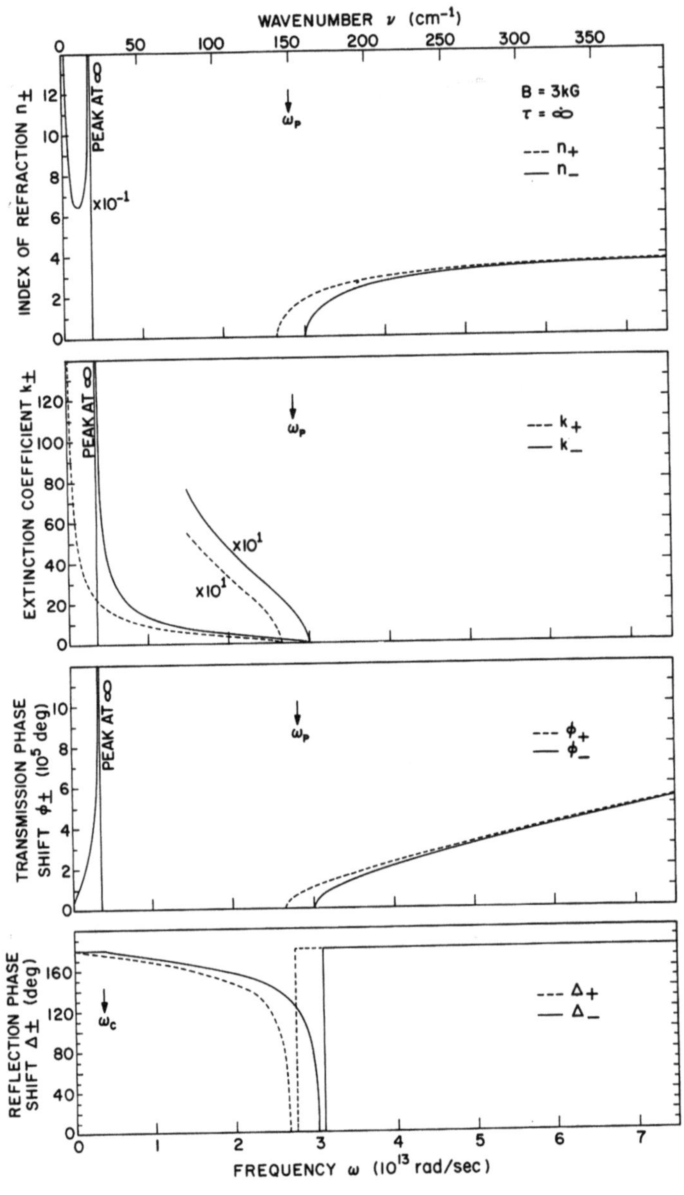

Figure 5. The parameters $n_\pm$, $k_\pm$, $\varphi_\pm$ and $\Delta_\pm$ for $\tau = \infty$ and $B_o = 3$ kG. The idealized InSb sample has $N = 6 \times 10^{16}$ cm^{-3}, $\epsilon_\infty = 16$, $m^* = 0.015\ m_o$ and $d = 1$ cm (where d is not specified otherwise).

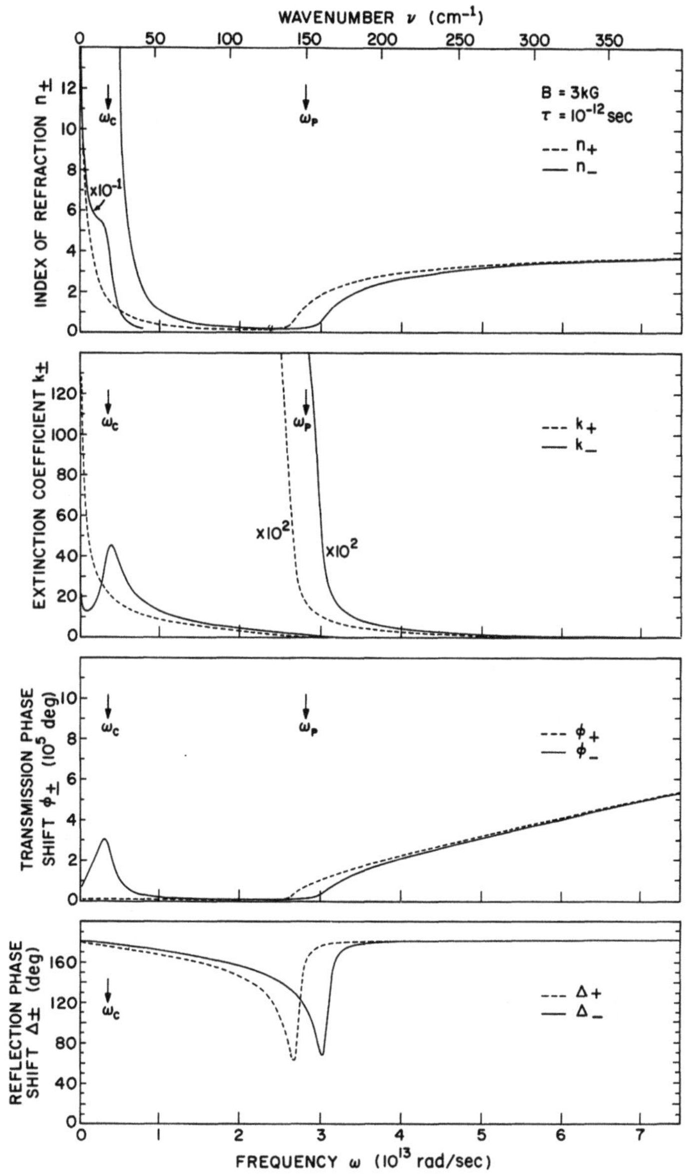

Figure 6. The parameters $n_{\pm}$, $k_{\pm}$, $\varphi_{\pm}$ and $\Delta_{\pm}$ for $\tau = 10^{-12}$ sec and $B_o = 3$ kG.

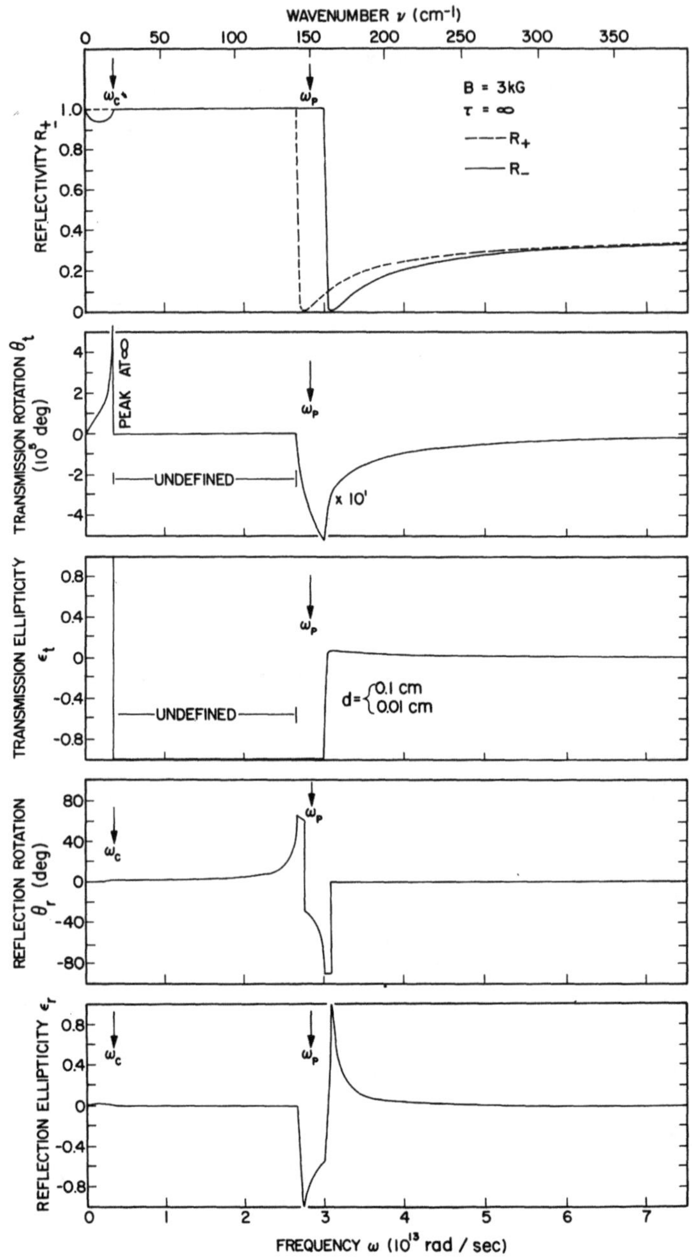

Figure 7. The parameters $R_{\pm}$, θ_t, ϵ_t, θ_r and ϵ_r for $\tau = \infty$ and $B_o = 3$ kG.

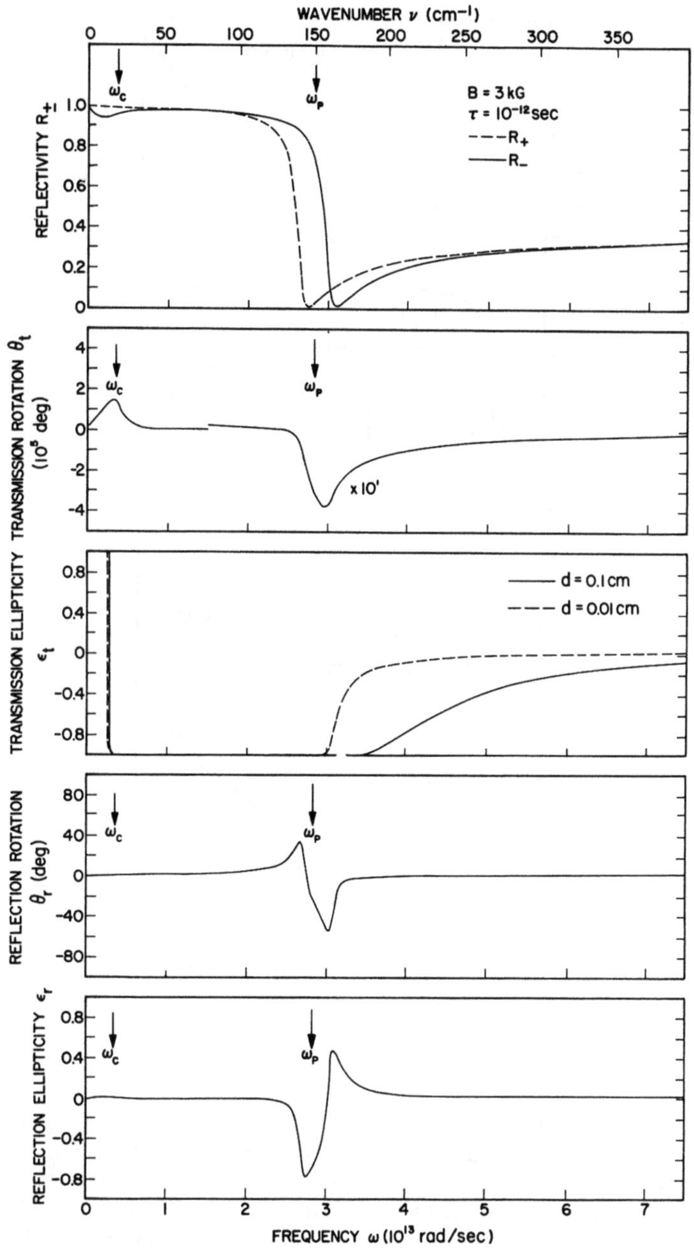

Figure 8. The parameters $R_{\pm}$, θ_t, ϵ_t, θ_r and ϵ_r for $\tau = 10^{-12}$ sec and $B_o = 3$ kG.

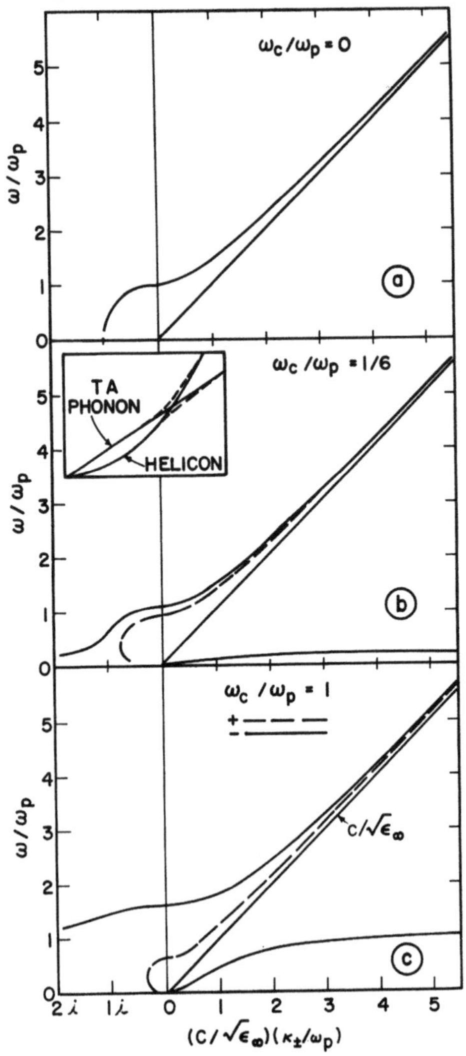

Figure 9. Dispersion curves for the Faraday geometry for $\nu = 0$. Figures a, b and c represent the cases $\omega_c/\omega_p = 0$, 1/6 and 1, respectively. The inset in b indicates how transverse acoustic phonons and helicons can interact at an intersection of their dispersion curves. The solid straight lines of slope unity are the dispersion curves of radiation in a medium of dielectric constant ϵ_∞.

transverse helicons[28]. When EM radiation enters a solid, classically we still think of the wave inside the sample described by the dielectric constant curves of Fig. 3. However, the externally applied EM wave has interacted with the charge system, exciting it and, in turn, being altered by the moving charges as dictated by Maxwell's equations. This excitation in the solid (wave and oscillators) is called a polariton[29] and has its energy distributed in both mechanical and electromagnetic parts. For example, in Fig. 9b, just below ω_c where $\varkappa$ is very large, most of the energy is in the mechanical motion of the charges and very little is in the form of EM radiation. For large ω the energy is almost all EM and the polariton is essentially an EM wave. We associate the wave vector with the polariton now, not just with the EM wave. A classical measurement of n and k over the entire spectral range would yield ϵ and the $\varkappa$ vs ω curves without ever having to consider the mechanical motion of the charges.

The time-average energy density of radiation in a dispersive medium for the lossless case is[30,31]

$$\overline{U} = \frac{c}{4\pi}\sqrt{\epsilon}\,\frac{d\varkappa}{d\omega}\,\overline{E^2} = \frac{1}{8\pi}\left[\left(\epsilon + \omega\frac{d\epsilon}{d\omega}\right)\overline{E^2} + \overline{H^2}\right] \tag{64}$$

where we remember that $\overline{H^2} = \epsilon\,\overline{E^2}$. A portion of this can be associated with the kinetic and potential energy in the mechanical motion of the charges, the remainder to the EM fields; or as in Eq. (64), all can be associated with the EM fields.

The mechanical portion may be expressed as

$$\overline{U}_M = \frac{1}{8\pi}\left(\epsilon - \epsilon_\infty + \omega\frac{d\epsilon}{d\omega}\right)\overline{E^2} \tag{65}$$

and the electromagnetic portion is given by

$$\overline{U}_{EM} = \frac{1}{8\pi}\left(\epsilon_\infty\overline{E^2} + \overline{H^2}\right) . \tag{66}$$

The mechanical portion is also expressible as

$$\overline{U}_M = \frac{1}{2}\,Nm\overline{v^2} + \frac{1}{2}\,Nm\,\omega_o^2\,\overline{r^2} \tag{67}$$

where $m\,\omega_o^2$ represents the force constant for a bound oscillator.

We guessed from Fig. 9b, and Eqs. (65) and (66) show, that near the resonant frequencies of transverse oscillators, most of the energy is mechanical, since $d\epsilon/d\omega$ tends to infinity. Far above ω_p, $d\epsilon/d\omega$ becomes zero and $\epsilon \rightarrow \epsilon_\infty$, so the mechanical energy goes to zero. We only consider regions where ϵ is positive, so that the polariton modes are well defined, although in regions of negative ϵ, there is also mechanical and EM energy in the sample. As an example, in Eqs. (13) and (16) we let $B_0 \rightarrow 0$ and $\nu \rightarrow 0$ to get

$$\epsilon = \epsilon_\infty\left(1 - \frac{\omega_p^2}{\omega^2}\right) \qquad x = -\frac{eE_x}{m^* \omega^2}$$

where we only look at the x (transverse) motion.

The mechanical energy density is just

$$\overline{U}_M = \frac{1}{2} Nm^* \overline{\dot{x}^2} = \frac{1}{8\pi} \frac{\epsilon_\infty \omega_p^2}{\omega^2} \overline{E_x^2} \, .$$

This is the same result as obtained from Eq. (65). The EM energy density is

$$\overline{U}_{EM} = \frac{1}{8\pi} (\epsilon_\infty + \epsilon) \overline{E_x^2} = \frac{\epsilon_\infty}{8\pi} \left(2 - \frac{\omega_p^2}{\omega^2}\right) \overline{E_x^2} \qquad \text{for } \omega \geq \omega_p \, .$$

Above ω_p we see that at high ω the mechanical portion goes to zero while the EM portion approaches the usual value $\epsilon_\infty \overline{E_x^2}/4\pi$. $\overline{E_x^2}$ must be calculated from $E_x = E_{xo} \exp i(\omega t - \varkappa z)$ and is found to be $\overline{E_x^2} = E_{xo}^2/2$.

For the case of $B \neq 0$, we get for the two circular motions

$$\overline{U}_{M\pm} = \frac{1}{2} Nm^* \overline{\dot{r}_\pm^2} = \frac{1}{8\pi} \frac{\epsilon_\infty \omega_p^2}{(\omega \mp \omega_c)^2} (\overline{E_x^2} + \overline{E_y^2})$$

which is just the result obtained from Eq. (65) and

$$\overline{U}_{EM\pm} = \frac{\epsilon_\infty}{8\pi} \left[2 - \frac{\omega_p^2}{\omega(\omega \mp \omega_c)}\right] (\overline{E_x^2} + \overline{E_y^2}) \qquad \begin{cases} \omega > \pm \frac{1}{2}\omega_c \pm \frac{1}{2}(\omega_c^2 + 4\omega_p^2)^{\frac{1}{2}} \\ \omega < |\omega_c| \text{ for lower sign} \end{cases}$$

where we remember the sign in ω_c, and the ranges of interest are bounded by $\epsilon_\pm = 0, \infty$. Note that we do not consider the longitudinal plasma motion along z as it is not driven by a transverse EM wave.

In the case of lattice vibrations in pure semiconductors the dielectric constant is given by Eq. (86) plus the background lattice term ϵ_∞. The relative lattice displacement $\vec{R}$ of two ions (see Eqs. (95)) can be used to write the kinetic and potential energy densities as

$$\overline{U}_M = \frac{1}{2}\, nM\, \overline{\dot{X}^2} + \frac{1}{2}\, nM\, \omega_T^2\, \overline{X^2}$$

where M is the reduced mass of the lattice ion oscillators (pairs of ions) and n is the density of these oscillators. Better yet, we can use Eq. (65) directly to get

$$\overline{U}_M = \frac{\epsilon_\infty(\omega_L^2-\omega_T^2)(\omega_T^2+\omega^2)}{8\pi(\omega_T^2-\omega^2)^2}\,\overline{E^2}\,.$$

The electromagnetic energy density is

$$\overline{U}_{EM} = \frac{1}{8\pi}\,\epsilon_\infty \left(2 + \frac{\omega_L^2 - \omega_T^2}{\omega_T^2 - \omega^2}\right)\overline{E^2}\quad.$$

Since transverse EM radiation does not interact with longitudinal oscillators, the polariton idea is not applicable. However, it is still possible to discuss the mechanical and electrical energy density. The electric field is that of the longitudinal oscillator and the magnetic field is H = 0.

The interaction of polaritons (helicons) and transverse acoustical (TA) waves is illustrated in the insert of Fig. 9b. The dispersion curve for a TA phonon of velocity v_s is linear near $\varkappa = 0$ and is given by $q = \omega/v_s$. When the energy and momentum conservation requirements are fulfilled at $\varkappa_H = q_{TA}$ and $\hbar\omega_H = \hbar\omega_{TA}$, an interaction occurs resulting in a repulsion of the curves. The TA phonon becomes helicon-like while the helicon has become TA phonon-like[28].

Polaritons associated with lattice vibrations have been measured as physical entities by small angle Raman scattering techniques in the case of pure semiconductors like GaP[32] and ZnO[33].

Another point of interest, which can be discussed in simple terms for the lossless case, by use of Fig. 9, is the group and phase velocities of the EM waves[34,35]. The phase velocity is given by $v_p = \omega/\varkappa$ while the group velocity is $v_g = d\omega/d\varkappa$. Thus, a line from the origin to any point $(\omega_1,\varkappa_1)$ on the dispersion curve has a slope equal to the phase velocity, while the tangent at that point has a slope equal to the group velocity. Thus in Fig. 9a for $\omega_c = 0$ and for $\omega > \omega_p$, we find that $v_p > c/\sqrt{\epsilon_\infty}$ and $v_g < c\sqrt{\epsilon_\infty}$ obtains. Also, at large $(\omega,\varkappa)$ values, group and phase velocities are nearly equal to $c/\sqrt{\epsilon_\infty}$. In Fig. 9b in the helicon region we also note that both velocities are very small and nearly equal. For the lossy case, where strong absorption takes place, the concept of phase and group velocity becomes complicated and is not treated in a simple manner.[36]

DISCUSSION OF DIELECTRIC CONSTANT IN THE VOIGT GEOMETRY

General Remarks

We had mentioned previously, that in the Faraday geometry, there is a transverse oscillator of resonant frequency ω_c and a longitudinal oscillator at ω_p. This is illustrated in Fig. 10

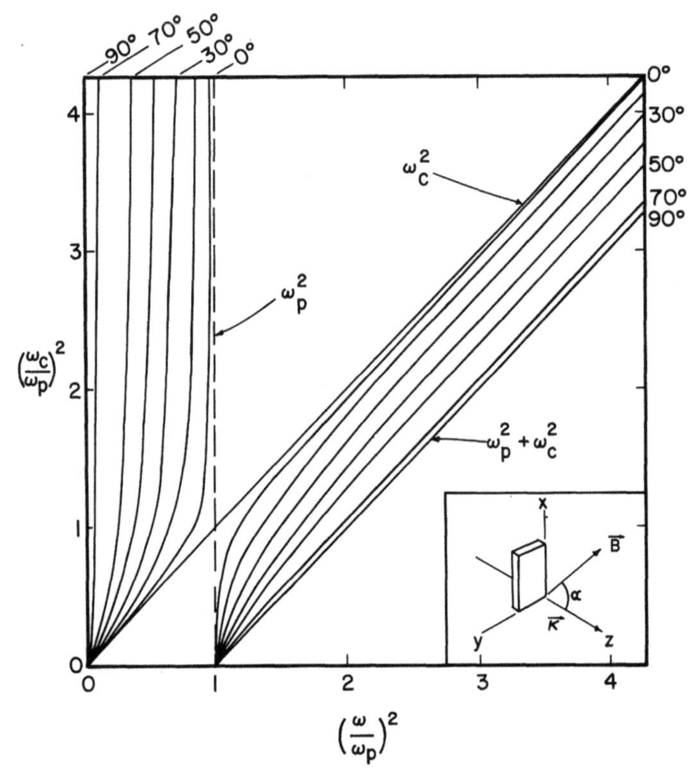

Figure 10. Shift of the resonant frequencies from the Faraday to the Voigt geometry for $\nu = 0$.

where we plot ω vs B_o in the form ω_c^2/ω_p^2 vs ω^2/ω_p^2. The dotted vertical line ω_p represents the plasmon resonant frequency which is not measured in a normal incidence experiment. The 45° line labeled ω_c is measurable as an absorption line. In the Voigt geometry, the plasmon and cyclotron oscillators are coupled together with resonant frequency $\omega = (\omega_p^2 + \omega_c^2)^{\frac{1}{2}}$ as given by Eq. (19) (when $\varepsilon \to \infty$). As the magnetic field is tipped from z to x, we should expect a smooth shift in the resonant frequencies as indicated in Fig. 10. For $\vec{\kappa}$ not parallel to $\vec{B}_o$ the resonances are associated with coupled plasmon-cyclotron oscillators (collective cyclotron excitations) with both transverse and longitudinal character as examination of Eqs. (22) will reveal for the Voigt geometry. We measure only the transverse aspect of the charge motion with transverse EM radiation. The intensity of the absorption line will be weak if most of the oscillator motion is along z (longitudinal) and strong if most of the oscillator motion is along y (transverse). We discuss the Voigt geometry first, and will return to the case of arbitrary direction of $\vec{B}_o$ shortly.

The solution of the Maxwell wave equation yields the wave vectors $\kappa_{\|,\perp}$ which show the resonances at $\omega = 0$ and $\omega = (\omega_p^2 + \omega_c^2)^{\frac{1}{2}}$ as indicated in Eq. (19). Cyclotron resonance is not measurable at ω_c in this geometry because the Hall field E_{2z} can be thought of as an additional restoring force which shifts the natural frequencies of the charge system (ω_p and ω_c) to $\omega = 0$ and $\omega = (\omega_p^2 + \omega_c^2)^{\frac{1}{2}}$. Another way to see why no resonant absorption occurs at ω_c is to examine Eqs. (22) when $\omega = \omega_c$ and $\nu = 0$. Then $E_{2z} = iE_{2y}$ represents a circularly polarized electric field in the y-z plane (the plane of cyclotron motion). However, the sense of this motion is opposite to the sense of the cyclotron motion so no resonant absorption takes place at ω_c.

The Drude model is a single particle model in which we simply multiply by N to get the effects of N particles. It has been shown[37] that N electrons will have N-1 single particle cyclotron modes at ω_c (to which EM radiation cannot couple) in the Voigt geometry and one collective mode at $(\omega_p^2 + \omega_c^2)^{\frac{1}{2}}$ (to which EM radiation can couple). Thus, if one illuminates a cube sample propagating the beam parallel to $\vec{B}_o$ and reflecting the transmitted beam to propagate through the cube perpendicular to $\vec{B}_o$, then one sees both resonant absorption lines.

General Magnetic Field Effects

We examine the lossless case for the Voigt geometry first for $\varepsilon_{\perp R} = n_\perp^2 - k_\perp^2$. The isoindex of refraction curves are shown in Fig. 11 for $\|$ and $\perp$ polarizations. Of special interest are

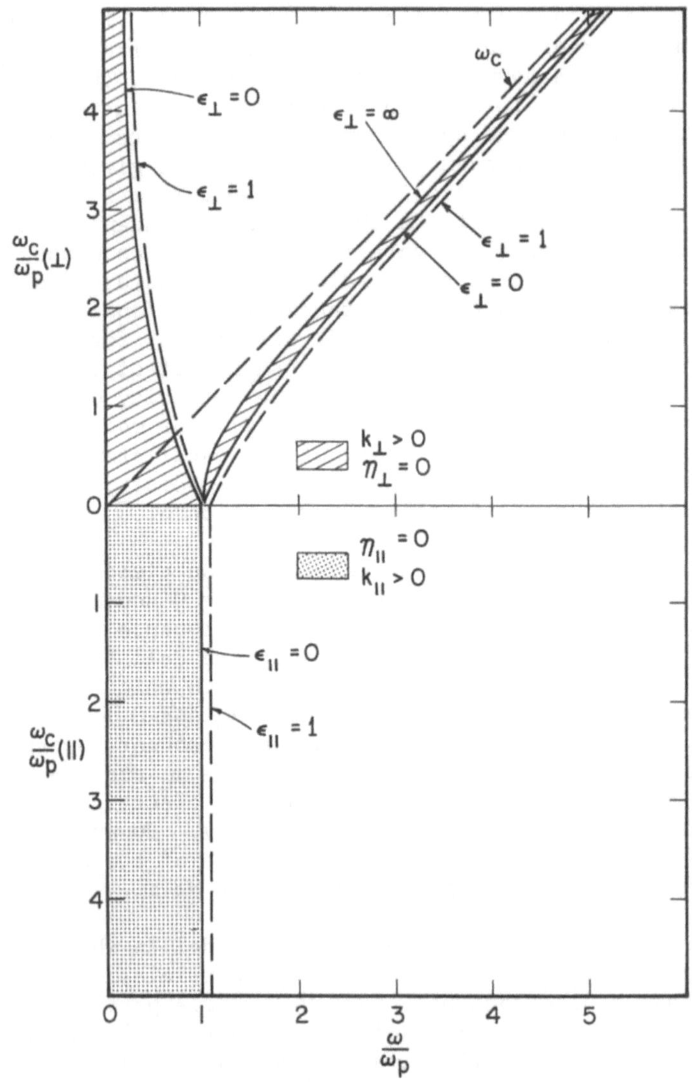

Figure 11. Index of refraction plot showing zeroes, unities and infinities in $n_{\parallel,\perp}^2 = \epsilon_{\parallel,\perp}$ for the lossless case $\nu = 0$, Voigt geometry. In shaded regions $k > 0$, $n = 0$ holds. The dotted line ω_c marks the position where cyclotron resonance would have occurred in the Faraday geometry.

the points where the dielectric constant is $\epsilon_{\perp R} = 0$, 1 and ∞. From Eq. (19) for $\nu = 0$ we get

$$\epsilon_{\perp R} = n_\perp^2 - k_\perp^2 = \epsilon_\infty \left[1 - \frac{\omega_p^2(\omega_p^2-\omega^2)}{\omega^2(\omega_p^2+\omega_c^2-\omega^2)}\right] . \qquad (68)$$

The zeroes occur at

$$\omega^2 = \omega_p^2 + \frac{\omega_c^2}{2} \pm \frac{\omega_c}{2} (\omega_c^2+4\omega_p^2)^{\frac{1}{2}} \qquad (69)$$

The infinities occur at

$$\omega = 0 \qquad \omega = (\omega_p^2+\omega_c^2)^{\frac{1}{2}} \qquad (70)$$

The unities occur at

$$(\epsilon_\infty-1)2\omega^2 = (2\epsilon_\infty-1)\omega_p^2+(\epsilon_\infty-1)\omega_c^2 \pm\{[(2\epsilon_\infty-1)\omega_p^2+(\epsilon_\infty-1)\omega_c^2]^2 - 4(\epsilon_\infty-1)\epsilon_\infty\omega_p^4\}^{\frac{1}{2}} \qquad (71)$$

The function $1/\epsilon_\perp$ which emphasizes the zeroes in $\epsilon_\perp$ is also useful but will not be given explicitly. Horizontal lines across Fig. 11 can be plotted as $\epsilon_\perp$ vs ω/ω_p. Four cases of interest, $\omega_c = 0$, $\omega_c \ll \omega_p$, $\omega_c = \omega_p$ and $\omega_c \gg \omega_p$, are illustrated in Fig. 12. The reflectivity is also shown. In Fig. 13 we show the effects on $\epsilon_{\perp R}$ and $\epsilon_{\perp I}$ as ν increases from zero for the case $\omega_c = (1/2)\omega_p$.

The reflection and transmission phase shifts and amplitude changes given by Eqs. (29)-(32) produce all the magnetooptical experiments of interest. In the Faraday geometry the phase shifts produced the rotation while the differential amplitudes produced the ellipticity as indicated by Eqs. (45), (46), (49), and (50). For the Voigt geometry, the opposite is true, at least at low fields. Generally, two linearly polarized waves described by

$$E_{x\|} = A_x \cos(\omega t + \Delta_1)$$
$$E_{x\perp} = A_y \cos(\omega t + \Delta_2) \qquad (72)$$

when combined vectorially give elliptically polarized radiation. The ellipse major axis makes an angle θ with the x axis, while the ratio of the minor to the major axis b/a is the ellipticity $\epsilon = b/a = \tan\gamma$. Then, for $\tan\eta = A = A_y/A_x$ and $\Delta_2-\Delta_1$, it can be shown that

$$\tan 2\theta = \tan 2\eta \cos\Delta$$
$$\sin 2\gamma = \sin 2\eta \sin\Delta \quad . \qquad (73)$$

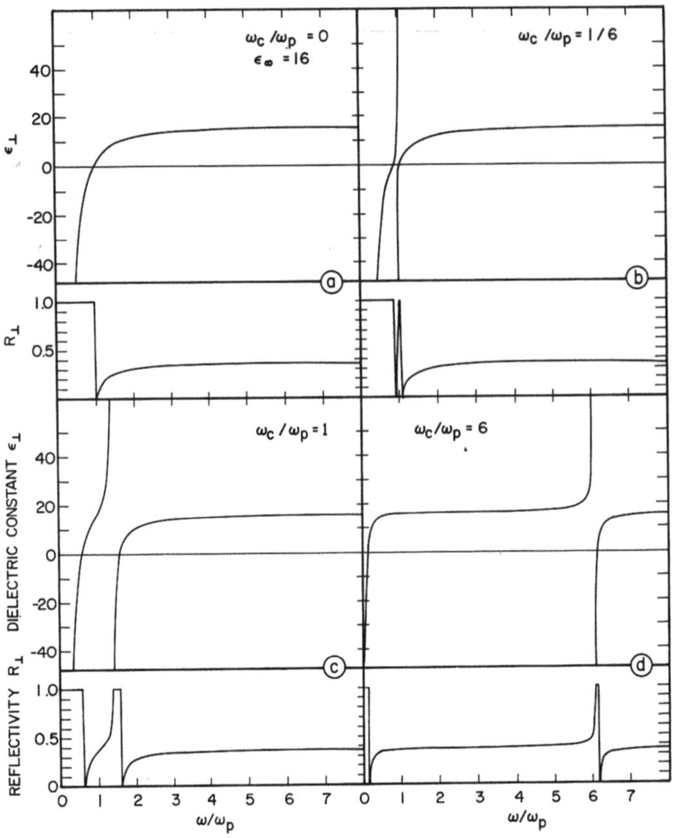

Figure 12. Dielectric constant $\epsilon_\perp$ and reflectivity $R_\perp$ for the Voigt geometry for the lossless case $\nu = 0$. Figures a, b, c and d represent the cases $\omega_c/\omega_p = 0$, 1/6, 1 and 6, respectively. The case $\omega_c/\omega_p = 0$ also gives $\epsilon_{||}$ and $R_{||}$ for all magnetic fields.

For a reflection experiment in the Voigt geometry, a linearly polarized beam at an angle of 45° to $\vec{B}_0$ is normally incident on a sample surface. This is equal components of $||$ and $\perp$ polarization. Upon reflection there are phase changes $\Delta_{||}$ and $\Delta_\perp$ and amplitude changes $r_{||}$ and $r_\perp$. The reflected beam is elliptically polarized and rotated as described by Eq. (73) if we let $\rho = |r_{||}|/|r_\perp|$ and $\Delta = \Delta_{||}-\Delta_\perp$. Here, we use the moduli of $r_{||}$ and $r_\perp$ to define ρ. Since tan $\eta = \rho$ and $\tan 2\eta = 2\rho/(1-\rho)^2$, it follows that

$$\tan 2\theta = \frac{2\rho}{1-\rho^2} \cos \Delta \qquad (74)$$

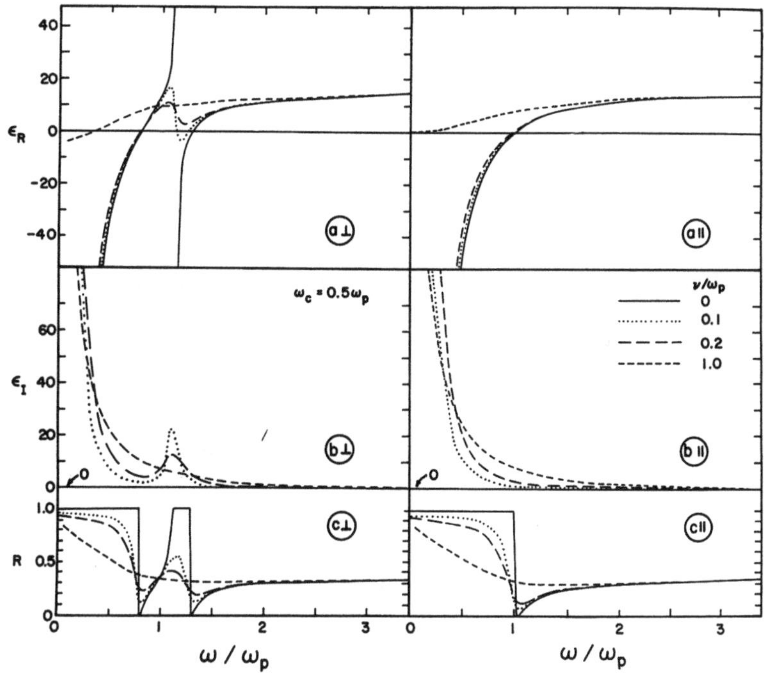

Figure 13. Real and imaginary parts of the dielectric constants $\epsilon_{\parallel,\perp}$ and reflectivities $R_{\parallel,\perp}$ for the Voigt geometry for ν/ω_p = 0, 0.1, 0.2 and 1 for $\omega_c/\omega_p = 0.5$.

$$\tan 2\gamma = \frac{2\rho}{1+\rho^2} \sin \Delta \tag{74}$$

This result shows that the rotation θ_r and the ellipticity $\epsilon_r = \tan \gamma$ depend on both reflectivity and phase shift. For small differential phase shift Δ, rotation θ_r primarily depends on reflection amplitudes while for small differential reflectivities ($\rho \approx 1$), ellipticity ϵ_r primarily depends on phase shift.

The same treatment holds for the transmission experiment. Here, the moduli of the transmission amplitudes are given by $|t_{\parallel,\perp}| = |1-r^2_{\parallel,\perp}| \exp(-2\pi k_{\parallel,\perp} d/\lambda)$. Then $t = |t_\parallel|/|t_\perp|$ and $\varphi = \varphi_\parallel - \varphi_\perp$ replace ρ and Δ in Eq. (74). The transmission experiment is the Voigt effect. Multiple reflection effects in thin samples alter these results and those in the Faraday geometry also[38,39].

Low Field Approximation

In the vicinity of ω_p for the conditions $\omega \approx \omega_p \gg \omega_c$, $\omega\tau \gg 1$, the zeroes in $\epsilon_{\perp R}$ are given by

$$\omega_\perp \approx \omega_p \pm \frac{\omega_c}{2} + \frac{\omega_c^2}{4\omega_p} + \dots \quad . \tag{75}$$

This splitting of the reflection edge is similar to the case for the Faraday geometry.

No helicon wave occurs at low frequencies in the Voigt geometry.

For the Voigt effect for the conditions $\omega \gg \omega_c$, ω_p, ν, the differential phase shift is given by

$$\delta = \varphi_{||} - \varphi_\perp = \frac{\omega d}{c}\,(n_{||} - n_\perp) \approx \frac{e^4 dN\lambda^3 B_o^2}{4\pi^2 c^6 n m^{*3}} \quad . \tag{76}$$

In this approximation the effect is independent of τ. If the Faraday and Voigt effects are measured on the same sample, we get

$$\frac{\delta}{\theta} = \frac{\omega_c}{\omega} \tag{77}$$

which indicates that δ is always much smaller than θ.

An analogy to the Hall effect may be made in this geometry. With B_o along z, and the radiation electric field given by $E_\perp = E_{2y}$, the ac Hall field E_{2z} is given by Eq. (23).

High Field Approximation

The transition to high magnetic field produces no new effects. The approximations just discussed break down and the full formulas are needed to analyze an experiment.

In passing we again note that reflectivity, as shown in Fig. 12, can be interpreted in terms of zeroes, unities and infinities in $\epsilon_{\perp R}$. The rapid rise to unity toward higher frequency locates the resonance $(\omega_p^2+\omega_c^2)^{\frac{1}{2}}$. The particular oscillator is both longitudinal and transverse in character, but we measure only its transverse properties. We note that the points $\epsilon_{\perp R} = 0$ do not locate resonant frequencies of longitudinal oscillators as examination of Eqs. (22) reveals. At $\omega = \omega_p$ between the two zeroes occurs the point $\epsilon_{\perp R} = \epsilon_\infty$. Examination of Eqs. (22) indicates that at this point for $\perp$ polarization the charge motion is entirely longitudinal (y=0). This is reasonable since if the wave "sees" a dielectric constant ϵ_∞, it doesn't "see" any free carrier transverse motion. Of course, the transverse EM wave does not "react" to the longitudinal free carrier motion.

Wave Vector Analysis

We plot $\varkappa_\perp$ vs ω for $\nu = 0$ in Fig. 14. Note that $\varkappa_\perp(B_o = 0)$ equals $\varkappa_\parallel(B_o)$. Again we can interpret these plots in terms of polaritons. The phase and group velocities can also be obtained.

LOW FIELD APPROXIMATION FOR CONDUCTIVITY

The current density can be expanded in the form

$$J_i = (\sigma_{ik} + \sigma_{ik\ell} B_\ell + \sigma_{ik\ell m} B_\ell B_m + \ldots) E_k \ . \tag{78}$$

In Eq. (7) the diagonal and off-diagonal terms may be expanded in powers of B as (we drop the zero subscript on B_o)

$$\sigma_{diag} = \frac{-i\omega_p^2 \epsilon_\infty}{4\pi(\omega - i\nu)} \left[1 + \frac{\omega_c^2}{(\omega - i\nu)^2} + \frac{\omega_c^4}{(\omega - i\nu)^4} + \ldots \right]$$

$$\sigma_{off\ diag} = \frac{-\omega_p^2 \epsilon_\infty \omega_c}{4\pi(\omega - i\nu)^2} \left[1 + \frac{\omega_c^2}{(\omega - i\nu)^2} + \frac{\omega_c^4}{(\omega - i\nu)^4} + \ldots \right] . \tag{79}$$

The first diagonal term in Eq. (79) is independent of B and is identified with the term σ_{ik} in Eq. (78). It is this term we associate with the B = 0 optical plasma effects and with the electrical conductivity. The first off-diagonal term is linear in B and can be associated with the optical Faraday effect (by use of Eq. (62) to get Eq. (52)) and the electrical Hall effect (Eq. (56)). The second diagonal term is quadratic in B and can be associated with the optical Voigt effect (by use of the definition of δ in Eq. (76)) and with the electrical magnetoresistance effect.

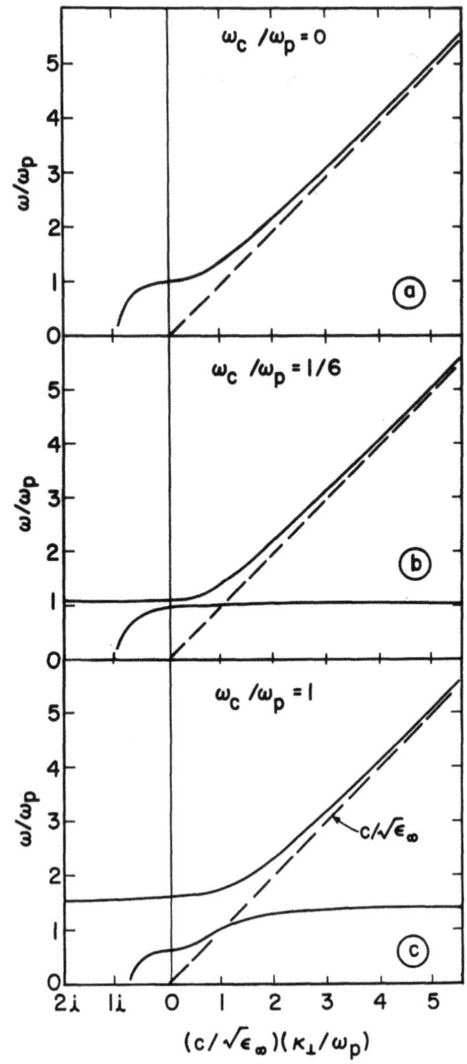

Figure 14. Dispersion curves for the Voigt geometry for $\nu = 0$. Figures a, b and c represent the cases $\omega_c/\omega_p = 0$, 1/6 and 1, respectively. The dashed lines are the dispersion curves of radiation in a medium of dielectric constant ϵ_∞.

This treatment has proved useful for treating semiconductors like Ge with anisotropic effective masses[7].

SEMIMETALS WITH EQUAL NUMBERS OF HOLES AND ELECTRONS

If the medium contains equal numbers of holes and electrons, each makes an additive contribution to the conductivity[21,40,41]. The conductivity components Eq. (7) become (for $N_e = N_h$, $\nu_e = \nu_h = \nu$)

$$\sigma_{xx} = \sigma_{yy} = -Ne^2 i(\omega - i\nu)\left\{\frac{1}{m_e^*}\frac{1}{[(\omega-i\nu)^2-\omega_{ce}^2]} + \frac{1}{m_h^*}\frac{1}{[(\omega-i\nu)^2-\omega_{ch}^2]}\right\}$$

$$\sigma_{zz} = \frac{-Ne^2 i}{(\omega-i\nu)}\left[\frac{1}{m_e^*}+\frac{1}{m_h^*}\right] \tag{80}$$

$$\sigma_{xy} = -\sigma_{yx} = Ne^2\left\{\left[\frac{-\omega_{ce}}{(\omega-i\nu)^2-\omega_{ce}^2}\right]\frac{1}{m_e^*} + \left[\frac{-\omega_{ch}}{(\omega-i\nu)^2-\omega_{ch}^2}\right]\frac{1}{m_h^*}\right\}$$

$$= \frac{-Ne^2(\omega-i\nu)^2[\omega_{ce}m_h^* + \omega_{ch}m_e^*]}{m_e^* m_h^*\,[(\omega-i\nu)^2-\omega_{ce}^2][(\omega-i\nu)^2-\omega_{ch}^2]}$$

where the sign convention for ω_{ce} and ω_{ch} is used. Thus, there is no dc Hall conductivity σ_{xy} for $\omega_c \gg \nu$. We can switch to $\epsilon_{1,2}(\nu=0)$ for the Faraday and Voigt geometries to get

$$\epsilon_{1,2F} = \epsilon_\infty\left\{1 - \frac{4\pi Ne^2}{\epsilon_\infty}\,\frac{m_h^*(\omega^2-\omega_{ch}^2)+m_e^*(\omega^2-\omega_{ce}^2) \pm \omega[\omega_{ce}m_h^*+\omega_{ch}m_e^*]}{m_e m_h(\omega^2-\omega_{ce}^2)(\omega^2-\omega_{ch}^2)}\right\}$$

$$\epsilon_{\|V} = \epsilon_\infty\left\{1 - \frac{4\pi Ne^2}{\epsilon_\infty \omega^2}\left(\frac{1}{m_e^*}+\frac{1}{m_h^*}\right)\right\} \tag{81}$$

$$\epsilon_V = \epsilon_\infty \left\{ 1 - \frac{\omega_p^2(\omega^2 - \omega_p^2 + \omega_{ce}\omega_{ch})}{(\omega^2-\omega_{ce}^2)(\omega^2-\omega_{ch}^2) - \omega_p^2(\omega^2+\omega_{ce}\omega_{ch})} \right\} \tag{81}$$

where $\omega_p^2 = (4\pi Ne^2/\epsilon_\infty)(1/m_e^* + 1/m_h^*)$.

In the Faraday geometry for the spectral region $\omega \ll \omega_c < \omega_p$, the fourth term is negligible compared to the second and third, so the two waves are nearly identical and linearly polarized. The sample is transparent below the lower of the two cyclotron frequencies. These waves are called Alphen waves with $\vec{\kappa} \parallel \vec{B}_o$. If $\vec{\kappa}$ makes an angle α to $\vec{B}_o$ in the xz plane (we go from Faraday to Voigt geometry), two distinct waves propagate, an ordinary wave with E_{yo} and $\kappa_o^2 = (\omega^2/c^2)\epsilon_{xx}$ and an extraordinary wave with E_{xe}, E_{ze} and $\kappa_e = \kappa_o/\cos\alpha$. Even at the Voigt geometry there still exists a low frequency region of transparency for E_{yo}. Other optical properties at higher frequencies can be deduced from study of Eqs. (81). We only point out that in the Voigt geometry, a hybrid resonance occurs at $\omega^2 = \omega_{ce}\omega_{ch}$ which dominates the individual cyclotron resonances at ω_{ce} and ω_{ch} for high carrier concentrations (large ω_p). This is a form of collective cyclotron excitations which we have already noted for the simpler case of a single type of free carrier.

ARBITRARY MAGNETIC FIELD ORIENTATION

We now turn to the general solution of the Maxwell wave equation for arbitrary angle α between the direction of propagation and magnetic field. Normal incidence is still maintained. This problem has been done for a gaseous plasma[9] and the results can be translated immediately to a non-polar semiconductor by the substitution of ϵ_∞ for unity as the background dielectric constant. The two propagating waves are "governed" by

$$\epsilon_{1,2} = \epsilon_\infty \left[1 - \frac{2\omega_p^2[\omega(\omega-i\nu)-\omega_p^2]}{\omega\{2(\omega-i\nu)[\omega(\omega-i\nu)-\omega_p^2]-\omega\,\omega_c^2\sin^2\alpha \mp \left(\omega_c^4\,\omega^2\sin^4\alpha + 4\omega_c^2[\omega(\omega-i\nu)-\omega_p^2]^2\cos^2\alpha\right)^{\frac{1}{2}}\}} \right] \tag{82}$$

In the limit $\alpha = 0$ (Faraday geometry) $\epsilon_{1,2} \rightarrow \epsilon_\pm$ holds while for

$\alpha = 90°$ (Voigt geometry) $\epsilon_{1,2} \rightarrow \epsilon_{\perp,\parallel}$ holds. The resonant frequencies occur when $\epsilon_{1,2} \rightarrow \infty$ occurs. For the lossless case $\nu = 0$, the resonant frequencies are given by

$$2\omega^2 = \omega_p^2 + \omega_c^2 \pm \left[(\omega_p^2+\omega_c^2)^2 - 4\omega_c^2\omega_p^2 \cos^2\alpha\right]^{\frac{1}{2}} . \qquad (83)$$

For the Faraday geometry, Eq. (83) gives two frequencies ω_p and ω_c, but the solution ω_p does not correspond to $\epsilon = \infty$ in Eq. (82). Fig. 10 was calculated by use of Eq. (83).

Another interesting feature occurs at $\omega = \omega_p$ regardless of the values of α and ω_c. Then we see that ϵ_1 equals ϵ_∞ while $\epsilon_2=0$.

The polarizations of the two waves are given by

$$\frac{E_{1,2x}}{E_{1,2y}} = \frac{-i2\omega_c\left[\omega(\omega-i\nu)-\omega_p^2\right]\cos\alpha}{\omega\omega_c^2\sin^2\alpha \pm \left\{\omega^2\omega_c^4\sin^4\alpha+4\omega_c^2\left[\omega(\omega-i\nu)-\omega_p^2\right]^2\cos^2\alpha\right\}^{\frac{1}{2}}} . \qquad (84)$$

For the Voigt geometry, the 1,2 notation here is the reverse of that arbitrarily defined as in Eqs. (15) and (16). For $\alpha = 0$, $E_{1,2x}/E_{1,2y} = \mp i$ holds and for $\alpha = 90°$ $E_{1,2x}/E_{1,2y} = 0$ and ∞ holds. Thus $E_{1x} = 0$ and $E_{1y} = E_\perp$ and $E_{2x} = E_\parallel$ and $E_{2y} = 0$ obtain. Generally, the waves are elliptically polarized.

Along the z axis the longitudinal electric field is given by

$$E_z = \frac{\omega_c^2\omega_p^2\cos\alpha\ \sin\alpha\ E_x - i\ \omega_c\omega_p^2(\omega-i\nu)\sin\alpha\ E_y}{\omega(\omega-i\nu)\omega_c^2-(\omega-i\nu)^2\left[\omega(\omega-i\nu)-\omega_p^2\right]-\omega_c^2\omega_p^2\cos^2\alpha} . \qquad (85)$$

E_z goes to zero for the Faraday geometry and to Eq. (23) for the Voigt geometry.

In Fig. 15 we plot $\epsilon_{1,2}$ for several angles for $\nu = 0$ and $\omega_c/\omega_p = 0.5$. We note that as α gets larger than zero, a collective cyclotron excitation resonance appears slightly above ω_p while another collective cyclotron excitation starting at ω_c drops in frequency. When α is 90°, this latter has gone to zero frequency. Note that a "helicon" region occurs for all angles $\alpha \neq 90°$ at low frequency. We leave as an exercise the construction of the reflection curves associated with the dielectric constants shown in Fig. 15.

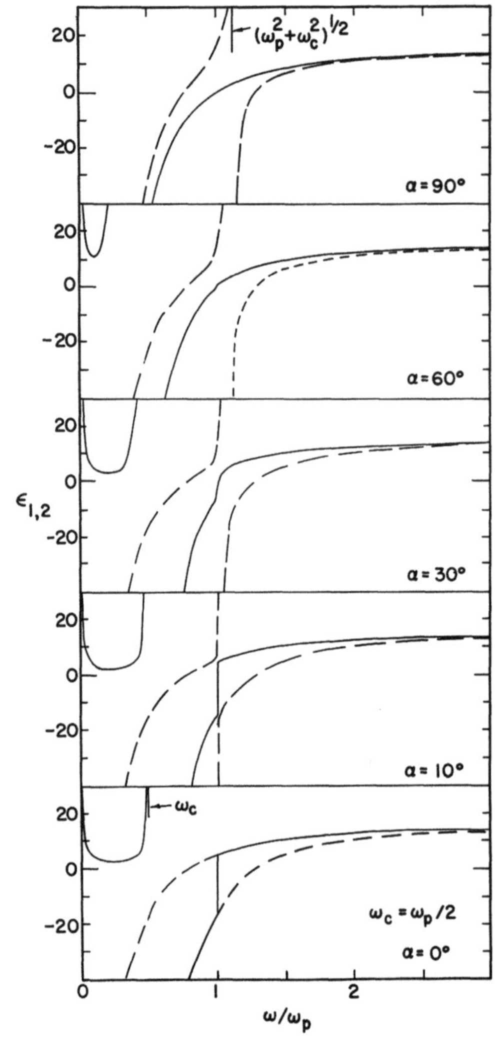

Figure 15. Dielectric constants $\epsilon_{1,2}$ for magnetic field tipped from Faraday (α = 0) to Voigt (α = 90°) geometry. The "helicon" dielectric constant shown in the upper left corner of each graph (below ω_c) should be multiplied by 10^2 to get the actual value of ϵ.

POLAR SEMICONDUCTORS

General Magnetic Field Results - Faraday and Voigt Geometries

The magnetooptical effects just discussed are altered when lattice vibrations are added to the crystal. The changes are largest in the vicinity of the optical phonon frequencies. The problem may again be solved by calculating the dielectric tensor for the medium and then solving the wave equation for the wave vectors of the two propagating modes. The lattice vibration contribution to each diagonal term of the dielectric tensor of Eq. (8) is[31]

$$\epsilon_\ell = \frac{\epsilon_\infty(\omega_L^2-\omega_T^2)}{\omega_T^2-\omega^2+i\Gamma\omega} \tag{86}$$

where ω_L is the longitudinal optical phonon frequency, ω_T is the transverse optical phonon frequency, Γ is the phonon damping constant, and exp $i\omega t$ dependences have been assumed. Then, when the free carrier density is zero, we get the usual diagonal dielectric tensor used to discuss reststrahlen effects in dielectric crystals.

For the Faraday geometry we may write the two dielectric constants as

$$\epsilon_\pm = \epsilon_\infty\left[1 - \frac{\omega_p^2}{\omega(\omega_\mp\omega_c-i\nu)} + \frac{\omega_L^2-\omega_T^2}{\omega_T^2-\omega^2+i\Gamma\omega}\right]. \tag{87}$$

Note that the lattice effects are simply additive because of the simple definition of $\epsilon_\mp = \epsilon_{xx} \pm i\epsilon_{xy}$. Resonances of transverse oscillators occur at ω_T and ω_c.

For the Voigt geometry since $\epsilon_\perp$ is given by[42]

$$\epsilon_\perp = \epsilon_{yy} + \frac{\epsilon_{yz}^2}{\epsilon_{yy}} = \frac{2\epsilon_+\epsilon_-}{\epsilon_++\epsilon_-}, \tag{88}$$

the lattice term gets mixed into $\epsilon_\perp$ in a complicated way. We get

$$\epsilon_\perp = \epsilon_\infty\left\{1 + \frac{\omega_T^2-\omega^2}{\omega_T^2-\omega^2+i\Gamma\omega} - \frac{\omega_p^2}{\omega(\omega-i\nu)}\left[1 - \frac{\omega_c^2\omega(\omega_L^2-\omega^2+i\Gamma\omega)}{(\omega-i\nu)D}\right]\right\} \tag{89}$$

$$\epsilon_{\|} = \epsilon_{\infty}\left[1 - \frac{\omega_p^2}{\omega(\omega-i\nu)} + \frac{(\omega_L^2-\omega_T^2)}{\omega_T^2-\omega^2+i\Gamma\omega}\right] \tag{89}$$

with

$$D = \left[\omega_p^2-\omega^2+i\nu\,\omega+\frac{\omega\omega_c^2}{(\omega-i\nu)}\right]\left[\omega_L^2-\omega^2+i\Gamma\omega\right] - \omega_p^2(\omega_L^2-\omega_T^2).$$

Resonances occur at

$$2\omega_{\pm}^2 = \omega_p^2 + \omega_L^2 + \omega_c^2 \pm \left[(\omega_p^2+\omega_L^2+\omega_c^2)^2 - 4(\omega_p^2\omega_T^2+\omega_c^2\omega_L^2)\right]^{\frac{1}{2}} \tag{90}$$

$$\omega = \omega_T .$$

A longitudinal Hall electric field arises in the z direction given by

$$E_z = \frac{i\omega_p^2\,\omega_c}{(\omega-i\nu)D}\left[\omega_T^2-\omega^2+i\Gamma\omega\right] E_y . \tag{91}$$

With no lattice effects Eq. (91) simplifies to Eq. (23). When the semiconductor becomes non-polar, effectively $\omega_T = \omega_L = 0$ holds and Eq. (90) reverts to the previously-discussed resonant frequency for the Voigt geometry $\omega^2 = \omega_p^2 + \omega_c^2$. We will deduce that the resonances in a polar semiconductor are associated with coupled plasmon, LO phonon, cyclotron oscillators, both transverse and longitudinal in character which we call coupled collective cyclotron excitation-LO phonon modes.

We could again make plots of ϵ vs ω for $\nu = 0$ to determine the optical properties, but we will only make the isoindex of refraction plots shown in Figs. 16 and 17 for the Faraday and Voigt geometries. The ϵ vs ω and R vs ω curves can be deduced from these. We have chosen two cases ($\omega_p = 0.5\,\omega_T$ and $\omega_p = 2\omega_T$) for illustration. The zeroes and infinities in ϵ are the easiest features to follow. Shaded regions, as usual, indicate non-propagating regions where $n = 0$ and $k > 0$. In the region between ω_T and ω_L, the graphs indicate that for $\nu = 0$ the reflectivity drops to zero near the frequency where the positive contribution to ϵ due to the ω_- resonance cancels the negative contribution due to the lattice vibrations. Wave vector properties follow from a study of ϵ vs ω.

Thin-Film Model for Voigt Geometry

While the results of the Maxwell wave equation treatment contain the physics of the coupled collective cyclotron excitation-LO

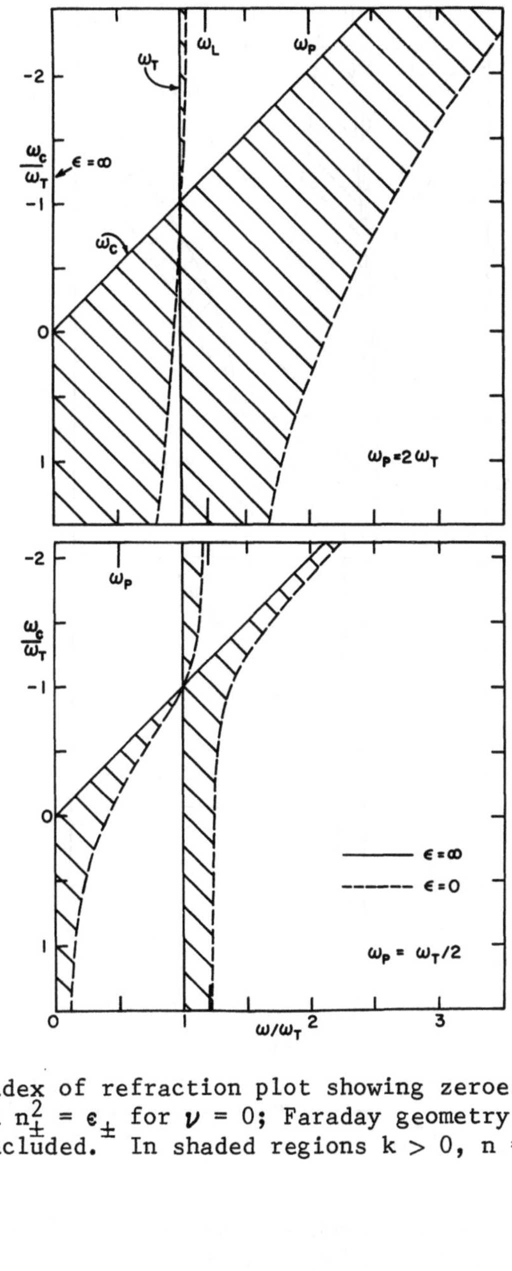

Figure 16. Index of refraction plot showing zeroes and infinities in $n_\pm^2 = \epsilon_\pm$ for $\nu = 0$; Faraday geometry with phonons included. In shaded regions $k > 0$, $n = 0$ holds.

phonon modes, it is informative to derive some of the results by use of a thin-film model[11,43,44]. This model gives information about the resonant frequencies of the charge system and the response of the system to an applied driving electric field but not about the eigenvectors of propagating radiation in bulk samples.

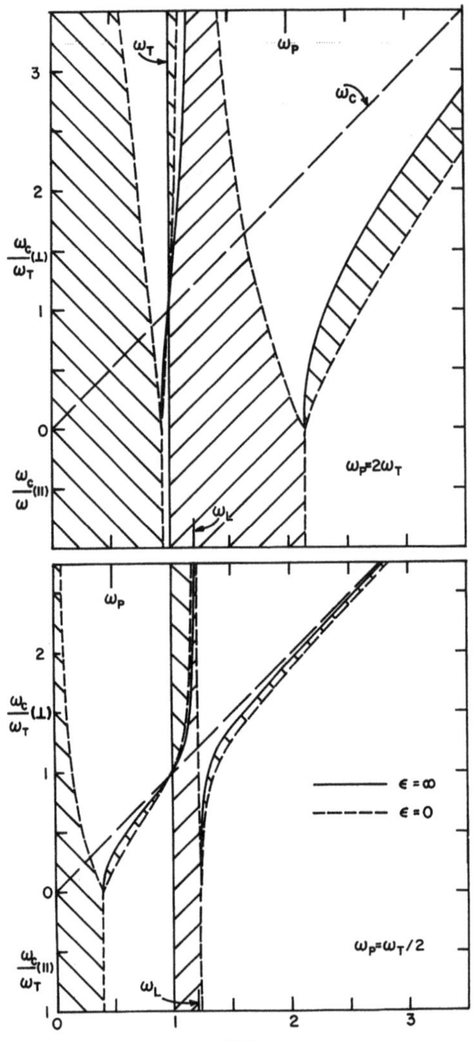

Figure 17. Index of refraction plot showing zeroes and infinities $n^2_{\parallel,\perp} = \epsilon_{\parallel,\perp}$ for $\nu = 0$; Voigt geometry with phonons included. In shaded regions $k > 0$, $n = 0$ holds.

In a thin-film sample such as shown in Fig. 1 for $\lambda \gg d$, the equations of motion describing lattice vibrations of wave vector $\vec{q} \approx 0$ and free carriers(45) are given by

$$\ddot{\vec{R}} + \Gamma \dot{\vec{R}} + \omega_T^2 \vec{R} = \frac{e^*}{M} \vec{E}_e \tag{92}$$

$$\ddot{\vec{r}} + \nu \dot{\vec{r}} - \frac{e}{c} (\dot{\vec{r}} \times \vec{B}) = \frac{d}{m^*} \vec{E}_e$$

where $\vec{R}$ is the relative displacement, e* is the effective charge, n is the density, and M is the reduced mass of the positive and negative ion oscillators, while $\vec{r}$ is the average displacement, e is the charge, N is the density, and m* is the effective mass of the free carriers. The electric field seen by a charge in the sample is $\vec{E}_e = \vec{E}_a - L\vec{P}$ where $\vec{E}_a$ is the externally applied electric field, and L is the depolarization factor (4π for the thickness direction and zero for directions parallel to the film surface). $\vec{P}$ is the long-wavelength macroscopic electric polarization given by

$$\vec{P} = ne^*\vec{R} + Ne\vec{r} + \alpha\vec{E}_e = \frac{ne^*\vec{R}+Ne\vec{r}+\alpha\vec{E}_a}{1 + \alpha L} \tag{93}$$

where $\alpha = (\epsilon_\infty-1)/4\pi$ is the background lattice electric susceptibility. Free carriers are not influenced by the Lorentz field[9], but its effect on bound ions is contained in the definition of ω_T[44]. Eqs. (92) are general and can be solved for arbitrary direction of magnetic field in the x,z plane of Fig. 1. The solution of these coupled equations for $E_a = 0$ give the resonant frequencies of the charge system. For $E_a \neq 0$, the solution also gives the response of the charge system to an applied electric field. Eqs. (92) are coupled because both lattice and free carriers see the same electric field E_e which is made up of terms which contain displacements of both types of charges. For the Voigt geometry of Fig. 1, the y and z equations of motion are coupled together by the Lorentz force, while the x equation is uncoupled. Also, the z and Z motion will be coupled because of the E_{ze} term. With exp $i\omega t$ dependences assumed, the equations of motion for free carriers and lattice ions, respectively, are

$$\begin{aligned}
-(\omega^2-i\nu\omega)x &= \frac{e}{m^*} E_{xa} \\
-(\omega^2-i\nu\omega)y - i\omega\omega_c z &= \frac{e}{m^*} E_{ya} \\
-(\omega^2-i\nu\omega)z + i\omega\omega_c y &= \frac{e}{m^*\epsilon_\infty} \left[E_{za}-4\pi(Nez+ne^*Z)\right] \\
-(\omega^2-i\Gamma\omega)X + \omega_T^2 X &= \frac{e^*}{M} E_{xa} \\
-(\omega^2-i\Gamma\omega)Y + \omega_T^2 Y &= \frac{e^*}{M} E_{ya} \\
-(\omega^2-i\Gamma\omega)Z + \omega_T^2 Z &= \frac{e^*}{M\epsilon_\infty} \left[E_{za}-4\pi(Nez+ne^*Z)\right]
\end{aligned} \tag{94}$$

where we ignore the effect of the Lorentz force on the ions directly.

The displacement coordinates are found to be

$$x = \frac{-e}{\omega(\omega-i\nu)m^*} E_{xa}$$

$$y = \left[\frac{-e}{\omega(\omega-i\nu)m^*} + (\omega_L^2-\omega^2+i\Gamma\omega)\left(\frac{\omega_c^2 e}{D(\omega-i\nu)^2 m^*}\right)\right] E_{ya} - (\omega_T^2-\omega^2+i\Gamma\omega) \cdot \left[\frac{i\omega_c e}{D(\omega-i\nu)m^*}\right] E_{za}$$

$$z = (\omega_L^2-\omega^2+i\Gamma\omega)\left[\frac{i\omega_c e}{D(\omega-i\nu)m^*}\right] E_{ya} + (\omega_T^2-\omega^2+i\Gamma\omega) \frac{e}{Dm^*\epsilon_\infty} E_{za}$$

$$X = \frac{e^*}{M(\omega_T^2-\omega^2+i\Gamma\omega)} E_{xa} \tag{95}$$

$$Y = \frac{e^*}{M(\omega_T^2-\omega^2+i\Gamma\omega)} E_{ya}$$

$$Z = \frac{-i\omega_c\omega_p^2 e^*}{M(\omega-i\nu)D} E_{ya} + \left[- (\omega-i\nu) + \frac{\omega_c^2\omega}{(\omega-i\nu)} \right] \frac{e^*}{DM\epsilon_\infty} E_{za}$$

$$D = \left[\omega_p^2-\omega^2+i\nu\omega + \frac{\omega\omega_c^2}{(\omega-i\nu)} \right] (\omega_L^2-\omega^2+i\Gamma\omega) - \omega_p^2\Omega_p^2 \ .$$

We may write the electric polarization in terms of electric susceptibility response functions as $\vec{P} = \overleftrightarrow{\chi}^r\vec{E}_a$. We can then define a response dielectric function as $\overleftrightarrow{\epsilon}^r = \overleftrightarrow{1} + 4\pi \overleftrightarrow{\chi}^r$. The present treatment is useful for determining the time average of power absorption per unit volume,

$$\wp = \frac{1}{2} \text{Re } \dot{\vec{P}}\cdot\vec{E}_a = \frac{\omega}{2} \text{Im } \vec{P}\cdot\vec{E}_a \ . \tag{96}$$

This equation is similar to Eq. (39), but in that case $\vec{J}$ and $\vec{E}$ are defined in the medium, and E is a solution of the wave equation.

The electric susceptibility response functions[43] are obtained from the electric polarization

$$P_x = ne^*X + Nex + \alpha E_{xe} = \chi^r_{11} E_{xa}$$

$$P_y = ne^*Y + Ney + \alpha E_{ye} = \chi^r_{22} E_{ya} + \chi^r_{23} E_{za} \tag{97}$$

$$P_z = ne^*Z + Nez + \alpha E_{ze} = \chi^r_{32} E_{ya} + \chi^r_{33} E_{za} \ .$$

Then the susceptibility response functions are

$$4\pi\chi^r_{11} = \epsilon_\infty - 1 - \epsilon_\infty \left[\frac{\Omega_p^2}{\omega_T^2-\omega^2+i\Gamma\omega} - \frac{\omega_p^2}{(\omega-i\nu)\omega}\right]$$

$$4\pi\chi^r_{22} = \epsilon_\infty - 1 + \epsilon_\infty \left[\frac{\Omega_p^2}{\omega_T^2-\omega^2+i\Gamma\omega} - \frac{\omega_p^2}{(\omega-i\nu)\omega} + \frac{\omega_c^2\omega_p^2(\omega_L^2-\omega^2+i\Gamma\omega)}{(\omega-i\nu)^2 D}\right]$$

$$4\pi\chi^r_{33} = \epsilon_\infty - 1 + \frac{1}{\epsilon_\infty D}\left[\omega_p^2(\omega_T^2-\omega^2+i\Gamma\omega) + \Omega_p^2\left(-\omega^2+i\gamma\omega + \frac{\omega_c^2\omega}{(\omega-i\nu)}\right)\right] \tag{98}$$

$$4\pi\chi^r_{23} = \frac{-i\omega_p^2\omega_c(\omega_T^2-\omega^2+i\Gamma\omega)}{(\omega-i\nu)D}$$

$$4\pi\chi^r_{32} = \frac{i\epsilon_\infty\omega_p^2\omega_c}{(\omega-i\nu)D}\left[-\Omega_p^2+(\omega_L^2-\omega^2+i\Gamma\omega)\right] .$$

Note that these are not necessarily the susceptibilities obtained from Eq. (8) from the definition $\epsilon_{ij} = 1 + 4\pi\chi_{ij}$ or the $\chi_{\|,\perp}$ obtained from $\epsilon_{\|,\perp} = 1 + 4\pi\chi_{\|,\perp}$. The χ's of Eq. (8) are defined in terms of the actual electric field in the medium, not just the applied electric field. For the Voigt geometry, since the propagating radiation is linearly polarized (as obtained from the wave equation solution) it happens that $\chi_\perp$ does equal χ^r_{22} so Eq. (89) is obtained from the thin-film model.

Eqs. (94) reveal many interesting aspects of the general problem of coupled modes in solids. The simplest case is B = 0 for a non-polar ($e^* = 0$) semiconductor. Then the z equation becomes

$$-(\omega^2-i\nu\omega)z = \frac{e}{m^*\epsilon_\infty}\left[E_{za} - 4\pi Nez\right]$$

which represents an oscillator which sees a damping force (ν term), an applied force (E_{za} term) and an additional force due to the electric polarization term which contains the charge displacement. This $-4\pi P_z$ term is a self field which acts on the oscillator and shifts its natural frequency from $\omega = 0$ to $\omega_p = (4\pi Ne^2/m^*\epsilon_\infty)^{\frac{1}{2}}$, the plasma frequency.

For the case of a polar semiconductor with no free carriers we may write

$$-(\omega^2 - i\Gamma\omega)Z + \omega_T^2 Z = \frac{e^*}{M\epsilon_\infty} E_{za} - 4\pi ne^* Z$$

which represents a bound oscillator of natural frequency ω_T which sees a damping force (Γ term), an applied driving force (E_{za} term) and a self-field force ($-4\pi P_z$ term). This self field shifts the natural frequency from ω_T to ω_L with ω_L, the longitudinal optical phonon frequency, given by[44]

$$\omega_L^2 - \omega_T^2 = \Omega_p^2 = \frac{4\pi ne^{*2}}{M\epsilon_\infty} . \qquad (99)$$

If the polar semiconductor has free carriers, the z and Z equations are coupled, because the free carrier plasma and the LO phonons see each others self fields. The resonant frequencies of the coupled plasmon-phonon modes are found to be Eq. (90) with ω_c set equal to zero, namely

$$2\omega_\pm^2(B{=}0) = \omega_p^2 + \omega_L^2 \pm [(\omega_p^2+\omega_L^2)^2 - 4\omega_p^2\omega_T^2]^{\frac{1}{2}} . \qquad (100)$$

Eqs. (95) give the displacements of the charges in response to the applied electric field. Thus, for B = 0, resonances occur in z and Z at $\omega_\pm$ as given by Eq. (100) if we can apply an E_{za} field. Of course, a resonance occurs at ω_T for applied E_{xa} and E_{ya} fields. In a normal incidence experiment, no E_{za} can be applied, so no absorption line will be observed at $\omega_\pm$ (zeroes in ϵ do occur at $\omega_\pm$, however). These modes may be measured in non-normal incidence experiments where linear polarization in the plane of incidence does have an E_{za} component[11,44,46] and in Raman scattering experiments[47].

In Fig. 18 the positions of $\omega_\pm(B{=}0)$ are plotted as a function of $\omega_p^2 \propto N$. At N = 0 two resonant longitudinal modes occur at $\omega = 0$ and $\omega = \omega_L$. At very large N, the upper mode is nearly at ω_p while the lower mode is near ω_T. Thus, there are two longitudinal oscillators, one nearly all plasmon in content and one almost all phonon. But the longitudinal phonon has a frequency ω_T! This can be understood by considering the contributions to

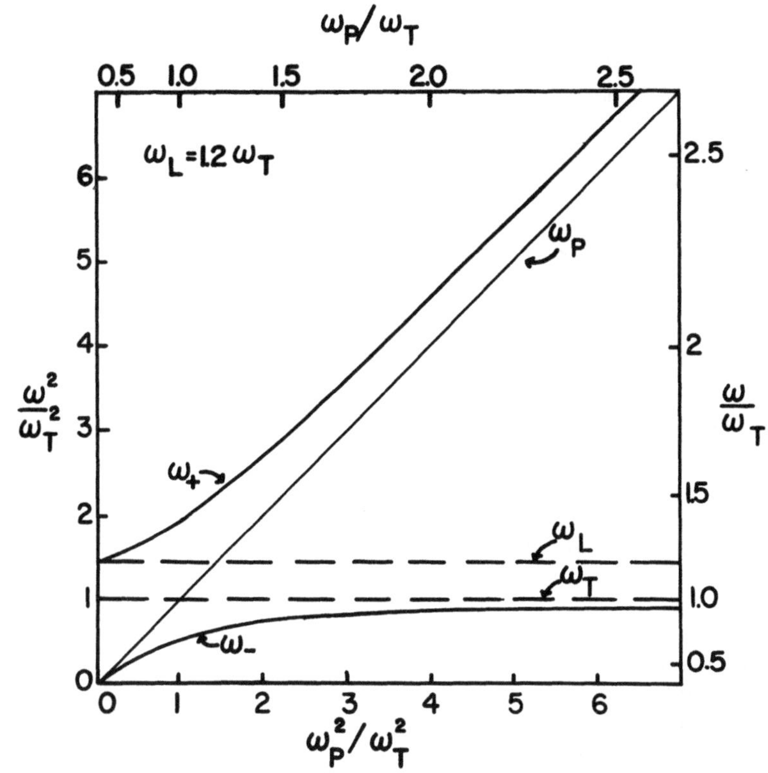

Figure 18. Frequencies of coupled plasmon-LO phonon modes as a function of carrier density for $\nu = 0$.

the self fields $-4\pi P_{z,Z}$. In the ω_+ mode, the contributions from free carriers and ions are in phase, while in the ω_- mode they are out of phase. As N increases (n is fixed), the longitudinal phonon oscillator self field is screened out by the plasma. Thus the lower mode will approach ω_T, the natural resonant frequency of the lattice as determined by the lattice spring constant while the upper mode will approach $(\omega_p^2 + \Omega_p^2)^{\frac{1}{2}}$.

The application of a magnetic field now introduces the Lorentz force which couples the cyclotron oscillator (y,z motion) into the coupled plasmon-phonon oscillators (z,Z motion) and the full Eqs. (94) and (95) now apply. Resonances occur at $\omega_\pm(B)$ as given by Eq. (90). The coupled oscillators are now both transverse and longitudinal in character, so that we can measure $\omega_\pm$ with radiation linearly polarized $\perp$ to B in a normal-incidence experiment. We are still not probing the longitudinal motion of the coupled oscillators. In Fig. 19 are plotted the positions of $\omega_\pm$ for

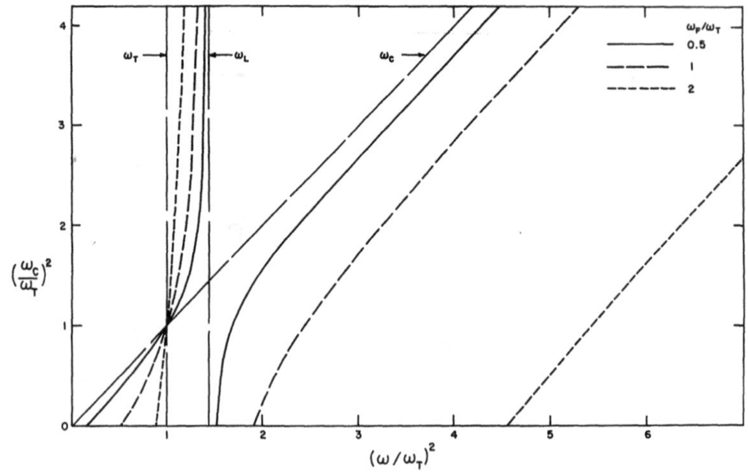

Figure 19. Magnetic field dependence of coupled cyclotron-plasmon-phonon modes for three samples with $\omega_p/\omega_T = 0.5$, 1 and 2 and $\nu = 0$.

several values of N. Some useful approximations for $\omega_\pm$ are: For small N the modes near B = 0 are $\omega_+ \approx \omega_L$ and $\omega_- = \omega_p\omega_T/\omega_L$ while for large B they are $\omega_+ \approx \omega_c$ and $\omega_- \approx \omega_L$. For large N the modes near B = 0 are $\omega_+^2 \approx \omega_p^2 + \Omega_p^2$ and $\omega_-^2 \approx \omega_L^2 - \Omega_p^2 = \omega_T^2$, while for large B, they are $\omega_+^2 \approx \omega_p^2 + \omega_c^2$ and $\omega_- \approx \omega_L$. Whenever a mode is near ω_L or ω_p, the resonant absorption line will not have much intensity as the y coordinate will get small while the z and Z coordinates will get large. The Y coordinate will be large near ω_T always and the transverse lattice oscillator will always be measurable.

Another interesting feature will be the half width of the resonant absorption lines. The half widths of $\omega_\pm$ are determined by complicated mixtures of ν and Γ, the free carrier and lattice damping frequencies(48). Generally it is true that $\nu \gg \Gamma$ with $\Gamma \approx 1\ \text{cm}^{-1}$. Thus in a pure "dielectric" semiconductor TO absorption lines are sharp and narrow, with half widths Γ. Cyclotron resonance and plasmon resonance line half widths are 2ν. This different dependence of half width on damping constant is due to the different forms of the resonant denominators in Eq. (87). For coupled modes, when $\omega_\pm$ is near ω_L, the line half widths will be predominantly determined by Γ while far from ω_L, the half widths depend strongly on ν. Thus, in a material with low mobility (large ν), while the usual cyclotron condition $\omega_c\tau > 1$ may not be fulfilled, the coupled mode line for the same B may be quite sharp if it is near ω_L (of course, it will be weak).

Another interesting point is shown in Fig. 19 where the ω_- mode always crosses ω_T at ω_c for any carrier density. At this point in frequency, the Hall field given by Eq. (91) is zero. This field has contributions from the lattice ions and free carriers and may be considered as a restoring force field which shifts the coupled plasmon-phonon oscillator frequencies $\omega_\pm(B = 0)$. For $\omega_- = \omega_T = \omega_c$, this field is zero so the lattice vibrates at its natural frequency ω_T for both the Y and Z motion, while the free carriers oscillate at their natural frequency ω_c (as determined by the Lorentz force) for the y and z motion. Note that the upper mode at ω_+ continues to see an additional magnetic field-dependent electric field.

Arbitrary Magnetic Field Orientation

When B is oriented at an arbitrary angle α in the xz plane but wave propagation is still along z (normal incidence) results are more complicated. The thin-film model gives the resonant frequencies as

$$\omega^6 - \omega^4(\omega_c^2+\omega_p^2+\omega_L^2) + \omega^2[\omega_c^2(\omega_p^2\cos^2\alpha+\omega_L^2)+\omega_p^2\omega_T^2] - \omega_T^2\omega_p^2\omega_c^2\cos^2\theta = 0 \quad (101)$$

but does not give the polarization. A complete Maxwell wave equation treatment gives the same resonant frequencies and the two dielectric constants $\epsilon_{1,2}$ for the two propagating EM waves[49]. Here we only give a plot of the resonant frequencies from which, however, much can be guessed about the modes. The results for two samples are calculated in Fig. 20. The transition from Faraday to Voigt geometry is smooth. Resonances near $\omega_\pm(B = 0)$ and ω_L should be weak because most of the coupled oscillator motion will be longitudinal. Also, the line half widths will depend on the LO phonon and electron content of the mode.

It is obvious that for arbitrary magnetic field orientation, there is no longer a simple Faraday rotation or Voigt effect. Indeed, it might be difficult to analyze the general ellipticity and rotation to extract out the usual quantities like effective mass and carrier density, especially in the vicinity of ω_L. Fortunately, in solids, we can first pick the Faraday and Voigt geometries to most easily obtain information. Physicists interested in the propagation of long wavelength radiation through the earth's magnetoplasma (ionosphere) and in Faraday rotation of synchrotron radiation propagating from a star to earth through the galactic magnetoplasma are not so fortunate. On the other hand, they do not need to consider lattice vibration effects.

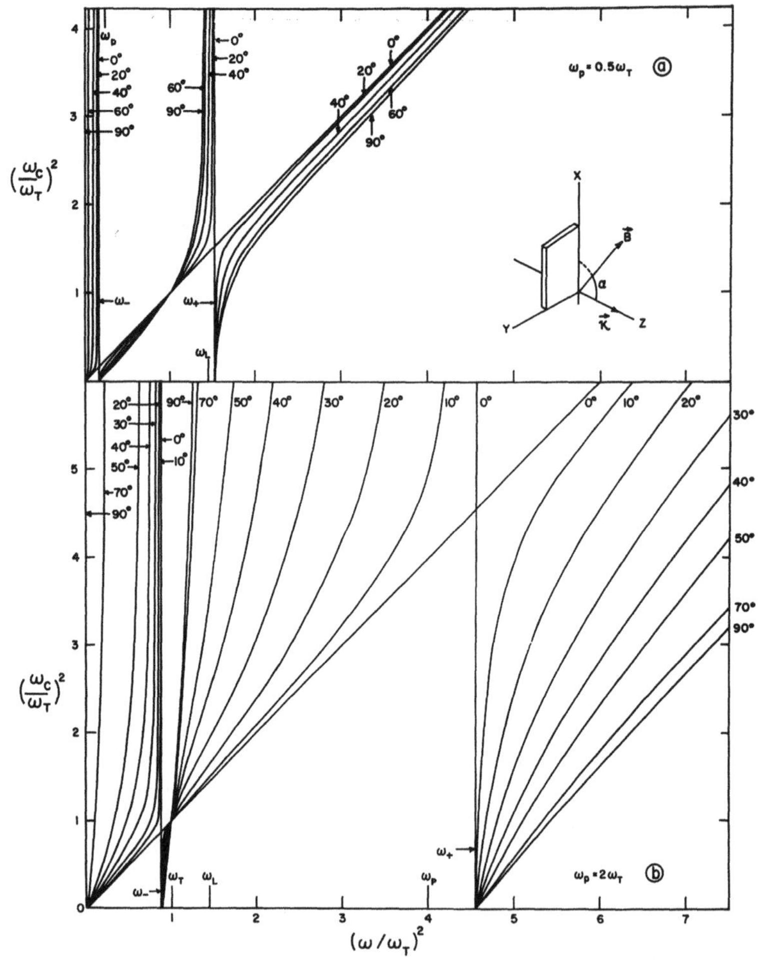

Figure 20. Shift of resonant frequencies from the Faraday to the Voigt geometry for two samples with ω_p/ω_T = 0.5 and 2 and ν = 0.

IMPURITY EFFECTS

Most free carrier effects in semiconductors are to be found in the infrared and microwave regions with some flexibility possible since $\omega_p^2 \propto N/m^*$. For InSb with N = 10^{14} cm^{-3} and m^*/m = 0.014, ω_p is about 6 cm^{-1} while for N = 10^{19} and m^*/m = 0.06, ω_p is about 1000 cm^{-1}. Non-parabolic energy bands cause the effective mass to be larger higher up in the band. Effects associated with lattice vibrations will usually occur in the far infrared since ω_T and ω_L are in this region for most materials.

There remains a class of far infrared experiments which has not been considered. When impurities such as group III and group V elements are put substitutionally into Ge and Si, at low temperature the impurity acts as a hydrogenic atom with a single electron or hole loosely bound to a positive or negative core. Because this "atom" is in a medium of dielectric constant $\gg 1$ and because the effective mass of the electron or hole is less than the free electron mass, the binding energy corresponds to far infrared energies $E_0 \approx 0.01$ eV. A system of hydrogen-like energy levels are formed just below the conduction band for donors and just above the valence band for acceptors. For donors, ionization of the impurity with increasing temperature or with infrared radiation would correspond to a transition from the ground state to the continuum conduction band. This is the mechanism for impurity photoconducting detectors(50). Electronic transitions from the ground state to excited states crudely correspond to Balmer-type series(51). Details of the electronic levels are clarified by the use of the Zeeman effect(52) and strain(51,53) to split degenerate levels.

For small magnetic field a classical-type Zeeman effect is observable(52). For large fields where distinct Landau levels are formed and $\hbar\omega_c \gg E_0$, the impurity levels "tag along" with Landau levels. This can give rise to an impurity transition which occurs near a free carrier cyclotron resonance line(54). At higher temperature the impurity electron is ionized off the parent atom to become a free carrier in the conduction band, so the impurity line weakens and disappears while the cyclotron line grows. The reverse happens as the temperature is lowered since free carriers become bound to ionized impurity atoms.

The far infrared spectral studies of impurities have been confined primarily to impurities in Ge, Si and InSb. A wide range of phenomena have been uncovered including polaron effects(55), phonon-assisted transitions(56) and transitions from the ground impurity level to impurity levels associated with various Landau levels(57). Polaron effects also influence free carrier effects near ω_L in low carrier density samples(58,59).

REFERENCES

1. P. Drude, The Theory of Optics, (Tuebner, Leipzig, 1900; Dover Publications, Inc., 1959).

2. H. A. Lorentz, The Theory of Electrons, (G. W. Stechert Co.,

New York, 1916; Dover Publications, Inc., 1952).

3. W. Voigt, Electro- and Magneto-Optics, (Tuebner, Leipzig, 1907).

4. B. Lax and S. Zwerdling, Progress in Semiconductors, (Wiley, New York, 1960), Vol. 5, p. 223.

5. B. Lax and J. G. Mavroides, Solid State Phys. 11, 261 (1960). Edited by F. Seitz and D. Turnbull (Academic Press, New York 1960), p. 261.

6. T. S. Moss, Phys. Stat. Solidi 2, 601 (1962).

7. B. Lax, International School of Physics "Enrico Fermi" Course XXII, Varenna, 1961 (Academic Press, New York,1963) p. 240.

8. E. D. Palik and G. B. Wright, Semiconductors and Semimetals, Edited by R. K. Willardson and A. C. Beer (Academic Press, New York, 1967), Vol. 3, p. 421.

9. V. L. Ginzburg, Propagation of Electromagnetic Waves in Plasma, (Gordon and Breach, Inc., New York, 1961).

10. S. D. Smith, Handbuch der Physik, (Springer-Verlag, Berlin, 1967), V. 25/2a, p. 234.

11. S. Iwasa, Y. Sawada, E. Burstein and E. D. Palik, J. Phys. Soc. Japan Suppl. 21, 742 (1966).

12. M. J. Stephen and A. B. Lidiard, J. Phys. Chem. Solids 9, 43 (1959).

13. S. Teitler and E. D. Palik, Phys. Rev. Letters 5, 546 (1960).

14. B. Lax and G. B. Wright, Phys. Rev. Letters 4, 16 (1960).

15. W. G. Spitzer and H. Y. Fan, Phys. Rev. 106, 882 (1957).

16. T. S. Moss, Optical Properties of Semiconductors, (Academic Press, Inc., New York, 1959), p. 25.

17. J. M. Bennett, J. Opt. Soc. Am. 54, 612 (1964).

18. W. R. Smythe, Static and Dynamic Electricity, (McGraw-Hill Book Co., New York, 1968).

19. S. D. Smith, T. S. Moss and K. W. Taylor, J. Phys. Chem.

Solids 11, 131 (1959).

20. See various articles in Plasma Effects in Solids, Proceedings of the 7th International Conference on the Physics of Semiconductors, Paris 1964 (Dunod, Paris and Academic Press, Inc., New York, 1965).

21. B. Ancker-Johnson, Semiconductors and Semimetals, Edited by R. K. Willardson and A. C. Beer (Academic Press, New York, 1966), Vol. 1, p. 379.

22. E. Burstein, P. J. Stiles, D. N. Langenberg and R. F. Wallis, Report of the International Conference on the Physics of Semiconductors, Exeter 1962 (The Institute of Physics and The Physical Society, London, 1962), p. 345.

23. E. D. Palik, Applied Opt. 2, 527 (1963).

24. J. K. Furdyna, Appl. Opt. 6, 675 (1967).

25. H. S. Bennett and E. A. Stern, Phys. Rev. 137, A448 (1965).

26. D. L. Mitchell and R. F. Wallis, Phys. Rev. 131, 1965 (1963).

27. J. K. Furdyna and M. E. Brodwin, Phys. Rev. 132, 97 (1963).

28. C. C. Grimes and S. J. Buchsbaum, Phys. Rev. Letters 12, 357 (1964).

29. J. J. Hopfield, J. Phys. Soc. Japan Suppl. 21, 77 (1966).

30. F. Stern, Solid State Physics 15, 299 (1963), Edited by F. Seitz and D. Turnbull, (Academic Press, New York, 1963).

31. M. Born and K. Huang, Dynamical Theory of Crystal Lattices, (Oxford University Press, London, 1954).

32. C. H. Henry and J. J. Hopfield, Phys. Rev. Letters 15, 964 (1965).

33. S. S. Porto, B. Tell and T. C. Damen, Phys. Rev. Letters 16, 450 (1966).

34. W. P. Allis, S. J. Buchsbaum and A. Bers, Waves in Anisotropic Plasmas, (MIT Press, Cambridge, Mass., 1963).

35. E. H. Holt and R. E. Haskell, Foundations of Plasma Dynamics, (Macmillan, Co., New York, 1965).

36. J. A. Stratton, Electromagnetic Theory (McGraw-Hill, Inc., New York, 1941), p. 333.

37. R. F. Wallis (private communication).

38. B. Donovan and T. Medcalf, Brit. J. Appl. Phys. 15, 1139 (1964).

39. J. Webster and B. Donovan, Brit. J. Appl. Phys. 16, 25 (1965).

40. S. J. Buchsbaum, Plasma Effects in Solids, Proceedings of the 7th International Conference on the Physics of Semiconductors, Paris 1964 (Dunod, Paris and Academic Press, New York, 1965) p. 3.

41. G. E. Smith, L. C. Hebel and S. J. Buchsbaum, Phys. Rev. 129, 154 (1963).

42. P. R. Wallace, J. Can. Phys. 46, 243 (1968).

43. R. Kaplan, E. D. Palik, R. F. Wallis, S. Iwasa, E. Burstein and Y. Sawada, Phys. Rev. Letters 18, 159 (1967).

44. E. Burstein, Dynamical Processes in Solid State Optics, Edited by R. Kubo and H. Kaminura (W. A. Benjamin, Inc., New York, 1967) p. 1.

45. G. Dresselhaus, A. F. Kip and C. Kittel, Phys. Rev. 100, 618 (1955).

46. D. W. Berreman, Phys. Rev. 130, 2193 (1963).

47. A. Mooradian and G. B. Wright, Phys. Rev. Letters 16, 999 (1966).

48. E. D. Palik, R. Kaplan, B. W. Henvis, J. R. Stevenson, S. Iwasa and E. Burstein, International Conference on the Physics of Semiconductors Moscow, 1968. (In press)

49. S. Teitler (private communication).

50. H. Levinstein, Appl. Opt. 4, 639 (1965).

51. A. Onton, P. Fisher and A. K. Ramdas, Phys. Rev. 163, 686 (1967).

52. S. Zwerdling, K. J. Button and B. Lax, Phys. Rev. 118, 975 (1960).

53. P. Fisher, A. K. Ramdas and J. H. Reuszer, Proceedings of the International Conference on Lattice Defects in Semiconductors, Edited by R. R. Hasiguti (University of Tokyo Press and The Pennsylvania State University Press 1968).

54. R. Kaplan, J. Phys. Soc. Japan Suppl. 21, 249 (1966).

55. B. D. McCombe and R. Kaplan, Phys. Rev. Letters 21, 756 (1968).

56. R. Kaplan and R. F. Wallis, Phys. Rev. Letters 20, 1499 (1968).

57. R. Kaplan, Phys. Rev. Letters 20, 329 (1968).

58. D. H. Dickey, E. J. Johnson and D. M. Larson, Phys. Rev. Letters 18, 599 (1967).

59. C. J. Summers, R. B. Dennis, B. S. Wherrett, P. G. Harper and S. D. Smith, Phys. Rev. 170, 755 (1968).

FAR INFRARED ABSORPTION IN ORDERED MAGNETIC SYSTEMS

M. Tinkham

Harvard University

Cambridge, Massachusetts

I. SPIN WAVES VS. LOCAL EXCITATIONS

The most elementary and familiar model for ordered magnetic systems is the so-called molecular field model, in which each spin is supposed to be acted on by a "molecular field" proportional to the magnetization or mean polarization of the other spins. This model leads to a second-order phase transition at a Curie temperature T_C above which there is no mean polarization. It gives qualitatively correct results, but is not quantitatively correct. The physical basis for the model is the exchange interaction, which introduces a coupling energy which depends on the relative spin orientations of nearby atoms. One expects the exchange coupling to fall off roughly exponentially with distance in insulating crystals, since it depends on the overlap of wavefunctions. Thus it is common to consider only nearest-neighbor coupling. Since the molecular field model essentially couples a given spin equally to <u>all</u> spins that participate in the average, it will give particularly bad results when each spin has only a few near neighbors.

For spectroscopic as well as thermodynamic purposes, the excitation spectrum of the system is the key property. In the molecular field model, the excitations are just those of single spins in an effective magnetic field, with energy

$$E_m = mg\beta H_{eff} = mg\beta\lambda M \tag{1}$$

where g is the spectroscopic splitting factor, β is the Bohr magneton, λ the molecular field parameter, M the magnetization, and m is the number of units of spin reversal. Such excitations are actually seen only in special situations, because in most cases they are poor approximations to the true elementary excitations of the systems. The reason for this is that such a spin flip at one site is coupled by the exchange Hamiltonian to an excited state of equal energy in which an adjacent spin is flipped instead. Because this coupling is between degnerate states, there is a zero-order effect on the eigenstates. In fact, the true eigenexcitation must reflect the translational symmetry of the crystal and hence be extended over the entire crystal. These are the spin waves, characterized by the crystal momentum k so that translation by one lattice spacing $\underline{a}$ introduces a phase factor e^{ika} in the wave function.

A simple model which illustrates these points is the ferromagnetic linear chain with nearest neighbor exchange coupling. For reasonable generality, we let the exchange be anisotropic. Then the Hamiltonian takes the form

$$\mathcal{H} = -2\Sigma_i(J_x s_{ix}s_{i+1,x} + J_y s_{iy} s_{i+1,y} + J_z s_{iz} s_{i+1,z}) \tag{2}$$

We assume $J_z > J_x, J_y$, so that the ground state has spins aligned along the z direction and introduce raising and lowering operators

$$s_{i\pm} = s_{ix} \pm is_{iy} \tag{3}$$

which connect states with s_{iz}=m to those with s_{iz}=m$\pm$1. Then (2) becomes

$$\mathcal{H} = -2\Sigma_i \Big[(J_z s_{iz} s_{i+1,z} + \frac{J_x+J_y}{4} (s_{i+}s_{i+1,-} + s_{i-}s_{i+1,+}) + \frac{J_x-J_y}{4} (s_{i+}s_{i+1,+} + s_{i-}s_{i+1,-}) \Big] \tag{4}$$

To start, let us assume there is axial symmetry about z, so that J_x=J_y, and the last term vanishes. In that case, $S_z=\Sigma_i s_{iz}$ is conserved and has a value M (in units of $\hbar$). In the ground state Φ_o, all spins are aligned, so S_z=Ns, where N is the number of spins in the chain, which for simplicity is considered to be closed into a ring. The corresponding energy is

$$E_o = -2NJ_z s^2 \tag{5}$$

The elementary excitations are states in which S_z=Ns-1, with the form

$$\Psi_k = \frac{1}{\sqrt{N}} \quad \Sigma_i \; e^{ikx_i}\Psi_i \tag{6a}$$

where

$$\Psi_i = \frac{s_{i-}\;\Phi_o}{\sqrt{s(s+1)-s(s-1)}} = \frac{s_{i-}}{\sqrt{2s}}\;\Phi_o \tag{6b}$$

is the state which differs from the ground state Φ_o by having s_z of the ith spin reduced by one unit. Evidently (6a) has the Bloch form appropriate to a periodic Hamiltonian. If the Hamiltonian (4) with J_x=J_y operates on this Ψ_k, we find

$$\mathcal{H}\Psi_k = \{-2J_zNs^2 + [4J_z-2(J_x+J_y)\cos ka]s\}\;\Psi_k \tag{7}$$

so that the spin wave state is indeed an eigenstate with excitation energy

$$\varepsilon_k = E_k-E_o = 4s\left[\;J_z-\left(\frac{J_x+J_y}{2}\right)\quad\cos ka\;\right] \tag{8}$$

$$\equiv \quad 4sJ_{\parallel}\left(1-\frac{J_{\perp}}{J_{\parallel}}\cos ka\right) \tag{8a}$$

where $J_{\parallel} = J_z$ and $J_{\perp} = J_x = J_y$.

If the exchange is isotropic, so that $J_{\parallel}$=$J_{\perp}$, then ε_k goes to zero as $k \to 0$. This gives the familiar result

$$\varepsilon_k = 8Js\sin^2 ka/2 \approx 2Jsa^2k^2 \qquad \text{for } ka<<1 \tag{9}$$

With any finite anisotropy, there will be a gap in spectrum, i.e., no energy levels below $4s(J_{\parallel} - J_{\perp})$. The other interesting limit is when $J_{\perp}/J_{\parallel} \to 0$. Then $\varepsilon_k = 4sJ_{\parallel}$ for all k, and the spin wave dispersion curve is flat. Also, in this limit, localized excitations in which a single spin is flipped are just as valid eigenstates as the spin waves. This is only exactly true in the limit, however; for any finite $J_{\perp}$ the only exact eigenstates immediately become the spin waves (6). A final remark is that, while $J_{\parallel}$ is all that plays a role in determining the ground state energy E_o, $J_\perp$ and $J_\parallel$ play an equal role in determining excitation energies ε_k.

Although we won't go into any detail here, if $J_x \neq J_y$, then S_z is not a good quantum number, and even the ground state contain admixture of order $[(J_x - J_y)/(J_\parallel - J_\perp)]^2$ of states with $S_z = (Ns-2)$, $(Ns-4)$,... This absence of S_z as a constant of the motion greatly complicates the analysis of excited states as well.

If one considers more highly excited states, with more than one unit of spin reversal, the situation again becomes more complicated. The simplest and most generally useful approximation is to treat the spin waves as independent non-interacting boson excitations, so that the energies are simply additive

$$E-E_o = \Sigma_k n_k \varepsilon_k \tag{10}$$

where n_k is the occupation number of the spin wave state k. In fact, spin waves do interact, so that this is not rigorously true. A dramatic example which we shall discuss later is the case of the completely anisotropic linear chain with $J_\perp = 0$. Then we can rigorously consider localized excitations. Two separate single excitations have energy

$$2\varepsilon_1 = 2(4sJ_\parallel) \tag{11}$$

as does a double excitation at a single site (for $s > 1/2$). If instead the two spin deviations are on adjacent sites, the excitation energy is

$$\varepsilon_2 = (4sJ_\parallel)(2 - \frac{1}{2s}) \tag{12}$$

lower than $2\varepsilon_1$ by $2J_\parallel$. In the extreme case of $s = 1/2$ corresponding to the Ising model, we see that $\varepsilon_2 = \varepsilon_1$. That is, it costs no more energy to flip over two adjacent spins than one. In fact, a cluster of any number of consecutive spins in the chain can be flipped for this same energy $4sJ_\parallel = 2J_\parallel$ which is just the energy associated with the "bonds" broken at the two ends of the cluster. As we shall see later, this situation is well approximated in $CoC\ell_2 \cdot 2H_2O$, but that is a rather special case.

If $s > 1/2$, if there are more than two neighbors to a given spin, or if the anisotropy is less complete, the interactions lead to less dramatic effects. The possibility of two-magnon bound states in the case of isotropic, or Heisenberg, exchange has been discussed by Wortis and Fukuda,[1] and Wortis has considered the effects of allowing $J_\perp < J_\parallel$. For a given system, the interaction effects are made most observable when the interaction producing the magnons preferentially flips adjacent spins. This effect, pointed out recently,[2] will be discussed later in connection with specific processes.

Having taken this brief look at some of its limitations, we now return to the approximation that the response of an ordered magnetic system to the far infrared radiation can be discussed in terms of exciting simple non-interacting spin waves. To conserve momentum, $\vec{k}$ of the spin wave must equal the $\vec{k}$ of the absorbed photon. In the far infrared, this is a typically $10^2 cm^{-1}$. This is negligible compared to the size of the Brillouin zone in k-space, namely, $\pi/a \sim 10^8 cm^{-1}$. Thus, for practical purposes, we can say that far infrared absorption probes the spin wave spectrum only at k=0. Only by considering processes in which a single photon creates more than one excitation (for example, two magnons) can the rest of the spin wave dispersion relation be investigated.

Returning to our ferromagnetic linear chain with uniaxially anisotropic exchange, the only excitation at k=0 has energy

$$\varepsilon_o = 4s(J_{\parallel} - J_{\perp}) \tag{13}$$

If a magnetic field H_o were applied along the z (or $\parallel$) axis, it would add a Zeeman term

$$\mathcal{H}_z = -g_{\parallel}\beta HS_z = -g_{\parallel}\beta H\sum_i s_{iz} \tag{14}$$

to the Hamiltonian. Since S_z is a good quantum number of the field free Hamiltonian $\mathcal{H}_o$, it remains one for $\mathcal{H}_o+\mathcal{H}_z$. Thus, the eigenvalues are just shifted by $-g\beta HS_z$. The ground state energy shifts by $-g_{\parallel}\beta HNs$, and each single spin wave state shifts by $-g_{\parallel}\beta H(Ns-1)$. Thus, the excitation energies become

$$\varepsilon_k = \varepsilon_k(H=0) + g_{\parallel}\beta H \tag{15}$$

In particular, the observable k=0, or uniform precession, spin wave has energy

$$\varepsilon_o = 4s(J_{\parallel} - J_{\perp}) + g_{\parallel}\beta H \tag{16}$$

rising linearly with H. [Of course, if the field were reversed the energies would decrease with H, but in that case the ground state would be metastable against a reversal of all the spins. Such a regime could be set up by cooling a sample through T_c in a field of one sense (to eliminate the ambiguity of sense when the magnetized state was set up) and then reversing the field while preventing the sample from rotating. Because of the energy barrier against a macroscopic reorientation of all the spins, it may be possible to observe spectra in the metastable configuration even over long periods of time.]

II. ABSORPTION INTENSITY

It is convenient to describe the intensity of the absorption associated with magnon creation in terms of the complex susceptibility χ of the material. The real and imaginary parts of χ are related by the Kramers-Kronig relations:

$$\chi'(\omega) = \frac{2}{\pi}\int_0^\infty \frac{\chi''(\omega_1)\omega_1 d\omega_1}{\omega_1^2-\omega^2}, \quad \chi''(\omega) = \frac{-2\omega}{\pi}\int_0^\infty \frac{\chi'(\omega_1)d\omega_1}{\omega_1^2-\omega^2} \tag{17a,b}$$

The first of these exactly parallels the structure of the expression for the susceptibility derived by perturbation theory in quantum mechanics, namely

$$\chi'_{ii}(\omega) = \frac{2}{\hbar}\sum_{\substack{n,m \\ \text{(pairs)}}} (f_n-f_m)\frac{\omega_{mn}}{\omega_{mn}^2-\omega^2}|(n|\mu_i|m)|^2 \tag{18}$$

where f_n is the fractional occupancy of system state n, and $(n|\mu_i|m)$ is the matrix element of the magnetic dipole operator connecting the states n and m. The normalization here is that χ is the ratio of the magnetic moment of the system to the applied field. This allows the identification

$$\chi''_{ii}(\omega)\Delta\omega|_{mn} = \frac{\pi}{\hbar}(f_n - f_m)|(n|\mu_i|m)|^2 \tag{19}$$

as the integrated intensity associated with the nm transition. The susceptibility per unit volume, χ_{vol}, enters into the attenuation coefficient α;

$$\alpha = 4\pi\chi''_{vol}\frac{\omega}{c}\sqrt{\varepsilon} \tag{20}$$

Applying this procedure to the case of our ferromagnetic system, at T=0 for simplicity, we need the matrix elements of μ_x(say) between the ground state Φ_o and the one magnon k=0 excited state Ψ_o. This is

$$(\Psi_o|\mu_x|\Phi_o) = (\Psi_o|\sum_j g_x\beta s_{jx}|\Phi_o)$$

$$= g_x\beta(\Psi_o|\sum_j \frac{s_{j+} + s_{j-}}{2}|\Phi_o)$$

But $s_{j+}\Phi_o = 0$ since all spins are aligned in the ground state. Thus

$$(\Psi_o|\mu_x|\Phi_o) = \frac{g_x\beta}{2}(\Psi_o|\sum_j s_{j-}|\Phi_o)$$

$$= \frac{g_x\beta}{2}\sqrt{2sN}\;(\Psi_o|\Psi_o)$$

$$= g_x\beta\sqrt{\frac{Ns}{2}}$$

Thus

$$\chi''_{vol}(\omega)\Delta\omega = \frac{\pi}{2\hbar}\frac{N}{V}g_x^2\beta^2 s \quad (21)$$

leading to

$$\alpha(\omega) \approx 2\pi^2 g_x^2 s\sqrt{\varepsilon}\,\frac{N}{V}\frac{\beta^2}{\hbar c}\frac{\omega}{\Delta\omega} \quad (22)$$

over the line width $\Delta\omega$.

Putting in reasonable values $g_x=2$, $s=1/2$, $\sqrt{\varepsilon} = 2$, $N/V=10^{22}$, we find

$$\alpha \approx 3(\frac{\omega}{\Delta\omega})\ \text{cm}^{-1} \quad (22a)$$

Thus for a fractional line width of 1%, the peak absorption is essentially total for samples more than a small fraction of a millimeter thick.

Although this purely quantum calculation of $\chi(\omega)$ is of some pedagogical interest in developing familiarity with magnon processes, the result is much more easily obtained in a semi-classical way using the Bloch torque equation

$$\frac{d\vec{M}}{dt} = \gamma\vec{M} \times \vec{H}_{eff} \tag{23}$$

for the total moment of the sample. (Here the gyromagnetic ratio $\gamma = g\beta/\hbar$). The anisotropic exchange favoring the z-direction can be replaced by a phenomenological anisotropy field $\vec{H}_A$ along z. If there is also an external static field H_o along z as well as the weak oscillating field $\vec{h}$ along x, we have three contributions to $\vec{H}_{eff}$, and

$$\frac{d\vec{M}}{dt} = \vec{M} \times (\vec{H}_A + \vec{H}_o + \vec{h}_x)$$

Assuming the driving field has frequency ω, this becomes in component form

$$i\omega M_x = \gamma M_y H_z$$

$$i\omega M_y = -\gamma M_x H_z + \gamma M_z h_x \tag{24}$$

where $H_z = H_A + H_o$. Neglecting h_x, the only frequency consistent with non-zero M_x or M_y is

$$\omega_o = \gamma H_z = \gamma H_A + \gamma H_o \tag{25}$$

corresponding exactly to the eigenenergies in (16). If h_x is included, we find

$$\chi'_{xx}(\omega) \equiv \frac{M_x}{h_x} = \frac{\gamma M \omega_o}{\omega_o^2 - \omega^2} \tag{26}$$

Comparing with (17a), we see that

$$\chi''_{vol}\Delta\omega = \frac{\pi}{2}\gamma M = \frac{\pi}{2}\left(\frac{g\beta}{\hbar}\right)\left(\frac{N}{V} g\beta s\right) = \frac{\pi}{2\hbar}\frac{N}{V} g^2\beta^2 s \tag{27}$$

in exact agreement with the quantum result (21).

Because of the great simplicity of this semiclassical approach, we shall usually prefer to use it whenever possible. It must be recalled, however, that some care will be required to find the appropriate phenomenological parameters to use. For example, in the present example we neglected the anisotropy of the g-tensor to get a simpler treatment. When in doubt, it is always necessary to return to a quantum mechanical treatment in which (23) is replaced by something of the form

$$\frac{ds_i}{dt} = \frac{i}{\hbar}[\mathcal{H}, s_i] \equiv \frac{i}{\hbar}[\mathcal{H}s_i - s_i\mathcal{H}] \tag{28}$$

In the present problem with $H_o = 0$, for example

$$\frac{dS_x}{dt} = \frac{d}{dt}\sum_i s_{ix} = \frac{i}{\hbar}[\mathcal{H}, \sum_i s_{ix}]$$

$$= \frac{-2i}{h}[\sum_i (J_y s_{iy} s_{i+1,y} + J_z s_{iz} s_{i+1,z}) \quad , \sum_j s_{jx}]$$

$$= \frac{-4i}{h}\{\sum_i J_y s_{i+1,y}[s_{iy}, s_{ix}] + J_z s_{i+1,z}[s_{iz}, s_{ix}]\}$$

so

$$\frac{dS_x}{dt} = -4\sum_i (J_y s_{i+1,y} s_{iz} + J_z s_{i+1,z} s_{iy})$$

In the linear spin wave approximation, all $s_{iz} = s$. Hence this reduces to

$$\frac{dS_x}{dt} = 4(J_z - J_y)s_z \sum_i s_{iy} = 4(J_z - J_y)s_z S_y$$

$$= 4(J_z - J_y)(\vec{S} \times \vec{s})_x$$

$$\equiv \gamma(\vec{S} \times \vec{H}_A)_x \tag{29}$$

where H_A is along z and has magnitude $4(J_z - J_y)s/\gamma$. With this indentification, (25) agrees with (13), as it must if the approaches are to be consistent. Since the value of γ cancels here, it is convenient to take it equal to $\gamma_{\parallel}$. Then the effect of an external field H_0 along z can be included by simply adding it to H_A, leading to

$$\frac{d\vec{S}}{dt} = \gamma_{\parallel} \vec{S} \times (\vec{H}_A + \vec{H}_0) \qquad (30)$$

III. ANTIFERROMAGNETIC RESONANCE

Consider first a classic two sublattice antiferromagnet with spins aligned along an easy axis. (Examples are FeF_2 and MnF_2, whose structure is shown in Fig. 1.) The ground state will be one with $S_z = 0$. In the simplest approximation, the Néel state, each sublattice is fully aligned with $S_{1z} = -S_{2z} = (N/2)s$, there N is the total number of spins. Unlike the ferromagnetic case, this is not an exact ground state. The true ground state has admixed in its components with $|S_{1z}|$ and $|S_{2z}| < Ns/2$, but with $S_z = S_{1z}+S_{2z}=0$ still. For most purposes we can neglect these admixtures, which typically amount to only a few percent, and think in terms of the Néel ground state. The elementary excitations of the system will have S_z values $\pm$ 1, but in general are not confined to one sublattice or the other. Rather, the expectation values of S_{1z} and S_{2z} are both changed, with a net change which must be exactly integral. These excitations were first predicted in 1951 by Kittel[3] and Nagamiya.[4]

If we confine our attention to simple antiferromagnetic resonance in which one magnon is excited at k=0, we can readily derive the resonance frequency by the phenomenological Bloch equation

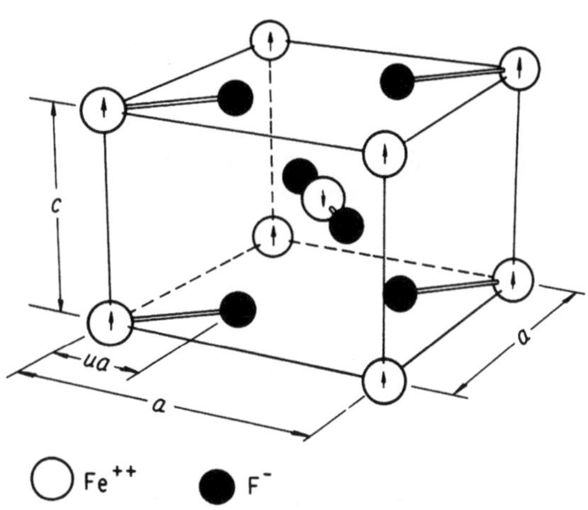

Fig. 1 Unit cell of rutile structure antiferromagnets such as FeF_2 and MnF_2.

approach. At k=0, all spins on each sublattice move as a unit, and we have only to consider the torques acting on them as a unit. If we consider an anisotropy field H_A on each sublattice, an external field H_o, and an exchange field $\vec{H}_{exi} = -\lambda\vec{S}_j$ acting between the sublattices, we have

$$\frac{d\vec{S}_1}{dt} = \gamma\vec{S}_1 \times (\vec{H}_{A1} + \vec{H}_o + \vec{H}_{ex1})$$
$$\frac{d\vec{S}_2}{dt} = \gamma\vec{S}_2 \times (\vec{H}_{A2} + \vec{H}_o + \vec{H}_{ex2}) \tag{31}$$

where $\vec{H}_{A1} = -\vec{H}_{A2} = H_A$. Assuming an $e^{i\omega t}$ time dependence, and taking components, we have

$$\frac{i\omega}{\gamma} S_{1x} = S_{1y}(H_A + H_o + H_{ex}) + H_{ex} S_{2y}$$
$$\frac{i\omega}{\gamma} S_{1y} = -S_{1x}(H_A + H_o + H_{ex}) - H_{ex} S_{2x}$$
$$\frac{i\omega}{\gamma} S_{2x} = S_{2y}(-H_A + H_o - H_{ex}) - H_{ex} S_{1y} \tag{32}$$
$$\frac{i\omega}{\gamma} S_{2y} = -S_{2x}(-H_A + H_o - H_{ex}) + H_{ex} S_{1x}$$

where we set $H_{ex} = \lambda S_{1z} = -\lambda S_{2z}$ for convenience. We see that these four coupled equations simplify to two on going to circular polarized components such as $S_- = S_x - iS_y$. Namely,

$$\frac{\omega}{\gamma} S_{1-} = (H_A + H_o + H_{ex})S_{1-} + H_{ex} S_{2-}$$
$$\frac{\omega}{\gamma} S_{2-} = (-H_A + H_o - H_{ex})S_{2-} - H_{ex} S_{1-}$$

or

$$\left(-\frac{\omega}{\gamma} + H_A + H_{ex} + H_o\right)S_{1-} + H_{ex}\,S_{2-} = 0$$

$$\left(-\frac{\omega}{\gamma} - H_A - H_{ex} + H_o\right)S_{2-} - H_{ex}\,S_{1-} = 0 \tag{33}$$

The eigenfrequencies are found by solving the secular equation for non-trivial solutions:

$$\begin{vmatrix} -\frac{\omega}{\gamma} + H_o + H_A + H_{ex} & H_{ex} \\ -H_{ex} & -\frac{\omega}{\gamma} + H_o - H_A - H_{ex} \end{vmatrix} = 0 \tag{34}$$

or

$$\left(-\frac{\omega}{\gamma} + H_o\right)^2 - (H_A + H_{ex})^2 + H_{ex}{}^2 = 0$$

leading finally to

$$\omega = \gamma\left[H_o \pm \sqrt{2H_{ex}H_A + H_A{}^2}\right] \tag{35}$$

for the two antiferromagnetic resonance frequencies.

If we carried this through with S_{i+} it would only change the sign of ω, so as to have the same physical result. Thus, we have two distinct modes with opposite sense of circular polarization (since $\sqrt{2H_{ex}H_A + H_A{}^2} > H_o$). They are degenerate at $H_o = 0$ and split symmetrically by $\pm\,\gamma H_o$ in a field. These two modes correspond to the two types of k=0 magnons that are possible, with $S_z = \pm\,1$, corresponding to the field dependent energies found here. The structure of the modes at resonance is indicated by the fact that

$$\frac{S_{1-}}{S_{2-}} = \pm\sqrt{2x + x^2} - x - 1 \tag{36}$$

where $x = H_A/H_{ex}$. This tends to -1 for x=0 and to 0,-∞ for x=∞. For the typical case x=0.1, S_{1-}/S_{2-} = -1.56, -0.64. Thus the ratio of the decreases in $|S_{iz}| \sim S_{i-}^2$ is 2.43:1. For one quantum of excitation, then, $\Delta S_{1z} = -1.7$ (or -0.7) while $\Delta S_{2z} = +0.7$ (or 1.7) so that both sublattices are substantially involved.

This result for the field-free resonance frequency can be put into a form which relates more directly to macroscopic properties by reexpressing H_{ex} and H_A. The perpendicular susceptibility of an antiferromagnet is a measure of these parameters since both of them resist the torque of the external field if it is oriented perpendicular to z. In equilibrium, the torque on a sublattice is zero, so the sum of fields perpendicular to each sublattice magnetization is zero. If each sublattice tips through an angle θ, this requirement leads to

$$H_{\perp} \cos\theta - H_{ex} \sin 2\theta - H_A \sin\theta = 0$$

or

$$\theta = \frac{H_{\perp}}{2 H_{ex} + H_A}$$

for small angles. Thus

$$\chi_{\perp} = \frac{M_{\perp}}{H_{\perp}} = \frac{2M\theta}{H_{\perp}} = \frac{2M}{2 H_{ex} + H_A} \tag{37}$$

where M is a sublattice magnetization. Using (37), (35) becomes (for $H_o=0$),

$$\omega = \gamma \left(\frac{2H_A M}{\chi_{\perp}} \right)^{1/2}$$

But $H_A M$ is equal to K, the anisotropy constant defined so that the energy increase on deviating $\vec{M}$ from equilibrium by θ is $E=1/2\ K\theta^2$. Thus, our final result , allowing for anisotropic g-values is

$$|\omega| = \gamma_{\parallel} \left(\frac{2K}{\chi_{\perp}} \right)^{1/2} \pm \gamma_{\parallel} H_o \tag{38}$$

[We do not consider in detail corrections to the field dependence that come in for $T > 0$, in which case $\alpha \equiv \chi_{\parallel}/\chi_{\perp} \neq 0$, so that H_o introduces an asymmetry in the sublattice magnetizations. The effect is essentially to replace H_o by $(1-\alpha)\ H_o$.] Because this form is in terms of quantities of physical significance (rather than model parameters like H_{ex}) it appears to have rather general validity. Since $\chi_{\perp}$ can be measured rather easily in a DC measurement, the contribution of a far-infrared spectroscopic measurement of $\omega(H_o)$

is to determine $\gamma_{\parallel}$ and the anisotropy constant K, quantities relatively inaccessible to static measurement.

Although we have not computed the high-frequency susceptibility, the order of magnitude of χ''_{max} must be $\chi_{\perp}(\omega/\Delta\omega)$ as is seen from the Kramers-Kronig relation (17a) after noting that the two AFMR modes give the only contribution to the integral.

The first far-infrared application of this theory was made in FeF_2 by Ohlmann and Tinkham.[5] This material has $T_N = 78^{\circ}$ K, and the resonance could be followed from $\sim 1.5^{\circ}$K up to $\sim 70^{\circ}$K. The resonance frequency dropped slowly from its value (52.7 cm^{-1}) at T = 0, reflecting the decrease expected in K(T). It also broadened rapidly, making observation difficult at the higher temperatures. The $g_{\parallel}$ value was found to be 2.25 $\pm$ 0.05 in agreement with earlier EPR work[6] on Fe^{++} as a paramagnetic impurity in $Zn\ F_2$, a host of similar crystal structure. The anisotropy constant K also agreed reasonably well with the large value expected from the single-ion anisotropy term DS_z^2 found in the earlier work. This anisotropy originates from the crystalline electric field via spin-orbit coupling. Thus, its origin is quite different in detail from the anisotropic exchange treated above, since the latter arises only in the interaction between two ions.

Another source of anisotropy energy is the dipolar interaction among the magnetic ions. If they are not arranged in a cubic array, the interaction energy depends on their absolute orientation with respect to crystal axes even if they stay parallel to each other. The magnitude of this term can be estimated quite well by suitable lattice sums. Its order of magnitude is limited by a rather weak magnetic interaction to an $H_A \sim 10^4$ Oe, corresponding to a frequency $\sim 1\ cm^{-1}$. In $Fe\ F_2$ this contribution to the anisotropy is small compared to the DS_z^2 term. On the other hand, in $Mn\ F_2$, the DS_z^2 term is negligible. As a result of having only dipolar anisotropy, the AFMR frequency in MnF_2 is only $\sim 9\ cm^{-1}$ compared to 53 cm^{-1} in FeF_2. Since both materials have the same structure and similar T_N (corresponding to similar H_{ex} or $\chi_{\perp}$), this difference rather directly reflects the different amounts of anisotropy.

A somewhat different case of AFMR is that typified by NiO and MnO. In these, there is an easy (111) plane with only weak anisotropy forces preferring one direction over another within the plane. Small oscillations of M_1 and M_2 about their equilibrium positions now sense very different "anisotropy fields" for their two orthogonal motions, in-plane and out-of-plane. Thus, there is no longer axial symmetry, and excitations need not have $S_z = \pm 1$. In fact it turns out that the two normal modes are non-degenerate and both have $S_z = 0$ at $H_o = 0$, so that the frequencies shift only quadratically with an applied field. The modes are linearly polarized, parallel and perpendicular to the hard axis. The anisotropies are largely dipolar in origin; hence H_A is small compared to H_{ex}. Dropping H_A compared to H_{ex}, the two modes have frequencies of the form

$$\omega_1 = \gamma\sqrt{2H_{ex}H_{A1}} \quad , \quad \omega_2 = \gamma\sqrt{2H_{ex}H_{A2}} \tag{39}$$

where H_{A1} and H_{A2} refer to the in-plane and out-of-plane anisotropy, respectively. Because H_{A1} is so small, ω_1, is expected to lie in the microwave region, but has not been observed. As shown by Sievers and Tinkham,[7] ω_2 at $T \approx 0$ falls at 27.5 cm^{-1} for MnO compared to a value of 29 cm predicted[8] from the measured $\chi_\perp$ and the calculated dipolar anisotropy. For NiO the corresponding measured[7] and theoretical[8] frequencies are 36 cm^{-1} and 43 cm^{-1}, again in reasonable agreement. In both materials, magnetic fields of 10 kOe produced no observable shift or broadening of the line. This is consistent with the fact that the excitations are linearly polarized and have $S_z = 0$.

Finally, the temperature dependence of the resonant frequencies was measured in both cases up to about 0.85 T_N. Since $\chi_\perp$ is essentially independent of temperature, $\tilde{\nu}(T)/\tilde{\nu}(0)$ should be a measure of $[K(T)/K(0)]^{1/2}$. According to the $\ell(\ell+1)/2$ rule first proposed by Zener[9] and discussed by many others, $K(T)$ should vary as $M(T)^{\ell(\ell+1)/2}$ where ℓ is the order of the spherical harmonic of the appropriate anisotropy. For this simple uniaxial anisotropy $\sim(3\cos^2\theta - 1)$, $\ell = 2$. Thus, we expect $\nu \sim K^{1/2} \sim M^{3/2}$. If one assumes $M(T)$ follows the appropriate Brillouin function, the agreement with experiment is very bad. Rather, $M(T)$ must fall much more slowly at first as T increases. Independent evidence both from the intensity of neutron scattering ($\sim M^2$) and from temperature-dependent lattice distortion (also $\sim M^2$) confirms that this is the case. In fact, if one compares the temperature dependence of $M^{3/2}$ from neutron data with the temperature dependence of ν_{AFMR}, one finds very satisfactory agreement both for MnO and NiO. This is illustrated in Fig. 2 showing data for MnO compared with theoretical curves based on $M(T)$ as determined from neutron scattering. The "Zener approximation" is a plot of $M^{3/2}$ whereas the "molecular field" curve is a plot of M itself. Note that the $M^{3/2}$ curve gives a good fit over a substantial range of temperature.

IV. EXCHANGE RESONANCE

In 1953 Kaplan and Kittel[10] pointed out that in a ferrimagnet (a material with opposite but unequal spins), there is an observable resonant mode with frequency directly proportional to the exchange constant. This contrasts with the situation in simple antiferromagnets where a geometric mean of H_{ex} and H_A is measured and with

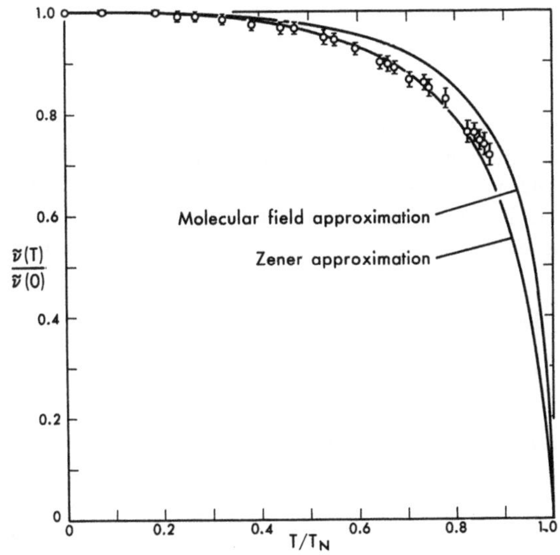

Fig. 2 Temperature-dependence of antiferromagnetic resonance frequency in MnO compared with the dependence predicted by the molecular field and Zener approximations. The square root of the neutron scattering intensity has been used for the temperature dependent of the sublattic magnetization.

simple ferromagnets where only H_A is measured.

To derive the resonance condition we start with coupled equations of motion similar to (31), namely, in zero field,

$$\frac{d\vec{M}_1}{dt} = \gamma_1 \vec{M}_1 \times (-\lambda \vec{M}_2)$$

$$\frac{d\vec{M}_2}{dt} = \gamma_2 \vec{M}_2 \times (-\lambda \vec{M}_1) \tag{40}$$

where we have shifted to the usual convention of expressing the molecular exchange field H_{ex} as λM_i, rather than λS_i. Taking an $e^{i\omega t}$ time dependence, and solving, one finds the eigenfrequencies given by

$$\omega = 0, \ \omega_e$$

where ω_e is the Kaplan-Kittel exchange resonance frequency

$$\omega_e = \lambda(\gamma_2 M_1 - \gamma_1 M_2) \tag{41}$$

The root at $\omega = 0$ refers to the ferrimagnetic resonance or uniform precession of the total moment. No exchange energy enters the frequency because the moments remain collinear. It has zero frequency here since we have included neither external field nor anisotropy. In the exchange resonance mode, one can easily see that

$$\frac{M_{1-}}{\gamma_1} + \frac{M_{2-}}{\gamma_2} = 0 \tag{42}$$

so the transverse angular momenta $\sim M_i/\gamma_i$ cancel. However the transverse <u>magnetic moment</u> does not cancel, since $\gamma_1 \neq \gamma_2$. In fact the oscillating moment and high frequency susceptibility are proportional to $(\gamma_1 - \gamma_2)^2$, the exact relation for the integrated absorption being

$$\chi''\Delta\omega = \frac{\pi}{2}\,\frac{\lambda M_1 M_2 (\gamma_1 - \gamma_2)^2}{\omega_e} \tag{43}$$

A. The Rare Earth Iron Garnets

The first system in which such resonance were studied extensively was the rare earth iron garnet system, with chemical formula $5Fe_2O_3 . 3R_2O_3$ where R is a trivalent rare earth ion. The actual crystallographic unit cell contains four such formula units. The ferric ions are located on 16 octahedral (a) sites and 24 tetrahedral (d) sites; the rare earth ions located on 24 (c) sites. Below the Curie point ($\sim 550°K$) the ferric ions order into opposing sublattices (a) and (d), with a net moment from 8 ions per unit cell or 2 per formula unit. Since the g value is 2, this leads to an iron magnetization of 10 β per formula unit. The rare earth ions are coupled by relatively weak antiferromagnetic exchange to the iron ions of sublattice <u>d</u>, tending to align their spins against the net moment of the iron. Because of the relative weakness of this coupling, significant rare earth moment develops only below $\sim 50°K$, at which temperatures the iron sublattice can be considered fully ordered.

Although the rare earth ions are very anisotropic, which plays a key role in some properties, for a first orientation we ignore this, treating them all as equivalent and isotropic. Then we can apply the simple exchange resonance model, treating the iron as one sublattice and the rare earths as the other. One then expects a resonance at

$$\omega_e = \lambda(\gamma_R M_{Fe} - \gamma_{Fe} M_R)$$

In the temperature range of interest, M_{Fe} is constant, but M_R falls away toward zero as T increases. Then, the resonance frequency should rise with increasing T, approaching the precession frequency of a single rare earth ion in the iron exchange field, namely $\gamma_R(\lambda M_{Fe})$. The total rise should be a fraction

$$\frac{\gamma_{Fe} M_R}{\gamma_R M_{Fe}} = \frac{(M_R/\gamma_R)}{(M_{Fe}/\gamma_{Fe})} = \frac{J_R}{J_{Fe}}$$

of the final frequency; that is, the ratio of the angular momenta. In the usual case, the rare earth is characterized by a spin of 1/2 per ion since a single Kramers doublet lies lowest. Since there are 6 R ions per formula unit, we have

$$\frac{J_R}{J_{Fe}} = \frac{6 \times 1/2}{2 \times 5/2} = \frac{3}{5}$$

for this ratio. In other words, $\omega_e(T=0)$ should be 2/5 of the limiting value of ω_e at high temperature.

A temperature dependent resonance of this sort was first observed by Sievers and Tinkham[11] in YbIG, some data being shown in Fig. 3. However, the temperature dependence was not so great as expected from this model, and other temperature independent lines were observed. These properties were explained by extensions of the model[12] to take account of anisotropy and of the many inequivalent rare earth sites. The anisotropy results from anisotropy in the exchange coupling and can be described for small displacements from the (111) easy axis by a simple uniaxial anisotropy energy, equivalent to an anisotropy field. This raises the resonant frequency for $T \approx 0$, but not for high temperatures, since the anisotropy energy falls as a power of the rare earth magnetization which goes to zero as T increases. From independent information of Henderson and White[13] we could estimate that this anisotropy correction raised $\omega_e(0)$ from $\sim 10\ cm^{-1}$ to the observed value of $14\ cm^{-1}$. The $10\ cm^{-1}$ value was then $\sim 2/5$ of the high frequency limiting value, estimated from the data to be $\sim 25\ cm^{-1}$.

The other two, temperature - independent, lines observed were identified as the single-ion frequencies of the two inequivalent rare earth sites appropriate to the equilibrium situation when the iron magnetization and exchange field were aligned along the (111) easy axis. The problem was to explain why these frequencies should

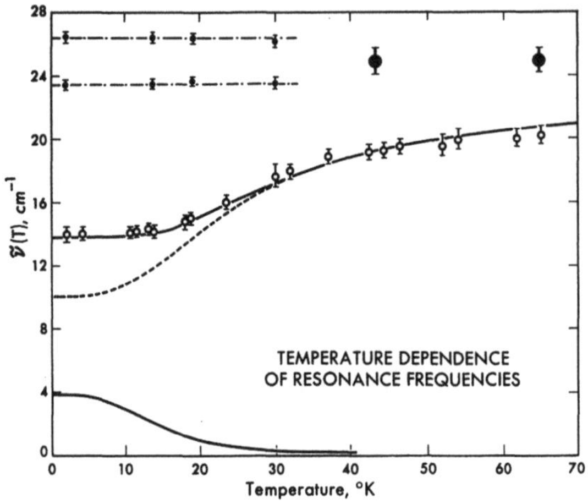

Fig. 3 Temperature dependence of the resonance frequencies in YbIG for the three strong lines. The two highest frequencies are temperature independent. Above 30°K only the center of gravity of the doublet could be measured. The dashed curve gives the calculated temperature dependence of the exchange frequency in the isotropic two-sublattice approximation. The solid curve demonstrates the good agreement between theory and experiment when anisotropy energy is introduced. The low-frequency curve is the predicted "ferri-magnetic" resonance with the same model.

be observable, since single ion transitions do not correspond to k=0 spin wave modes. The key observation for resolving this question is that with 24 rare earth ions per crystallographic unit cell, there are many "optical" k=0 modes in the exact solution. Even without attacking such a formidable problem, considerable insight can be obtained by solving a simple three sublattice linear chain problem. The three sublattices correspond to the iron and the two types of rare earth sites which have different g values along the (111) direction. The three k=0 spin wave solutions correspond to 1) the uniform precession ferromagnetic resonance at $\omega = 0$ (in absence of anisotropy), 2) the Kaplan-Kittel exchange resonance with averaged rare earth parameters, and 3) a new mode at $\omega = \bar{\gamma}\,\bar{\lambda}\,M$; the average single ion precession frequency.

We now generalize the model further by noting that the <u>transverse</u> g-values of all ions having the same longitudinal g-value need <u>not</u> be the same. Assuming there are two different transverse g-values at alternate sites along the chain, the true unit cell now contains 6 ions and has 6 spin waves at k=0. But if successive ions differ only in transverse g-value, it is reasonable to assume that the frequencies of these new k=0 modes are essentially the same as the old modes at $k=\pi/a$, modes which "fold back" to k=0 in the new smaller k-space cell. These three "new" modes have frequencies given approximately by

$$\Omega_{\pi/a}, \quad \omega_2, \quad \omega_3$$

where $\Omega_{\pi/a}$ in the very high frequency zone boundary frequency characteristic of the iron-iron exchange, and ω_2 and ω_3 are the single ion frequencies of the two types of rare earth ions in the iron exchange field. The intensity of all these modes near the single ion frequencies depends on the difference in transverse g-factors. Although no quantitative test was possible, this picture is confirmed qualitatively by the absence of any observable resonances in GdIG. This is a special case because both Gd^{3+} and Fe^{3+} are S-state ions, with g=2 and isotropic. Thus, neither the Kaplan-Kittel resonance, nor the "single-ion" transitions should be observable in GdIG, since both require differences in transverse g-values to be observable. The absence of any resonance in this material thus confirms the general picture we have developed here.

V. TWO MAGNON EXCITATIONS

In 1965 Halley and Silvera[14] discovered that absorption processes in which two magnons were simultaneously created could be observed in FeF_2 at 154 cm^{-1}. This frequency is much higher than the ordinary AFMR at 53 cm^{-1} corresponding to the creation of a single k=0 magnon. In fact it is more than $2\tilde{\nu}_{AFMR}$, but less than $3\tilde{\nu}_{AFMR}$, so that the excitation could not simply be a multiple quantum or harmonic of the k=0 magnon. Moreover, polarization studies showed that the transition was an <u>electric</u> dipole transition, contrary to the magnetic dipole nature of all the simple magnetic resonances discussed above. Another characteristic of the observed absorption was that the frequency did not vary appreciably with applied fields, whereas the AFMR frequency in FeF_2 varies linearly with the field component along the c-axis. This latter property suggests that the excitation being created consists of two magnons, with $\Delta S_z = +1$ and -1 respectively, so that $\Delta S_z|_{tot} = 0$. The frequency (154 cm^{-1} compared to $\tilde{\nu}_{AFMR} = 53$ cm^{-1}) suggests that the two magnons are being produced out near the Brillouin zone boundary, where $\tilde{\nu}_k > \tilde{\nu}_o \equiv \tilde{\nu}_{AFMR}$. Because the radiation field has $\vec{k}=0$, the two magnons must have essentially equal and opposite $\vec{k}$ to conserve momentum in the process. Since the amount of phase space volume per unit $|k|$ goes as k^2, one would expect the part of the Brillouin zone near the zone boundary to dominate in a strong absorption peak although a continuous weak absorption should start already at $2\tilde{\nu}_o$ (106 cm^{-1} for FeF_2 or 18 cm^{-1} for MnF_2).

This simple discussion accounts for many of the salient features of the process. It leaves open several questions, however, An obvious one is the question of coupling mechanism; how at all does an electric dipole transition, requiring a parity change, create two magnons? More detailed questions are involved in working out absolute intensities and line shapes, taking account of the full symmetry of the problem in a group theoretical way.

To start this more careful investigation, we first consider the nature of the spin waves for the whole Brillouin zone (BZ) rather than confining our attention to k=0 as we did in discussing ordinary AFMR. In these rutile structure antiferromagnets each magnetic ion is surrounded by an octet of opposite spin ions as shown in Fig. 1. Three sets of exchange parameters are apparently required for a good fit to all available data; namely, J_1, J_2, and J_3 to the first 3 sets of neighbors. The corresponding distances are

$$R_1 = c \approx 3.3 \text{ Å}$$

$$R_2 = \sqrt{2a^2+c^2}/2 \approx 3.7 \text{ Å}$$

$$R_3 = a = 4.7 \text{ Å}$$

J_2 is antiferromagnetic, whereas J_1 and J_3 are ferromagnetic. For modes at k=0, J_1 and J_3 play no role, since they connect spins in the same sublattice. Thus, our H_E was determined by J_2 alone. However, for k≠0, J_1 and J_3 do play a role, since they connect spins which now are inclined to one another. This comes into the equations by having J_1 and J_3 enter with coefficients of the form

$$\sum_j (e^{i\vec{k}\cdot\vec{\rho}_j} - 1)$$

where the sum runs over the appropriate set of neighbors of a given ion and $\vec{\rho}_j$ is the position of the jth neighbor with respect to the ion in question. Evidently each term of the sum vanishes for k=0 so these interactions obey no role in determining the frequency of k=0 magnons. However once k gets a significant way out in the BZ, the terms attain order unity, and play a large role. Although we won't go into details here, the effect is to make magnon energies rise to values dependent directly on exchange rather than on a geometric mean of exchange and anisotropy. Also, the eigenmodes change and approach the simple form of being localized to excitations purely on one or the other sublattice. More formally, we can write the magnon creation operator for excitations with $\Delta S_z = -1$, for example, as

$$\alpha^{+}_{\downarrow\vec{k}} = (2S_N)^{-1/2}\{u_k \sum_i e^{i\vec{k}\cdot\vec{r}_i} S_{i-} - v_k \sum_j e^{i\vec{k}\cdot\vec{r}_j} S_{j+}\} \qquad (44)$$

where i runs over the "up" sublattice and j runs over the "down" sublattice. For any $S_z = -1$ magnon, $u_k^2 - v_k^2 = 1$. We noted earlier in equation (36) that for k=0, the ratio of u_k^2 to v_k^2 was typically of order unity. However as the zone boundary is approached, $u_k^2 \to 1$ and $v_k^2 \to 0$, so that the magnons are truly confined to one sublattice rather than only having a larger excitation on one than on the other.

For any given set of values for J_0, J_1, J_2 and H_A, one can compute the complete $\hbar\omega_k$ spectrum over the BZ. The parameters can be fitted from experimental data such as the AFMR frequency, $\chi_\perp$, and, if available, neutron data on $\hbar\omega_k$ at some points in the BZ. From this, the density of magnon states in energy can be computed. The result for MnF_2 is shown in Fig. 4, with an inset showing the labelling of characteristic points in the Brillouin zone. Note that the lower peak in N(E) is associated with the longitudinal midplane of the tetragonal BZ, whereas the upper peak is associated with the top plane. This is reasonable in view of the fact that

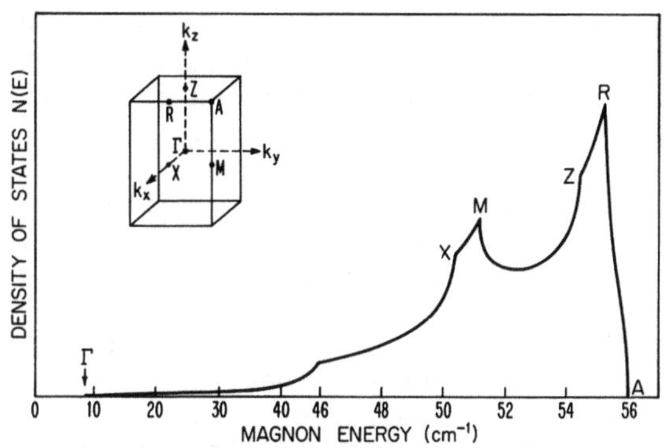

Fig. 4. Magnon density of states in MnF_2. (The apparent kink at 46 cm^{-1} arises from a change in horizontal scale at that point.) The inset shows the labelling convention for the Brillouin zone. The Γ point magnon is the usual AFMR resonance mode at k=0. Such a plot for FeF_2 would be qualitatively similar, but the Γ-point would be shifted up to 53 cm^{-1}, and the zone boundary peak to ~77 cm^{-1}, corresponding to the much higher anisotropy in FeF_2.

$|k|$ is greater on the top because of the small c/a ratio in this structure.

Now let us consider the nature of the <u>two</u> magnon states of k=0 which can be set up. There are four types:

$$
\begin{aligned}
|0,-\rangle &= |\uparrow\vec{k},\downarrow-\vec{k}\rangle - |\downarrow\vec{k},\uparrow-\vec{k}\rangle \\
|0,+\rangle &= |\uparrow\vec{k},\downarrow-\vec{k}\rangle + |\downarrow\vec{k},\uparrow-\vec{k}\rangle \\
|2,+\rangle &= |\uparrow\vec{k},\uparrow-\vec{k}\rangle \\
|2,+\rangle &= |\downarrow\vec{k},\downarrow-\vec{k}\rangle
\end{aligned}
\qquad (45)
$$

The numbers in the kets on the left side indicate ΔS_z and the $\pm$ indicates the parity, whereas the notations in the kets on the right hand side refer to the results of application of two operators of the form (44). The parity assignments can be verified by noting that inversion in one of the magnetic sites (which all are crystallographic centers of symmetry) reverses $\vec{k}$ but leaves the spins unchanged. Thus,

$$
I|\uparrow\vec{k},\downarrow-\vec{k}\rangle = |\uparrow-\vec{k},\downarrow\vec{k}\rangle = |\downarrow\vec{k},\uparrow-\vec{k}\rangle \qquad (46)
$$

for example, and the parity assignments given in (45) follow immediately. Since an electric dipole absorption must involve a parity change, only excitations of the $|0,-\rangle$ type can be produced in that case. Thus it is no accident that the observed absorption has the insensitivity to magnetic field associated with a $\Delta S_z = 0$ transition. We also see that the interaction creating the excitation must involve both sublattices equally.

More detailed selection rules apply at points, lines, and planes of special symmetry in the BZ. Loudon[15] has worked this out in detail. He finds in particular that two-magnon zero wave-vector excitations are forbidden for $\vec{k}$ at the Γ, M, Z, and R points of the zone; and that for $\vec{E}||\vec{c}$, only the corner point A contributes, whereas for $\vec{E}\perp\vec{c}$, only X contributes. When this sort of information is folded into the two-magnon density of states (which has the same shape as the one-magnon one, but with doubled energy scale), rather definite line shape predictions can be made. Also, it can be seen that the peak absorption should occur at a higher energy for $\vec{E}||\vec{c}$, where the A point contributes, than for $\vec{E}\perp\vec{c}$, where X is the highest energy critical point that contributes. These predictions are generally in good agreement with experiment, although some significant discrepancies occur in line shapes. At present it is thought that these may result from magnon-magnon interaction[2] rather than from the more extended range exchange interaction which was originally postulated.

The above discussion has skirted the question of the actual mechanism for exciting the two-magnon excitations, relying on general arguments which are valid independent of this mechanism. Three mechanisms have been proposed: 1) an electric quadrupole-dipole interaction relying on spin-orbit coupling, due to Halley[14,17]; 2) an exchange coupling scheme due to Tanabe et al.[18] and 3) a scheme based on infrared active phonons.[19] Orders of magnitude are hard to estimate with confidence. However, mechanism 1) seems unable to explain the strong two-magnon absorption in MnF_2 where spin-orbit coupling is unimportant, although it may be important in FeF_2, the material for which it was introduced. Each of these mechanisms must involve <u>two</u> ions, since parity would be a rigorous quantum number for a transition occuring on a single ion at a center of symmetry. Moreover, there must be no center of symmetry which interchanges the two ions. Given these conditions, there is no reason of symmetry to forbid the excitation, and one simply must estimate the magnitudes of the various possible processes. In any case, the coupling operator reduces to a form

$$\mathcal{H}_{ij}^{2\text{-magnon}} = \vec{E}\cdot(\vec{\pi}_{ij}+\vec{\pi}_{ji})S_{i-}S_{j+} \tag{47}$$

where i and j are ions on the two sublattices, and $\vec{\pi}_{ij}$ is the dipole moment associated with the odd-parity excited states of ion i whereas $\vec{\pi}_{ji}$ involves those of ion j. Summing this over all ion pairs, one is led to the full coupling term which excites the two-magnon electric-dipole absorption. Because $\vec{\pi}_{ij}$ falls off rapidly if ions i and j are not adjacent, this excitation process favors magnon interactions by flipping adjacent spins. As suggested by Elliott et al.[2], this will probably affect the line shape of the two magnon lines in such a way as to eliminate the need to invoke an ad hoc extended range interaction as was done by Allen et al.[16]

In conclusion, then, it appears that the theory of the two-magnon absorption is quite successful, and this sort of far-infrared absorption can give information about magnon energies near the zone boundary which complement the AFMR data on k=0 magnon energies.

VI. LOCALIZED PAIR AND CLUSTER EXCITATIONS

The most extreme case of magnon interaction was mentioned above in connection with (12). It was noted that in a linear chain of s=1/2 ions with completely anisotropic exchange coupling ($J_\perp = 0$) (i.e., an Ising model), the energy to flip any number of adjacent spins in a cluster is $2J_\parallel$. If a magnetic field is applied, these excitation energies shift by $g_\parallel \beta Hm$, where m is the number of flipped spins in the cluster. Thus, m can be readily determined by measuring the slope

$$d(h\nu)/dH = g_{\parallel} \beta m$$

where $h\nu$ is the observed excitation energy.

Although such an isolated Ising chain is rather idealized, in $CoCl_2 \cdot 2H_2O$ we have a system with approximately these features under certain conditions. Cluster excitations up to 5-fold have been observed in this system, with energies little different from single spin flips, but with 5 times the field dependence.

This material has $T_N = 17^oK$, below which it is antiferromagnetic, with moments along the b-axis. In magnetic fields along b at low temperatures it undergoes two successive metamagnetic transitions. At $H_{c1} \approx 32kOe$ it goes into a ferrimagnetic configuration with twice as many + spins as - ones. Then at $H_{c2} \approx 46kOe$ it goes into a ferromagnetic configuration with all spins parallel. The strongest exchange coupling, J_0, is along linear chains in the c-direction. This coupling is ferromagnetic, and in the metamagnetic transitions, these chains of spins always turn as units. There are two other important (antiferromagnetic) couplings J_1 and J_2 between adjacent chains. All the couplings are very anisotropic and the Co^{++} ions have a single low-lying Kramers doublet, so that $s=1/2$. Thus to a good first approximation the simple Ising model of interactions is applicable. This allows us to think in terms of localized excitations in the form of spin reversals.

Such an analysis has been applied to microwave experiments by Date and Motokawa,[20] and to the far infrared by Torrance and Tinkham.[21] The most striking feature is that for $H = H_{c2}$, the external field just cancels the molecular field of the antiferromagnetic neighbor coupling. This effectively reduces each chain to an isolated ferromagnetic linear chain in zero field. Then, as remarked above, the energy for one, two, three, or more adjacent spin flips is the same: $2J_o$. For fields above H_{c2}, these split and give a spectrum

$$h\nu = 2J_o + g_{\parallel} \beta m(H-H_{c2}) \quad (H > H_{c2}) \tag{48}$$

Similar simple relations govern the excitation energies in the other phases below H_{c2}. The results of this simple model are compared with the data in Fig. 5. The qualitative agreement is evident, and the quantitative agreement is not bad considering that after setting H_{c1} and H_{c2} to agree with experiment, the only remaining free parameters are $g_{\parallel}$ and J_o.

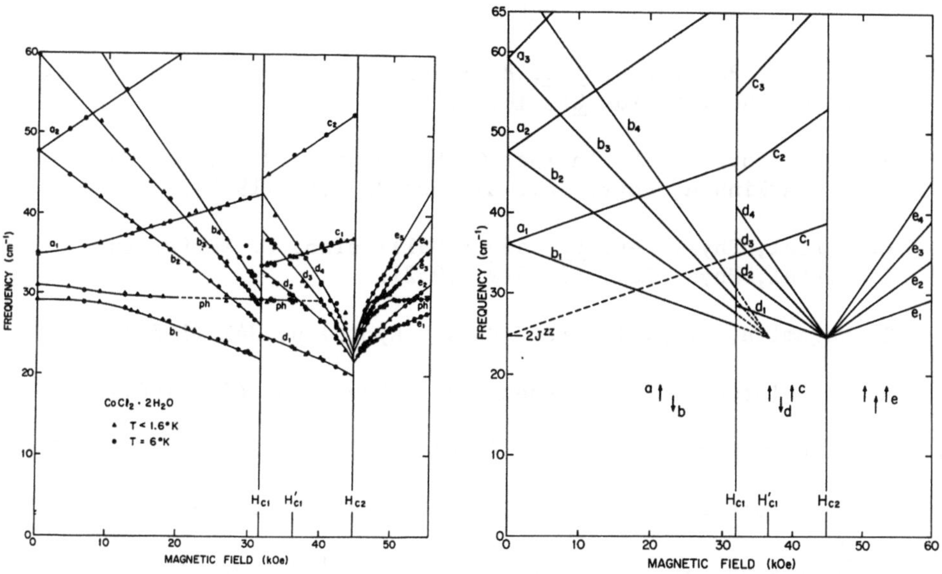

Fig. 5 (a) Measured frequencies of absorption lines in $CoCl_2 \cdot 2H_2O$. (b) Frequencies predicted by simple Ising model of interactions.

More recent better data (unpublished) has allowed a more critical comparison with theory. We find clear evidence (in an improved fit) for the residual effects of the non-zero, but small, values of $J_\perp$. These can be treated as perturbations on the simple Ising picture. Also we find evidence of a strong magnon-phonon interaction with a phonon near 30 cm^{-1}, which only gives absorption to the extent that magnon character is coupled into it. By taking all these refinements into account an excellent fit to the data can be obtained, but we won't go into that here.

In summary, this system with almost Ising linear chains shows an extremely exaggerated form of magnon-magnon interaction. Two or more magnons may be created in a strongly bound state, well approximated by a k=0 Bloch function of simple localized clusters of spin reversals. Although this is an extreme case, it does illustrate the fact that magnon-magnon coupling can produce dramatic new features in the spectrum.

REFERENCES

1. M. Wortis, Phys. Rev. 132, 85 (1963); N. Fukuda and M. Wortis, J. Phys. Chem. Solids 24, 1675 (1963).

2. R. J. Elliott, M. F. Thorpe, G. F. Imbusch, R. Loudon, and J. B. Parkinson, Phys. Rev. Letters 21, 147 (1968).

3. C. Kittel, Phys. Rev. 82, 565 (1951); also, F. Keffer and C. Kittel, Phys. Rev. 85, 329 (1952).

4. T. Nagamiya, Prog. Theoret. Phys. (Kyoto) 6, 342 (1951).

5. R. C. Ohlmann and M. Tinkham, Phys. Rev. 123, 425 (1961).

6. M. Tinkham, Proc. Roy. Soc. (London) A236, 535 (1956).

7. A. J. Sievers, III, and M. Tinkham, Phys. Rev. 129, 1566 (1963).

8. F. Keffer and W. O'Sullivan, Phys. Rev. 108, 637 (1957).

9. C. Zener, Phys. Rev. 96, 1335 (1954).

10. J. Kaplan and C. Kittel, J. Chem. Phys. 21, 760 (1953).

11. A. J. Sievers and M. Tinkham, Phys. Rev. 124, 321 (1961).

12. M. Tinkham, Phys. Rev. 124, 311 (1961).

13. J. W. Henderson and R. L. White, Phys. Rev. 123, 1627 (1961).

14. J. W. Halley and I. Silvera, Phys. Rev. Letters 15, 654 (1965).

15. R. Loudon, Adv. in Phys. 17, 243 (1968).

16. S. J. Allen, Jr., R. Loudon, and P. L. Richards, Phys. Rev. Lett. 16, 463 (1966).

17. J. W. Halley, Phys. Rev. 149, 423 (1966).

18. Y. Tanabe, T. Moriya, and S. Sugano, Phys. Rev. Letters 15, 1023 (1965).

19. J. W. Halley, Phys. Rev. 154, 458 (1967).

20. M. Date and M. Motokawa, Phys. Rev. Lett. 16, 1111 (1966).

21. J. B. Torrance, Jr., and M. Tinkham, J. Appl. Phys. 39, 822 (1968).

FAR INFRARED ABSORPTION IN SUPERCONDUCTORS

M. Tinkham

Harvard University

Cambridge, Massachusetts

I. REVIEW OF ELECTROMAGNETIC RESPONSE IN THE NORMAL STATE

A convenient characterization of the response of a metal to electromagnetic fields is the complex conductivity:

$$\sigma(\omega,q) = \sigma_1(\omega,q) - i\sigma_2(\omega,q) = \frac{i\omega\varepsilon(\omega,q)}{4\pi} \tag{1}$$

giving the current per unit electric field of frequency ω and wave number q. (We consider only the transverse conductivity, where E and J are perpendicular to q.) As indicated in (1), an equivalent description in terms of a complex dielectric constant ε is also possible. For metals in the normal state $\sigma_n(\omega,q)$ can be calculated in the free-electron approximation by applying Fourier analysis to the well-known Chambers integral expression for the current:

$$\vec{J}(\vec{r},t) = \frac{3\sigma_o}{4\pi\ell} \int \frac{\vec{R}[\vec{R}\cdot\vec{E}(\vec{r}',t')]}{R^4} e^{-R/\ell} d\tau' , \tag{2}$$

where $\sigma_o = ne^2\ell/mv_o$, $\vec{R} = \vec{r}' - \vec{r}$, $t' = t - (R/v_o)$, ℓ is the mean free path, and v_o is the Fermi velocity. At $\omega = 0$, for example, one finds $\sigma_2 = 0$, and the q-dependence of σ_1 is:

$$\sigma_{1n}(0,q) = \sigma_o f(q\ell), \tag{3}$$

where

$$f(q\ell) = \frac{3}{2q^3\ell^3} [(1 + q^2\ell^2) \tan^{-1} q\ell - q\ell] \tag{4}$$

which has the asymptotic forms:

$$f(q\ell) \to 1 - \frac{1}{5} q^2\ell^2 + \ldots \qquad (q\ell \to 0) \tag{4a}$$

$$\to \left(\frac{3\pi}{4q\ell}\right)\left(1 - \frac{4}{\pi q\ell} + \ldots\right) \quad (q\ell \to \infty) . \tag{4b}$$

Using this result one can derive the Fuchs-Sondheimer relation for the effective conductivity of the metal in a thin film.

For $\omega \neq 0$, one finds:

$$\sigma_n(\omega,q) = \frac{3\sigma_o}{4iq\ell} [(z^2 - 1) \ln \frac{z-1}{z+1} + 2z] \tag{5}$$

where $z = (\omega/qv_o)-(i/q\ell)$. Although this expression (5) is awkward, we recover a familiar elementary result by considering the limit at $q = 0$, where

$$\sigma_{1n}(\omega,0) = \frac{\sigma_o}{1 + \omega^2\tau^2} \qquad (\tau = \ell/v_o), \tag{6a}$$

$$\sigma_{2n}(\omega,0) = \frac{\sigma_o(\omega\tau)}{1 + \omega^2\tau^2} . \tag{6b}$$

Another simple result emerges if we let $\ell \to \infty$. In that case

$$\sigma_{1n}(\omega,q)\Big|_{\ell=\infty} = \frac{3\pi ne^2}{4mqv_o}\left(1 - \frac{\omega^2}{q^2v_o^2}\right) \quad \omega < qv_o, \tag{7a}$$

$$= 0 \qquad\qquad \omega > qv_o,$$

$$\sigma_{2n}(\omega,q)\Big|_{\ell=\infty} = \frac{3ne^2}{4mqv_o}\left\{\frac{2\omega}{qv_o} + \left(\frac{\omega^2}{q^2v_o^2} - 1\right) \ln\left|\frac{qv_o-\omega}{qv_o+\omega}\right|\right\} . \tag{7b}$$

The cutoff of σ_{1n} at $\omega = qv_o$ occurs because any higher frequency would imply a wave with velocity greater than v_o, which could not transfer energy to any electron while conserving momentum. It is readily verified that both (6a) and (7a) satisfy the general oscillator strength sum rule

$$\int_0^\infty \sigma_1(\omega,q)\, d\omega = \frac{1}{2}\pi\frac{ne^2}{m} \quad . \tag{8}$$

Knowing the general complex conductivity expression (5), one can in principle solve any problem of the electromagnetic response of a homogeneous medium by using Maxwell's equations and a suitable Fourier analysis of the fields involved. For example, the surface impedence of a bulk sample filling a half-space x > 0, is defined by

$$Z = R + iX = \frac{E_y\big|_{x=0}}{\int_0^\infty J_y(x)dx}. \tag{9}$$

The fractional absorption of normally incident radiation is equal to cR/π (for a metal, where $c|Z| << 1$). For diffuse surface scattering, Z is given by:

$$Z(\omega) = \frac{4\pi^2 i\omega}{c^2}\left\{\int_0^\infty \ln[1 + 4\pi i q^{-2}c^{-2}\omega\, \sigma(\omega,q)]dq\right\}^{-1}, \tag{10a}$$

and for specular surface scattering by

$$Z(\omega) = \frac{2}{\pi}\int_0^\infty \left[\sigma(\omega,q) - \frac{iq^2c^2}{4\pi\omega}\right]^{-1} dq \quad . \tag{10b}$$

Unfortunately these general expressions cannot be evaluated analytically even for the normal metal, although extensive numerical results have been tabulated. A useful limit for which an explicit solution is available is the <u>extreme anomalous limit</u>, which obtains when $\ell >> \delta$, the skin depth. In this case the important Fourier components have $q \sim 1/\delta >> 1/\ell$, and it is reasonable to approximate $\sigma(\omega,q)$ by its asymptotic form for $q\ell \to \infty$. Since σ_2 vanishes as q^{-2} (so long as $\omega << v_o/\delta$), it is neglected, and only $\sigma_{1n}(\omega,q) \cong 3\pi ne^2/4mqv_o$ is used for all q. The result is:

$$Z_{n\infty} = 3^{\frac{1}{6}}\, \pi^{\frac{1}{3}}\, (1 + \sqrt{3}\, i)\left(\frac{\omega^2\ell}{c^4\sigma_o}\right)^{\frac{1}{3}} \tag{11}$$

for diffuse surface scattering and $\frac{8}{9}$ times this for specular surface scattering. Thus, there is actually little difference in the final result for the two assumptions about the surface scattering.

The other useful simple limit is the classical skin effect limit, in which $\sigma(\omega,q)$ is treated as independent of q, having the value appropriate to $q = 0$. This limit will be valid when $q \sim 1/\delta$ is small compared to $1/\ell$ and ω/v_o. Both of these requirements are met if ℓ is small enough, since for small ℓ, $1/\delta \sim \ell^{1/2}$. Thus, this is often called the extreme dirty limit. According to (6), in this limit as $\ell \to 0$, $\sigma_{2n}/\sigma_{1n} = \omega\tau = \omega\ell/v_o \to 0$. Thus, for the normal metal, we can again neglect σ_2 compared to σ_1. This allows us to write, for specular surface scattering

$$Z = \frac{2}{\pi}\int_0^\infty \frac{dq}{\sigma_1(\omega) - i(q^2c^2/4\pi\omega)}$$

$$= \frac{i}{\pi\alpha}\int_{-\infty}^{\infty} \frac{dq}{(q-q_o)(q+q_o)}$$

where $\alpha = c^2/4\pi\omega$ and $q_o^2 = -i4\pi\omega\sigma_1/c^2$. By a simple contour integration, we find

$$Z_{n,c\ell} = (1 + i)(2\pi\omega/\sigma_1c^2)^{1/2} \tag{12}$$

Since we have assumed $\omega\tau \approx 0$, $\sigma_1 \approx \sigma_o$, the dc conductivity.

Comparing (11) and (12), we note that $Z_{n\infty} \sim \omega^{2/3}$ whereas $Z_{n,c\ell} \sim \omega^{1/2}$. We also note that $Z_{n\infty}$ is independent of ℓ (since $\sigma_o \sim \ell$), whereas $Z_{n,c_\ell} \sim \ell^{-1/2}$.

II. THE ABSORPTION EDGE IN SUPERCONDUCTORS

In the superconducting state, a complex conductivity $\sigma_s(\omega,q)$ may again be defined, but it is not given by any simple closed expression such as (5). Rather, it is given by a complicated integral expression. However, in the extreme anomalous limit, i.e. $q \to \infty$, (where $\sigma_n(\omega,q) \approx \sigma_{1n}(q) \sim 1/q$ is independent of ω, as noted above) Mattis and Bardeen[1] noted that $\sigma_s(\omega,q)$ also varied as $1/q$. Thus a well-defined ratio

$$\frac{\sigma_s(\omega,q)}{\sigma_n(\omega,q)} = \frac{\sigma_1(\omega)}{\sigma_n} - \frac{i\sigma_2(\omega)}{\sigma_n} \tag{13}$$

exists, which is dependent only on frequency, not on q (so long as q is large enough).

The response of a pure superconductor to a high-frequency field was also treated by Abrikosov et al.[2], using a powerful Green's function formulation. They treated the extreme anomalous("Pippard") limit for many limiting cases, and also some examples of the "London" ($qv \ll \Delta$) limit. No explicit results for intermediate cases are given. This treatment was extended to the case of finite mean free path by Abrikosov and Gor'kov.[3] In the limit $\ell \ll \xi_o$, they derived explicit analytic expressions equivalent to those of Mattis and Bardeen. Thus the same simple $\sigma_s(\omega)/\sigma_n$ results can be used in both our simple limits.

Numerical values of these ratios computed from the BCS theory were given by Miller[4] and more extensive calculations have been made by Waldram.[5] Confining our attention now to the absolute zero, these ratios can be expressed simply in terms of elliptic integrals as pointed out by Mattis and Bardeen. Naturally, $\sigma_1 = 0$ for $h\omega < E_g$, since quanta smaller than the energy gap cannot be absorbed by the metal. At $\omega = \omega_g = E_g/\hbar$, (σ_1/σ_n) rises from zero with finite slope such that

$$\left.\frac{d(\sigma_1/\sigma_n)}{d(\omega/\omega_g)}\right|_{\omega_g} = \frac{1}{2}\pi , \tag{14}$$

and also $\sigma_2(\omega_g)/\sigma_n = 1$. The form of these functions is shown in Figure 1a. For frequencies well above the gap, $\sigma_1/\sigma_n \to 1$ and $(\sigma_2/\sigma_n) \to 0$ as expected from the fact that the high frequency optical properties are known to be unaffected by the transition to the superconducting state.

For frequencies well below the gap, $\sigma_2(\omega)$ rises as ω^{-1}. The physical meaning of this term in $\sigma_2 \sim 1/\omega$ can be made clear by noting that it corresponds to the accelerative supercurrent described by the London equation

$$\vec{E} = \frac{\partial}{\partial t}(\Lambda \vec{J}_s) , \quad \Lambda = \frac{m_s}{n_s e^2} = \frac{4\pi}{c^2}\lambda_L^2 \tag{15}$$

where m_s and n_s are the mass and density of the superconducting electrons and λ_L is the London penetration depth. From this

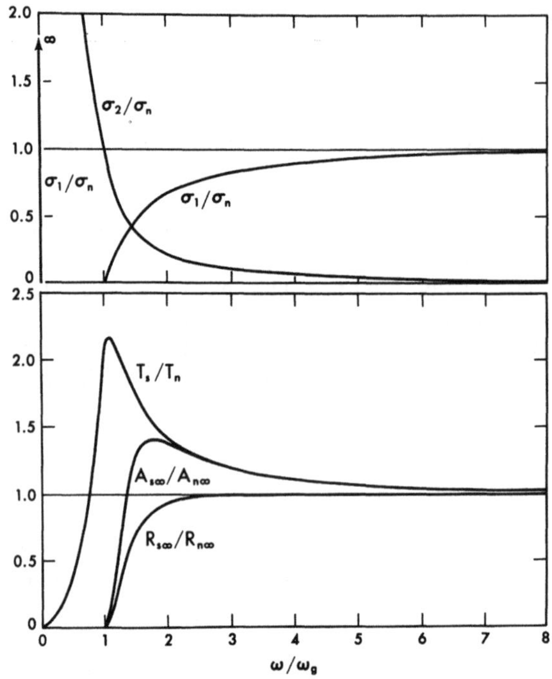

Fig. 1 (a) Ratio of complex conductivity in superconducting state $\sigma_1 - i\sigma_2$ to the real constant normal state conductivity in the extreme anomalous limit, as calculated by Mattis and Bardeen. These results are also valid in the extreme dirty limit. (b) $R_{s\infty}/R_{n\infty}$ is the ratio of the superconducting to normal surface resistance of a bulk sample, eq. (19); $A_{s\infty}/A_{n\infty}$ is the ratio of the superconducting to normal absorptivity for thin low-impedance films, T_s/T_n is the ratio of superconducting to normal transmissivity of a typical very thin film. All three curves are calculated using the extreme anomalous limit approximation of Fig. 1a.

relation,

$$\sigma_2 = \frac{1}{\Lambda\omega} = \frac{n_s e^2}{m_s \omega} \tag{16}$$

so that the $\sigma_2 \sim 1/\omega$ represents conductivity limited by inertia, without damping. Now Ferrell, Glover, and Tinkham[6] showed that a sum-rule argument allows determination of the strength of this supercurrent response from $\sigma_1(\omega)$. Namely, if $\int_{0+}^{\infty} \sigma_1(q,\omega)d\omega$ does not attain the full sum rule value $\pi ne^2/2m$, referred to in (8), then the missing area A under the integral of $\sigma_1(\omega)$ appears at $\omega = 0$ as $A\,\delta(\omega)$. The Kramers-Kronig transform of this in σ_2 is $2A/\pi\omega$. The missing area is easily estimated by inspection of

Fig. 1a and by noting that the normal conductivity σ_{1n} also must satisfy the sum rule. Thus, the missing area

$$A = \int_{0+}^{\infty} [\sigma_{1n}(\omega) - \sigma_{1s}(\omega)]d\omega = \tilde{\omega}_g \sigma_{1n},$$

where $\tilde{\omega}_g$ is of order ω_g, but somewhat larger. Thus, $\sigma_{2s}/\sigma_{1n} = 2\tilde{\omega}_g/\pi\omega \sim \omega_g/\omega$. In fact, the exact theory (in the extreme anomalous limit) gives

$$\sigma_2/\sigma_n \rightarrow (\pi/2)(\omega_g/\omega) \qquad (\omega << \omega_g) \tag{17}$$

To the extent that either the extreme anomalous limit or the extreme dirty limit is adequate, one can simply use the Mattis-Bardeen $\sigma_s(\omega)/\sigma_n$ ratios to calculate the ratio of the surface impedance of a superconductor to that of the normal state of the same metal. In the extreme anomalous limit, one finds

$$\frac{Z_{s\infty}}{Z_{n\infty}} = \left(\frac{\sigma_1 - i\sigma_2}{\sigma_n}\right)^{-\frac{1}{3}}. \tag{18}$$

From this, one easily finds that the ratio of surface resistances is

$$\frac{R_{s\infty}}{R_{n\infty}} = \frac{2 \sin \frac{1}{3}\theta}{[(\sigma_1/\sigma_n)^2 + (\sigma_2/\sigma_n)^2]^{\frac{1}{6}}}, \quad \theta = \tan^{-1} \sigma_1/\sigma_2, \tag{19}$$

from which the normalized absorption curve can be computed. The result is plotted in Fig. 1b. Inserting the values of σ_1 and σ_2 near ω_g mentioned above, we find

$$\left.\frac{d(R_{s\infty}/R_{n\infty})}{d(\omega/\omega_g)}\right|_{\omega_g} = \frac{2}{3}\,\frac{d(\sigma_1/\sigma_n)}{d(\omega/\omega_g)}\left(\frac{\sigma_2}{\sigma_n}\right)^{-\frac{4}{3}} = \frac{2}{3}\cdot\frac{1}{2}\,\pi = \frac{1}{3}\pi = 1.047 \tag{20}$$

If, instead, the extreme dirty limit is appropriate, one finds

$$\frac{Z_{s,\,dirty}}{Z_{n,dirty}} = \left(\frac{\sigma_1 - i\sigma_2}{\sigma_n}\right)^{-\frac{1}{2}} \tag{21}$$

so that

$$\frac{R_{s,dirty}}{R_{n,dirty}} = \frac{\sqrt{2}\quad \sin\theta/2}{[(\sigma_1/\sigma_n)^2 + (\sigma_2/\sigma_n)^2]^{\frac{1}{4}}}, \quad \theta = \tan^{-1}\sigma_1/\sigma_2 \tag{22}$$

Although not shown in Fig. 1b for clarity, this curve rises slightly more steeply at ω_g than the extreme anomalous limit curve, and it overshoots to a value a few percent above unity for $\omega/\omega_g \sim 3$. Taking the values for σ_1 and σ_2 near ω_g mentioned above, we find

$$\left.\frac{d(R_s/R_n)_{dirty}}{d(\omega/\omega_g)}\right|_{\omega_g} = \frac{1}{\sqrt{2}}\frac{d(\sigma_1/\sigma_n)}{d(\omega/\omega_g)}\left(\frac{\sigma_2}{\sigma_n}\right)^{-\frac{3}{2}} = \frac{\pi}{2^{3/2}} = 1.11 \tag{23}$$

which is only a few per cent greater than (20).

In contrast to these results, the experimentally measured absorption edges are considerably steeper, values of 4 to 7 for the normalized slope at ω_g being more typical in the case of pure lead. Since any strains, inhomogeneities, or anisotropies in the gap could only broaden the absorption edge, this steepness discrepancy cannot be explained away trivially by invoking sample imperfection. Rather, the source of at least part of the discrepancy resides in the inapplicability of these simple limits. Even for pure metals in which $\ell >> \delta$, the superconducting coherence length $\xi_o = 2v_o/\pi\omega_g$ may be comparable with the high frequency skin depth δ, which at ω_g is nearly equal to the superconducting penetration depth λ in both superconducting and normal states. To the same degree, ω_g will be of the same order as v_o/δ. This signals an error of the same order of magnitude (because of retardation effects) in the use of the extreme anomalous limit in the normal state at these high frequencies. These two effects add rather than cancel. Thus the true ratio R_s/R_n for a pure metal near ω_g will differ from (19) by an amount which increases with (λ/ξ_o).

Miller[4] attacked this problem, carrying out numerical calculations for the cases of tin and aluminum, under the simplifying assumption that $\ell = \infty$. This is an adequate approximation for pure samples with carefully prepared surfaces, but of course not for alloys. He found that even for Al $(\lambda/\xi_o \approx 0.03)$ the correction to the ratio R_s/R_n near ω_g reaches 30% at low temperatures, while for Sn $(\lambda/\xi_o \approx 0.2)$ the correction is about a factor of two. That is, the two absorption edges are steepened by factors of 1.3 and 2.0, respectively, compared to the extreme anomalous limit approximation. If one extrapolated these results to lead ($\lambda/\xi_o \approx 0.5$), one might expect a steepening by a factor of ~ 3 compared to the extreme anomalous limit, still short of the experimentally observed values.

It is of interest to try to identify the physical origin of the steepening of the absorption edge compared to the extreme anomalous (or extreme dirty) limit. We can get some insight by noting the forms of (20) and (23). These show that the edge will be steepened by having a steeper rise for σ_1/σ_n at ω_g or by having a lower value of σ_2/σ_n there. Thus, if one tries to qualitatively carry over these results to intermediate cases by replacing the extreme anomalous limit values of σ_s/σ_n by suitable weighted averages of σ_s/σ_n over q, this indicates what the effects of various changes will be.

The initial slope of σ_1 at ω_g is given by a simple exact result of Leplae[7], namely,

$$\left.\frac{d\sigma_{1s}(\omega,q)}{d(\omega/\omega_g)}\right|_{\omega_g} = \frac{\pi}{2}\ \sigma_o f(q\ell) = \frac{\pi}{2}\ \sigma_{1n}(0,q)$$

Then

$$\left.\frac{d[\sigma_{1s}(\omega,q)/\sigma_{1n}(\omega,q)]}{d(\omega/\omega_g)}\right|_{\omega_g} = \frac{\pi}{2}\ \frac{\sigma_{1n}(0,q)}{\sigma_{1n}(\omega,q)} \geq \frac{\pi}{2}$$

This reduces to $\frac{\pi}{2}$ in the extreme anomalous limit ($q \to \infty$) and the dirty limit ($\ell \to 0$), but otherwise exceeds $\pi/2$. Some indication of this behavior for intermediate cases is given in Fig. 2, showing the results of unpublished calculations of Strassler and Fulde for q = 0 and various mean free paths. [Qualitatively similar results would be expected if one took $\ell = \infty$ and replaced $\omega_g\tau$ by $(q\xi_o)^{-1}$.] We see that in extreme cases, σ_{1s} can even exceed σ_{1n}. Stated more physically, some oscillator strength from near $\omega = 0$ is elevated above the gap in the superconducting state, resulting in a very steep rise in absorption above ω_g. The physical reason for this elevation is that the superconducting condensation alters the energy-momentum relations for electrons near the Fermi surface. For example, if we consider the extreme case of $\ell = \infty$ and $q \to 0$, in the normal state the entire $\sigma_1(\omega,q)$ leading to the sum rule value lies essentially at $\omega = 0$ (below $\omega = qv_o$). In the superconducting

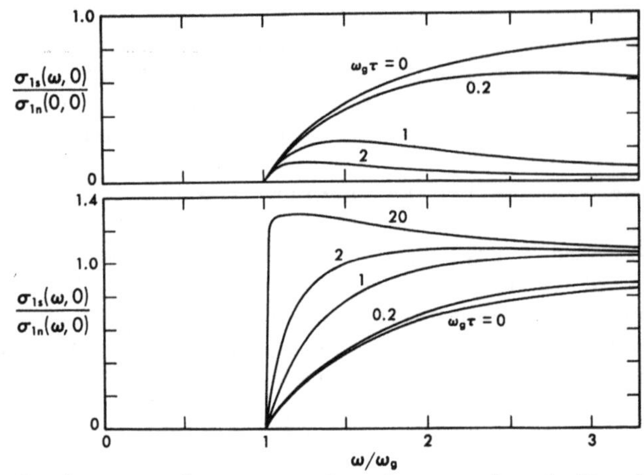

Fig. 2 Real part of superconducting conductivity for q = 0 Fourier component for various mean free paths, according to unpublished calculations of Strassler and Fulde. In (a), normalization is to σ_{1n} at $\omega = 0$, whereas in (b), the normalization is to σ_{1n} at ω. Note that with normalization (a), all curves start with same slope at ω_g as required by the general result of LePlae. However, with normalization (b), the onset becomes very steep as the mean path increases. $[\omega_g\tau = 2\ell/\pi\ \xi_o]$. These curves will also be qualitatively correct for $\ell = \infty$, and $\omega_g\tau$ replaced by $(q\xi_o)^{-1}$.

state, however, a $q \not\approx 0$ photon can be absorbed at $\omega = \omega_g$ because energy can be expended in depairing without momentum change.

Summing up, in all cases except the extreme anomalous limit or the extreme dirty limit we expect the rise of σ_{1s} at ω_g, normalized to σ_{1n} at ω_g, to be steeper than in the extreme anomalous limit because of the elevation of oscillator strength above the gap by the superconducting interactions. This effect will be most prominent for clean samples ($\ell >> \xi_o$) of materials in which λ/ξ_o is not too small, because then Fourier components for which the steepening is greatest will play an important role in determining the surface impedance.

The question of the observed "excess" steepness has been investigated particularly carefully by Ginsberg.[8] He notes that anisotropy of the energy gaps in pure samples may interfere with quantitative comparisons, and suggests that homogeneous alloys are more suitable because in them the gap is made isotropic by an averaging process described by Anderson.[9] Moreover, careful quantitative data are available for a number of lead alloys from the work of Leslie and Ginsberg.[10] Ginsberg carried through the machine calculations required to compute surface resistances from the exact results (10a) and (10b), for the appropriate mean free paths, and without any restriction on q values. The comparison with experiment is shown in Fig. 3, taken from his paper. The plotted quantity P is essentially a normalized difference $(R_n - R_s)$, whose fall above ω_g corresponds to the rise in $R_s(\omega)$ there. Note the small differences between the specular and diffuse scattering approximation, which become completely negligible for the dirtiest samples, as expected. Also note that all the exact theoretical curves (solid) fall well below either of the extreme limit approximate curves. Even for the sample with 6 percent impurity, the exact curve falls well below the extreme dirty (or local) limit curve, which shows how hard this limit is to actually reach. The general conclusion is that taking the q-dependence of σ into account greatly improves the agreement between theory and experiment, but that a significant "excess" steepness remains. As we shall describe in more detail below, it now appears that this residual discrepancy is due to a reduction of σ_2/σ_n below the Mattis-Bardeen value by a strong-coupling effect in lead.

III. FAR INFRARED PROPERTIES OF THIN FILMS

Although studies of absorption by bulk samples are preferable from the point of view of sample perfection and purity, studies of thin films offer compensating advantages in sensitivity and ease of quantitative interpretation. Let us see how this comes about.

We restrict our attention to films of thickness $d \ll \lambda$, the penetration depth, so that fields and currents are essentially constant through the thickness of the film. It is then convenient to define a thickness-dependent effective conductivity σ_{eff} such that the admittance per square, Y (which is the quantity actually measured) is given by

$$Y = \sigma_{eff} d = \frac{\bar{J} d}{E} = \frac{1}{E} \int_{-d/2}^{d/2} J(x)dx = \frac{1}{E} \int_{-d/2}^{d/2} \int_{-\infty}^{\infty} J(q) e^{iqx} dq dx$$

$$= \frac{1}{E} \int_{-d/2}^{d/2} \int_{-\infty}^{\infty} \sigma(q) E(q) e^{iqx} dq dx$$

$$= \frac{1}{2\pi} \int_{-d/2}^{d/2} \int_{-\infty}^{\infty} \int_{-\infty}^{\infty} \sigma(q) \frac{E(x')}{E} e^{-iqx'} e^{iqx} dx dx' dq$$

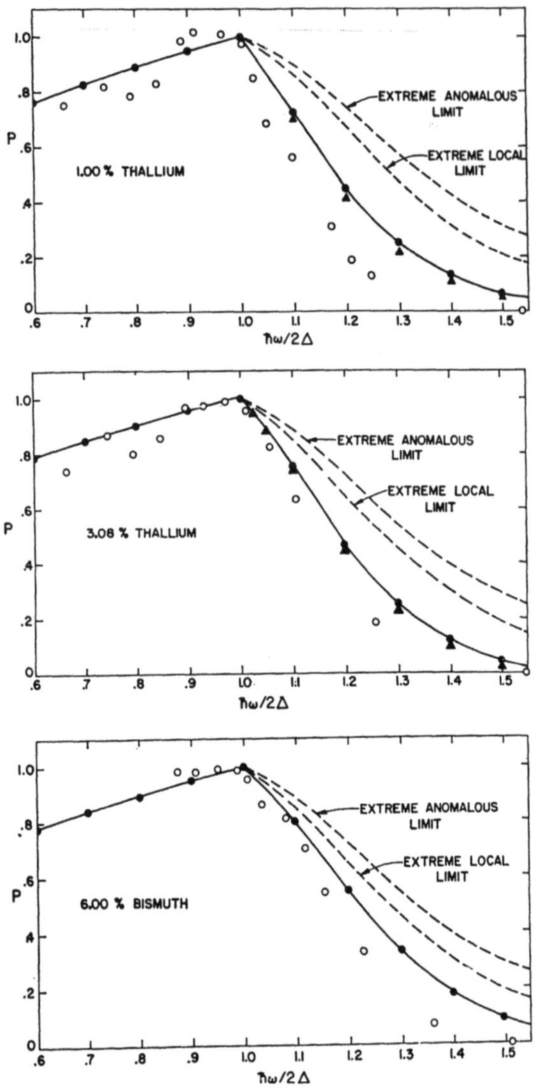

Fig. 3 Absorption curves (from ref. 10) for lead with various amounts of impurities, as indicated. P is a normalized difference in surface resistance of superconducting and normal states. The open circles are the experimental points. The solid circles and the triangles are Ginsberg's theoretical points for diffuse and specular scattering of the electrons at the surface, respectively. At frequencies below the gap edge, the diffuse and specular scattering are too close together to show separately; this is also true at frequencies above the gap edge for the most impure alloy. The extreme anomalous limit and the extreme local limit are explained in the text.

Now, diffuse scattering at boundaries is introduced into the Fourier analysis treatment by putting $E(x')=E$ for $|x'| \leq d/2$, and $E(x')=0$ otherwise. This prescription takes account of the fact that electrons are assumed to loose all memory of previously experienced fields in a diffuse wall collision. Hence the film is equivalent to a similar volume in an infinite medium but subjected to this localized field $E(x')$. [If the reflection were specular, the boundary would not have this effect, but experimental evidence indicates that the actual situation is closer to the diffuse approximation.] Putting in this assumption, then,

$$Y = \sigma_{eff} d = \frac{1}{2\pi} \int_{-\infty}^{\infty} \sigma(q) \int_{-d/2}^{d/2} e^{-iqx'} dx' \int_{-d/2}^{d/2} e^{iqx} dx dq$$

$$= \frac{2}{\pi} \int_{-\infty}^{\infty} \sigma(q) \frac{\sin^2 qd/2}{q^2} dq \qquad (24)$$

Evidently σ_{eff} is just a weighted average of $\sigma(q)$, with weight proportional to $|E(q)|^2$, which gives heaviest weight to a broad band of q-values extending out to $q= \pm 2\pi/d$. Thus, if d is small enough compared to ξ_o the majority of the weight will come for q values in which the extreme anomalous limit applies. Accordingly, we expect the extreme anomalous limit to be a better approximation for a thin film than for even a pure bulk sample, if $d \ll \lambda$, as we are assuming throughout. This should allow us to compare measured frequency dependences directly with the Mattis-Bardeen results without needing to worry much about small q-values. Evidently this offers a considerable simplification in interpretation.

For these thin films, the actual thickness d is not known to any great accuracy, and there may be granularity, etc. Thus, as remarked above, σ_{eff} itself is not measured. What one really measures is the admittance per square $Y=\sigma_{eff}d$. In the normal state, this is just the reciprocal of the resistance per square $R_n=1/\sigma_n d$, i.e., the resistance of a square (of any size) of the film material measured between electrodes along the length of opposing edges, a readily measured quantity. So long as the film is thin enough, σ_n (or Y_n) is constant from dc even up to far infrared frequencies. (We have verified this to a few per cent.) Thus it forms a convenient normalization, and we can simply equate

$$R_n Y_s = Y_s/Y_n = \sigma_s/\sigma_n \qquad (25)$$

Thus uncertainties in d cause no difficulty in interpretation.

Since the film thickness is negligible compared to a wavelength of far infrared radiation, the film may be replaced by an equivalent film of equal admittance but infinitesimal thickness in a "lumped constant" electromagnetic analysis. It is convenient to normalize to the impedance of free space, $Z_o = 377\ \Omega/\text{square} = 4\pi/c(\text{esu})$. Thus our dimensionless normalized film admittance is

$$y = \frac{4\pi}{c}\sigma_{eff}d \tag{26}$$

and

$$y_n = 377/R_n(\text{ohms}).$$

From simple transmission line theory, the reflection coefficient for power is

$$R = \left|\frac{1-(y+1)}{1+(y+1)}\right|^2 = \left|\frac{y}{y+2}\right|^2 = \frac{y_1^2+y_2^2}{(y_1+2)^2+y_2^2} \tag{27}$$

Similarly, the transmission coefficient is

$$T = \left|\frac{2}{y+2}\right|^2 = \frac{4}{(y_1+2)^2+y_2^2} \tag{28}$$

and the absorption coefficient is

$$A = 1-R-T = y_1T = \frac{4y_1}{(y_1+2)^2+y_2^2} \tag{29}$$

Of course all these results are changed significantly when one takes account of the substrate, multiple reflection, etc., but the qualitative features stay the same.

In the normal state, y_n = real, so these results simplify somewhat to read

$$\begin{aligned} R_n &= y_n^2/(y_n+2)^2 \\ T_n &= 4/(y_n+2)^2 \\ A_n &= 4y_n/(y_n+2)^2 \end{aligned} \tag{30}$$

all of which are independent of frequency if the film is thin enough.

In the superconducting state, one must determine two quantities y_1 and y_2 at each frequency. This requires two independent measurements at each frequency, such as transmission and reflection. In the early work of Ginsberg and Tinkham[11] only transmission data were taken, and the Kramers-Kronig relations used to supply the second relation. However this is not straight forward, and it now appears that significant errors were made in the assumed extrapolation necessary for use of the integral transforms. Thus, we shall pass immediately to the more recent work of Palmer and Tinkham[12] on very thin lead films in which precise measurements of both transmission and reflection were made simultaneously.

The experimental arrangement is shown in Fig. 4 (rotated by 90° from its orientation in use). Radiation from a grating monochromator was incident through the light pipe at left, reflected from spherical mirrors C and D into a gently convergent beam incident on the sample film at I. The transmitted and reflected beams were detected by bolometers M and E, aided by condensing cones. By rotating a turntable, the sample could be replaced by a blank quartz disc or else by a mirror. Also, the sample could be heated above T_c to make it normal conducting. In this way, it was convenient to normalize superconducting signals against normal ones and also to make "absolute" measurements by comparison against a "perfect" reflector or a bare quartz substrate. The films were evaporated in situ onto a quartz substrate cooled to $77^{\circ}K$ to minimize agglomeration. The optimum film resistance is of order $Z_0 \sim 377\Omega$. This implies that they must be very thin. In fact, they turn out to have "resistive thickness" d $\sim$ 10-15Å. This thickness

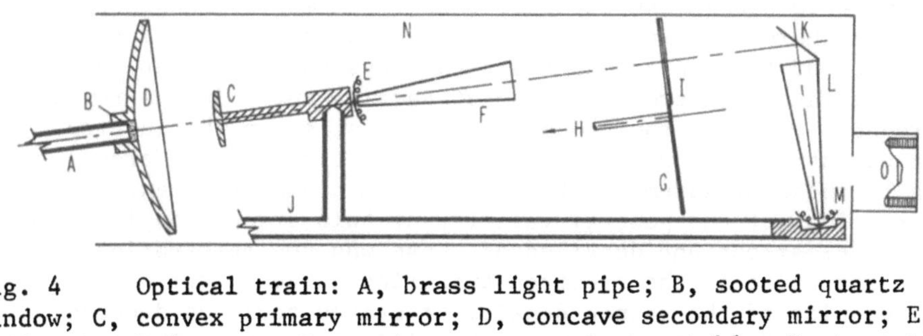

Fig. 4 Optical train: A, brass light pipe; B, sooted quartz window; C, convex primary mirror; D, concave secondary mirror; E, reflection bolometer; F, condenser cone; G, turntable; H, to rotor shaft and heat switch; I, quartz substrate; J, tube connnected to helium reservoir; K, right-angle plane mirror; L, condenser cone; M, transmission bolometer; N, low-temperature radiation shield; O, molybdenum boat Pb evaporation source.

estimate is determined from the temperature dependent part of the film resistance, assuming Matthiessen's rule that the residual resistivity due to defects and surfaces is simply additive to the phonon-induced part. A more realistic "mass thickness" is probably greater, perhaps as much as 50Å. In any case, the thickness is surely much less than λ or ξ_o, so the extreme anomalous limit should be very closely approximated.

From the measured values of R and T at each frequency, values of y_1 and y_2 were found by inverting (27) and (28), or more precisely, the analogs of these equations taking account of substrate and multiple internal reflections. These are readily converted to the desired conductivity ratios since $\sigma_i/\sigma_n = y_i/y_n = y_i R_n/377$. The results for σ_1/σ_n and σ_2/σ_n at 2°K are shown in Figures 5 and 6. As is evident from Fig. 5, the σ_1/σ_n results agree with the Mattis-Bardeen predictions to within a scatter of $\pm$ 5%. However, Figure 6 shows that σ_2/σ_n is systematically much less than the BCS result of Mattis and Bardeen. In fact, we found the coefficient of $1/\omega$ in σ_2/σ_n for frequencies well below the gap to be reduced by about 25% from the MB result, which leads to σ_2/σ_n at $\omega = 2\Delta$ being 40% below the MB value of 1.0. This discrepancy was puzzling at first because it appeared to violate the Kramers-Kronig relations connecting σ_2 and σ_1. That is, since $\sigma_2(\omega)$ is determined by causality once $\sigma_1(\omega)$ is known, it would be impossible for $\sigma_2(\omega)$ to disagree with MB if $\sigma_1(\omega)$ did not.

This dilemma has since been completely resolved in terms of the strong-coupling theory of the superconductivity of lead. In this theory, discussed by Scalapino, et al.[13], the electron-phonon interaction is treated dynamically and this leads to an energy gap parameter which is both complex and frequency-dependent. The imaginary part of the gap parameter represents the fact that electronic excitations in the superconductor can decay in a finite lifetime into phonon excitations; thus it is maximum at energies for which there is a large phonon density of states. The frequency dependence of the gap arises from the physical fact that the phonon-mediated electron-electron attraction turns into a repulsion at frequencies above the phonon frequencies, since the lattice deformations then become out of phase with the driving forces. Previous to these measurements there had been a number of experimental confirmations of the general validity of this strong-coupling theory. The most clear-cut and detailed of these was the observation of phonon structure in the electron tunneling curves, from which a detailed plot of the phonon density of states could be deduced, including van Hove singularities at critical points. These phonon anomalies lead to deviations from simple BCS density of states curves which reach maxima of some 5%, at energies of order the Debye energy. Another consequence of the strong coupling in lead is the anomalously high ratio of the energy gap at T=0 to T_c, namely, 4.5 rather than the BCS value of 3.5.

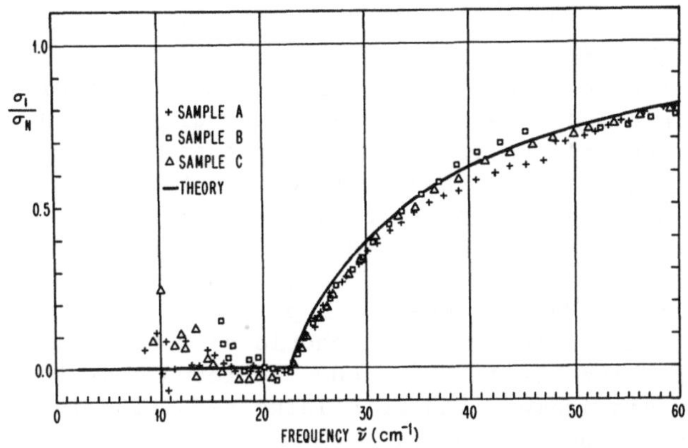

Fig. 5 Results of measurements of the real part of the normalized conductivity of three thin lead films at 2°K, compared with Mattis-Bardeen theory with gap frequency fitted to 22.5 cm^{-1}. To reduce the clutter in the figure, only about one fourth as many points are shown as were taken and recorded. The points shown are selected typical points above the gap and local averages below the gap.

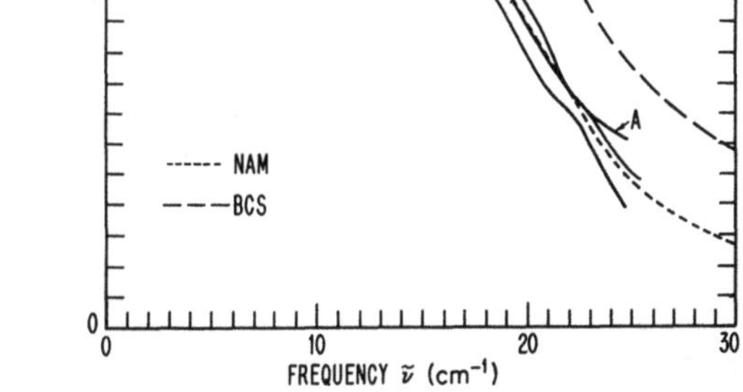

Fig. 6 Smoothed results of measurements of the imaginary part of the normalized conductivity of three lead films (A,B, and C) at 2°K. Curve labeled BCS is the weak-coupling result of Mattis and Bardeen, while that labeled Nam presents his corrected version. In both cases, the gap frequency was taken to be 22.5 cm^{-1}.

In a pair of papers, Nam[14] had considered the modification to be expected in the Mattis-Bardeen results as a result of using the Scalapino-Schrieffer-Wilkins results for $\Delta(E)$ in computing the response of superconducting lead to electromagnetic radiation. He found changes in $\sigma_1(\omega)$ of a few percent extending over a wide frequency range, even well above the Debye frequency. These were comparable in amplitude to the phonon effects in tunneling, mentioned above. Such small changes spread over a wide range in frequency were just at the limit of the accuracy of our data in Fig. 5, and hence our results can be considered consistent either with the MB weak-coupling or the Nam strong-coupling results. In these papers, Nam found that the changes in $\sigma_2(\omega)$ should also be small, but on subsequently checking his calculations, he found that the strong-coupling theory did indeed lead to a large reduction in σ_2 of just the size observed experimentally.

In fact, this correction had been anticipated, since, using his strong-coupling $\sigma_1(\omega)$ curve together with the sum rule and Kramers-Kronig relations we saw that the relatively small theoretical changes in $\sigma_1(\omega)$ extending over a wide frequency range could be accumulated to cause a large change in $\sigma_2(\omega)$ at low frequencies. The point is that in a superconductor any "shortage" A of area under the $\sigma_1(\omega)$ curve for $\omega > 0$ (see Fig. 1) compared to the sum rule value (or normal state value, since they must be equal) $(\pi/2)ne^2/m$ appears as a δ-function in σ_1 at $\omega = 0$. The Kramers-Kronig transform of this is the $1/\omega$ supercurrent term in $\sigma_2(\omega)$. Hence the strength of the supercurrent response is proportional to the integrated shortage of $\int\sigma_1(\omega)d\omega$. Now, it turns out that in the strong-coupling treatment of lead, $\sigma_1(\omega)$ exceeds the MB weak-coupling $\sigma_1(\omega)$ over a range of frequencies extending well beyond the typical phonon frequencies and hence far beyond the frequencies (up to only a few times the gap) over which measurements were made. Although such small changes were not observable with our limited accuracy, the accumulated effect on the δ-function and on σ_2 at low frequencies was very conspicuous.

Recently, Shaw and Swihart[15] have shown that this sum rule-dispersion theory approach leads to results for $\sigma_2(\omega)$ which are completely consistent with the (corrected) direct calculation of Nam and with the experiments. Moreover, they showed that 99% of the change in σ_2 in the low frequency range was due to the change in strength of the δ-function and only 1% to the change in distribution of the oscillator strength remaining above the gap frequency. This result illustrates the usefulness of the sum rule technique in isolating the major effects on the supercurrent response of a superconductor.

Because the 25% reduction in what is effectively n_s/m_s is comparable in order of magnitude to the increase in the electronic effective mass in lead due to "phonon-dressing" effects, it is tantalizing to wonder whether a more direct relation could be found between these two related electron-phonon effects.

As a final remark, it should be pointed out that the decrease in $\sigma_2(\omega)$ described here leads to a steeper absorption edge at the gap in experiments on bulk samples. Since we know the size of the strong-coupling correction only in the extreme anomalous limit, we can use it only to correct the surface resistance predictions in the extreme anomalous and extreme dirty limits. From (20) and (23) we see that the steepness of the absorption edge in these two limits will be increased by factors of $(\sigma_2/\sigma_n)_{\omega g}^{-4/3}$ and $(\sigma_2/\sigma_n)_{\omega g}^{-3/2}$ respectively, since $(\sigma_2/\sigma_n)_{\omega_g} = 1$ in the MB theory. Since $(\sigma_2/\sigma_n)_{\omega g} \approx 0.6$, these increases are factors of 1.98 and 2.15 respectively. Such changes are in the direction and of the order of magnitude required to explain the excess steepness mentioned above in connection with Fig. 3. Thus it seems reasonable to expect that, if the strong-coupling conductivity of lead were available for all q-values, quantitative agreements with the observed absorption edge shape in bulk lead would also be achieved.

IV. THE EFFECT OF MAGNETIC PERTURBATIONS

The work of Abrikosov and Gor'kov[16] showed that magnetic impurities in a superconductor have a profoundly different effect from ordinary non-magnetic scattering centers. Whereas Anderson had shown that non-magnetic impurities average out anisotropies, leading to a more sharply defined average BCS energy gap, Abrikosov and Gor'kov showed that magnetic impurities cause a large decrease in the gap and in T_c, together with a marked smearing out of the peak in the density of states curve at the gap edge. In fact, for concentrations above 0.91 of the critical concentration at which T_c would be decreased to zero, there is no gap at all, although there is still a substantial degree of pair correlation in the superconducting state. We denote the actual gap in the spectrum by $\hbar\omega_g$, while Δ measures the pairing potential. Unlike in BCS, $\hbar\omega_g$ is no longer equal to 2Δ. For example, in the gapless regime, $\omega_g = 0$, while $\Delta \neq 0$.

Subsequently, Maki[17] and others have shown that in a thin film of a dirty superconductor a parallel applied magnetic field or a current have the same effect as paramagnetic impurities, with the equivalent impurity concentration being proportional to H^2 or p^2, where p is the pairing momentum. The equivalence of various pair-breaking mechanisms has been considered critically by Maki and Fulde.[18] Skalski, et al.,[19] have calculated many properties of such systems, including the penetration depth and σ_1/σ_n for q=0 in the extreme dirty limit $\ell \ll \xi_o$. The results for σ_1/σ_n are shown in

Fig. 7. They showed that there is a Meissner effect (with increased penetration depth) and hence persistent currents, even in the gapless regime. However, $\sigma_1 > 0$ for <u>all</u> frequencies in the gapless regime, leading to absorption for all non-zero frequencies. The absorption approaches zero as $\omega \to 0$, since σ_1 is finite (apart from the δ-function at $\omega = 0$) while $\sigma_2 \sim 1/\omega$. Thus the real part of the impedance (i.e., the resistance) vanishes as $\sigma_1/(\sigma_1^2 + \sigma_2^2) \sim \omega^2$, as $\omega \to 0$.

The theoretical tunneling density of states has been quite well confirmed by Woolf and Reif[20] for the case of paramagnetic impurities and by Levine and by Millstein and Tinkham[21] for the case of magnetic fields.

Far infrared absorption experiments offer the possibility of a further test of the theory, since matrix elements are involved in addition to the density of states, and also because electrons moving parallel to the film plane dominate the far infrared properties whereas those moving perpendicular to the surface dominate tunneling. (According to the work of the Orsay school and of Tewordt and coworkers, the effect of the field is quite different on these two groups of electrons.) Accordingly it is of interest to consider the results of Martin and Tinkham[22] on the absorption of far infrared radiation by lead films $\sim$100Å thick in strong magnetic fields.

These films are thick enough so that most of the radiation is reflected (>90%), some absorbed (<10%), and very little transmitted. To build up a measurable signal from the small change in reflectance between the superconducting and normal state, the film

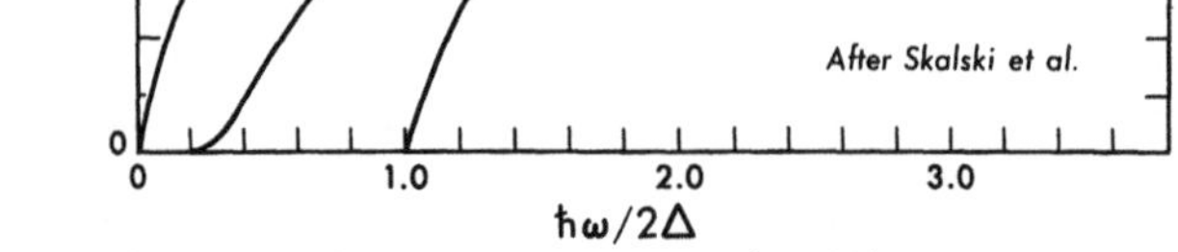

Fig. 7 Results of Skalski, et al. (1964) for real part of conductivity for q = 0 at T = 0, for several concentrations of magnetic impurities $\sim\Gamma$. The curve $\Gamma/\Delta = 1$ represents the transitional case in which ω_g has just barely been reduced to zero. The Mattis-Bardeen curve shown in Fig. 1a results when $\Gamma/\Delta = 0$.

is mounted in a cavity arrangement, allowing multiple reflections from the film before the radiation reaches the detector. By means of a series of control experiments and suitable data analysis, it is possible to extract a reasonably reliable result for the ratio A_s/A_n of the absorptivity in the superconducting and normal states as a function of frequency and magnetic field. Some data for a 100 Å film are shown in Fig. 8.

Note first that in zero field, the virgin curve is qualitatively similar to the BCS theoretical curve of Fig. 1b, including the marked overshoot where $A_s/A_n > 1$. Quantitative agreement is not to be expected, since corrections are required from the extreme-anomalous weak-coupling limit, from the low-impedance limit, and for non-normal incidence. Also note that if the film has been exposed to a field, there is a residual "precursor" absorption which remains after the applied field has been removed. This effect is presumed due to trapped flux.

Next let us consider the curves taken in the presence of a field. The field is nominally parallel to the film, and has been aligned to a precision of about 1/2° by means of the dc resistive transition. However, the existence of a trapped flux precursor effect suggests that the small perpendicular component of the field

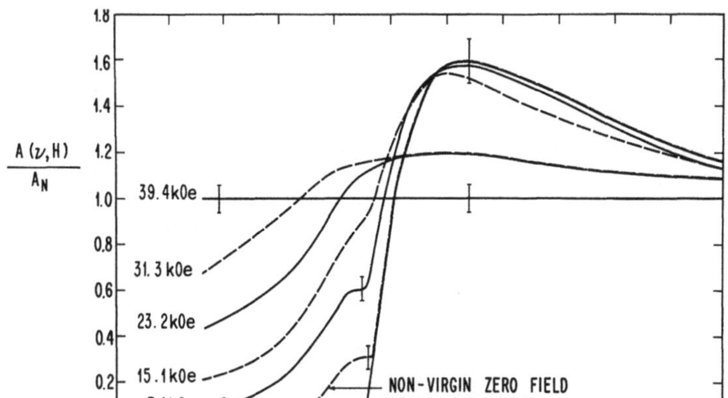

Fig. 8 Experimental results for the absorptivity ratio A_s/A_n for a lead film $\sim$ 100 Å thick, with $\ell \gtrsim$ 300 Å, at T = 1.3°K, in nominally parallel field, estimated to be misaligned by $\sim$0.5°. Note that the virgin curve resembles the theoretical curve in Fig. 1b, but is steeper and higher due to corrections to the weak-coupling extreme anomalous limit. The non-virgin H = 0 curve shows a per-cursor absorption due to trapped flux. Other curves show absorption at all frequencies due to core regions, and progressive downward shift of gap, as discussed in text. (H_c = 44.7 kOe.)

is playing a significant role. Such a component would punch through the film in an array of quantized vortices, some of which would remain trapped in the film after the external field was reduced to zero. In such a vortex, we know from the work of Abrikosov[23] that the order parameter goes smoothly to zero at the center. The complete theory of the actual absorption spectrum of such a spatially inhomogeneous system is not available. However, for clean superconductors Caroli, et. al.[24] have shown that with each vortex there is associated a density of low-lying excited states equal to that in a cylinder of normal metal of radius $\sim\xi$, the coherence length. This agrees qualitatively with the result of applying the Skalski curves to compute a local density of states or local value of σ_1 in the dirty limit. Thus in general it seems clear that the presence of a number of vortices proportional to the perpendicular magnetic field will lead to some field-dependent absorption at all frequencies, with the details depending on the model. For $\omega \ll \omega_g$, this effect will completely predominate over the effect of the much larger parallel component, since that simply decreases the gap by an amount proportional to $h_\parallel^2 = (H_\parallel/H_{c\parallel})^2$. Of course at frequencies near ω_g, the parallel field effect is important. However, because $h_\perp = (H_\perp/H_{c\perp})$ enters to the first power and produces an effect for all frequencies, it generally dominates the qualitative behavior even for a nominally parallel field. This $h_\perp$ effect complicates the theoretical interpretation of the data, but serves to point up rather dramatically the sensitivity of all thin film experiments to even a minimal perpendicular field component.

The model which we use for qualitative discussion of our results distinguishes three types of region in the film in the presence of vortices: 1) an effectively normal core of radius $\sim\xi \approx \sqrt{\xi_0 \ell}$, where $\omega_g = 0$ (even if $\Delta \neq 0$ except at the center); 2) a background region between vortices where ω_g has a value reduced by $H_\parallel$ to a value roughly approximated by $\omega_g(h_\parallel) \approx \omega_g(0)(1-h_\parallel^2)$; 3) a transition region, in which ω_g increases smoothly from zero to this outer value. If $h_\perp \ll 1$, the vortices are independent, producing an area of type (1) in proportion to their number, and hence to $h_\perp$. In fact, since at the transition the whole sample must be normal, we can normalize by saying that the fraction of "normal" area is given approximately by

$$f_n \approx h_\perp/(1-h_\parallel^2), \tag{31}$$

which leads to $f_n = 1$ when $h_\perp + h_\parallel^2 = 1$, the correct expression for the angular dependence of the critical field of a thin film.[25]

At the high frequencies of a far-infrared experiment and particularly because the field is nominally parallel, we must take account of the absorption in all three regions. In the absence of a detailed theory, we simply note the following: In low fields we see absorption ~ H throughout the gap due to the core and transition regions, with a sharp onset of additional absorption at the background gap $\omega_g(h_{\parallel})$. As the field is increased, one reaches a "packed vortex" regime in which the background region (2) has been "squeezed out" by the vortices. At this point, the shoulder at $\omega_g(h_{\parallel})$ washes out because there is no particular dominant gap value left. With further increase in field, there is a continuous approach of A_s toward A_n at all frequencies, as the order parameter and ω_g are everywhere driven toward zero. This behavior is evident in the data of Fig. 8.

REFERENCES

1. D. C. Mattis and J. Bardeen, Phys. Rev. 111, 412 (1958).
2. A. A. Abrikosov, L. P. Gor'kov, and I. M. Khalatnikov, Zh. Experim. i Teor. Fiz. 35, 265 (1958); English translation: Soviet Physics JETP 8, 182 (1959).
3. A. A. Abrikosov and L. P. Gor'kov, Zh. Experim. i Teor. Fiz. 35, 1558 (1958); English translation, Soviet Physics JETP 8, 1090 (1959).
4. P. B. Miller, Phys. Rev. 118, 928 (1960).
5. J. R. Waldram, Advan. Phys. 13, 1 (1964).
6. R. A. Ferrell and R. E. Glover, III, Phys. Rev. 109, 1398 (1958); M. Tinkham and R. A. Ferrell, Phys. Rev. Letters 2, 331 (1959).
7. L. Leplae, Ph.D. Thesis, University of Maryland, Technical Report 283, 1962.
8. D. M. Ginsberg, Phys. Rev. 151, 241 (1966).
9. P. W. Anderson, J. Phys. Chem. Solids 11, 26 (1959).
10. J. D. Leslie and D. M. Ginsberg, Phys. Rev. 133, A362 (1964).
11. D. M. Ginsberg and M. Tinkham, Phys. Rev. 118, 990 (1960).
12. L. H. Palmer and M. Tinkham, Phys. Rev. 165, 588 (1968).
13. D. J. Scalapino, J. R. Schrieffer, and J. W. Wilkins, Phys. Rev. 148, 263 (1966).
14. S. B. Nam, Phys. Rev. 156, 470 (1967); 156, 487 (1967).

15. W. Shaw and J. C. Swihart, Phys. Rev. Letters 20, 1000 (1968).

16. A. A. Abrikosov and L. P. Gor'kov, Zh. Eksperim. i Teor. Fiz. 39, 1781 (1960). (English translation: Soviet Physics JETP 12, 1243 (1961).

17. K. Maki, Prog. Theoret. Phys. (Kyoto) 31, 731 (1964).

18. K. Maki and P. Fulde, Phys. Rev. 140, A1586 (1965).

19. S. Skalski, O. Betbeder-Matibet, and P. R. Weiss, Phys. Rev. 136, A1500 (1964).

20. M. Woolf and F. Reif, Phys. Rev. 137, A557 (1965).

21. J. L. Levine, Phys. Rev. 155, 373 (1967); J. Millstein and M. Tinkham, Phys. Rev. 158, 325 (1967).

22. W. S. Martin and M. Tinkham, Phys. Rev. 167, 421 (1968).

23. A. A. Abrikosov, Zh. Eksperim, i. Teor. Fiz. 32, 1442 (1957), (English translation: Soviet Physics JETP 5, 1174 (1957)).

24. C. Caroli, P. G. deGennes, and J. Matricon, Phys. Letters 9, 307 (1964).

25. M. Tinkham, Phys. Rev. 129, 2413 (1963).

FAR INFRARED DISPERSION AND THE RAMAN SPECTRA OF FERROELECTRIC CRYSTALS

A. S. Barker, Jr.

Bell Telephone Laboratories, Murray Hill, New Jersey

INTRODUCTION

In the past eight years, the understanding of the ferroelectric transition has increased enormously. This jump forward was caused by the discovery that lattice vibrations play a central role in the transition in certain ferroelectric crystals. Cochran and Anderson proposed that on cooling towards the transition, the frequency of a particular lattice vibration mode decreases to the point where the mode becomes unstable. The present paper discusses the spectroscopy of such unstable modes both in ferroelectrics and antiferroelectrics. We list below the topics to be covered.

A. Lattice Vibrations in the NaCl Type Crystal with Uncharged Atoms
B. Lattice Vibrations with Charged Ions
C. Lattice Mode Instabilities - The Ferroelectric Case
D. Lattice Mode Instabilities - The Antiferroelectric Case
E. The Infrared Response - The Dielectric Function
F. The Dielectric Function with Damping for Several Ferroelectrics
G. The Raman Spectra of Ferroelectric Crystals
H. Field Induced Raman Spectra

Since the original papers of Cochran and Anderson, two review books have appeared.[1,2] While the present paper attempts to be self-contained, the reader is referred to these books for references to original papers and for additional background reading. References 3 and 4 predate the Cochran and Anderson work but contain very detailed and important material on crystallographic and thermodynamic behavior of ferroelectric crystals.

A. LATTICE VIBRATIONS IN THE NaCl TYPE CRYSTAL

Uncharged Atoms and Nearest Neighbor Forces Only

Figure A1 shows the ion positions in an NaCl type crystal. We first consider longitudinal vibrations along x (see Fig. A1) of wave vector k_x. Describing the nearest neighbor inter-ion potential by the two parameters A and B and indexing the planes of constant phase by n we obtain for the equations of motion

$$\text{A1)} \quad \begin{aligned} m_1\ddot{x}_1(n) &= -A\{2x_1(n)-x_2(n+1)-x_2(n-1)\} - 4B\{x_1(n)-x_2(n)\} \\ m_2\ddot{x}_2(n) &= -A\{2x_2(n)-x_1(n+1)-x_1(n-1)\} - 4B\{x_2(n)-x_1(n)\}. \end{aligned}$$

We have used $x_\ell(n)$ for the amplitude of ion type ℓ at the nth position in the lattice. We now try as trial solutions waves of the form

$$\text{A2)} \quad x_j(n) = x_j e^{i\omega t} e^{inka},$$

where a is half the X-ray cell parameter as shown in Fig. A1. Substituting A2 into A1 and making use of identities such as

$$\text{A3)} \quad \begin{aligned} x_2(n+1)+ x_2(n-1) &= (x_2 e^{i(n+1)ka} + x_2 e^{i(k-1)ka}) e^{i\omega t} \\ &= x_2 e^{inka} e^{i\omega t} (e^{ika} + e^{-ika}) \\ &= x_2 e^{inka} e^{i\omega t} (2\cos ka), \end{aligned}$$

we obtain the secular equation

$$\text{A4)} \quad \begin{vmatrix} -\omega^2 m_1 + 2A + 4B & -4B - 2A\cos ka \\ -4B - 2A\cos ka & -\omega^2 m_2 + 2A + 4B \end{vmatrix} = 0.$$

Kittel[5] has discussed this set of equations for the case B = 0. The left side of Fig. A2 shows the dispersion curves, i.e., the frequencies allowed for each wave vector k. If B = 0 there is an instability at $k = \pi/a$ in the lower branch. To avoid collapse of the crystal or distortion to a new crystal structure we must therefore have $B > 0$. In Fig. A2 we denote the upper branch or locus of solutions as the optic branch and the lower, the acoustic branch.

We now consider transverse vibrations for the same wave vector shown in Fig. A1. Ion motion is now along y. The equations of motion are

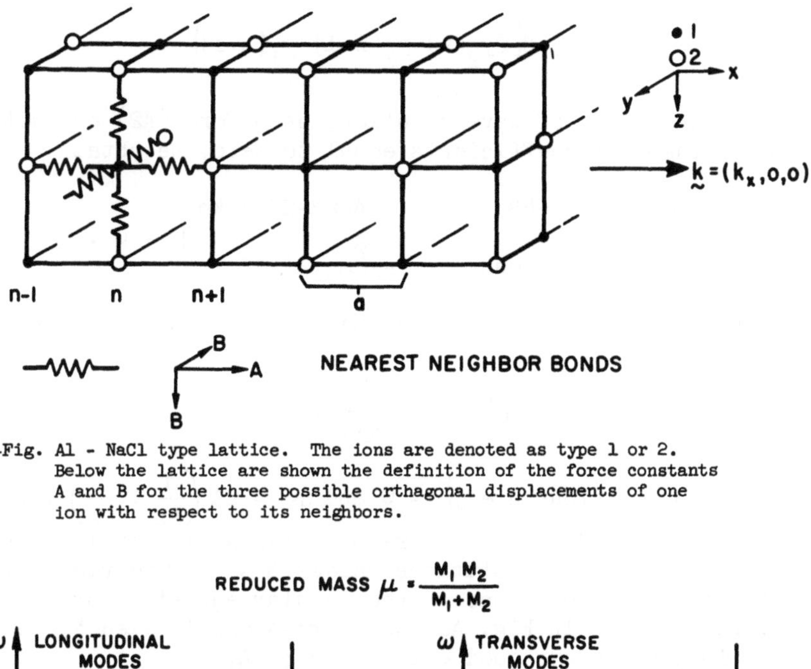

Fig. A1 - NaCl type lattice. The ions are denoted as type 1 or 2. Below the lattice are shown the definition of the force constants A and B for the three possible orthagonal displacements of one ion with respect to its neighbors.

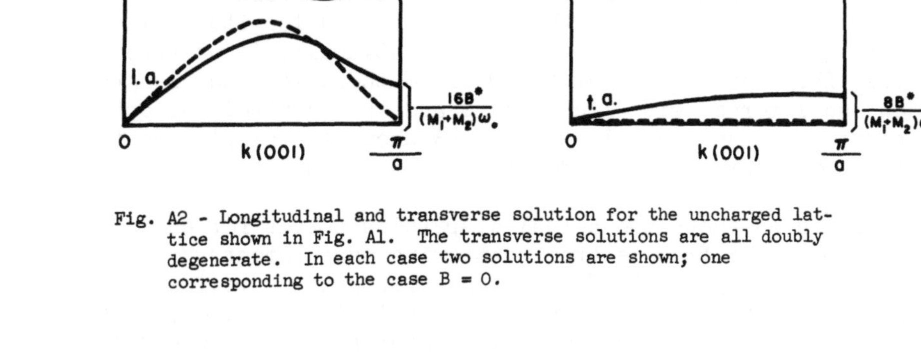

Fig. A2 - Longitudinal and transverse solution for the uncharged lattice shown in Fig. A1. The transverse solutions are all doubly degenerate. In each case two solutions are shown; one corresponding to the case B = 0.

A5)
$$m_1\ddot{y}_1(n) = -2A\{y_1(n)-y_2(n)\} - 2B\{y_1(n)-y_2(n)\}$$
$$-B\{2y_1(n)-y_2(n-1)-y_2(n+1)\}$$
$$m_2\ddot{y}_2(n) = -2A\{y_2(n)-y_1(n)\} - 2B\{y_2(n)-y_1(n)\}$$
$$-B\{2y_2(n)-y_1(n-1)-y_1(n+1)\} \ .$$

Using the same type of wave solutions as before (A2) and setting the determinant of coefficients equal to zero, we obtain

A6)
$$\begin{vmatrix} -\omega^2 m_1+2A+4B & -2A-2B-2B\cos ka \\ -2A-2B-2B\cos ka & -\omega^2 m_2+2A+4B \end{vmatrix} = 0 \ .$$

For this transverse case we find that if B = 0 the secular equation simplifies to

$$\omega^4 - \omega^2(2A) \cdot \left(\frac{1}{m_1} + \frac{1}{m_2}\right) = 0 \ .$$

The two solutions are $\omega^2 = 0$ and $\omega^2 = 2A\left(\frac{1}{m_1} + \frac{1}{m_2}\right)$; i.e., they are k independent. We therefore have an instability of the entire lower branch of solutions unless $B > 0$. The right side of Fig. A2 shows the solutions of the secular equation for the transverse modes. In Fig. A2 the expressions labeled * are approximations which are good only for $B \ll A$.

The equations we have used thus far neglect the charges on the ions and the long range coulomb forces resulting from these charges. A consequence of this neglect is the result shown in Fig. A2 that at k = 0 both the transverse and longitudinal optic mode frequencies are equal. We will continually focus attention on the vibration modes near k = 0 since these long wavelength modes can interact with light waves and thus control the optical properties of the crystal. Figure A3 shows the displacement patterns for some of the modes for k near 0 and π/a.

Figure A4 shows the actual dispersion curves for an NaCl type crystal obtained from neutron scattering measurements.[6] Except for the splitting apart of the transverse optic (t.o.) and longitudinal optic (l.o) modes (which we shall discuss later) we see that our model reproduces the main features of the dispersion curves for k along the x-axis. It is an instructive exercise to show that the t.a. (transverse acoustic) frequency at $\underset{\sim}{k} = (\pi/a,0,0)$ is the same as the l.a. frequency at $\underset{\sim}{k} = (\pi/a,\pi/a,0)$ for our model, as shown experimentally in Fig. A4.

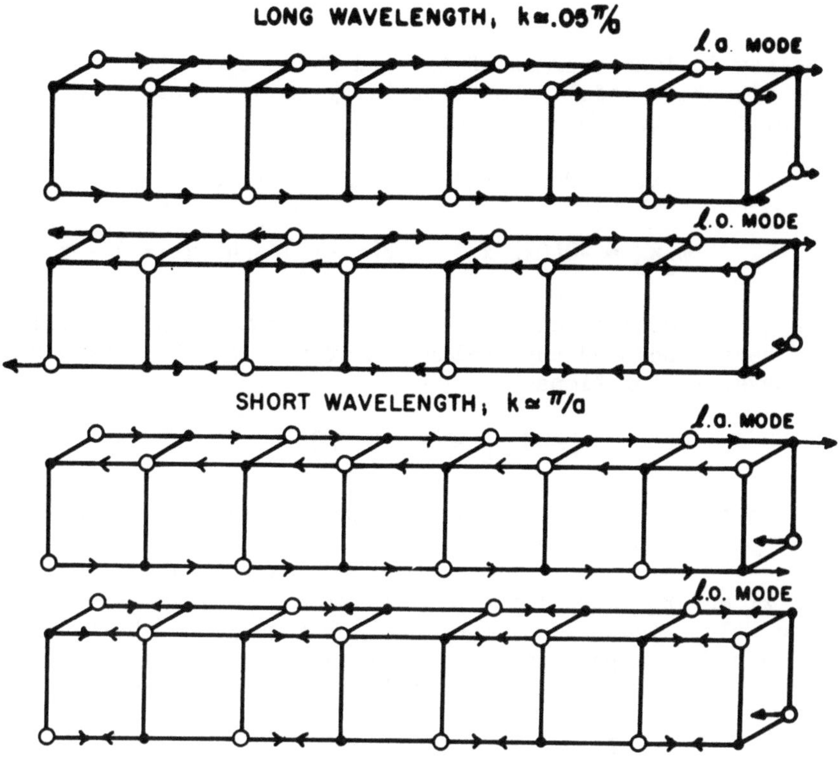

Fig. A3 - Eigenvectors for some longitudinal modes taken from Fig. A2. The upper two displacement patterns are for modes near wavevector k = o. The lower two correspond to solutions near the zone boundary.

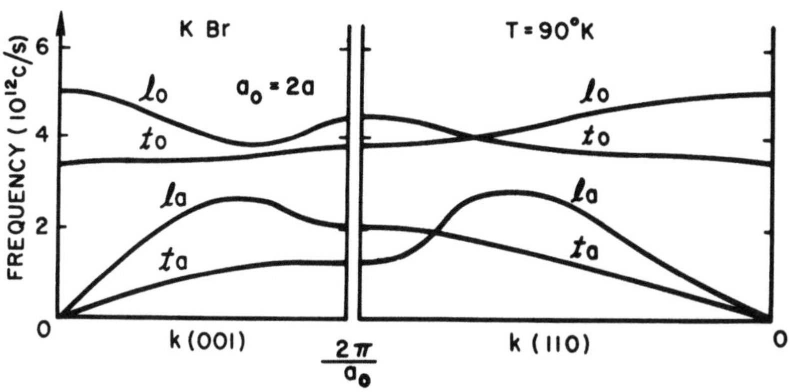

Fig. A4 - Dispersion curves for KBr from Cowley et al., Ref. 6. Here the t.o. and l.o. modes are split well apart even at k = o. This is to be contrasted with the result shown in Fig. A2 for the uncharged lattice.

B. CHARGED IONS

We now include charges of $+z_1$ and $z_2 = -z_1$ on the ions in our lattice vibration model. For simplicity we now restrict ourselves to $k \approx 0$, i.e., to long wavelength vibrations. We write

B1)
$$m_1\ddot{x}_1 = -2A(x_1-x_2) - 4B(x_1-x_2) + z_1E_{loc}$$
$$m_2\ddot{x}_2 = -2A(x_2-x_1) - 4B(x_2-x_1) - 4B(x_2-x_1) - z_1E_{loc}$$

where E_{loc} is local electric field acting on the ion. Kittel[5] and others[7] have discussed the evaluation of the local electric field. For transverse vibrations

B2)
$$E_{loc} = E + 4\pi P/3$$

where E is the transverse externally applied field. P, the macroscopic polarization, has two contributions. First the ion motions x_1 and x_2 cause ionic polarization $\frac{x_1z_1}{v} + \frac{x_2(-z_1)}{v}$. Here v is the volume occupied by two ions. Secondly the charge cloud around each ion is polarized by the local field. This contribution we write as $\alpha\, E_{loc}/v$. The total polarization is therefore

B3)
$$P = z_1(x_1-x_2)/v + \alpha\, E_{loc}/v \ .$$

Since we are interested only in the optical modes at $k \sim 0$ we define the optic mode coordinate $x = x_1-x_2$. The two equations B1 may be subtracted to obtain the single equation

B4)
$$\frac{m_1m_2}{m_1+m_2}\,\ddot{x} = -x(2A+4B) + z_1E_{loc} \ .$$

The polarization equation becomes

B5)
$$P = z_1x/v + \alpha\, E_{loc}/v \ .$$

Eliminating E_{loc} and P using B2 and B5 we obtain from B4 the equation of motion

B6)
$$\frac{m_1m_2}{m_1+m_2}\,\ddot{x} = -x\left[2A+4B - \frac{\frac{4\pi z_1^2}{3v}}{1-\frac{4\pi\alpha}{3v}}\right] + \frac{z_1}{1-\frac{4\pi\alpha}{3v}}\,E \ .$$

The square bracketed expression is the effective force constant for the $k \sim 0$ t.o. mode. For uncharged ions ($z_1=0$) we obtain the same answer as previously. The effect of the charges and the local fields is to reduce the effective force constant for the transverse mode, lowering its frequency. It is the lowering of this frequency to zero (or to some low value) which gives long wavelength ($k \sim 0$) optic mode instability and can lead to a transition to the ferroelectric state.[8] Instabilities at other values of k can arise also. The particular instability which leads to an antiferroelectric transition is discussed below. In NaCl or KBr the third term in the bracket (Eq. B6) is about 30 percent of the first two. Thus these materials have moderately high frequency, stable, t.o. modes. Cochran,[8] and others,[9] have discussed the special situation which enhances the local field effect in the perovskite lattice which can cause the third (local field) term to be as large as the first two terms giving an unstable optic mode.

For the natural t.o. mode we set E = 0 and use our trial solution A2 in Eq. B6 to obtain:

B7)
$$\omega_{t.o.}^2 = \frac{m_1+m_2}{m_1 m_2}\left[2A+4B-\frac{4\pi z_1^2/3v}{1-4\pi\alpha/3v}\right] .$$

For completeness we show now that the l.o. mode at $k \sim 0$ is affected in a much different manner by local field effects. For longitudinal fields, the local field is

B8)
$$E_{loc} = E - 8\pi P/3 .$$

Following the same procedures as above we obtain for the natural l.o. mode frequency:

B9)
$$\omega_{l.o.}^2 = \frac{m_1+m_2}{m_1 m_2}\left[2A+4B+\frac{8\pi z_1^2/3v}{1+8\pi\alpha/3v}\right] .$$

This frequency is always larger than $\omega_{t.o.}$ and there is no cancellation of the short range forces A and B by the local field effects.

To summarize, we have derived the dispersion curves for a simple NaCl type model with short range forces between uncharged particles. For this model the t.o. and l.o. modes are degenerate at $k \simeq 0$. When charge effects are included $\omega_{t.o.}^2$ is pushed down and $\omega_{l.o.}^2$ up by terms proportional to z_1^2/v. There is therefore the possibility that even if the uncharged lattice is stable, the t.o. mode in the charged lattice could have an instability. The unit cell volume, v, decreases with decreasing temperature in most solids. We can picture this type of instability occurring on cooling to a certain temperature where the local field term becomes as large in magnitude as the short range force term in Eq.B7. (See however the discussion in the summary).

All of the infrared optical effects we will discuss come from equation B5 which shows that for a crystal with charged ions the optic-mode ion motions have an associated macroscopic polarization P. Since this polarization vibrates at frequency $\omega_{t.o.}$ and $\omega_{l.o.}$, we might expect interaction of infrared light with the crystal at these natural mode frequencies. If the crystal has an unstable optic mode, an infrared study will show one of the frequencies to be anomalously low.

C. LATTICE MODE INSTABILITIES

The Ferroelectric Case

We now examine in more detail a case where a transverse optic mode decreases in frequency to the point where the lattice becomes unstable and makes a transition to a new crystal structure. Much of what will be discussed is parallel to the macroscopic Devonshire theory[10] of ferroelectric phase transitions. The approach here however follows Cochrans theory in that the transition is described as a t.o. mode with associated polarization $P = zx/v$ becoming unstable. In Devonshire's theory P appears explicitly but no connection is made with lattice vibrations. The relevance of the Cochran theory is that infrared,[9] neutron,[11] and Raman spectrocopy[12] have shown that a lattice vibration mode does indeed drop markedly in frequency as a crystal is cooled towards its ferroelectric transition.

From Eq. B6 we note that the effective restoring force for our model of the t.o. ($\underline{k} \sim 0$) mode is

C1) $$k_{eff} = 2A + 4B - \frac{4\pi z_1^2/3v}{1-4\pi/3v}.$$

If this force constant becomes zero because the last term is large, the natural motion of the t.o. mode is the solution of:

C2) $$\frac{m_1 m_2}{m_1+m_2}\ddot{x}=0.$$

The solution is of course

C3) $$x = at + b ,$$

where a and b are determined by some initial conditions. We recall that the optic mode coordinate $x = x_1 - x_2$ is the separation of the positive and negative ions from their equilibrium positions. Equation C3 says the ions move apart linearly in time. This motion must be modified very quickly however by the close approach to other ions which brings in strong repulsive forces. To properly handle the transition we must put in these higher order forces from the start. To do this we rewrite the t.o. mode equation as:

C4)
$$\frac{m_1 m_2}{m_1+m_2}\ddot{x} = (-xk_{eff}+\beta x^3-\gamma x^5) + zE.$$

On comparing with our previous equation of motion B6, we have lumped all the previously defined forces into k_{eff}. k_{eff} may be positive zero or negative depending on the size of the local field effects. We also introduce z for the local-field-corrected z_1 for simplicity. The β and γ terms are nonlinear force terms which become important for large x or if k_{eff} goes to zero. The new restoring force $k_{eff}x - \beta x^3 + \gamma x^5$ may be integrated twice to give the potential shown in Fig. C1. We have chosen the parameters β and γ to fit the case of $BaTiO_3$. In Fig. C1 we choose k_{eff} to be linearly dependent on temperature. As explained earlier this effect can arise from v being dependent on temperature. At the top of Fig. C1 we show the high temperature case where k_{eff} is quite large. The system vibrates about the minimum in the well with the frequency given in the figure. As k_{eff} is lowered, the frequency drops and and the side dips caused by the β and γ terms become relatively more important. The figure shows the frequencies for these side wells, however the system will remain in the center well until we cool to T = 120°C. At 120°C a first order transition takes place to a new tetragonal phase. The ions prefer not to remain in the old equilibrium positions but move to the new position $x = x_s$. Since the ionic part of the polarization is zx/v the crystal developes a spontaneous polarization $P_s = zx_s/v$. At the same time the system vibrates in the new displaced potential well and thus has a new frequency. Figures C2-3 show plots of the mode frequency squared and of the ion shift as the crystal is cooled through T_t, the transition temperature. T_c, the Curie temperature, can be obtained by extrapolating in Fig. C2 and is defined as the temperature where the mode would have gone to zero frequency if the crystal had not changed phase. The connection with the more conventional definition of a Curie temperature will be made when we discuss the dielectric response function and its temperature dependence.

There are several points to be made about the transition we have derived. First, the crystal below T_c is no longer cubic. Because x_s has a static non-zero value, the crystal has distorted

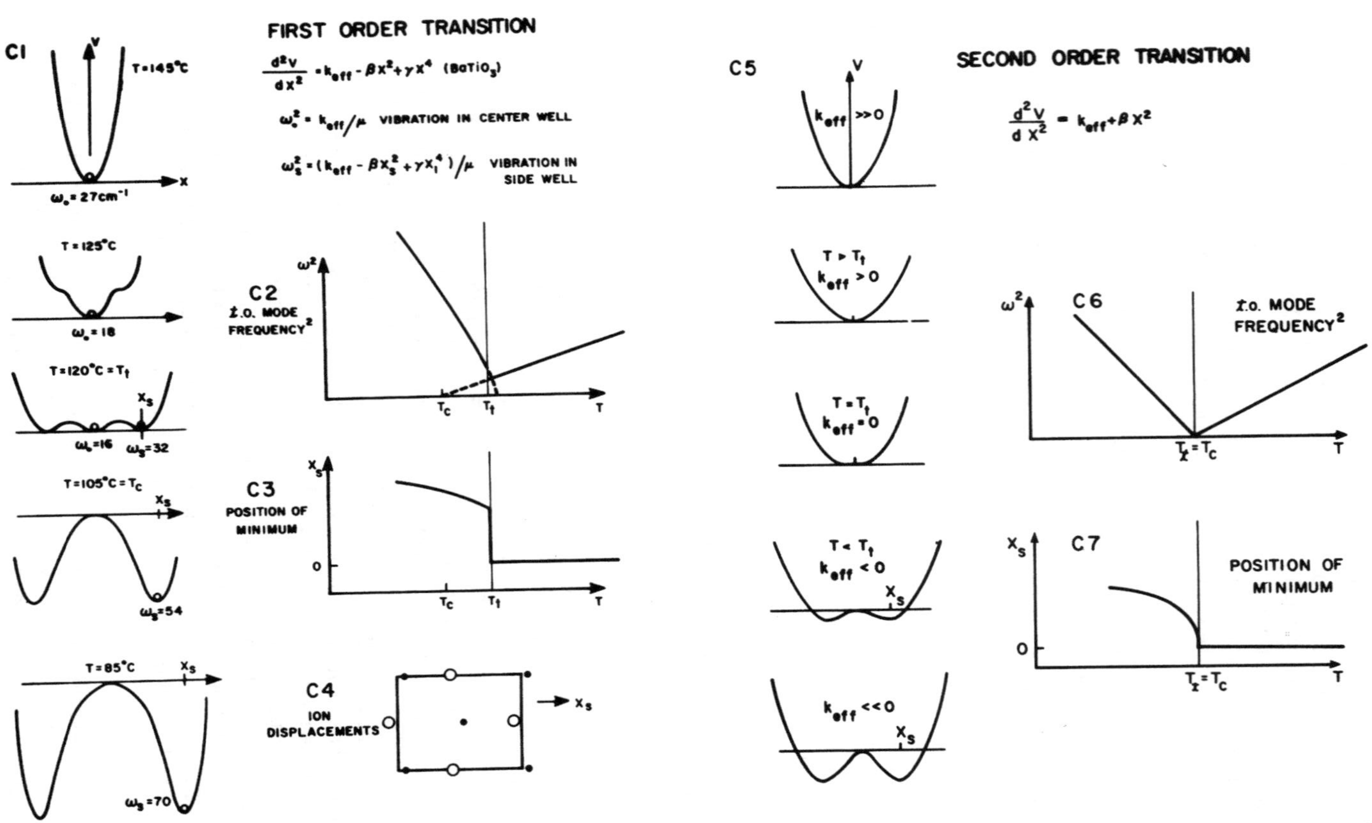

Fig. C1-C7. Potential wells, mode frequencies and spontaneous ion shifts for a first and second order ferroelectric transition. In C1 and C5 as the crystal is cooled we move from the top of the page downwards. The transition occurs at T_t, however, in C1, the sample may have to be cooled below T_t in practise to induce the ion to overcome the potential hill. This causes the hysteresis typical of first order transitions.

to a tetragonal phase with the c-axis along x_s. Figure C4 shows one plane of ions for $T < T_t$. The instability can of course develop not only along +x (as shown in Fig. C4) but also along -x, or ±y or ±z. In fact a real crystal shows this ambiguity by choosing different directions for x_s in different regions of the crystal giving the characteristic domain pattern of a ferroelectric. Application of an electric field along the x axis obviously can force the system into one large domain with all ions choosing the same-side minimum well in the low temperature phase.

A second point is that the crystal no longer has a center of inversion on each ion. This is obvious from comparing Fig. A1 with C4. The lack of inversion symmetry below T_t causes the new t.o. mode to be Raman active as well as infrared active for this model. We will see other examples of mode activity changing at a transition in later discussions.

A third point worth emphasizing now is that while we have focused upon an unstable mode, $\omega_{t.o.}$, at $k \simeq 0$, the mathematics could be applied to other modes. For example, the optic mode at $k = \pi/a$ in Fig. A2 will also have local field corrections which could cause it to go towards zero frequency on cooling. First principles calculations have in fact suggested that this mode is more likely to go unstable than the one at $k \simeq 0$ in $BaTiO_3$ though this result is contrary to experiment.

In Figs. C5-7 we sketch the potential well, mode frequency and static ion displacement x_s for a case of a second order transition. While we won't discuss this case in detail, we note that the mode actually does go to zero frequency (in this ideal case) and x_s does not make a jump at T_t but rises smoothly. In the ideal cases presented here we see that the ferroelectric instability may or may not have a mode drop to zero frequency.

As a final point in discussing ferroelectric transitions we recall that it was the close cancellation of the short range force by the local field correction which made a transition possible. The unit cell volume v which enters the effective restoring force may then swing the balance, making a transition occur on cooling. In practice the cancellation may take place at temperatures above the melting point of the crystal so that it is already ferroelectric with a distortion x_s when it is grown from the melt. $LiNbO_3$ comes very close to this situation. Other crystals may be approaching a ferroelectric transition on cooling but the volume does not decrease enough. Since the decrease in volume (thermal contraction) saturates in most crystals on cooling to low temperatures, a crystal may appear to be heading towards an instability but never make it. $SrTiO_3$ and $KTaO_3$ are good examples of such "nearly ferroelectric" crystals. Another possibility is that two or more different lattice vibration frequencies may be tending towards zero. For example the "ferroelectric" $\omega_{t.o.}$ vibration may be going to zero but the $k = \pi/a$ vibration reaches zero and the crystal goes to a new non-ferroelectric phase. The resultant crystalline distortion puts new forces on the $\omega_{t.o.}$ mode raising its frequency enough that further cooling does not cause it to become unstable. Examples of all of these types of behavior exist. Our knowledge of actual short range forces and local field corrections is still too poor however to allow predictions of such behavior in advance.

D. LATTICE MODE INSTABILITIES

The Antiferroelectric Case

We consider briefly now additional cases of instabilities to help understand some experimental results. Figure D1 shows a linear chain model with first, second, third and fourth-neighbor forces. We work with effective force constants only; we allow some of these force constants to be temperature dependent without specifying what provides the mechanism. We also assume linear forces, omitting higher order terms of the type included in Eq. C4. If we find an instability, we will assume that the ions shift in the obvious directions to new minima in potential wells appropriate to the neglected nonlinear terms. The second neighbor force constant k_1^2 which links type 1 ions is of particular interest. In Fig. D2-3 we see that for the force constant changes shown, an instability develops at wavevector $k = \pi/a$. The displacement pattern for this mode is shown in Fig. D4 and below it (Fig. D5) we show the instability actually accomplished with new ion positions. We now have a new crystal structure with the repeat distance 2a rather than a. Figure D6 shows the new lattice vibration frequencies assuming reasonable force constants for the new crystal structure. The mode displacement patterns for the distorted crystal show that we now have at $k = 0$ two infrared active modes and one Raman active mode instead of the single infrared active mode which exists for temperatures above the transition temperature T_t (Fig. D2). In addition, contrary to the case of the ferroelectric transition where the distortion x_s carried charge z to give a static dipole moment $P_s = x_s z$ in the distorted phase, here we find the polarization $P_s = x_s \cdot z_1 + 0 \cdot z_2 - x_s \cdot z_1 + 0 \cdot z_2 =$ zero. The "frozen in" displacement pattern in the distorted phase is antiferroelectric with two cancelling dipoles in each of the new unit cells.[5] To summarize, the transition discussed here is connected with a mode near $k = \pi/a$ becoming unstable. There is no spontaneous polarization, however new $k \sim 0$ optic modes appear below the transition. A more realistic model with Lorentz correction included for all forces shows that the same volume changes that cause the zone boundary mode to go unstable, also lower the frequency of the $k \sim 0$ zone center optic mode slightly. For the simple antiferroelectric

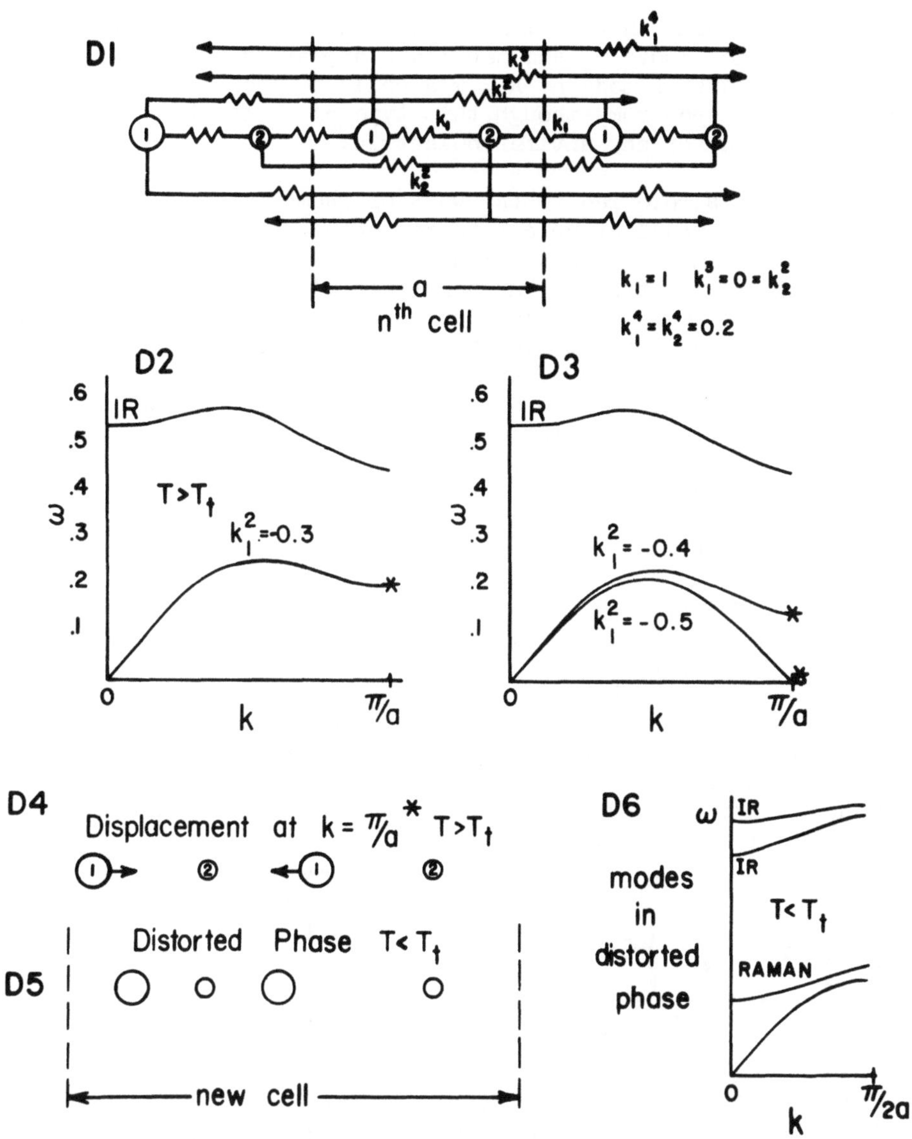

Fig. D1-D6 - One dimensional model of an antiferroelectric. D1 shows the ions, bonds and unit cell above the transition. D2 shows the corresponding lattice modes. In D3 the restoring force k_1^2 has been lowered to -.4 and then to the critical value -.5. D4 shows the ion displacements for the unstable mode. Below are shown the new ion positions and in D6 the new lattice modes. A detailed model including anharmonicity shows that the new Raman mode in D6 rises from very low frequencies as the crystal is cooled below T_t.

transition discussed here which doubles the unit cell, first order infrared or Raman spectra above T_c will show no modes becoming unstable contrary to the ferroelectric case. We can expect however that the spectra will show a slight softening (Frequency lowering) of the optic modes above T_t and new modes appearing below T_t on cooling. The unstable mode at $k = \pi/a$ could be seen optically in the second order spectrum, though this type of observation has not yet been reported for any antiferroelectric.

The reader is referred to the work of Thomas and Muller for a three dimensional model of an antiferroelectric transition.[13] Their lattice vibration model is appropriate to the perovskite crystal structure and is geometrically much more complicated than the model presented here. The same general features are found to hold however. Cochran and Zia have also studied the problem of antiferroelectricity in several real crystals.[14] They study the frozen in displacement pattern below T_t and from this deduce the types of mode which can be unstable above T_t. This latter work emphasizes the complication which can arise in a real crystal where several modes may go unstable at the antiferroelectric transition.

In the next chapter we define the dielectric response function at $k = 0$ in terms of the $k = 0$ polarization. The ferroelectric instability is signaled by the divergence of the $k = 0$ response function at the transition i.e., by the appearance of a $k = 0$ polarization wave in response to a negligible driving force. It is worth noting here that the antiferroelectric transition can be thought of as the divergence of a $k \neq 0$ response function. In Fig. D3-4 it is the longitudinally polarized wave at $k = \pi/a$ which has a divergent response as $T \rightarrow T_t$. Fig. D7 shows the transversly polarized $k = \pi/a$ (1,1,1) wave which goes unstable in $SrTiO_3$ at the 110°K antiferroelectric transition.[15]

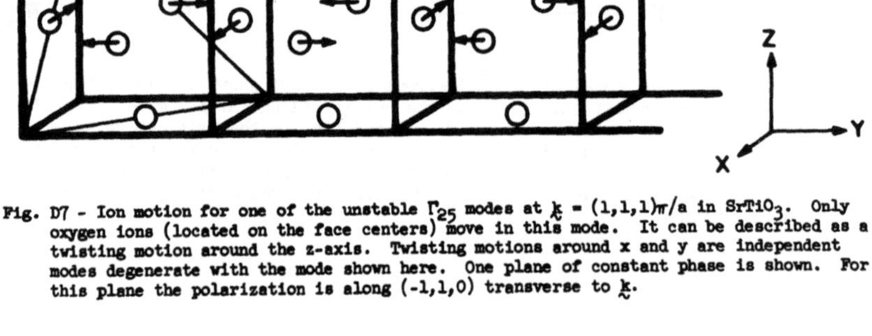

Fig. D7 - Ion motion for one of the unstable Γ_{25} modes at $\underset{\sim}{k} = (1,1,1)\pi/a$ in $SrTiO_3$. Only oxygen ions (located on the face centers) move in this mode. It can be described as a twisting motion around the z-axis. Twisting motions around x and y are independent modes degenerate with the mode shown here. One plane of constant phase is shown. For this plane the polarization is along (-1,1,0) transverse to $\underset{\sim}{k}$.

E. THE INFRARED RESPONSE - THE DIELECTRIC FUNCTION

We must now return to our original equations for the t.o. lattice vibration mode to discuss the effects of these vibrations on the optical properties. We will find strong effects in the absorption and reflection of light near $\omega_{t.o.}$ and $\omega_{l.o.}$. Since typically $\omega_{t.o.}$ and $\omega_{l.o.}$ lie in the frequency range 3×10^{12} to 3×10^{13} cps, we use infrared techniques for the direct observation of the mode frequencies. The infrared behavior is most conveniently described in terms of the dielectric function $\varepsilon(\omega)$. For our model of the transverse lattice vibrations we found

E1)
$$\frac{m_1 m_2}{m_1+m_2}\ddot{x} = -k_{eff}\, x + zE$$

and

E2)
$$P = zx/v + \alpha E/v.$$

In these equations k_{eff}, α, and z have all been modified by the local-field correction. Trying $e^{i\omega t}$ time dependences for all quantities

E3)
$$x = x_o e^{i\omega t}$$
$$P = P_o e^{i\omega t}$$
$$E = E_o e^{i\omega t},$$

we find

E4)
$$x_o = \frac{zE_o/m}{k_{eff}/m-\omega^2},$$

E5)
$$P_o = \frac{z^2E_o/mv}{k_{eff}/m-\omega^2} + \alpha E_o/v .$$

We have simplified notation by using m for the effective mass $m_1m_2/(m_1+m_2)$. Using the definition

$$\varepsilon = 1 + 4\pi P/E,$$

we find

E6)
$$\varepsilon(\omega) = 1 + 4\pi\alpha/v + \frac{4\pi z^2/mv}{k_{eff}/m-\omega^2} .$$

This is the standard classical oscillator dispersion relation.[9] Since for very high frequencies $\varepsilon \to 1 + 4\pi\alpha/v$, we call these constants the high frequency dielectric constant,

$$\text{E7)} \qquad \varepsilon_\infty \equiv 1 + 4\pi\alpha/v.$$

ε has a resonance at the frequency $\sqrt{k_{eff}/m}$. This is the t.o. mode frequency (Eq. B7). We can write

$$\text{E8)} \qquad \varepsilon(\omega) = \varepsilon_\infty + \frac{4\pi z^2/mv}{\omega_{t.o.}^2 - \omega^2}.$$

For an isotropic crystal, the index of refraction n is just

$$\text{E9)} \qquad n(\omega) = \sqrt{\varepsilon(\omega)}.$$

The problem of infrared waves incident on a sample crystal is a boundary value problem in which the fields are matched using E8. We produce one standard formula here for the power reflectivity of a very thick sample at normal incidence,

$$\text{Refl} = \left| \frac{n-1}{n+1} \right|^2 .$$

Figure E1-3 shows the three most useful kinds of experimental measurement and the results for the case of $\varepsilon(\omega)$ appropriate to a KBr crystal. In this figure we have included some damping in (Eq. E8) which is typical of lattice vibrations and which will be discussed below. From Fig. E4 we see that the response function has one peak and one zero. The peak comes of course at $\omega_{t,o.}$. It is left as an exercise to show that the zero comes at $\omega_{\ell.o.}$, the longitudinal optic mode frequency. The response function $1/\varepsilon$ which is sometimes useful in describing a material is of course peaked at $\omega_{\ell.o}$ and has a zero at $\omega_{t.o.}$

The first measurements of unstable ferroelectric modes consisted of observing the lower reflectivity shoulder (c.f. Fig. E1) of $SrTiO_3$ drop in frequency when the crystal was cooled. The higher frequency shoulder did not shift with temperature. These results show that $\omega_{\ell.o.}$ is stable but $\omega_{t.o.}$ is unstable.[9] $\omega_{t.o.}$ has been observed to drop to less than one fifth of its room temperature value in $SrTiO_3$ and $KTaO_3$.[12]

F. THE DIELECTRIC FUNCTION WITH DAMPING

In any real crystal, the t.o. or ℓ.o. mode amplitudes decay if they are initially excited. This leads to finite linewidths for the modes and imaginary parts in both $\varepsilon(\omega)$ and the index $n(\omega)$.

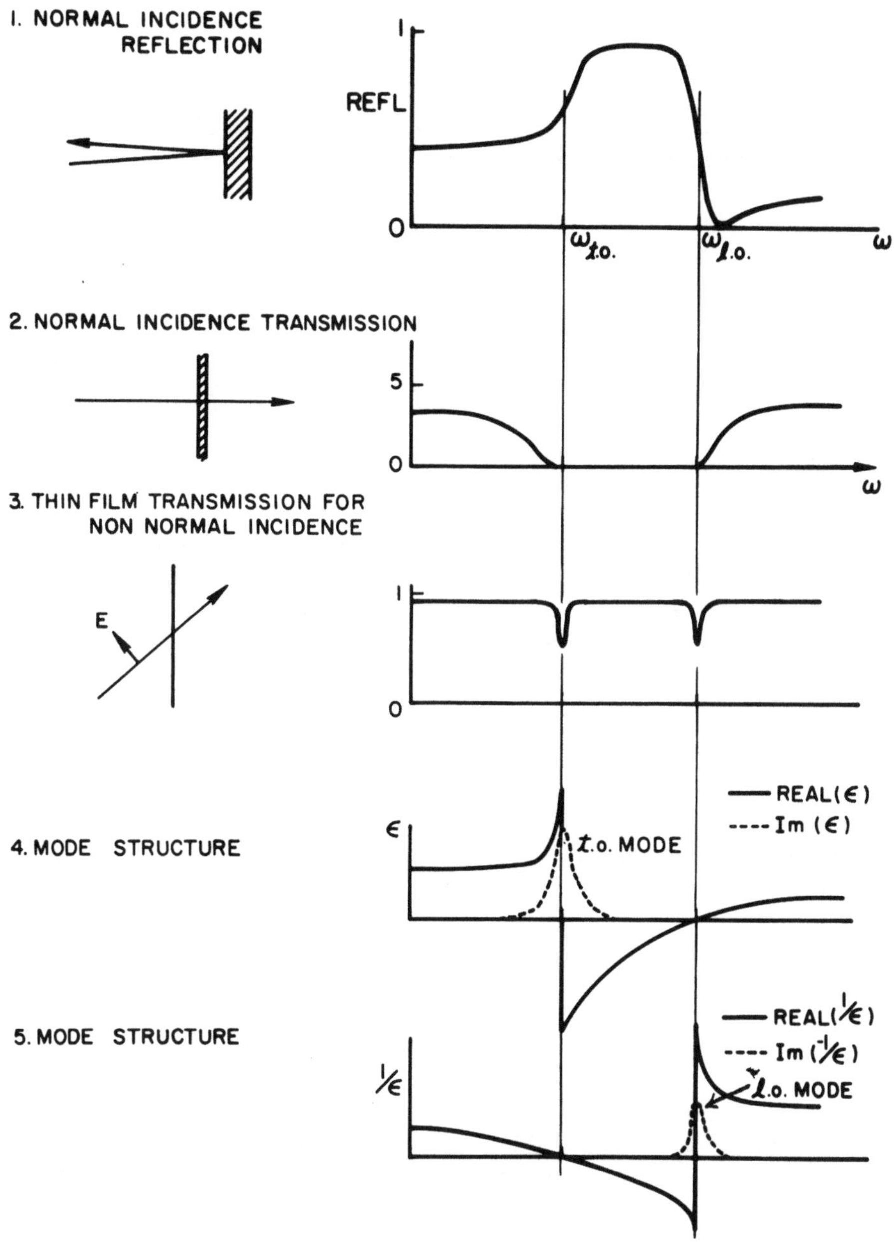

Fig. E1-E5 - Infrared spectra for three types of sample configuration. The sample is taken to have the idealized 1 oscillator mode structure given at the bottom of the figure with one t.o. frequency, one l.o. frequency and a constant (frequency independent) damping factor. Any real crystal will require a frequency dependent damping factor for a detailed fit to its spectra.

The simplest model for damping consists of putting a velocity dependent viscous damping force in the equation of motion

F1) $$m\ddot{x} = -k_{eff}x - m\gamma\dot{x} + zE ,$$

where γ describes the strength of the damping. Following our previous derivation for $\varepsilon(\omega)$ we now obtain

F2) $$\varepsilon(\omega) = \varepsilon_\infty + \frac{4\pi z^2/mv}{\omega_{t.o}^2 - \omega^2 + i\omega\gamma}$$

$$= \varepsilon'(\omega) - i\varepsilon''(\omega),$$

where the second line merely defines the notation for the real and imaginary parts ε' and ε''. Figures E4,5 show the real and imaginary parts of $\varepsilon(\omega)$ for a case where $\gamma/\omega_{t.o.} \sim 0.1$.

It was pointed out previously that the peak in ε occurred at $\omega_{t.o.}$ and the zero at $\omega_{\ell.o.}$. We must reexamine these statements in the presence of damping. It is useful before proceeding to introduce some new notation. We note that $4\pi z^2/mv$ has the dimensions of ω^2. Since this quantity represents the strength of the lattice vibration resonance, we define the dimensionless strength

F3) $$S = \frac{4\pi z^2}{mv\omega_{t.o.}^2} .$$

In terms of this strength, the low frequency (d.c.) dielectric constant is

$$\varepsilon_o = \lim_{\omega \to 0} \varepsilon(\omega) = \varepsilon_\infty + S.$$

We can now write Eq. F2 as

F4) $$\varepsilon(\omega) = \varepsilon_\infty + \frac{S\omega_{t.o.}^2}{\omega_{t.o.}^2 - \omega^2 + i\omega\gamma} = \varepsilon_\infty \frac{\frac{\varepsilon_o}{\varepsilon_\infty}\omega_{t.o.}^2 - \omega^2 + i\omega\gamma}{\omega_{t.o.}^2 - \omega^2 + i\omega\gamma} .$$

The zero of ε occurs where the numerator is zero and the peak or pole occurs where the denominator is zero. The frequencies where these occur are complex. We define the phonon frequencies as the real quantities $\omega_{t.o.}$ and

F5)
$$\omega_{\ell.o.} = \sqrt{\frac{\varepsilon_o}{\varepsilon_\infty}}\,\omega_{t.o.}\,.$$

It is clear that, since $4\pi z^2$ cannot be measured in an independent way, we might as well regard $\omega_{t.o.}$, $\omega_{\ell.o.}$, ε_∞ and γ as the basic parameters of our lattice vibration model. Figure F1 shows plots of contours of the imaginary parts of ε and $1/\varepsilon$. The peaks of these functions in complex frequency space are usually called "modes." Since the infrared beam has a real frequency, we need only evaluate ε along the real ω axis; however, it is instructive to see the actual poles and zeros (p and z) spaced back $\gamma/2$ back from the real axis. Our definitions of $\omega_{t.o.}$ and $\omega_{\ell.o.}$ correspond to the distance from the origin to the pole and zero respectively.

For our model of the lattice vibrations we have from F5

F6)
$$\frac{\omega^2_{\ell.o.}}{\omega^2_{t.o.}} = \frac{\varepsilon_o}{\varepsilon_\infty}\ . \ \text{L.S.T. relation}$$

General proofs of this Lyddane-Sachs-Teller relation have been given before.[9] The importance of this relation for ferroelectric crystals arises from the Curie Law behavior for ε_o. For ferroelectrics above their transition temperature it is found that[1-4]

F7)
$$\varepsilon_o \simeq \frac{C}{T - T_c}\ .$$

As T is lowered towards T_c, ε_o becomes very large. Experimentally ε_∞ is found to be practically unaffected by the temperature. From the Lyddane-Sachs-Teller relation (F6), we wee that either $\omega_{t.o.} \rightarrow o$ or $\omega_{\ell.o.} \rightarrow \infty$. From our earlier discussion of local field effects it is obvious that the former is more likely and this is born out by experiment. Combining F6 and F7, we obtain

F8)
$$\omega^2_{t.o.} = \omega^2_{\ell.o.}\,\frac{\varepsilon_\infty}{C}\,(T-T_c).$$

We find that, if $\omega_{\ell.o.}$ and ε_∞ are temperature insensitive, then $\omega^2_{t.o.}$ decreases linearly with decreasing temperature. This is a general result, however, we foresaw this behavior with our particular model utilizing the linear volume expansion and local field corrections. Figure F2 shows the mode behavior for a hypothetical ferroelectric at three temperatures above T_c.

$SrTiO_3$

Figure F3 shows the dielectric function of $SrTiO_3$ determined

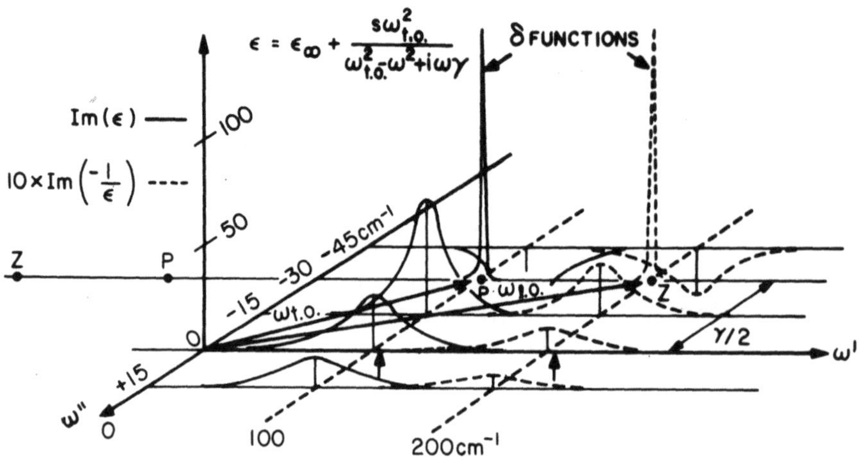

Fig. F1 - Contours of Im(ε) and Im(-1/ε) in the complex frequency plane. ε has a t.o. mode at 104.4 cm^{-1} and an l.o. mode at 202.2 cm^{-1}. The mode frequencies defined in the text are the distances from the origin to the pole and zero shown as heavy arrows. The mode structure seen along the real frequency axis (ω′) is usually obtained from analysis of an infrared spectrum. The numbers $\omega_{t.o.}$ and $\omega_{\ell.o.}$ have also been measured off along ω′ and are marked by short vertical arrows. Note that along the ω′ axis the mode peaks are damped and shifted below the mode frequencies.

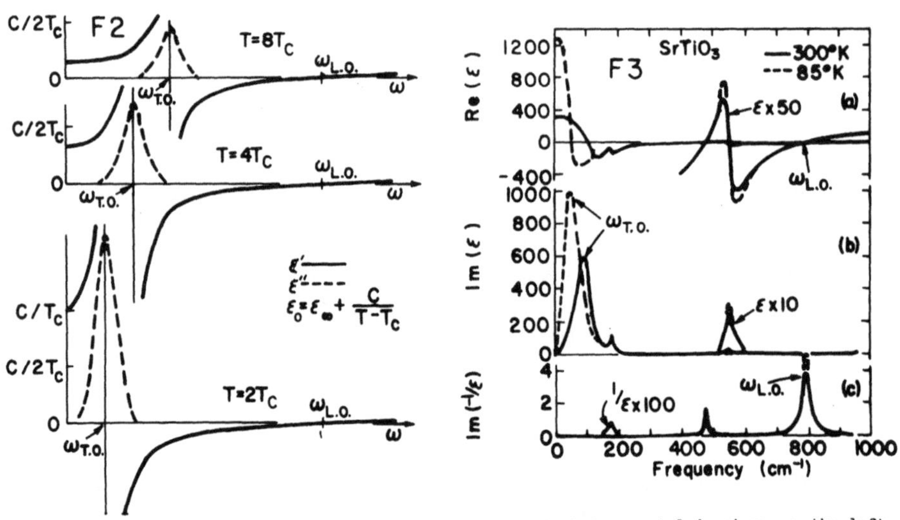

Fig. F2-3 - The model behavior of a one-mode ferroelectric crystal is shown on the left. On cooling towards T_c the transverse optic mode drops in frequency while the longitudinal optic mode remains fixed. Note that ϵ_0, the d.c. dielectric constant rises as $\omega_{t.o.}$ drops. On the right, the spectrum of $SrTiO_3$ is shown for two temperatures. At the lower temperature, the lowest t.o. mode has dropped, while the five other modes remain fixed.

by infrared reflection. $SrTiO_3$ is not ferroelectric but comes close since a k = 0 polar mode becomes quite soft. This material has three t.o. and three ℓ.o. modes. The Lyddane-Sachs-Teller relation still predicts that a t.o. mode approaches zero and this is observed.[9] The figure shows the mode shift on cooling from 300°K to 85°K. With further cooling ε_0 deviates below $\frac{C}{T-T_c}$ (T_c = 34°K) and finally saturates. Correspondingly, (Eq. F6) $\omega_{t.o.}$ deviates above the value predicted by Eq. F8 and also finally saturates,[12] reaching a limiting value of about 10 cm^{-1}.

$BaTiO_3$

In $BaTiO_3$ the picture of mode instability is much less clear because of very large damping. Figure F4 shows the best fits of ε'' using an equation like F4. These fits to the original reflectivity data have been discussed extensively elsewhere.[9] The peak of ε'' does not shift in accordance with the Lyddane-Sachs-Teller relation. This is because of the large γ. $\omega_{t.o.}$ is shown separately for the three temperatures which have been fitted. While there is a large uncertainty in each $\omega_{t.o.}$, this parameter does shift in agreement with the Curie law confirming the Cochran-type soft or unstable mode.

Hydrogen-Bonded Ferroelectrics

The hydrogen-bonded ferroelectrics (eg. KH_2PO_4) are known to have quite different thermodynamic properties from the perovskite ferroelectrics ($BaTiO_3$, $KNbO_3$, $PbTiO_3$, etc.). The hydrogen-bonded

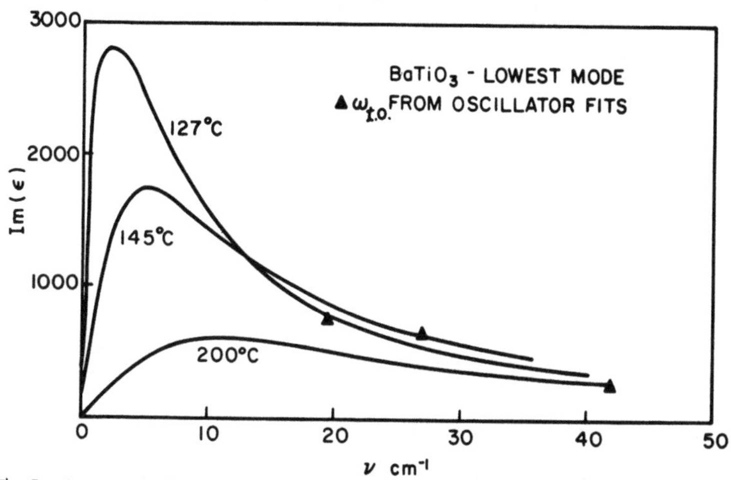

Fig. F4- Imaginary part of ε for $BaTiO_3$ taken from infrared reflection data. The lowest t.o. mode shown here is highly damped. The mode frequency $\omega_{t.o.}$ (solid triangles) is very temperature dependent and shifts in accordance with the Curie law and Lyddane-Sachs-Teller relation.

ferroelectrics have much larger entropy changes, much larger peak values for the dielectric constant at the ferroelectric transition, and smaller Curie constants than the ferroelectrics with ionic binding. Though the large entropy change suggests an order-disorder transformation, Cochran has suggested that, at least in KH_2PO_4, the transition involves an unstable optic mode, as it does in the perovskites.

There have been many conflicting reports of "ferroelectric behavior" of infrared modes particularly in KH_2PO_4 and KD_2PO_4. There must be a region of intense absorption in the infrared or microwave region connected with the large value of ε_0. The absorption may or may not be a phonon mode in the usual sense. A study of the infrared spectrum of triglycine sulfate (TGS) has shown that the strong "ferroelectric mode", i.e., the absorptive mode connected with the spontaneous polarization which appears below T_c must lie at frequencies below 14 cm^{-1}. For KH_2PO_4 at room temperature ($T-T_c$ = 300-123 = 177°K), a large part of the ferroelectric mode was observed as a broad Debye type peak.[9] However, near T_c this shifted downwards in frequency showing that most of the absorption must occur below 1.25 cm^{-1} for temperatures near T_c. Hill and Ichiki have investigated the dielectric spectrum of TGS and of KD_2PO_4 in the radio and microwave region and find the ferroelectric mode.[9] Figure F5 shows their result for ε'' of TGS replotted with the data smoothed somewhat. The data shown in Fig. F5 resembles a Debye dispersion function where

$$\varepsilon(\text{Debye}) = \varepsilon_\infty + \frac{S_D}{1+i\omega\tau} . \qquad \text{F9)}$$

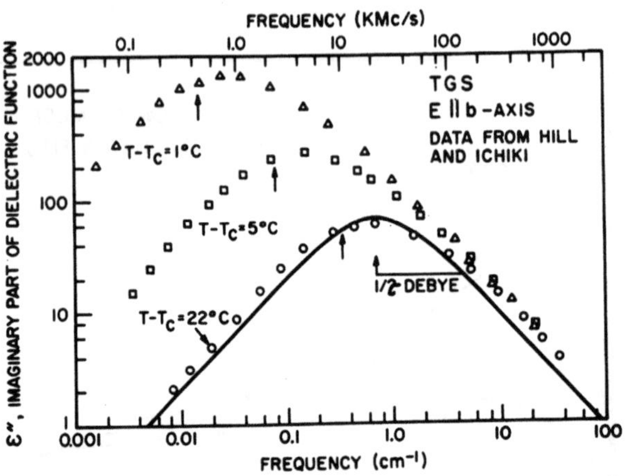

Fig. F5 - Broad ferroelectric absorption in triglycine sulfate for three temperatures above T_c. This type of mode is best described by the relaxation frequency defined by Hill and Ichiki and marked by the vertical arrows. The solid curve is a simple Debye function fit using the relaxation frequency $1/\tau$.

As the temperature is lowered towards T_c, S_D increases and the frequency where ε'' peaks, decreases. Hill and Ichiki fit their data not with Eq. (F9) but with a Gaussian distribution of Debye functions and find that τ_0, the relaxation time associated with the width of the distribution, varies as

F10) $$\tau_o \propto \frac{1}{T-T_c} .$$

Actually for rough purposes, a single Debye relaxation can be fit to the data of Fig. F5 (see solid curve) though again it is found that the relaxation time must obey Eq. F10, which replaces Eq. F8 in determining the characteristic mode softening for these very broad-mode situations.[9] While Eq. F10 looks quite different than Eq. F8 for mode instability, the picture in terms of the motion of the pole of ε in the complex plane is identical. In Fig. F1 it is the motion of P towards the origin which constitutes an instability. If its path is close to the imaginary frequency axis, the mode is highly damped. F10 is then a more appropriate description of the instability and ε obeys the Debye relation (F9). A new L.S.T. relation can also be derived in this extreme high damping limit.[9]

Cochran has treated a model of the ferroelectric mode in a piezoelectric material based on the zinc-blende structure. The model suggests that while the optic mode does drop in frequency with cooling, an acoustic mode also drops and actually becomes unstable first.[16] The recent results of Gammon and Cummins[2] in detecting unstable acoustic modes by Brillouin scattering suggests this approach may be more useful than measuring the broad ferroelectric mode. In the next chapter we show that Raman scattering can also give a measure of the infrared dispersion and show the instability of the optic mode.

The Antiferroelectric Crystal $NH_4H_2PO_4$

In Section D it was pointed out that for an antiferroelectric above T_t, there can be a softening of the optic modes on cooling, though an instability will occur for a zone boundary mode (or perhaps a zone center, infrared inactive mode) first. This softening has been measured by the present author in $NH_4H_2PO_4$ by the Raman effect. Since the lowest frequency mode is both Raman and infrared active, a comparison is possible with the 300°K infrared spectrum of Kawamura et al.[17] Both types of spectra show a broad mode at 300°K with ε_a'' peaked near 60 cm^{-1}. The Raman spectrum shows that the peak moves somewhat lower on cooling; but not nearly to as low a frequency as the low frequency peak in ε_c'' for ferroelectric KH_2PO_4.[18]

Summary of Dielectric Response

We have emphasized the importance of the Lyddane-Sachs-Teller relation in connecting Curie Law behavior of a ferroelectric with mode instability. Equation F6 shows that for $\omega_{t.o.} \to 0$, the response ε_o must diverge. Although we started with a particular model there are more general model independent arguments which relate the instability of a pole or peak in the response function to the strong temperature dependence of the dielectric response.[9] Measurements of the infrared reflectivity yield valuable information on the position and temperature dependence of the pole. Neutron diffraction can also show the low frequency pole and in addition its dispersion in k-space.[11] The next section discusses the more recent Raman measurements which with the advent of laser sources give in many cases spectra with better signal to noise allowing more detailed studies of unstable lattice modes.

G. RAMAN SPECTRA OF FERROELECTRIC CRYSTALS

We first develop a simple expression for the Raman scattering induced by a laser beam incident on a crystal with a thermally excited lattice vibration which is both Raman and infrared active. Many ferroelectric crystals have such optic modes. Recently Fleury and Worlock have induced Raman activity in Raman-inactive, infrared active modes using a large d.c. electric field.[12,15] The present treatment which focuses on the infrared dielectric function ε, has application to both natural and induced activity.

For a Raman active mode the thermal vibrations of the lattice at frequency ω disturb the electronic polarizability α, causing scattering of an incident laser beam and a shifting of its frequency by $\pm\omega$. Our reason for presenting the following argument is to allow analysis of the scattering in terms of $\varepsilon(\omega)$. For sharp δ-function lattice modes one just picks the lattice mode out of the scattered light (Raman) spectrum. The mode appears as a pair of side bands spaced by ω above and below the laser line. With very broad modes we find we must fit the spectrum to a model to analyze for the lattice mode ω. We can not merely take the peaks of the spectrum when looking for unstable modes which are broad.

The fluctuation-dissipation theorem or Nyquist's theorem relates the thermal fluctuation in a variable to the response function for that variable.[19,20] For example the fluctuating (noise) voltage across a resistor is related to the resistance. The fluctuating current is related to the conductance. We start with the high temperature (classical limit) form of Nyquist's theorem familiar in circuit theory. For a conductor of conductance G at temperature T the current fluctuation power spectrum is[19,20]

$$\overline{I^2}(\omega) = 4kT\,\mathrm{Re}(G). \tag{G-1}$$

To relate this circuit approach to our dielectric crystal, we consider a slab shaped crystal of area A and length ℓ, then

$$G = \sigma A/\ell \tag{G-2}$$

where σ is the conductivity function. The conductivity is just

$$\sigma = i\omega\varepsilon \tag{G-3}$$

so we can relate G on the right side of Eq. G1 to the dielectric function of the crystal. We next consider the current fluctuations $\overline{I^2}$. The current in the sample comes from the charge motion associated with the polarization P. Since the current density associated with the ion position x is $J = z\dot{x}/v$ we have

$$I = z\dot{x}A/v = i\omega zxA/v \tag{G-4}$$

The ionic polarization is $P = zx/v$ so

$$\overline{I^2} = \omega^2 A^2 \, \overline{P^2}. \tag{G-5}$$

Substituting G5 and G2-3 into G1, we obtain

$$\overline{P^2}(\omega) = \frac{4kT\,\mathrm{Im}(\varepsilon)}{\omega v} \tag{G-6}$$

as the Nyquist theorem relating the thermal fluctuation power spectrum of the polarization to the dielectric function. Of particular importance is the factor ω in the denominator. We will show below that the Raman effect usually measures $\overline{P^2}$, so that a peak seen in $\overline{P^2}$ is not the same peak as seen by measurements which give $\mathrm{Im}(\varepsilon)$. For broad peaks the distortion caused by the $1/\omega$ factor can be appreciable.

We must now make a basic assumption about the mechanism of the Raman scattering. The usual procedure is to use the first term in a power expansion of the electronic polarizability induced by lattice polarization and we follow this course.

$$\alpha = \alpha_o + \alpha' P \tag{G-7}$$

Equation G7 says that the electrons have a polarizability α_o in the absence of P and, when they are squeezed by a lattice distortion P, the polarizability changes. The coefficient α' is zero for Raman inactive crystals for which the electronic polarizability α is unaffected by the lattice vibrations. If we apply an electric field (from a laser) $E_\ell \cos(\omega_\ell t)$, there will be an induced polarization

$$P^i = \alpha E_\ell \cos(\omega_\ell t) \tag{G-8}$$

We are now neglecting explicit mention of local-field effects. We assume the crystal has a thermally vibrating polarization P cos ωt. Using Eq. G7 for α and retaining only the α' term, the polarization due to P and E_ℓ is

$$(G\text{-}9) \qquad P^i = \frac{\alpha' P E_\ell}{2} \cos(\omega_\ell + \omega)t + \frac{\alpha' P E_\ell}{2} \cos(\omega_\ell - \omega)t.$$

This equation shows that the induced polarization has sidebands proportional to α', P, E_ℓ centered about ω_ℓ. The laser signal E_ℓ cos $\omega_\ell t$ is present also due to the α_o term in Eq. G7. Equation G9 is for one spectral component of lattice polarization vibrating at frequency ω. If there is an envelope of lattice vibrations with frequencies near ω, then the sidebands in P^i will be envelopes of the same shape centered near $\pm\omega$ about ω_ℓ. To make the final connection with the linear response function ε, we note that the induced polarization has a power spectrum proportional to $(P^i)^2$. Defining the Raman signal, S, as the scattered power per unit of incident laser power we obtain:

$$(G\text{-}10) \qquad S(\omega_\ell \pm \omega) \sim (P^i)^2/E_\ell^2$$

$$\sim \alpha'^2\, \overline{P^2(\omega)}$$

$$\sim \frac{\alpha'^2\, kT\, \mathrm{Im}(\varepsilon)}{\omega}\ .$$

The full derivation of S to obtain all the numerical factors involves substituting P^i into the formula for an oscillating dipole to find the radiated field.[21,22] In addition the factor kT must be replaced by the quantum factor $\hbar\omega/(\exp(\hbar\omega/kT)-1)$ for down-scattered or Stokes light $(\omega_o - \omega)$. For the up-scattered or anti-Stokes light $(\omega_o + \omega)$ the correct thermal factor consists of the replacement $kT \rightarrow \hbar\omega/(1-\exp(-\hbar\omega/kT))$. Both of these factors approach kT in the classical limit of $kT \gg \hbar\omega$.

Figure G1 shows the phonon sidebands on each side of the laser line for $LiNbO_3$.[23] The lines on the low frequency side are stronger because at the temperature of the experiment the classical form of Eq. G6 and G10 is not correct. The thermal factor kT must be replaced by the quantum factors above and the asymmetry seen in the spectrum results. The lattice vibration mode marked by an arrow in Fig. G1 is analogous to the unstable mode in $BaTiO_3$ below T_t. Though $LiNbO_3$ has hexagonal crystal form both above and below T_t, it is closely related to the cubic perovskite form of some other ferroelectric crystals. The temperature dependence of the designated

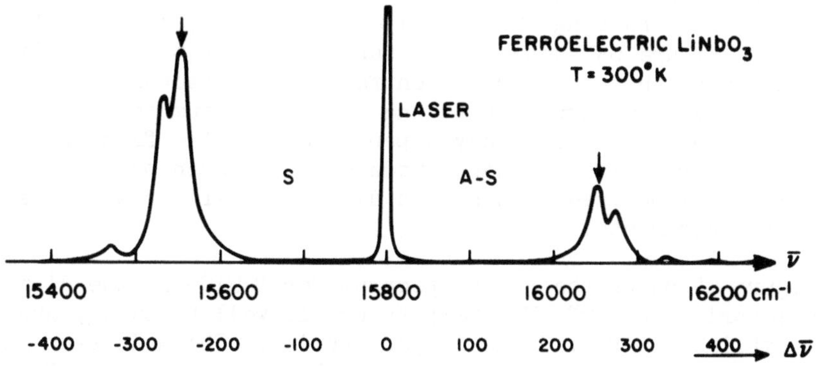

Fig. G1 - Raman spectrum of $LiNbO_3$ taken below T_t using a He-Ne laser. Both Stokes (S) and antistokes (AS) shifts are shown. The arrow marks the temperature dependent mode connected with the transition.

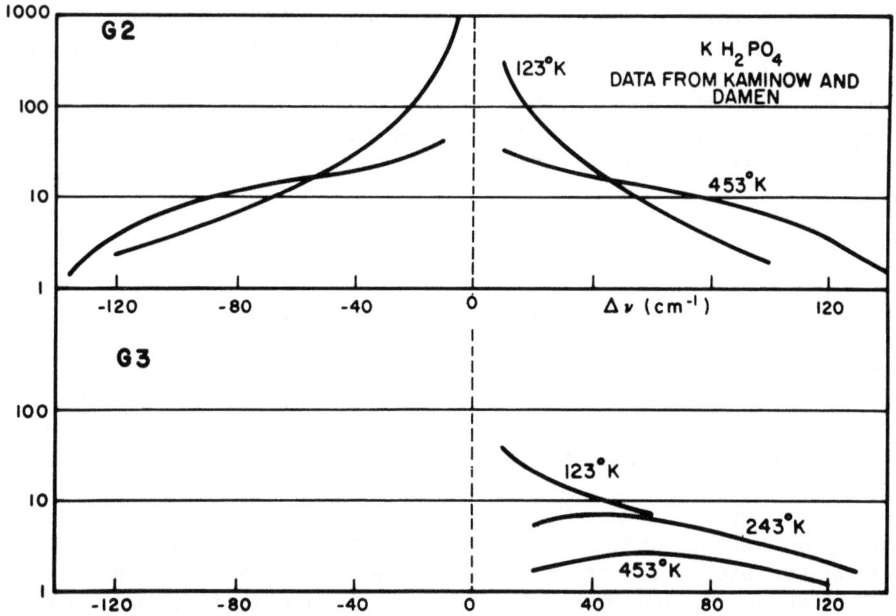

Fig. G2-3 - The Raman spectrum of KH_2PO_4 above the transition temperature. The lower figure shows the signal corrected for the ω^{-1} factor and quantum factor to give the dielectric response function which is very broad.

mode and its symmetry type show that it is related to the unstable mode above T_t which causes the ferroelectric transition.[23]

Figure G2 shows a Raman spectrum of KH_2PO_4.[18] Here the modes are of such low frequency and so broad that the $1/\omega$ correction is important in extracting the actual mode shape Im(ε) from the Raman signal S. Figure G3 sketches the subtraction of the laser line from the data, then the correction for the $1/\omega$ factor. Another correction is necessary because the Stokes quantum factor used in place of kT also has frequency dependence. In the final stage in Fig. G3 the broad low frequency mode of a hydrogen bonded ferroelectric is recognized as has been discussed earlier with the infrared experiments.

Figure G4 shows the Raman spectrum for $BaTiO_3$ in the distorted ferroelectric phase.[24] The temperature is well below T_t, and the model outlined in Fig. C3 no longer applies. The mode seen in Fig. G4 in the ferroelectric phase is also soft because $BaTiO_3$ makes another transition to another ferroelectric phase at T = 0° C.[25] The crystal distortion near T_t = 120°C makes this mode Raman active allowing direct measurement by Raman scattering. Cochran[8] has discussed the form of the short range forces and their anharmonicities needed in BaTiO to cause three modes to become unstable in succession, causing the transitions near 120° C, 0° C and -80° C. Figure G4 shows the position of the mode in the temperature range 10 to 60° C. For $BaTiO_3$ the $1/\omega$ factor and large damping parameter contained in $\varepsilon(\omega)$ remove all peaked structure from the Raman signal.

Alternate Derivation of Raman Response

For completeness we now give another argument for the form of the polarization fluctuations which will include both the transverse and longitudinal modes in a dielectric. For any transport quantity there are two important descriptive properties; the driving force which couples to the flow or current and the response function, Y, which is the ratio of the flow to the driving force. We give an entirely heuristic argument and develop the driving force by considering the stored energy H′. H′ is the perturbation term in the Hamiltonian due to the forces. This simple argument cannot be made rigorous in the general case of a lossy dielectric since a Hamiltonian formation is inapporpriate when there is dissipation. We will proceed by analogy with the high temperature (classical) form of Nyquist's theorem for a circuit element. To prevent cumbersome expressions, we use very loose notation omitting the taking of real parts of complex quantities, and using proportionalities rather than equalities.

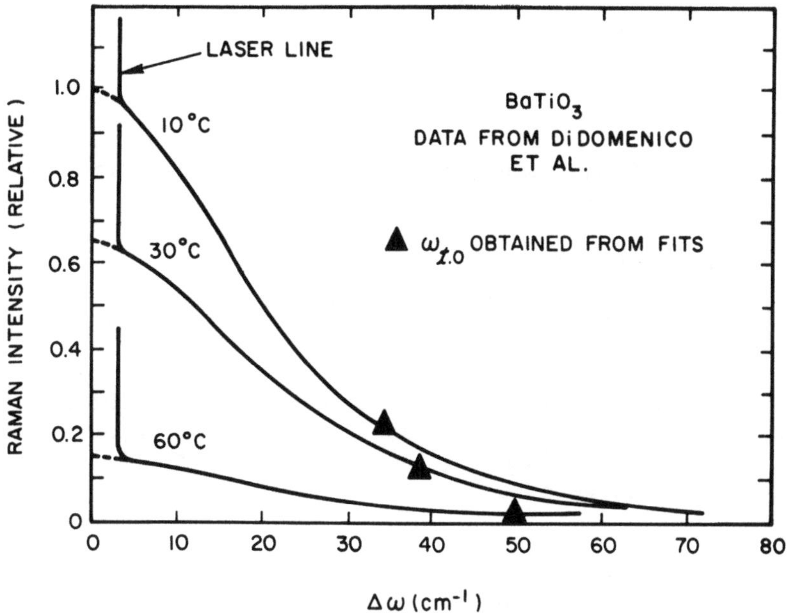

Fig. G4 - Raman spectrum of $BaTiO_3$ in the ferroelectric phase. The sample is being cooled towards the second phase transition near 0°C. Analysis of these spectra give the mode frequency shown.

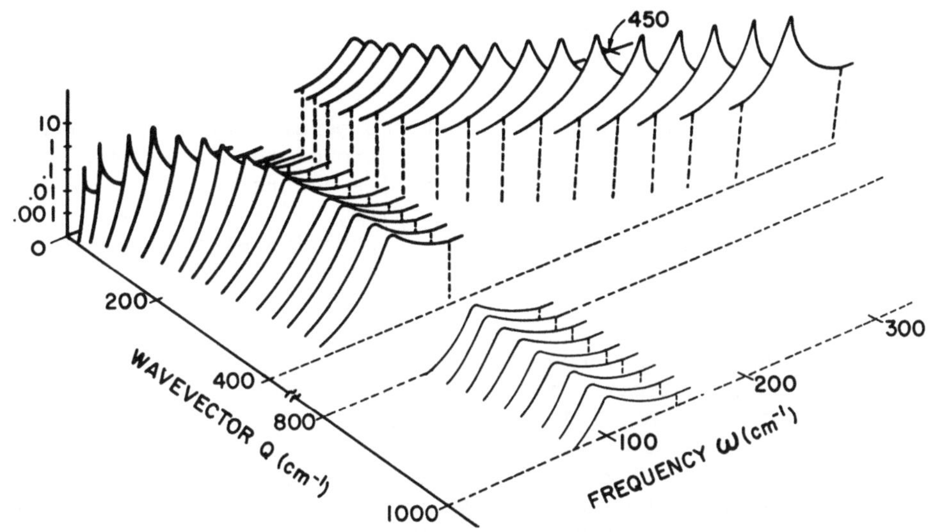

Fig. G5 - Imaginary part of the electric field response function $\overline{E^2}(\omega,Q)$ arbitrarily normalized. The vertical scale is logarithmic. The response function applies to a simple 1-oscillator dielectric with $\omega_{t.o.} = 100$ cm^{-1}, $\omega_{\ell.o.} = 200$ cm^{-1} and $\gamma/\omega_{t.o.} = 0.15$. The peaks show that most of the response occurs near the usual polariton curve defined by $\omega = Qc/\sqrt{\varepsilon}$ where all quantitites are taken to be real.

The response function is the admittance $Y = I/V$ and the Nyquist's theorem for the current power spectrum is[19,20]

$$\overline{I^2} \sim kT \, \mathrm{Re}(Y)$$

For the dielectric, the stored energy is $E_{app} \cdot P$. We wish to bring the polarization P into the problem as the variable corresponding to current. The obvious correspondences are

$$\left.\begin{array}{l} I \leftrightarrow i\omega P \\ V \leftrightarrow E_{app} \end{array}\right\} \quad \text{so that } \overline{I^2} \sim kT \,\mathrm{Re}\, Y \text{ becomes} \quad \overline{P^2} \sim \frac{kT}{\omega^2} \mathrm{Re}\left(\frac{i\omega P}{E_{app}}\right) = \frac{kT}{\omega} \mathrm{Im}(P/E_{app})$$

At optical frequencies we must recognize that any vector fields such as P have independent solenoidal and irrotational components which are both significant. Denoting these components by t(transverse) and ℓ(longitudinal) we have

(G-11)

$$\frac{t}{E_{in} = E_{app}} \qquad\qquad \frac{\ell}{E_{in} = E_{app}/\varepsilon}$$

$$\frac{P}{E_{app}} = \frac{P}{E_{in}} = \frac{\varepsilon-1}{4\pi} \qquad\qquad \frac{P}{E_{app}} = \frac{P}{\varepsilon E_{in}} = \frac{\varepsilon-1}{4\pi\varepsilon}$$

(G-12)

$$\overline{P^2} \sim \frac{kT}{\omega} \mathrm{Im}\left(\frac{\varepsilon-1}{4\pi}\right) + \frac{kT}{\omega} \mathrm{Im}\left(\frac{\varepsilon-1}{4\pi\varepsilon}\right)$$

(G-13)

$$\overline{P^2} \sim \frac{kT}{\omega} \mathrm{Im}(\varepsilon - \frac{1}{\varepsilon})$$

For the dielectric, Eq. G13 shows that the polarization fluctuation spectrum has peaks at both the pole and zero of the dielectric function ε. The minus sign in front of $1/\varepsilon$ occurs naturally since if our convention for ε makes Im ε positive, $\mathrm{Im}(1/\varepsilon)$ will be negative. Both peaks must appear with a positive sign in the power fluctuation of a passive medium. A $1/\omega$ factor appears as it did earlier.

By solving Maxwell's equations for longitudinal and transverse plane waves in a dielectric medium characterized by dielectric function ε one finds natural modes which are transverse polarization waves at the pole of ε and which can be driven by a transverse electric field. At the zero of ε longitudinal polarization waves occur which can be driven by a longitudinal displacement (D) field. For circuit components (and low frequencies) $E_{app} = E_{in}$ so we can drive only the transverse modes. Hence Eq. G1 misses the longitudinal modes. Fits to Raman spectra using both ε and $1/\varepsilon$ terms have been carried out in GaP[26] where the modes are so narrow that the frequency dependences of the $1/\omega$ factor and the quantum thermal factor can be neglected compared with the frequency dependencies of ε and $1/\varepsilon$.

The simplest approach to the Raman spectra is to assume that the polarizability of the dielectric is modulated by the polarization fluctuations as before. This is reasonable since these fluctuations reflect ion and electron motion which can modulate the polarizability. In Eq. G7 above we used

$$\alpha = \alpha_o + \alpha' P \tag{G7}$$

There is no reason why α' should be constant - in fact fairly simple models show that it is frequency dependent and complex. We proceed however with our simplifing assumption and take α' to be constant over the width of the individual line we are studying. For the case of one infrared active mode we have for the laser induced polarization

$$P_t^i(\omega_\ell \pm \omega) = \alpha_t' E_\ell(\omega_\ell) P_t(\omega)$$

$$P_\ell^i(\omega_\ell \pm \omega) = \alpha_\ell' E_\ell(\omega_\ell) P_\ell(\omega)$$

the subscripts t and ℓ refer to transverse and longitudinal fields.

The same arguments used with Eqs. G7-9, the induced polarizations given above, and Nyquist's theorem for the fluctuations in $P(\omega)$ predict:

$$S_t(\omega_\ell \pm \omega) \sim \alpha_t^2 \frac{kT}{\omega} \, \mathrm{Im}(\varepsilon(\omega)) \tag{G14}$$

$$S_\ell(\omega_\ell \pm \omega) \sim \alpha_\ell^2 \frac{kT}{\omega} \, \mathrm{Im}(-\, 1/\varepsilon(\omega)) \tag{G15}$$

Eq. G14 has been used to analyze for ω_{to} in the case of large damping for $BaTiO_3$[24,27] and KH_2PO_4.[18,27] Equation G15 shows that if the damping is large in the region of the longitudinal mode, then its spectrum is very flat and featureless similar to the transverse mode. To examine this behavior we take the limit of very large damping of a classical oscillator.

$$\varepsilon = \lim_{\substack{\Gamma \to \mathrm{large} \\ \omega < \omega_o}} \left\{ \varepsilon_\infty + \frac{S\omega_o^2}{\omega_o^2 - \omega^2 + i\omega\Gamma} \right\} = \varepsilon_\infty + \frac{S}{1 + i\omega\tau}$$

For this Debye response function, if we assume frequency independent polarizabilities α_t' and α_ℓ', the transverse and longitudinal modes give scattering

$$S_t \sim \frac{kT}{\omega}\,\mathrm{Im}(\varepsilon) \sim \frac{kTS\tau}{1+\omega^2\tau^2}$$

$$S_\ell \sim \frac{kT}{\omega}\,\mathrm{Im}(-1/\varepsilon) \sim \frac{kTS\tau/(\varepsilon_\infty+S)^2}{1+\omega^2\left(\frac{\tau\,\varepsilon_\infty}{S+\varepsilon_\infty}\right)^2}$$

These spectra have no peaks, they are flat out to some characteristic frequency then fall off as ω^{-2}. The high frequency roll-off will be modified above the frequency $kT/\hbar$ since in this region the classical thermal factor kT must be replaced by the correct quantum form.

Finally we must caution that throughout this discussion we have used approximations which neglect retardation. Equations G14 and G15 can give a good representation of Raman scattering under the usual 90° scattering conditions. For forward scattering however, scattering from polaritons can be detected. Here fluctuations in the electric field can be important in addition to the polarization fluctuations. The response function connected with the electric field fluctuation has both frequency and wavevector dependence. For polariton scattering it is not as useful to introduce a parameter α' or even two parameters α_t' and α_ℓ'. Polariton spectra usually cover a wide frequency range hence frequency dependences of α become important. Loudon,[28] and Henry and Garrett[29] have examined detailed models for Raman scattering which specify the polarizability by parameters in an energy expansion. They obtain an expression for the Raman scattering explicitly. Without going to such a model, we can only specify the linear response function part of the scattering as indeed has been our aim throughout this section. Using the appropriate response function[30] we obtain

$$\overline{E^2} = \frac{kT}{\omega}\,\mathrm{Im}\left\{\frac{4\pi\omega^2}{c^2Q^2-\omega^2\varepsilon}\right\} \qquad \text{(G16)}$$

for the classical form of the transverse electric field fluctuations. We now use Q for the wavevector to avoid confusion with the Boltzman constant k. The Raman scattering depends on a second order response function times this linear response function contained in the curly braces. If the second order response is fairly well behaved we expect $S_t(\omega_0 \pm \omega, Q) \sim \overline{E^2}$. i.e., the linear response function in Eq. G16 will give us the main details of the scattering efficiency. Eq. G16 shows that there will be Raman peaks near the usual polariton position, i.e. where $c^2Q^2 = \omega^2\varepsilon$. Fig. G5 shows the response function contained in Eq. G16. Electric field induced polariton peaks[31] have been detected in $SrTiO_3$ and $KTaO_3$ and appear to follow Eq. G16 to within the experimental uncertainty showing that under the conditions of the experiments the nonlinear terms have no rapidly varying frequency dependences.

In summary, we have seen that the Raman process brings in an additional frequency dependent response function α. In most cases however, α appears to be smooth enough that the Raman spectrum can be analysed to give the conventional linear response ε. The linear response in the cases of the ferroelectrics which have been studied shows the characteristic low frequency soft mode connected with the ferroelectric instability. In the case of $SrTiO_3$, modes connected with the antiferroelectric transition have also been detected below the antiferroelectric transition temperature.[15]

H. FIELD INDUCED RAMAN STUDIES OF FERROELECTRICS

In a series of recent experiments Fleury and Worlock have obtained new spectra of the unstable modes in $SrTiO_3$ and $KTaO_3$.[12] These crystals have the perovskite structure with 5 ions per unit cell. For small wavevector there are 3 distinct infrared active optical modes and 1 infrared inactive optic mode. All of the optic modes are Raman inactive. A Raman spectrum of $KTaO_3$ measured in the conventional way is shown in the top of Fig. H1. The features are all second-order. None of the broad peaks corresponds to the t.o. or ℓ.o. modes which have been measured previously by infrared reflection spectroscopy.[9] Fleury and Worlock apply a large biasing electric field E_o to $KTaO_3$. This field distorts the crystal slightly causing the infrared modes to become Raman active. It turns out that the induced activity (i.e. the effective α') is strongest for the mode of most interest, the lowest frequency unstable t.o. mode. The lower part of Fig. H1 shows the Raman spectrum with E_o applied. In this experiment E_o is actually a low frequency alternating field. By only recording spectra which alternate, the large constant background of second order scattering shown at the top of the figure is suppressed. Using such a "lock-in" technique, the gain can be increased over that used to record the second order spectrum. The induced first order spectrum shown reveals the t.o. mode at 45 cm^{-1} with better resolution than provided by the direct infrared measurements.

We will develop a simple model of the induced Raman effect. This model is a direct extension of the previous lattice vibration models. The theory is the same as for the Raman effect in any crystal, however the near cancellation of the short range forces by the effective field correction causes instability and enhances the Raman effect for the ferroelectric crystals we consider. To get the Raman effect at all, we must include nonlinearities in the models we have considered. The electronic polarizability α

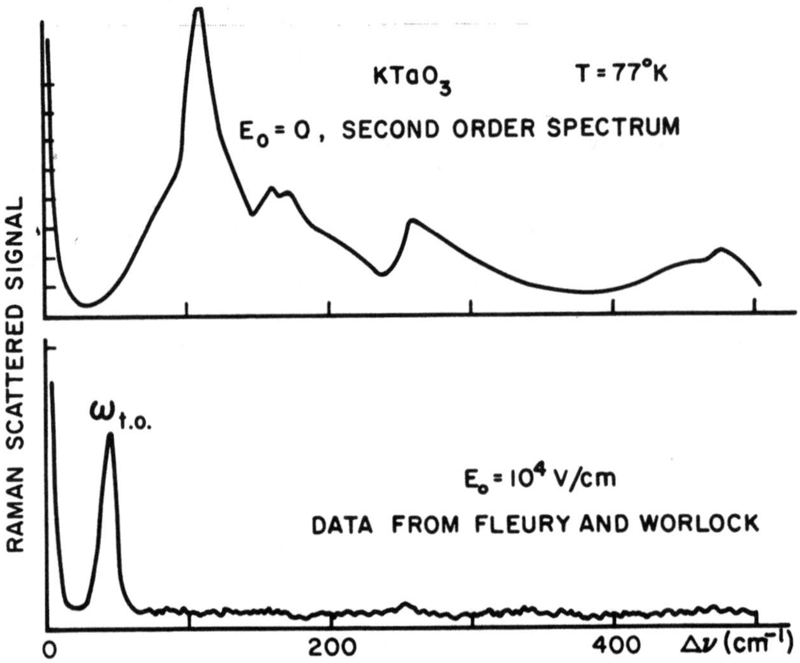

Fig. H1 - Raman spectra of $KTaO_3$ without and with an electric field E_0. E_0 is a low frequency alternating field and a lock-in technique is used to eliminate the steady second-order scattering shown at the top. This reveals the field-induced t.o. scattering.

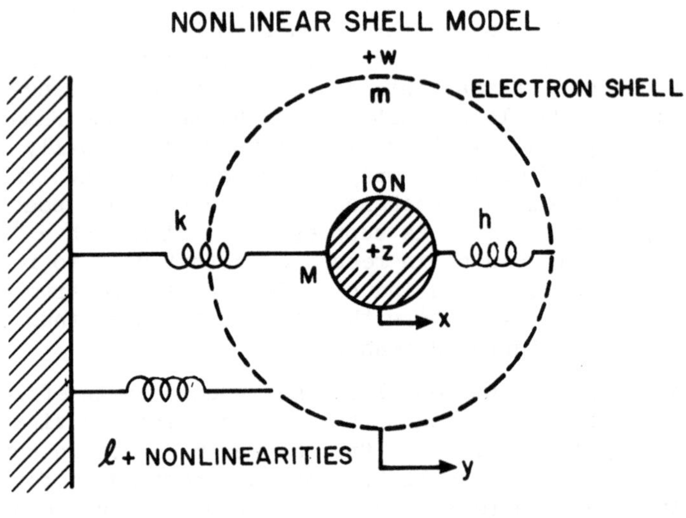

Fig. H2 - One ion-one electron shell model for the long-wavelength lattice vibrations. The unit shown here is positioned at each lattice point of a cubic crystal.

introduced earlier in a rather ad hoc manner must now be put into the lattice from the beginning. We do this as Garret has done using a shell model.[32] The model is shown schematically in Fig. H2. Only the long wavelength ($k \sim 0$) optic mode is considered. Our earlier equation for the optic mode

$$\frac{m_1 m_2}{m_1+m_2} \ddot{x} = -k_{eff}\, x + zE$$

corresponds to the motion of one particle of reduced mass $M = \frac{m_1 m_2}{m_1+m_2}$ shown in the figure. To this we add an electron shell of charge w, mass m, and displacement y from equilibrium. This shell is linked to its own ion and to the other ions by the force constants k and ℓ, respectively. For simplicity we make only ℓ nonlinear. The equations of motion and polarization are:

H1) $$M\ddot{x} = -kx - h(x-y) + z\,E_{loc}$$

H2) $$m\ddot{y} = -\ell y - h(y-x) + w\,E_{loc} - ay^2 - by^3$$

H3) $$P = zx/v + wy/v.$$

The terms with coefficients a and b represent the nonlinearities in the bond ℓ. The term ay^2 will give the regular Raman effect and the term by^3 the induced Raman effect.

Since the remainder of the derivation is mostly analysis of Eqs. H1-3, it is worthwhile to make some points here regarding these equations. First, we have introduced our nonlinearity in the equation of motion. Others have introduced the nonlinearity into the polarization equation. They note that when the local field is expanded fully it contains quadrupole and higher multipole terms which depend on x^2, x^3, etc. At present we do not have enough fundamental knowledge of the various microscopic parameters to decide which method is correct. Kelly has made estimates of the nonlinear dipole moment terms in the zinc blende structure.[33] Probably the present method is a good approximation in some crystals, and the nonlinear dipole moment method better in others.

A second point is that we treat thermal expansion in an approximate way. The nonlinear parameters in our model would control thermal expansion. For simplicity we have discarded all modes but the $k \simeq 0$ optic mode. By retaining only the optic coordinate, we cannot evaluate thermal expansion. To keep the model simple we retain v, the unit cell volume, in all local field expressions. We allow this volume to be temperature dependent according to the experimentally measured thermal volume expansion. This method gives mode instability but is incorrect for predicting volume effects (see Summary). A related problem is our neglect of electrostriction. The real crystal will dilate under high electric fields.[34] This effect is controlled in a complicated way by nonlinearities and by the optic mode frequencies. Electrostriction is important here as we wish to derive both temperature and electric field dependence of the soft mode frequency. We will neglect electrostriction consistent with our approximation of working only with the $k = 0$ optic mode and will comment on its effect where appropriate below.

Finally, we have made only one bond nonlinear. The other bonds may also be nonlinear; however, the derived equations become very cumbersome when they are included. Without any real justification we avoid these complications. Garrett[32] has given a clear discussion of equations like H1-3 modified to include the nonlinear force $(x-y)^2$. He derives the linear electro-optic effect as well as the Raman and parametric gain. We make only the y bond nonlinear but go to third order (y^3) since y^2 is not allowed in the centrosymmetric cubic perovskite crystals we wish to study. Our main justification for the use of Eqs. H1-3 is that they are simple enough that some feeling as to the ion and electron motions is retained even in the presence of the coupling and nonlinear terms which cause variations with temperature and electric field.

We now give a preliminary and somewhat sketchy derivation of the Raman effect and the electric field induced Raman effect. With a and b equal to zero, Eq. H1-2 have a combined resonance (lattice vibration) in which both x and y vibrate because of the coupling h. At temperature T this mode vibrates at frequency $\omega_{t.o.}$ and has amplitude x_t, y_t such that the mode possesses energy kT in the classical limit.

We consider only the y (electron) motion, $y_t\cos(\omega_{t.o.}t)$. If a laser field is applied, the electron also responds to that field as $y_\ell\cos\omega_\ell t$. We now allow a to be finite so the right side of the electron equation has a new driving term

$$\text{H4)} \qquad -a(y_t\cos\omega_{t.o.}t + y_\ell\cos\omega_\ell t)^2.$$

Expanding the square, one term is

$$\text{H5)} \quad -2ay_ty_\ell\cos(\omega_{t.o.}t)\cos(\omega_\ell t) = -ay_ty_\ell\{\cos(\omega_\ell-\omega_{t.o.})t+\cos(\omega_\ell+\omega_{t.o.})t\}.$$

There are therefore new driving terms in Eq. H2 with the Stokes and anti-Stokes frequencies $\omega_\ell-\omega_{t.o.}$ and $\omega_\ell+\omega_{t.o.}$. These driving terms cause a response in y at these frequencies. Since y vibrates with these frequencies, so does the polarization P through Eq. H3. We therefore have Raman signals in the crystal. The final step is just the creation of two electro-magnetic waves from the Raman polarization via the dipole radiation formula.

We can make contact with the discussion in the previous chapter on the Raman effect by noting that under appropriate conditions, the high frequency polarizability of the present model is

$$\alpha = \frac{w^2}{\ell+h} - \frac{2aw^2y}{(\ell+h)^2}$$

H6) $$= \frac{w^2}{\ell+h} + \frac{aCP}{(\ell+h)^2} \equiv \alpha_o + \alpha' P$$

where C is a complicated combination of all the parameters which relate the displacement y in the ionic vibration mode to the total polarization P. We therefore recover Eq. G7. In addition the present model has a dielectric function with a delta function peak at $\omega_{t.o.}$ so we have exactly the same situation as in Chapter G, except that the parameter α' there is here replaced by the more specifically defined nonlinear parameter a.

If a = o then there is no Raman effect for the lattice vibration. A d.c. (or low frequency alternating) electric field E_o can be applied, however, which will cause a static (or low frequency alternating) displacement y_o (even if b = o). In addition, we take y to have terms vibrating at the laser frequency and the phonon frequency as before. Inserting these terms into the non-linear part of Eq. H2, we obtain

H7) $$-b(y_o + y_\ell \cos\omega_\ell t + y_t \cos\omega_{t.o.} t)^3.$$

One of the terms resulting from the cube is

H8) $$\frac{-6b\, y_o y_\ell y_t}{2} \left\{\cos(\omega_\ell - \omega_{t.o.})t + \cos(\omega_\ell + \omega_{t.o.})t\right\}.$$

Again we can get terms on the right side of Eq. H2 which cause Stokes and anti-Stokes frequencies to appear in y and in the polarization. On comparing H8 with H5, we see that $3by_o$ plays the same role as a. This is, the distortion y_o (caused by the electric field) coupled with the third order term is exactly equivalent to having a second order term in the shell-ion restoring force as far as Raman scattering is concerned. We say that the distortion y_o induces Raman activity in the optic mode. In practice, for the largest d.c. fields obtainable, the 'a' term of a naturally Raman-active crystal will usually be larger than the by_o term in a Raman inactive crystal. The reverse can be true however for the very low frequency ferroelectric modes.

The Driven Response

We now solve Eq. (H1-2) using the local field appropriate to a diatomic cubic crystal Eq. (B2) . We will attempt to choose parameters so that the response corresponds to that of $SrTiO_3$. Since $SrTiO_3$ has 5 ions per unit cell, the correspondence will only be qualitative. For example, the electron shell in Eq. (H2) controls the high frequency dielectric response and the electron part of the local field which enters the equations of motion. In $SrTiO_3$, different electrons can be responsible for these phenomena allowing more parameters. We obviously have enough parameters already for the initial investigation of the model. Table I gives the model parameters chosen and Fig. H3 shows the driven response for a transverse electric wave incident on a slab of crystal. There is a low frequency (infrared) resonance at 93 cm^{-1}, the lattice vibration, and an electron resonance at 9.5×10^4 cm^{-1} in the ultraviolet.

Table I - Shell Model Parameters

	Ion	Electron
Mass	$M = 15$ amu	$m = 0.005$ amu
Charge	$z = 1.603\,\|e\|$	$w = 4.046\,\|e\|$
Force Constant	$k = 2.583\times10^5$ dynes/cm	$\ell = 2.57\times10^5$

linear coupling force $h = 2.57\times10^5$ dynes/cm
nonlinear force constant $b = 18\times10^{20}$ dynes/cm^3

Derived Parameters for $v = 64.2\ 10^{-24}\ cm^3$

Low Frequency Resonance	High Frequency Resonance	Dielectric Constants
$\omega_{t.o.} = 93\ cm^{-1}$	$\omega_{t.o.} = 9.5\times10^4 cm^{-1} = 11.8$ eV	$\varepsilon_0 = 291$
$\omega_{\ell.o.} = 810\ cm^{-1}$	$\omega_{\ell.o.} = 18.4\times10^4 cm^{-1} = 22.9$ eV	$\varepsilon_\infty = 3.75$
	for $v = 63.55\times10^{-24} cm^3$	
$\omega_{t.o.} = 11\ cm^{-1}$	$\omega_{t.o.} = 9.45\times10^4 cm^{-1}$	$\varepsilon_0 = 24{,}000$
$\omega_{\ell.o.} = 811\ cm^{-1}$	$\omega_{\ell.o.} = 18.45\times10^4 cm^{-1}$	$\varepsilon_\infty = 3.81$

Because of the electron-core coupling, there is electron <u>and</u> ion motion at each resonance. The ratio of electron-to-ion motion is listed in the figure. Table I lists also the longitudinal modes of the system. These may be found by locating the zeros of the dielectric function $\varepsilon(\omega)$, or by solving the equations of motion with the local field appropriate to longitudinal plane waves. The longitudinal lattice vibration resonance is at 810 cm^{-1}. At this resonance the ratio of electron-to-ion motion is very small. This

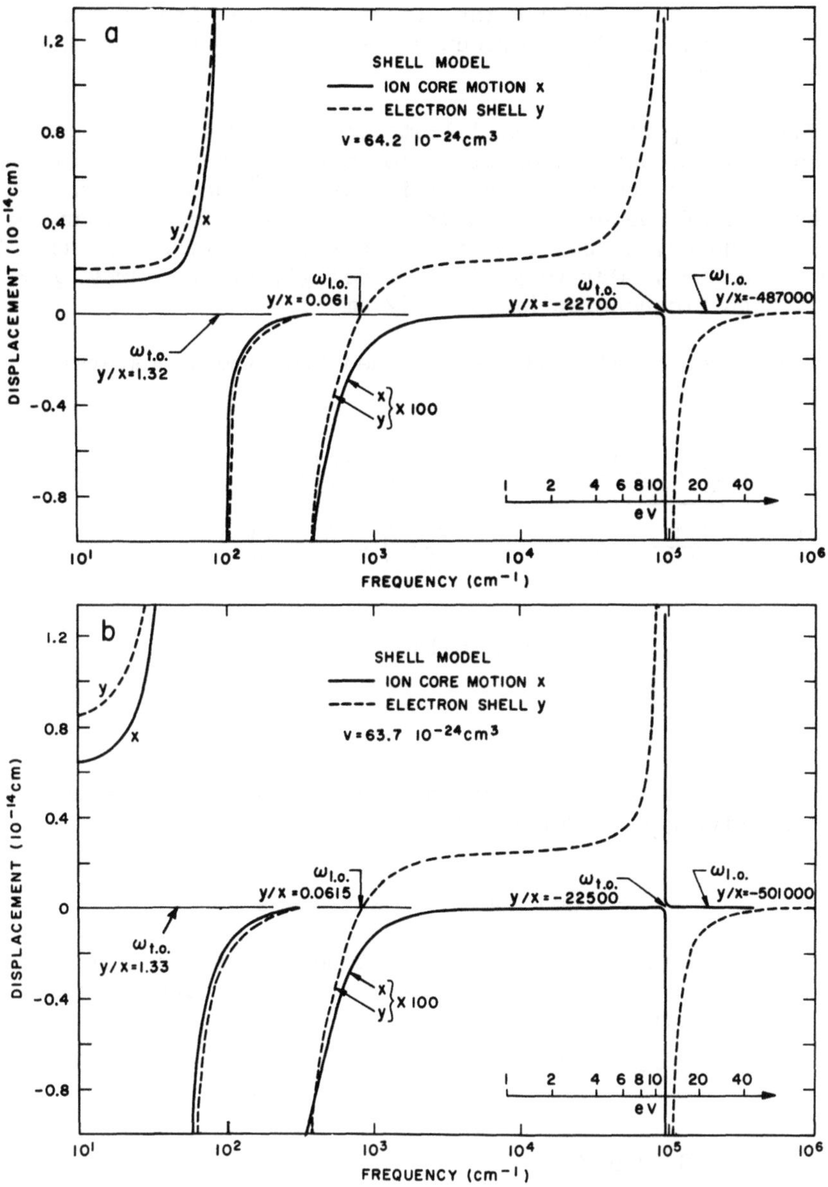

Fig. H3 - Ion and electron displacements for a one volt/cm driving field applied to the shell model. In a the model approximately reproduces the strongest modes in $SrTiO_3$ at 300°K. In b we reduce the cell volume v to simulate cooling to 100°K. The t.o. mode drops considerably, and the d.c. response increases while the other modes remain approximately stationary.

result will give weak ℓ.o. mode scattering relative to the t.o. scattering in the induced Raman effect. A mode with no electron motion will show no Raman effect since we have put all nonlinearities in the electron bond ℓ.

The lower part of Table I lists the mode frequencies and the high and low frequency dielectric constants ε_∞ and ε_o. Note that ε_∞ is defined in the manner usual in lattice dynamics[8]. It specifies the response of the crystal at frequencies much larger than the highest frequency phonon mode but less than the first electron band gap or resonance. Our very simple model gives, of course, $\varepsilon = 1$ at very high frequencies ($\omega \gg 2\times10^5$ cm^{-1}).

The curves in Fig. H3 are obtained from the transverse solutions:

H9)
$$x = \frac{\left[z\left(\ell + h - \frac{4\pi w^2}{3v} - m\omega^2\right) + w\left(h + \frac{4\pi wz}{3v}\right)\right]E}{\left(k + h - \frac{4\pi z^2}{3v} - M\omega^2\right)\left(\ell + h - \frac{4\pi w^2}{3v}\right) - \left(h + \frac{4\pi wz}{3v}\right)^2}$$

H10)
$$y = \text{same with } w \leftrightarrow z, \quad \ell \leftrightarrow k, \quad m \leftrightarrow M$$

In the figure the particle motion is given in 10^{-14} cm units for an external driving field of 1 volt/cm.

Some Numerical Estimates

To obtain some idea of the importance of various terms which are encountered when expansions are carried out, typical magnitudes are listed below:

H11) $v \sim 64\times10^{-24}$ cm^3

$h, k, \ell \sim 10^5$ dyne/cm

$|z| \sim |w| \sim |e| = 4.8\ 10^{-10}$ esu

$|a| \sim 2\times10^{12}$ dyne/cm^2 (zero for cubic perovskites and other crystals with centers of inversion)

$|b| \sim 10^{20}$ dyne/cm^3 (negative for $BaTiO_3$ but positive for $SrTiO_3$ and $KTaO_3$)

(a and b are obtained by setting the anharmonic energy equal to the harmonic energy at a displacement of one cell length.)

$\sqrt{\overline{x^2}} \sim 6\times10^{-10}$ cm (root mean square thermal motion at 300°K)

$x(E_o) \sim 16\times10^{-18}$ cm (static displacement for 1 volt/cm applied field)

$E_{laser} \sim 10^3$ volt/cm (2 watt laser focussed to 0.01 mm^2)

$E_{infrared} \sim 10^{-5}$ volt/cm (typical for infrared spectrometer beam)

From the last four quantities, it is obvious that the thermal displacements are larger than any displacement that we can induce by a d.c. field, infrared beam, or laser beam.

The Low-Frequency Response

We now take a = 0, b = 18×10^{20} dyne/cm^3 and choose the thermal expansion in the model to approximate that of $SrTiO_3$,

$$v = 63.45\ (1+3.95\times10^{-5}T)\text{Å}^3. \qquad \text{(H12)}$$

These estimates as well as others listed in Table I have not been carefully chosen to fit $SrTiO_3$ data, since the model cannot reproduce exactly this data at all temperatures. Rough correspondence is all that is desired. Figure H4 shows the t.o. lattice vibration mode as a function of electric field and volume (or temperature using Eq. (H12)). The mode is extremely soft, dropping from 93 cm^{-1} to zero (for $E_o = 0$) as v decreases. The volume effect is rather obvious - as volume decreases the terms like $-4\pi z^2/3v$ in Eq. (H9) increase negatively, cancelling the short-range forces k, ℓ, and h thereby lowering the mode frequency. The field-dependent effect comes from using $E = E_o + Ee^{i\omega t}$ in the equations of motion and allowing x and y to have d.c. components. The bond ℓ will stiffen to the new value $\ell + b(y_o)^2$ raising all mode frequencies when E_o is present.

Figure H5 shows the results of Fleury and Worlock[12] for the field-induced Raman effect in $SrTiO_3$. For small E_o the mode gives very weak Raman scattering; however its frequency can be obtained by an extrapolation procedure. This extrapolated frequency is given by the lowest curve, labeled $E_o \approx 0$. Higher static fields

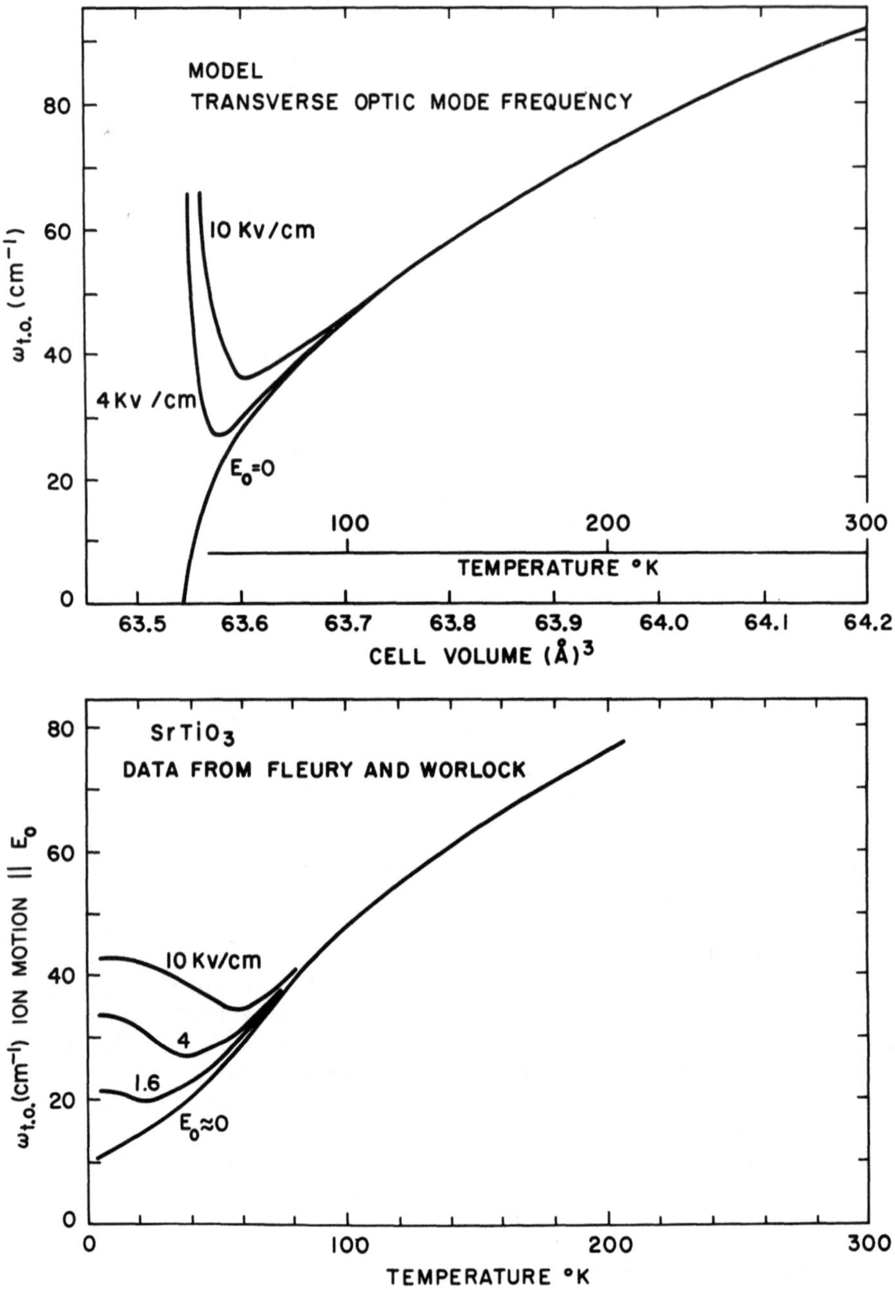

Fig. H4-5 - Transverse lattice vibration frequency given by the shell model (Parameters in Table I) and as measured in $SrTiO_3$.

increase the strength of the scattering and raise the frequency of the mode as shown. The agreement of the model with $SrTiO_3$ is gratifying.

There are two differences between the curves which must be discussed. First the $E_o = 0$ curve for $SrTiO_3$ shows a tendency to saturate, rather than continuing to drop and approach zero frequency near 35°K (the extrapolated Curie temperature). In the model, we have included only the linear part of the thermal volume expansion of $SrTiO_3$. Including the measured nonlinear terms corrects the model in the right direction but in fact overcorrects. The detailed fitting of $\omega(T,E=0)$ will involve quantum effects and may involve the effect of the phase transition at 110°K. Such fitting will not be pursued here with the present model.

The second discrepancy between the model and $SrTiO_3$ centers on the saturation of the contours of constant E_o as temperature is lowered. In the model there is no such saturation. An estimate using the measured electrostriction[34] in $SrTiO_3$ shows that at low temperatures with E_o = 10 kV/cm the elongation of the unit cell is equal in magnitude to the thermal contraction on cooling through a 100 degree temperature range. Electrostriction can therefore counteract the thermal contraction in producing mode shifts. Silverman[34] has shown that the electrostrictive parameter depends on the optic mode frequency as $\omega_{t.o.}^{-4}$. We obtain a rough measure of the effect of electrostriction by writing the cell volume as:

$$v = v_c(1 + \alpha(T-T_c) + \beta E_o^2/w_{t.o.}^4), \tag{H13}$$

where α is the linear volume expansion coefficient we have used all along to make the model simulate $SrTiO_3$. The β term arises from electrostriction. For $E_o = 0$, Fig. H4 shows the $\omega_{t.o.}$ frequency which is determined by v through the equations of motion. As E_o is increased at a fixed temperature, the $\omega_{t.o.}$ given by our model rises. The electrostriction will start to saturate the field dependence when

$$\beta E_o^2/\omega_{t.o.}^4 = \alpha(T-T_c). \tag{H14}$$

Since $\omega_{t.o.}^2 \propto (T-T_c)$ in the low-field region, the field where saturation begins is

$$E_o(\text{Saturation}) \propto (T-T_c)^{3/2}. \tag{H15}$$

The constant of proportionality depends on the electrostrictive constant. If we take the lower temperature inflection point as the place where electrostriction becomes important, then the $SrTiO_3$ data (Fig. H5) agrees qualitatively with Eq. (H15); however, whether the power is 3/2 or not cannot be determined accurately.

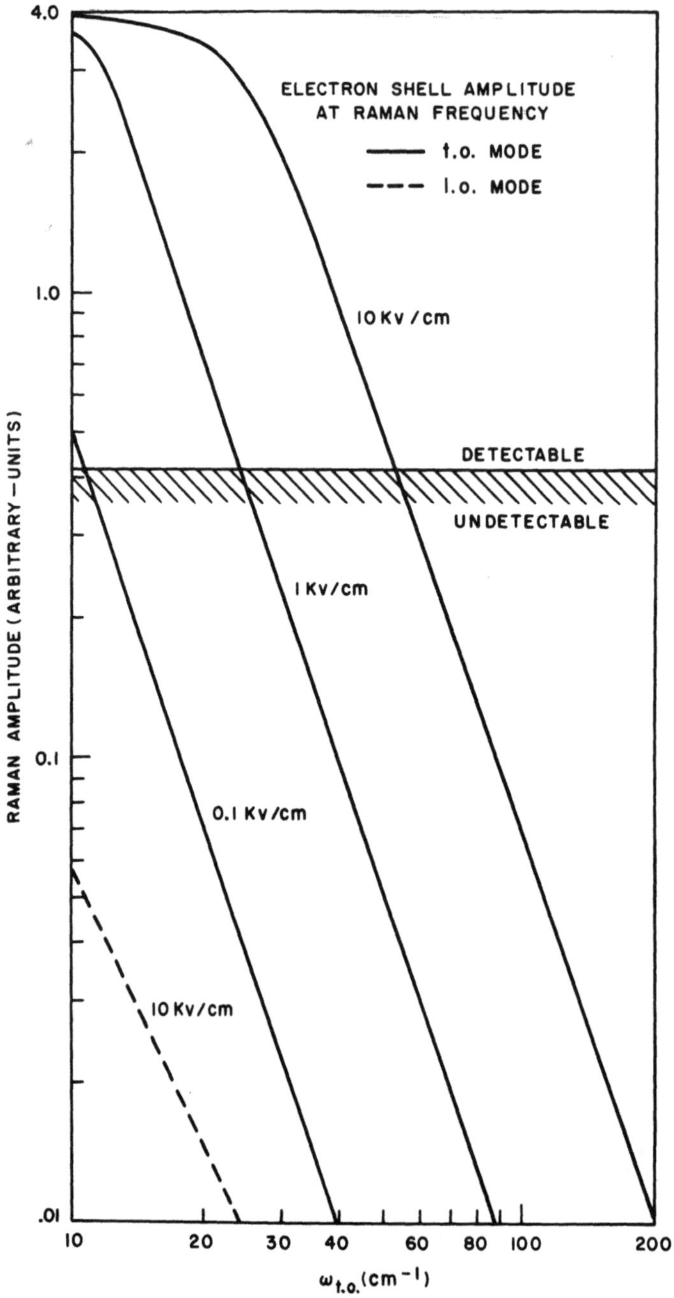

Fig. H6 - Displacement amplitude of the electron shell at the Raman frequency given by the shell model. The laser power and lattice mode energy are kept fixed while volume is varied to obtain the graph. The actual Raman line intensity depends on the square of the amplitude given here times the lattice mode excitation kT (for the classical case).

We now examine the Raman scattering strength of the $\omega_{t.o.}$ and $\omega_{\ell.o.}$ lattice vibrations. Up to this point we have been solving the equations of motion (H1-3) for an infrared and a d.c. field present. We now add a laser field $E_\ell \cos(\omega_\ell t)$. All variables are expanded to include terms oscillating at the lattice frequency, the laser frequency, and at certain sum and difference frequencies. The estimates given above (H-11) are used to eliminate most of the small terms. Given the thermal amplitude of the core and shell (x_t and y_t) at the lattice frequency, and the size of E_o and E_ℓ, we evaluate the electron motion at the Raman frequency. We take fixed thermal energy of the lattice mode, fixed laser power, and consider Raman frequencies well below the electron resonance at 9.4×10^4 cm^{-1} (nonresonant Raman scattering). Figure H6 shows the electron vibration amplitude at the Raman frequency. The scattered Raman signal will be proportional to the square of this amplitude times the appropriate quantum factor for Stokes or anti-Stokes scattering. In the present model the eigenvector (relative core to shell motion) changes less than 1 percent over the whole range of volume and electric field considered. For fixed energy of the t.o. mode therefore, the electron amplitude y_t varies as x_t over this range. Equation (H8) suggests that the electron motion at the Stokes frequency depends on $y_o \times y_\ell \times y_t$. Since $y_t \sim x_t \sim 1/\omega_{t.o.}$, y_ℓ = constant, and $y_o \sim 1/\omega^2_{t.o.}$ we would estimate a Raman amplitude proportional to $1/\omega^3_{t.o.}$. This third power dependence gives the straight line behavior shown in Fig. H6. Consideration of the complete solutions provides the saturation at low temperatures. The approximate limit of sensitivity of present-day Raman equipment is illustrated in Fig. H6 showing the necessity of high fields and a soft mode for the electric-field-induced Raman effect.

In Fig. H6 the Raman amplitude of the $\omega_{\ell.o.}$ mode is shown as a dashed line. The scattering is very weak since the electron motion in the ℓ.o. lattice vibration is much smaller than in the t.o. vibration (see y/x in Fig. H3). In addition, the frequency $\omega_{\ell.o.}$ is practically temperature independent so we drop from $1/\omega^3_{t.o.}$ to $1/\omega^2_{t.o.}$ dependence. For both the $\omega_{t.o.}$ and $\omega_{\ell.o.}$ modes, the actual Raman amplitude will have slight additional frequency dependence arising from the quantum factors. From Fig. H6 we conclude that the ℓ.o. mode would be too weak to detect.

The High Frequency Response

For frequencies between about 2000 and 20,000 cm^{-1}, the dielectric response of the present model is fairly flat (Fig. H3). The dielectric function in this region is dominated by the electron-shell whose resonance occurs at higher frequencies. Figure H7 shows this t.o. resonance frequency as a function of cell volume for the model. In the lower part of the figure, the index of refraction at

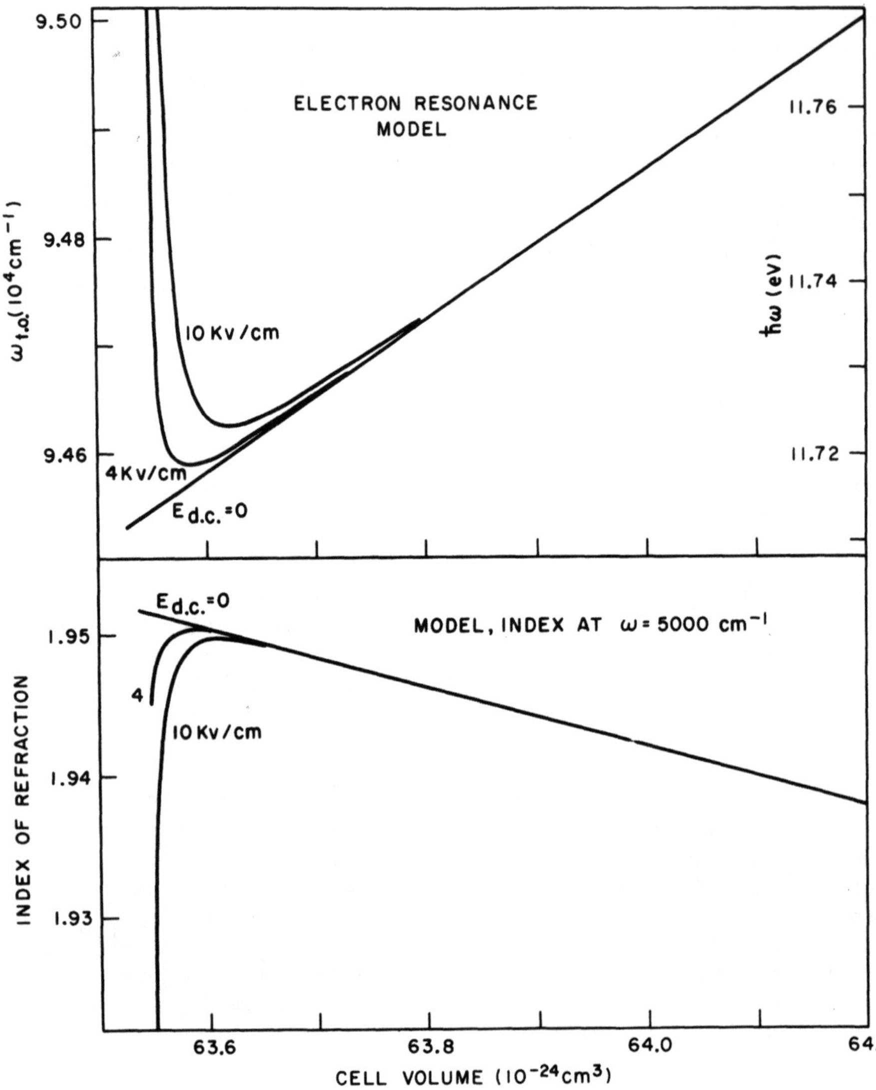

Fig. H7 - High frequency resonance and index of refraction for the shell model. The t.o. mode (electron resonance) rises with increasing volume and the index drops. About two thirds of the decrease in the index comes from the increase in t.o. mode frequency and the remainder from the decrease in oscillator strength or equivalently the decrease in l.o. mode frequency.

5000 cm^{-1} is shown. The index depends on the frequency and strength of the electron resonance and has volume and field dependence arising from the local field effects and the nonlinearity b. In a complete theory of the crystal, the electron resonance of our model would be an electron band transition. Wemple and DiDomenico have developed a phenomenological theory of the Raman effect using the shift of the electron band resonance with polarization or field as the starting point.[35] Butcher and McLean have discussed the microscopic approach to such a band theory.[36] The phenomenological theory[35] contains one basic nonlinear parameter similar to the b parameter we have used in the present model. Since it is a macroscopic theory, no attempt is made to evaluate the basic ion and electron motion; however, useful estimates can be made of the linear and quadratic electrooptic effects[35] and the Raman scattering strength[35,37] for many perovskite type crystals using essentially this one parameter. The electric field dependence shown in the top of Fig. H7 agrees with the band shift parameter of Wemple and DiDomenico.

Summary

In this chapter we have illustrated the electric-field-induced Raman effect using a shell model. One nonlinear bond has been included to give the Raman scattering. The essential temperature dependence is taken to come from the volume dependence of the local field. It must be remarked that this is only a heuristic way of obtaining the necessary temperature dependent effective force constant. Silverman has shown that a more complete model obtains this temperature dependence by thermal averages over modes other than the optic modes.[38] Our model, of course, neglects all other modes. The unstable optic mode frequency of a paraelectric (or ferroelectric above T_c) at constant pressure is:[8]

H16) $$m\omega^2_{t.o.} = |a| T - b.$$

We have obtained this result (Fig. H4) by expanding the volume dependence of the local field terms. In a real crystal the short range forces h, k, ℓ are volume and temperature dependent also.[38,39] A more general expression for the optic mode frequency including volume and temperature effects has the form:

$$m\omega^2_{t.o.} = |C_1| T - |C_2| v + C_3$$

This equation shows that specific volume effects actually counteract the purely thermal effects at constant volume.[39] We therefore emphasize that the procedure used in this and earlier chapters of considering only thermal expansion is heuristic and serves the purpose of introducing the temperature dependence but cannot be extended to predict the volume dependence of $\omega_{t.o.}$.

For simplicity the model has been based on a diatomic NaCl type crystal. To achieve the large local field corrections in terms like $4\pi w^2/3v$, large charges have been used (Table I). In the actual perovskite structure of $SrTiO_3$, the structure itself enhances the local field correction.[9] We conclude by illustrating the importance of the local field correction with an example. We take the model in Table I and reduce the charge w from 4.046 $|e|$ to 3.3 $|e|$. Table II shows the result. The model is now insensitive to volume (and temperature) and conforms roughly to the behavior of a cubic material like LiF. There is no soft mode; the electric field induced scattering is too small to be observed. The electron motion in the ℓ.o. mode is relatively much larger and can be made larger than the t.o. electron motion by choosing w negative. The Raman strength curves plotted against temperature are horizontal instead of rapidly rising at low temperatures. The high-frequency dielectric constant has also dropped. We can physically picture this difference in models occurring even in the perovskite system by comparing an oxide perovskite ($SrTiO_3$) with a fluoride perovskite ($KMgF_3$). The fluorine ion is less polarizable (i.e. has smaller w) than the oxygen ion. $KMgF_3$ is found to have no soft mode and very little temperature dependence of any other dielectric parameter. Its electric field induced Raman scattering is unmeasurably small. The comparison of the models in Tables I and II shows the importance of the shell charge or electronic polarizability.

Table II - Parameters for Second Model

All parameters the same as Table I except w = 3.3 $|e|$

Derived parameters for v* = 64.2×10^{-24} cm^3

Low Frequency Resonance	High Frequency Resonance	Dielectric Constants
$\omega_{t.o.} = 411\ cm^{-1}$, y/x =.97	$10.8\times10^4\ cm^{-1}$ = 13.4 eV	$\epsilon_o = 9.2$
$\omega_{\ell.o.} = 810\ cm^{-1}$, y/x =.11	$16.9\times10^4\ cm^{-1}$ = 20.9 eV	$\epsilon_\infty = 2.4$

*Note - All derived quantities except ϵ_o vary less than 1 percent when v is lowered to 63.55. ϵ_o rises 2 percent.

ACKNOWLEDGMENTS

The author acknowledges with pleasure helpful discussions with M. DiDomenico, Jr., R. Loudon, D. E. McCumber and J. M. Worlock.

REFERENCES

1. E. Fatuzzo and W. Merz, Ferroelectricity (North Holland Publishing Co., Amsterdam, 1967).
2. Ferroelectricity, ed. E. F. Weller (Elsevier Publishing Co., Amsterdam, 1967).
3. W. Känzig, Solid State Physics Vol. 4, ed. F. Seitz and D. Turnbull (Academic Press, New York, 1957).
4. H. D. Megaw, Ferroelectricity in Crystals (Methuen, London, 1957).
5. C. Kittel, Introduction to Solid State Physics (John Wiley and Sons, New York, 1953).
6. R. A. Cowley, W. Cochran, B. N. Brockhouse, and A. D. B. Woods, Phys. Rev. 131, 1030 (1963).
7. See H. D. Megaw (Ref. 4), Chs. 9 and 10.
8. W. Cochran, Advan. Phys. 9, 387 (1960).
9. A. S. Barker, Jr., Ferroelectricity, see Ref. 2.
10. See Ref. 1, Ch. 3.
11. R. A. Cowley, Phys. Rev. 134, 981 (1964).
12. P. A. Fleury and J. M. Worlock, Phys. Rev. Letters 18, 665 (1967), also Phys. Rev. 174, 613 (1968).
13. H. Thomas and K. A. Müller, Phys. Rev. Letters 21, 1256 (1968).
14. W. Cochran and A. Zia, Phys. Stat. Sol. 25, 273 (1968).
15. P. A. Fleury, J. F. Scott and J. M. Worlock, Phys. Rev. Letters 21, 16 (1968).
16. W. Cochran, Advan. Phys. 10, 401 (1961).
17. T. Kawamura, A. Mitsuishi and H. Yoshinaga, Jap. J. Appl. Phys. 7, 1303 (1968).
18. I. P. Kaminow and T. C. Damen, Phys. Rev. Lett. 20, 1105 (1968).
19. M. Lax, Rev. Mod. Phys. 32, 25 (1960).
20. D. K. C. MacDonald, Noise and Fluctuations (John Wiley and Sons, Inc., New York, 1962).
21. M. Born and K. Huang, Dynamical Theory of Crystal Lattices (Clarendon Press, Oxford, 1954), Sec. 19, 49.
22. R. Loudon, Advan. Phys. 13, 423 (1964).
23. A. S. Barker, Jr. and R. Loudon, Phys. Rev. 158, 433 (1967).
24. M. DiDomenico, Jr., S. H. Wemple, S. P. S. Porto and R. P. Bauman, Phys. Rev. 174, 522 (1968).
25. S. H. Wemple, M. DiDomenico and I. Camlibel, J. Phys. Chem. Solids 29, 797 (1968). These authors find that melt-grown $BaTiO_3$ has somewhat higher transition temperatures than those given here.
26. A. S. Barker, Jr., Phys. Rev. 165, 917 (1968).
27. Light Scattering in Solids, ed. G. B. Wright (Springer Verlag, New York, 1969). This conference proceedings contains several papers on ferroelectrics.
28. R. Loudon, Ref. 27.
29. C. H. Henry and C. G. B. Garrett, Phys. Rev. 171, 1058 (1968).

30. Methods of Quantum Field Theory in Statistical Physics, A. A. Abrikosov, L. F. Gorkov and I. E. Dzyaloshinski, (Prentice Hall, Inc., New Jersey, 1963), Sec. 6.
31. J. F. Scott, P. A. Fleury and J. M. Worlock, Phys. Rev. 177, (1969).
32. C. G. B. Garrett, IEEE J. Quantum Electron. QE-4, 70 (1968).
33. R. L. Kelley, Phys. Rev. 151, 721 (1966).
34. B. D. Silverman, Phys. Rev. 131, 2478 (1963).
35. S. Wemple and M. DiDomenico, Ref. 27, also Phys. Rev. (to be published).
36. P. N. Butcher and T. P. McLean, Proc. Phys. Soc. 81, 219 (1963).
37. J. M. Worlock, Ref. 27.
38. B. D. Silverman, Phys. Rev. 135, 1596 (1964).
39. A. J. Bosman and E. E. Havinga, Phys. Rev. 129, 1593 (1963).

MODIFICATION OF LATTICE VIBRATIONS BY IMPURITIES

C.W.McCOMBIE

J.J.Thomson Physical Laboratory

University of Reading, U.K.

1. Introduction

The interaction of lattice vibrations with a point imperfection has two aspects. In the first place the imperfection will modify the normal modes of vibration of the lattice. In the second place these modified modes will affect processes, such as optical absorption, taking place at the imperfection. These two aspects can hardly be separated since it is, as we shall see, frequently most convenient to treat the effect of the modified modes on a process at the imperfection without ever explicitly examining the modified normal modes themselves. It will, therefore, be necessary for us to consider both aspects although the main object of these lectures is to show how to take account of the modification of the modes.

We shall start by revising some very elementary matters concerning the classical dynamics of a system of particles which will crop up frequently during the lectures. These classical considerations are relevant even when the problem of interest requires the lattice to be treated quantum mechanically, as one then usually starts with a classical discussion of the normal modes and then quantises later.

2. Generalised Co-ordinates

We are concerned with a system of N particles each with a well defined equilibrium position. A generalised displacement of the system may be specified by giving the vector displacement of each atom from its equilibrium position i.e. by the

N 3-dimensional vectors $\underset{\sim}{a}_r$ where $\underset{\sim}{a}_r$, r = 1 ....N, is the displacement of the rth atom. The generalised displacement will be denoted by $\underset{\sim}{A}$, where $\underset{\sim}{A}$ may be thought of as a vector with 3N components but its precise specification need not concern us.

We attach a meaning in an obvious way to forming linear combinations of generalised displacements $\underset{\sim}{A}_s$ (associated displacements of particles $\underset{\sim}{a}_{sr}$). The linear combination $\Sigma\alpha_s\underset{\sim}{A}_s$ is defined to be the generalised displacement in which the displacement of the rth atom is $\Sigma\alpha_s\underset{\sim}{a}_{sr}$.

It is possible to choose (in many ways) a set of 3N generalised displacements $\underset{\sim}{A}_s$ which are independent (i.e. such that no one of them can be represented as a linear combination of the others). Since it is not possible to have a set of 3N+1 independent generalised displacements it follows that any generalised displacement $\underset{\sim}{A}$ can be represented as a linear combination of the $\underset{\sim}{A}_s$

$$\underset{\sim}{A} = \sum_s \alpha_s\underset{\sim}{A}_s \qquad (1)$$

Equation (1) is, of course, identical with

$$\underset{\sim}{a}_r = \Sigma\, \alpha_s\underset{\sim}{a}_{sr}, r = 1 \,....N \qquad (2)$$

and we shall use it in this form in calculations. It is easy to see that for fixed $\underset{\sim}{A}$ (and, of course, for fixed $\underset{\sim}{A}_s$, s = 1....3N) the α_s are unique. Thus the coefficients α_s are determined by, and conversely uniquely determine the generalised displacement $\underset{\sim}{A}$. They can therefore be taken to be a set of 3N coordinates specifying the configuration of the system and they will be termed generalised coordinates(although coordinates so defined are in fact only a restricted class of generalised coordinates: they do not include, for example, coordinates which consist of the polar coordinates of all the particles). The generalised displacements $\underset{\sim}{A}_s$ are said to be the basic displacements in terms of which the generalised coordinates α_s are defined.

It is often more convenient to describe the configuration of the particle system in terms of such generalised coordinates than to define it by giving, say, the Cartesian components of the displacement of each particle from its equilibrium position. (We may note in passing that this set of Cartesian components is the set of generalised coordinates which result if we take each basic displacement $\underset{\sim}{A}_s$ to represent a configuration in which just one particle is displaced and that by a unit displacement along one of the Cartesian axes.)

The kinetic and potential energies of the system of particles can be expressed as functions of the generalised coordinates α_s and their time derivatives $\dot{\alpha}_s$. These expressions lead us, by way of Lagrange's equations, to the equations of motion satisfied by the generalised coordinates α_s.

3. Small Displacements : the Linear Approximation

We shall be interested in small displacements of the system from its equilibrium configuration. By this we mean motions small enough for all but linear terms in the equations of motion (and so, as follows from Lagrange's equations, all but quadratic terms in the kinetic and potential energies) to be neglected. These small motions have the vital property that they can be superposed. To be more precise, if we have a number of time dependent generalised displacements $\underset{\sim}{B}_n(t)$ each of which satisfies the equations of motion (for the system without any applied forces) then any linear combination of these $\Sigma\ \beta_n \underset{\sim}{B}_n(t)$, with the β_n independent of time, will also satisfy the equations and so represents a possible motion.

A related superposition principle applies for small displacement when the system is acted on by external time-dependent forces, in which case the equations of motion become inhomogeneous. If $\underset{\sim}{C}(t)$ represents a possible motion of the system under the action of the applied forces (i.e. is a solution of the inhomogeneous equation) and the $\underset{\sim}{B}_n(t)$ as before represent motions of the system in the absence of external forces, then $\underset{\sim}{C}(t) + \Sigma\beta_n \underset{\sim}{B}_n(t)$, with any values for the constant β_n, also represents a possible motion of the system under the action of the applied forces. This corresponds to the familiar mathematical result that an arbitrary solution of the corresponding homogeneous equation can be added to the solution of the inhomogeneous equation.

It must, of course, be recognised that for realistic force laws the potential energy will contain cubic and higher terms, and some phenomena can be discussed only by taking these into account . Even then, however, the usual procedure is to start by neglecting these higher terms and determining the normal modes of the resulting linear system: one then quantises and subsequently takes the higher terms into account by quantum mechanical perturbation theory or some other quantum mechanical procedure. Thus the classical discussion of normal modes to which we now proceed is still relevant.

4. Normal Co-ordinates

We now turn to considering the consequences of making a very special choice for the 3N basic displacements. In the absence of any external forces there will be certain motions of the particle system in which the particles execute very small oscillations about their equilibrium positions, all oscillating with the same frequency and passing simultaneously through their equilibrium positions. These motions of the system (of which there are 3N, including some with zero frequency) are termed the normal modes. The sth normal mode will be characterised by giving its angular frequency ω_s and the displacement $\underset{\sim}{x}_{sr}$ (3-dimensional vector) of each of the particles $r = 1 \ldots N$, when all particles have their maximum displacements. These maximum displacements are, of course, undetermined to within a common multiplying constant and it will be convenient to think of them as having been chosen to satisfy the normalising condition

$$\Sigma_r m_r \underset{\sim}{x}_{sr}{}^2 = 1 \tag{3}$$

It can be shown that, with suitable choice in the case of some arbitrariness which arises when some of the modes have the same angular frequency (are degenerate), the displacements for any two distinct modes s and s' are orthogonal in the sense

$$\Sigma_r m_r \underset{\sim}{x}_{sr} \cdot \underset{\sim}{x}_{s'r} = 0 \tag{4}$$

The N displacements $\underset{\sim}{x}_{sr}$ associated with the sth mode define a generalised displacement $\underset{\sim}{X}_s$ (3N-dimensional vector) which may be described as the sth normal displacement. The 3N normal displacements $\underset{\sim}{X}_s$ can be taken to be basic displacements and an arbitrary generalised displacement $\underset{\sim}{A}$ can then be written

$$\underset{\sim}{A} = \Sigma_s \xi_s \underset{\sim}{X}_s \tag{5}$$

or

$$\underset{\sim}{a}_r = \sum_s \xi_s \underset{\sim}{x}_{sr} \tag{5'}$$

The generalised coordinates ξ_s thus defined in terms of this particular set of basic displacements are termed normal coordinates. They have a number of properties which render them extremely useful.

(1) In terms of the normal coordinates the kinetic and potential energies T and V for small displacements take

the form of sums of square terms

$$T = \tfrac{1}{2} \sum_s \dot{\xi}_s^{\,2} \qquad (6)$$

$$V = \tfrac{1}{2} \sum_s \omega_s^{\,2} \xi_s^{\,2} \qquad (7)$$

(2) As a consequence of property (1) the equation of motion satisfied by ξ_s is independent of all other ξ's and takes the form

$$\ddot{\xi}_s + \omega_s^{\,2} \xi_s = 0 \qquad s = 1 \ldots 3N \qquad (8)$$

These equations have the solutions

$$\xi_s(t) = \xi_s^{(0)} \cos (\omega_s t + \varepsilon_s) \qquad s = 1 \ldots 3N \qquad (9)$$

with the $\xi_s^{(0)}$ and ε_s arbitrary constants. The motion represented by this when all the $\xi_s^{(0)}$ are zero except $\xi_{s'}^{(0)}$ is, as is seen by substituting in (5),

$$\mathbf{A}(t) = \xi_{s'}^{(0)} \cos (\omega_{s'} t + \varepsilon_{s'}) \, \mathbf{\chi}_{s'} \qquad (10)$$

which is the s'^{th} normal motion with a specific choice for amplitude and phase. In general the solution of the set of 3N equations (9) yields when substituted in (5) just a superposition of all the normal motions with specific amplitudes and phases and the motion of the linear system in the absence of external forces may be thought of in this way. All this is, of course, just what was to be expected from the definition of a normal mode from which we started, taken together with the superposition property.

(3) If the system is acted upon by a number of external time dependent forces these give rise to a generalised force on each mode. The generalised force $\mathfrak{F}_s$ (a scalar) on the sth mode is defined by the requirement that if the sth normal coordinate ξ_s is increased by an amount $d\xi_s$, all other normal coordinates being fixed, then the work done by the applied forces in the resulting displacements of the particles is

$$\mathfrak{F}_s \, d\,\xi_s .$$

If the external force on the rth particle is $\mathbf{F}_r^{(e)}$ the work done by this in the displacement $d\xi_s \mathbf{\chi}_{sr}$ will be $\mathbf{F}_r^{(e)} \cdot \mathbf{\chi}_{sr} \, d\xi_s$ and summing the works done over the particles gives at once

$$\mathcal{F}_s(t) = \Sigma \mathbf{F}_r^{(e)}(t) \cdot \mathbf{x}_{sr} \qquad (11)$$

Thus the generalised forces can be obtained immediately when we know the applied forces and also the normal displacements (i.e. the forms of the modes).

In the presence of the applied forces the equations of motion satisfied by the normal coordinates become

$$\ddot{\xi}_s + \omega_s^2 \xi_s = \mathcal{F}_s(t) \quad s = 1 \ldots 3N \qquad (12)$$

and we notice that the motions of the various normal modes are still uncoupled. These equations imply that once the normal modes have been determined (so that we know the ω_s^2 and the $\mathbf{x}_{sr}$) the problem of determining the response of the system of particles to any system of forces varying sinusoidally with time is essentially solved. To see this let us suppose that the force on the rth particle is $\mathbf{F}_r^o \cos \omega t$ (13)
and so

$$\xi_s(t) = -\frac{1}{(\omega^2 - \omega_s^2)} \Sigma_r \mathbf{F}_r^o \cdot \mathbf{x}_{sr} \cos \omega t + C_s \cos(\omega_s t + \varepsilon_s) \qquad (14)$$

the values of the constants C_s and ε_s being determined by the initial conditions

Substitution in (5') gives

$$\mathbf{a}_r(t) = -\sum_{r',s} \frac{\mathbf{F}_{r'}^o \cdot \mathbf{x}_{sr'}}{\omega^2 - \omega_s^2} \mathbf{x}_{sr} \cos \omega t + \sum_s C_s \mathbf{x}_{sr} \cos(\omega_s t + \varepsilon_s) \qquad (15)$$

This expression (with the omission of the transient terms in the C_s) deals satisfactorily with the problem of determining the steady state response of a crystal to forces varying sinusoidally with time, provided the frequency ω of the applied forces lies outside the frequency range (or ranges) covered by the normal mode frequencies of the crystal. For applied frequencies which lie in normal mode ranges further discussion is needed and a major objective of what follows is to provide a simple way of handling this situation.

5. Damping

We shall find it convenient in connection with the

problem mentioned at the end of the last section to suppose that the motion of the particles suffers a small damping, and then to let this damping tend to zero at a suitable point in the calculation. The precise form of the damping introduced is arbitrary and it is convenient to suppose that if $\underset{\sim}{v}_r$ is the velocity of the rth particle the damping force which acts on it is $-\kappa m_r \underset{\sim}{v}_r$, κ being the same for all particles. In terms of the generalised coordinates

$$\underset{\sim}{v}_r = \sum_{s'} \dot{\xi}_{s'} \underset{\sim}{\chi}_{s'r} \tag{16}$$

and the generalised force on the sth mode exerted by the damping forces will be given by

$$\mathfrak{F}_s^{(d)} = -\kappa \sum_{rs'} m_r \underset{\sim}{\chi}_{sr} \cdot \underset{\sim}{\chi}_{s'r} \dot{\xi}_{s'} = -\kappa \dot{\xi}_s \tag{17}$$

where the last step follows from the normalisation condition (3) and the orthogonality condition (4). Thus the particular damping considered does not couple different normal modes and the damping force on a mode takes a very simple form.

If we have in addition external forces $\underset{\sim}{F}_r^o \cos \omega t$ as before equation (14) is replaced by

$$\ddot{\xi}_s + \kappa \dot{\xi}_s + \omega_s^2 \xi_s = \sum_r \underset{\sim}{F}_r^o \cdot \underset{\sim}{\chi}_{sr} \cos \omega t \tag{18}$$

It is convenient to start solving this by writing $Re^{i\omega t}$ for $\cos \omega t$ (R denotes "real part of ") and, solving (18) and substituting in (5'), one finds for the steady state solution which holds after the decay of transients

$$\underset{\sim}{a}_r(t) = -R \sum_{r',s} \frac{\underset{\sim}{F}_{r'}^o \cdot \underset{\sim}{\chi}_{sr'}}{\omega^2 - \omega_s^2 - i\kappa\omega} \underset{\sim}{\chi}_{sr} e^{i\omega t} \tag{19}$$

This completes our preliminary outline of the main properties of normal coordinates. As has already been stressed the results which have been given are easy to apply if we know the forms and frequencies of the normal modes. We shall now discuss briefly the way in which this information is obtained for the normal modes of vibration of a perfect crystal: it turns out that a knowledge of these perfect lattice modes (which, as we shall see, is easily obtained) forms a convenient starting point for the treatment of the lattice dynamics of imperfect crystals.

6. Normal Modes of a Perfect Crystal

The determination of the normal modes of a system of particles starts with introducing convenient basic displacements and corresponding generalised coordinates. The coefficients in the quadratic expressions for the kinetic and potential energies are then calculated and the equations of motion of the generalised coordinates are obtained from them by using Lagrange's Equations. All this is as already described briefly at the end of section 2. One then seeks solutions of these equations in which all the generalised coordinates vary with a common frequency. The corresponding motions are, of course, normal modes. The details of this method are set out in elementary texts and we shall not reproduce it here.

The method may clearly be applied to systems in which the number of particles N is small, since the problem reduces to investigating 3N linear equations in 3N unknown. When N is large (or even infinite) the method may still be usable if there is sufficient symmetry. Symmetry considerations, which may be useful even for a small number of particles, help by guiding us to a specially advantageous choice of basic displacements with which to start. With this choice the 3N basic displacements fall into sets which considerations of symmetry alone show to have the following property. If we suppose the system constrained so that the only allowed displacements are superpositions of the basic displacement in one of the sets and work out the normal modes of this constrained system, then the normal modes obtained will in fact be normal modes of the unconstrained system. By considering all the sets in turn we obtain all the 3N normal modes without having to solve a normal mode problem for a system with number of degrees of freedom greater than the number of basic displacements in the largest set.

Perhaps the most striking example of such simplification by symmetry occurs for the case of the perfect crystal, where the symmetry concerned is translational (described by the operations of translation which send the infinite crystal into itself). To be specific let us talk of NaCl. A set of generalised displacements of the kind under discussion may be chosen as follows. One displacement in the set is such that only chlorine ions are displaced, and these only in the direction of the x-axis; the magnitude of the displacement depends on the position of the ion, being given by sin $\underset{\sim}{k}.\underset{\sim}{R}$, where $\underset{\sim}{k}$ is an arbitrary fixed vector (the wave vector) and $\underset{\sim}{R}$ is the position vector of the equilibrium position of the ion relative to the equilibrium position of a certain chlorine ion which is chosen as origin. It should be noted that this puts the origin at a centre of symmetry and this is important for the discussion which we are giving: if there is no centre of symmetry in the lattice slight modifications are necessary, but we shall not concern ourselves with these. Two other displacements are obtained by replacing x by y and then by z in the above specification, the sin $\underset{\sim}{k}.\underset{\sim}{R}$ variation of magnitude with position remaining the same. Finally three further are obtained by similar displacements in which only the sodium ions are displaced all along the same axis, the magnitude of the displacement varying as sin $\underset{\sim}{k}.\underset{\sim}{R}$ where $\underset{\sim}{R}$ is the position vector of a sodium ion relative to the same origin as before. It can be shown that the six normal modes calculated on the assumption that the only displacements possible are superpositions of the six just mentioned are in fact normal modes of the unconstrained system. Each of the six normal modes is specified by giving its frequency and the coefficients of the six basic displacements in the linear superpositions which represent the extreme position in a normal mode.

The normal displacements obtained in the way just described are even, in the sense that the displaced configurations of the ions which they represent go into themselves under inversion in the origin. For each value of $\underset{\sim}{k}$ one obtains also a set of odd modes in which sin $\underset{\sim}{k}.\underset{\sim}{R}$ is replaced by cos $\underset{\sim}{k}.\underset{\sim}{R}$, the coefficients in the linear combination which gives the normal displacement remaining the same.

When the normal modes are investigated in this way for NaCl the squares of the frequencies of the modes and the sets of coefficients specifying the corresponding normal displacements emerge as the eigenvalues and eigenvectors of a 6 x 6 matrix (the dynamical matrix) which depends on $\underset{\sim}{k}$ and, of course, on the model assumed for the forces in the crystal. For this reason one talks of giving the eigenfrequencies and eigenvectors for the modes. We shall not discuss the calculations further as to do this adequately would take us too far afield.

The question of the range of values of $\underset{\sim}{k}$ for which results are to be given is of some interest to us. Two vectors $\underset{\sim}{k}$ which differ by a reciprocal lattice vector (i.e. a vector such that its scalar product into any displacement vector connecting two equivalent ions is a multiple of 2π) will yield exactly the same set of basic displacements. For this reason we can restrict our interest to a region of $\underset{\sim}{k}$ space termed the first Brillouin zone, which can be chosen to have cubic symmetry. The cubic symmetry of the crystal means further that we have complete information about the modes if we have investigated only those with k-values in a region occupying 1/48 of the volume of the zone: the modes with $\underline{k}$-values in the rest of the zone can be determined from these by symmetry. If one investigates the modes with k-values uniformly distributed in k-space one is in effect determining the modes of a certain portion of perfect crystals. With modern computer techniques one can investigate a dense mesh of k-values and so obtain the normal modes of a portion of crystal containing perhaps 10^5 atoms. A main objective of what follows is to see how to avail ourselves of this great ease in determining perfect lattice modes even when we are interested in the dynamics of a lattice with point imperfections.

7. Symmetry

It is an easy matter to verify by direct dynamical calculation that the assertions in the last section are correct and in fact the problem is usually treated in essentially the way described without explicit reference to symmetry properties as such. In general, however, the choice of the advantageous sets of basic displacements requires the use of Group Theory, which provides a systemmatic way of constructing such sets. While we shall try to keep the main exposition in these lectures independent of Group Theory it will be necessary to make some use of this subject and at this point it will be worth making some of our earlier statements more precise by using the terminology of Group Theory. For simplicity we shall confine our attention to cubic crystals.

To exploit symmetry one starts the normal mode investigation by constructing symmetry adapted generalized displacements, i.e. displacements belonging to definite rows of definite irreducible representations of the symmetry group concerned. All of the displacements so formed which belong to a particular row of a particular irreducible representation form one of the sets mentioned. In our example we made use of the translation group combined with the inversion group and we were able to construct symmetry adapted basic displacements with real components: if we had not been able to use the inversion symmetry the basic displacements would have had to be complex. Of course, once the normal modes have been obtained in complex form one can get real normal modes simply by taking the real part of the displacement.

If an operation of rotation or reflection which sends the crystal into itself is applied to a k vector, the normal modes associated with the k vector so obtained will clearly be degenerate with those associated with the first k vector. From a general k vector it is possible to obtain 47 others in this way, so there is considerable degeneracy among the modes.

One is therefore free to form alternative normal displacements by taking linear combinations of the degenerate plane wave modes.

While we shall seldom have occasion to work explicitly with such linear combinations of plane wave modes the fact that linear combinations with other symmetry properties can be found will be important for some of our arguments. This situation arises when we are more concerned with the cubic symmetry of the lattice about a particular ion than with the translational symmetry. The normal modes of the perfect lattice can be chosen (by forming linear combinations of the plane waves) to belong to definite rows of definite irreducible representations of the cubic point group. The modes so formed are no longer suited to exploiting translational symmetry (as we might wish to do if treating, for example, an anharmonic perturbation which also had this symmetry), but since this translational symmetry is to be removed as soon as a point defect is introduced this is not important. If the defect introduced has cubic symmetry these new modes allow us to exploit this symmetry in a manner to be explained later.

8. Nature of Modified Modes

As is well known the frequencies of the normal modes of an infinite perfect crystal cover one or more continuous ranges lying between zero and some maximum frequency. If there is more than one continuous range (and this is the exception rather than the rule) we shall talk of gaps in the frequency spectrum of the perfect crystal. In the case of a macroscopic portion of perfect crystal the continua of normal mode frequencies are replaced by discrete but extremely closely spaced frequencies and we talk of quasi-continua.

When an imperfection is introduced into the crystal it can modify the modes in two ways; it will always modify the form of the modes with frequencies in the continuous frequency ranges of the perfect crystal modes and it may, in some cases, introduce "localised" modes at quite different frequencies lying either above the maximum normal mode frequency of the perfect crystal or in gaps in the perfect crystal frequency spectrum. Before we take up the detailed investigation of these two types of modified modes some further preliminary remarks on them may be useful.

The modifications of the continuum modes may be viewed in two different ways, depending on the physical phenomena with which we are concerned. If we are interested in the effect of the lattice vibrations on optical absorption or other processes at the imperfection itself, then we shall be interested in the modification of the form of the normal displacement in the vicinity of the imperfection, as the magnitudes of the displacements of atoms near the imperfection will determine the coupling of the mode to the process concerned. On the other hand if we are interested in the lattice thermal conductivity, or other processes involving phonon transport, we shall think in terms of the perfect lattice modes being scattered by the imperfection. Of course the distinction between the two aspects of the modification of the modes is not as clear cut as this description might suggest and, as we might expect, substantially the same technique may be used in treating either type of problem .

The localised modes outside the ranges of perfect lattice frequencies are in many respects much easier to deal with than the modifications to the continuum modes. These localised modes are closely related to electronic energy levels in the energy gaps between allowed bands in solids. The localised mode occurs at a frequency at which a disturbance cannot propagate through the crystal but dies away exponentially in one direction and therefore increases exponentially in the other. In the same way the localised electronic state occurs at an energy for which a solution of Schrodinger's equation dies away exponentially in one direction and increases in the other. In neither case do we get a satisfactory solution in the infinite perfect crystal but if there is a point imperfection of the right kind it is possible to construct solutions which die away exponentially on going out from the centre. We shall not have occasion to exploit this formal analogy but it should be noted that the methods we shall develop for the lattice defect problem can be applied to treatments of the electronic states at a defect.

Before we take up the methods of handling modified mode problems which will be the main theme of our talks we shall discuss a more direct method which is in some respects crude, but which within its limitations gives us very clear and convincing information. In particular it will demonstrate that the effect of the modes on processes at the defect can often be changed by factors of ten or more when the modifications of the modes are taken into account.

9. Direct Treatment of Normal Modes: 11 shell calculation.

We shall consider first the modification of continuum modes in NaCl by the presence of a F-centre. As we shall discuss more fully later the temperature dependence of the breadth of the F-band suggested that the coupling of low frequency modes to the electronic transition associated with the F-band was greater than would be calculated for unmodified modes. The approach to be described was applied to testing this conjecture.

The idea of the method is to reduce the problem to a straightforward normal mode calculation by considering only a small portion of crystal and making full use of symmetry. To be more specific one considered the motion of ions out to the eleventh nearest neighbour of the F-centre, all other ions in the infinite lattice being held fixed. This introducing of the infinite fixed lattice round the position being considered was convenient and amounts merely to a special choice of boundary conditions.

Even the small portion of crystal which we have described contains 171 atoms and it was essential to make use of symmetry. Attention was fixed on modes which have, roughly speaking, spherical symmetry about the F-centre as it was felt that these would probably make a major contribution to the broadening of the F-band. To be more precise the normal modes considered are all those which could be formed from the largest possible set of independent generalised displacements which went into themselves under all the symmetry operations of the full cubic group (i.e. from the symmetry adapted displacements belonging to the identity representation of the group). Clearly the normal modes formed in this way will also go into themselves under all the operations (i.e. they belong to the identity representation).

The required generalised displacements can be chosen in such a way that each involves only the displacements of a set of atoms which can be sent into one another by symmetry operations. For example, the nearest neighbour sodium ions clearly form such a set and there is only one corresponding generalised displacement with the required symmetry, namely a radial displacement of all six atoms through equal distances (a "spherically symmetric" displacement). Similarly the second, third and fourth neighbours each form such a set and in each case there is only one generalised displacement (namely the "spherically symmetric" one) of the required symmetry associated with each. The fifth neighbours (coordinate of a typical ion (2a,a,0) also form such a set but in this case there are two generalised displacements of the required symmetry, the "spherically symmetric" one and another involving displacements at right angles to the radius vectors of the ions. In this displacement all ions in the z = 0 plane remain in that plane and the way in which they are displaced is shown in Fig. 1.

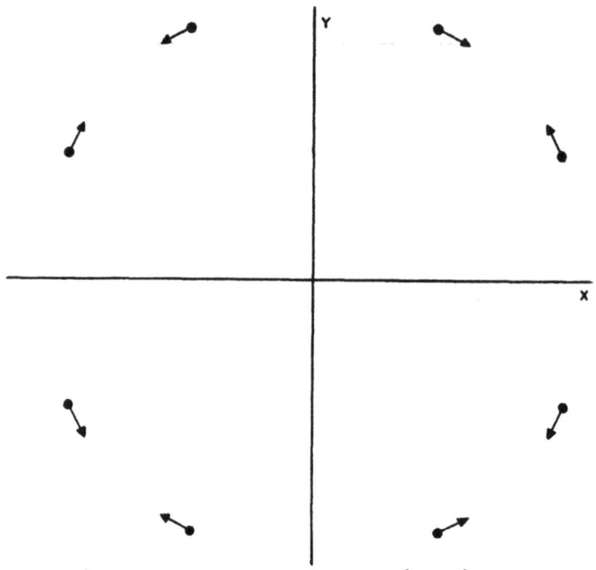

Fig.1. Non-spherical completely symmetric displacement of fifth neighbour shell (only one plane of atoms shown)

The same diagram applies, of course, to the x = 0 and y = 0 planes. Another feature arises with the eight nearest neighbours in that these include two different sets of atoms unrelated by symmetry: one set contains the atom (2a,2a,a) and the other contains the atom (3a, 0,0). One talks of the eighth shell of neighbours being formed of two sub-shells. The first sub-shell contributes two generalised displacements with the required symmetry, the second contributes one.

Proceeding in this way one finds that the eleven shells considered contribute in all sixteen generalised displacements of the required symmetry. The sets of coefficients for the kinetic and potential energy coefficients (these can conveniently be written as two 16 x 16 matrices) have to be calculated. In practice one considers separately the electrostatic interaction and the short-range interactions and adds the coefficients obtained. The determination of the normal modes reduces to finding the eigenvalues and eigenfrequencies of a 16 x 16 matrix. As already discussed in the case of the perfect lattice modes the eigenvalues give the squares of the normal mode frequencies and the eigenvectors give the forms of the normal displacements.

The model used for the crystal was that of Kellerman in which the ions are treated as rigid and non-polarisable, in a sense which will be made precise a little later. In this model the potential energy of the crystal is obtainable from the interaction potential of two ions as a function of their distance of separation. The form of this interaction potential will depend only on whether the two ions considered are both anions, both cations or one of each. If, as in the calculation under discussion, the repulsive interaction is neglected for all except nearest neighbour ions, one merely introduces this nearest neighbour repulsive interaction in the form $\lambda_{\pm}\ \exp(-r/\rho)$, where the two constants $\lambda_{\pm}$ and ρ are determined from the equilibrium lattice spacing and the compressibility. In addition one has to consider the Coulomb interaction.

The procedure for determining the potential and kinetic energies in terms of the generalised coordinates associated with the symmetry adapted displacements was straightforward although lengthy. The calculation was done both for the central atom in position and for an F-centre at the central position. In the latter case it was necessary to make estimates of the changes in the potential field in which ions near the F-centre moved as a result of replacing the central ion by the F-centre. Estimates were made for these charges which implied that the F-centre was "soft" in the sense that the restoring forces on the neighbouring ions were reduced. The estimates were, however, based on rather crude arguments and we shall not elaborate on them here. Our concern is to demonstrate the effect of the soft centre on the modes and whether or not the F-centre is in fact soft is, for the moment, of secondary importance.

The complete form of the modes obtained will not be shown as this aspect of the results will emerge more convincingly and strikingly in an example involving a larger portion of crystal which we shall take up later. We shall fix our attention on the frequencies of the normal modes and on the displacement of the nearest neighbour shell of ions. More precisely we shall consider for each normal displacement (normalised in the sense of (3)) the square of the displacement of the nearest neighbour shell divided by the frequency of the mode, and the frequency itself. As will emerge later the former quantity gives a measure of the relative contribution of the mode to the mean square breadth of the F-band at $0\,^{\circ}$K, on the assumption that the energy difference between the ground and excited electronic states of the F-centre depends only on the position of the nearest neighbour ions.

In Fig.2 the position of a line gives the frequency of the corresponding mode and the height gives (in arbitrary units) the square of the displacement of the nearest neighbour ions divided by the frequency in the associated normal displacement. Fig.2(a) gives the results obtained when the central atom is in position (it is, of course, not itself displaced in the "spherically

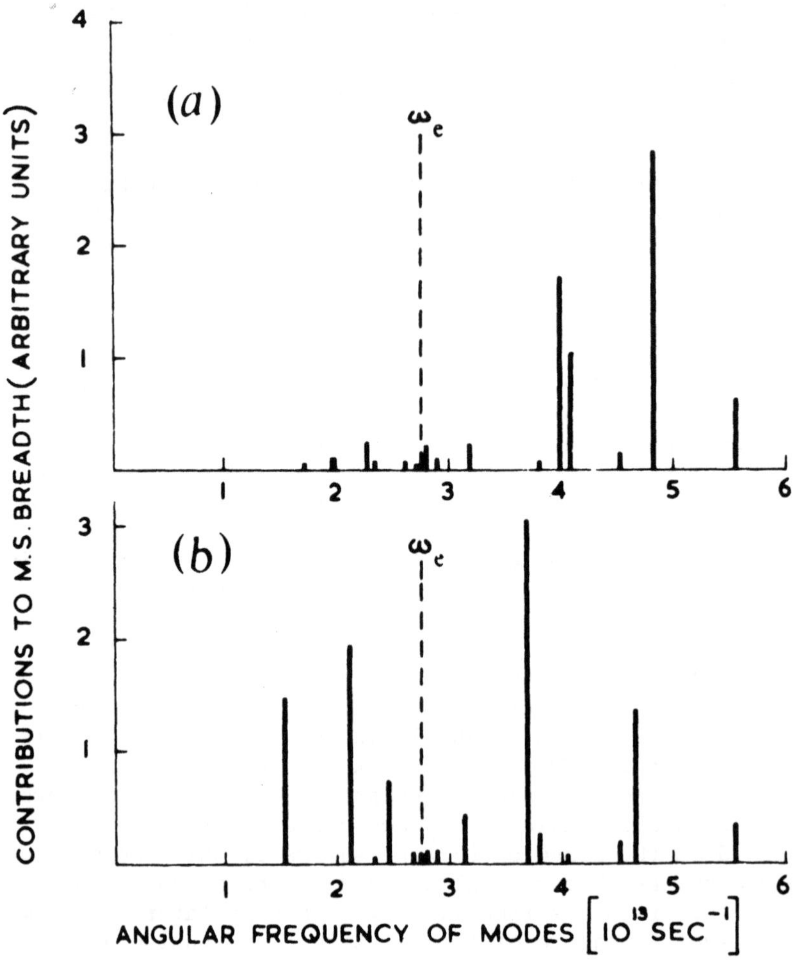

Fig.2.Eleven Shell NaCl Model. Square of displacement of nearest neighbours divided by frequency for each of the 16 spherically symmetric modes (a) with central atom present (b) with vacancy at centre.

symmetric" motion) and Fig. 2(b) gives the results found when the central ion is replaced by the "soft" F-centre. One sees that taking account of the modification of the modes due to the introduction of the F-centre greatly enhances the contributions to the broadening from low frequency modes : this bears out at least to some extent, the conjecture which initiated the investigation. The fact that the amplification is very large should be noted. It should also be observed that the frequencies of the modes are not changed greatly.

Before proceeding to another more ambitious application of the direct method of treating the modified modes it may be worth emphasising that the Kellerman model of the NaCl crystal employed in the above calculation is relatively crude compared with the models required to account for neutron and X-ray scattering results obtained for the perfect crystal. To account for the dispersion curves (curves of frequency against $\underset{\sim}{k}$-vector for a particular type of mode)obtained by these methods it has been necessary to go beyond treating the ions as rigid and non-polarisable as in the Kellerman model. Clearly one should take account of the fact that an electric field at an ion (such as will arise from the displacement of other ions in the generalised displacement being considered) will cause the ion to polarise. The electric field resulting from these dipoles must be taken into account in calculating the forces on ions (and so will affect the potential energy), and it will also contribute to the polarisation of other ions so that a self-consistency calculation is required. One must also take into account the fact that the forces exerted on a given ion by its neighbours may distort the electronic wave functions and so give rise to a dipole moment even if there is no electric field (one talks of the deformation dipole). In such work the ions are treated as non-rigid and polarisable. To our knowledge there has been no direct calculation of the type under discussion here which has embodied these refinements although such a calculation presents no difficulties in principle. As already indicated, however, these refinements have been incorporated into treatments of the vibrations of the perfect lattice, as in the work of Hardy and Karo. Thus the calculations on modified modes to be described later, which start from the vibrations of the perfect lattice, embody a more sophisticated and adequate model of the crystal than that used in the direct calculations.

The direct method can be applied to determining the normal modes with all symmetries. To illustrate the sizes of the matrices with which one has to work after full use is made of symmetry let us consider the vibrations of just four shells of atoms round the F-centre. These four shells contain 32 ions and so there are 96 normal modes in all. After using symmetry the largest matrix which has to be considered is 10 x 10. If only the "spherically symmetric" modes were required one would need only a 4 x 4 matrix.

10. Direct Method : Diamond Calculation

The eleven shell calculation carried out for NaCl gives only a crude picture of the forms of the modes near the F-centre (the boundaries are too near) and moreover the lowest frequency "spherically symmetric" mode which one gets has a frequency which is rather more than one quarter of the maximum continuum frequency. One gets no information at all about "spherically symmetric" modes with lower frequency. This was not too important for the problem to which the modified NaCl modes were being applied, namely the temperature dependence of the breadth of the F-band. The form of this temperature dependence gives, roughly speaking, a typical frequency for the modes mainly responsible for the broadening and this typical frequency, as deduced from experimental results, was comfortably higher than the minimum frequency obtained for a "spherically symmetric" mode in the eleven shell model. There was therefore no reason to believe that using a larger portion of crystal to introduce lower frequency modes would substantially alter the situation.

The situation is quite different for the optical absorption associated with certain imperfections in diamond. There one gets, not a broad smooth band like the F-band, but a very sharp absorption line with attendant absorption. The attendant absorption has quite a complicated structure and it yields rather detailed information about lattice vibrations. In particular it contains some features which suggest that modifications to quite low frequency modes may be important.

With this in mind the "spherically symmetric" vibrations round a vacancy in diamond were investigated allowing atoms out to the 60th nearest neighbour to vibrate. The problem was simplified somewhat by ignoring all generalised displacements with the required symmetry other than those in which all atoms at a given distance from the centre had displacements which were all radial and equal. It seems unlikely that this simplification will greatly influence the forms obtained for the modes. Even with it one is concerned with 60 x 60 matrices for the potential and kinetic energies. The whole operation of calculating the elements of the matrices and then solving the eigenvalue problem was programmed for an Elliott 803 Computer. The interactions of each atom with all its first, second and third nearest neighbours were considered. The results were insensitive to which of the available force constants were used.

If the "spherically symmetric" generalised displacements

are suitably normalised the elements of the corresponding eigenvectors give directly the radial displacements of successive shells in the various normal displacements. Some of the normal modes, calculated both for the central atom present (perfect lattice) and for it missing are shown in Figures 3 to 9. The assumptions made about the changes in force constants resulting from introducing the vacancy were somewhat arbitrary, but they were plausible in terms of the force constant model used. As one would expect the model made the vacancy "soft".

The vertical lines in the Figures represent the magnitudes of the (radial) displacements of successive shells: the distance of a line from the origin corresponds to the distance of the corresponding shell from the central site. It will be noted that the forms of the perfect lattice modes are close to what one would expect, this being particularly true of the lowest frequency mode Fig.3. (which one expects to be the longest wavelength acoustic mode consistent with the boundary conditions) and the highest frequency mode Fig.9 (which one expects to be the longest wavelength optical mode consistent with the boundary conditions). The enhancement of the displacement of the shell of atoms nearest to the central site when the central atom is removed is well marked for low frequency modes. The effect, moreover, is clearly resonant, being most marked in the present case for the mode which is third lowest in frequency. This resonant behaviour is, perhaps, suggested more clearly by Fig. 10 in which the squares of the displacements of the nearest neighbour ions divided by the frequency cubed (in the normalised normal displacements) are displayed as a function of frequency. The length of the line represents in arbitrary units the quantity being plotted and the position of the line gives the frequency of the corresponding mode. Fig.10 therefore corresponds to Fig. 3 which gave the NaCl results in essentially the same way: the difference in the powers of ω appearing in the quantity plotted in the two cases arises because Fig.10 relates to mean number of phonons created in each mode when the electronic transition takes place, while Fig. 3 relates to the contribution of each mode to broadening.

It is reasonable to assume that if we could increase greatly the number of shells considered, and so increase correspondingly the number of lines in Fig.10 these lines would have an envelope retaining the peak suggested by the rather small number of lines already there. This is equivalent to saying that we are regarding the relatively small number of modes in our 60 shell model as a sample of continuum modes. By investigating the modifications of the modes in the relatively small sample we get an idea of how the modification of continuum

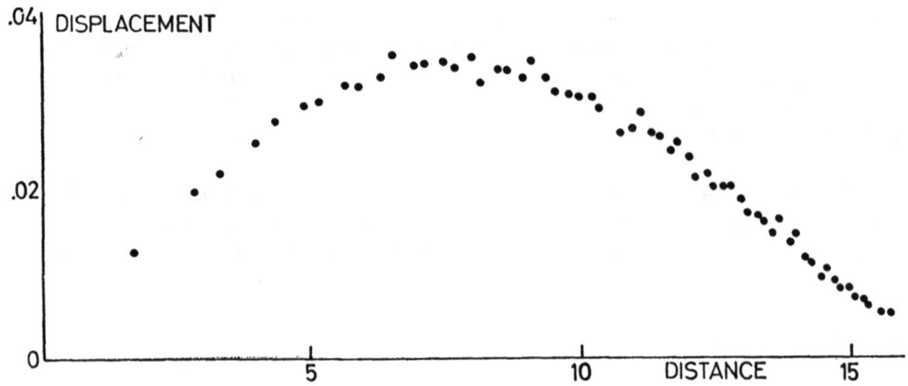

Fig.3. Sixty Shell Diamond Model. Lowest frequency mode - central atom present.

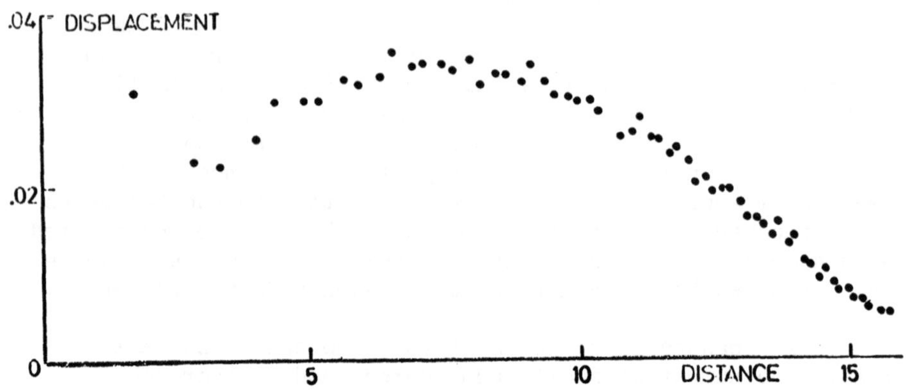

Fig. 4. Sixty Shell Diamond Model. Lowest frequency mode - central atom absent.

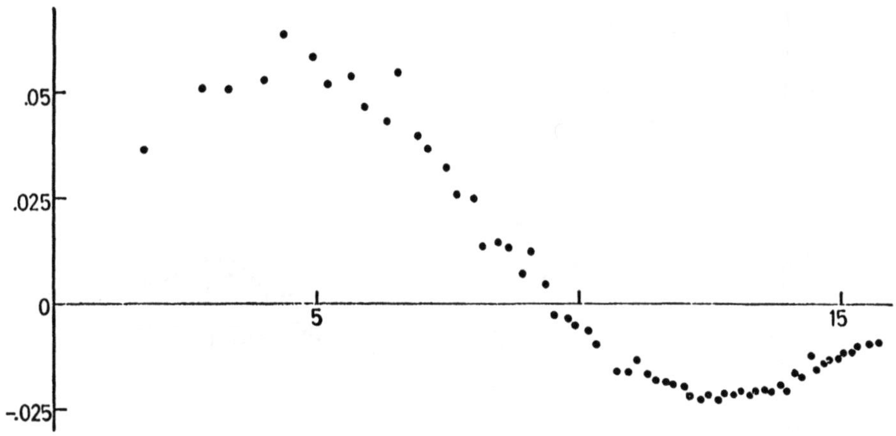

Fig.5. Sixty Shell Diamond Model. Second lowest frequency mode - central atom present.

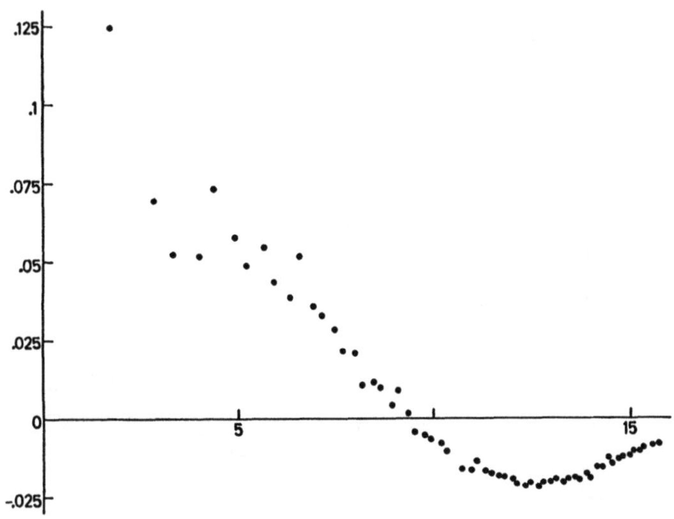

Fig.6. Sixty Shell Diamond Model. Second lowest frequency - central atom absent.

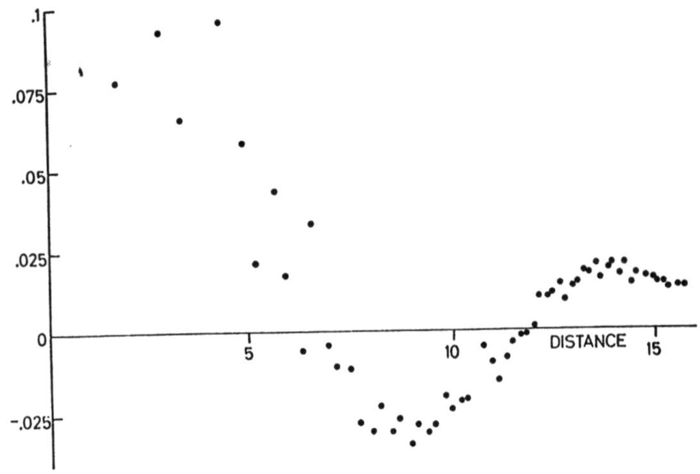

Fig.7. Sixty Shell Diamond Model. Third lowest frequency mode - central atom present.

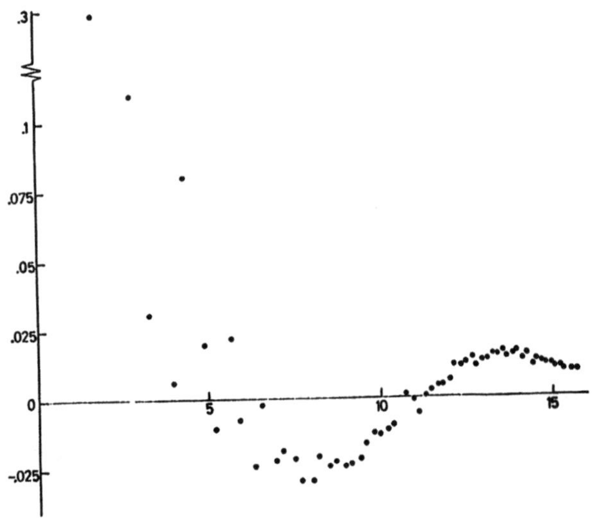

Fig.8. Sixty Shell Diamond Model. Third lowest frequency mode - central atom absent.

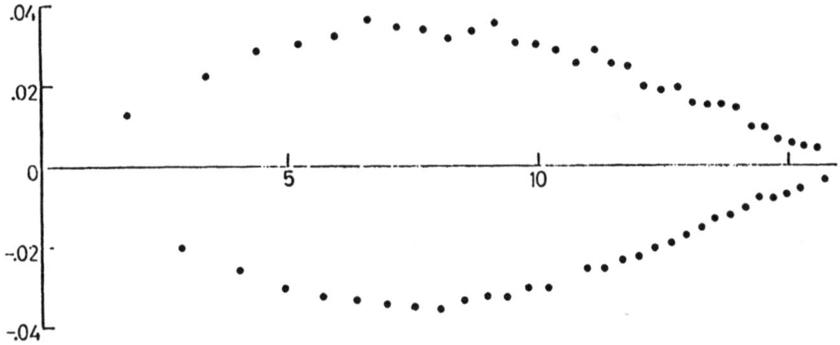

Fig.9. Sixty Shell Diamond Model. Highest frequency mode - central atom present.

modes of the symmetry under consideration varies with the frequency of the modes. In particular by sketching in the envelope as indicated in Fig. 10 we get an estimate, which is admittedly rather crude, for the magnitude of nearest neighbour displacements for modes (of the given symmetry) with all frequencies in the continuum range. This will allow us to deal in a rough and ready manner with problems involving the coupling of modes of this symmetry to processes at the defects.

Up to this point we have been concerned entirely with the modifications of continuum modes and we have not considered localised modes. These will not arise for the "soft" imperfections which we have been treating. To show exactly what we mean by a localised mode it is convenient to consider a hypothetical substitutional impurity in diamond which is "hard" in the sense that it increases the restoring forces on neighbouring atoms. In Fig.11 we show the form of the highest frequency mode obtained from the 60 shell modes, when the nearest neighbour restoring forces were increased sufficiently to make the angular frequency of this mode 2.526×10^{14} sec^{-1} as compared with 2.51×10^{14} sec^{-1} for the maximum frequency of the perfect lattice. The highly

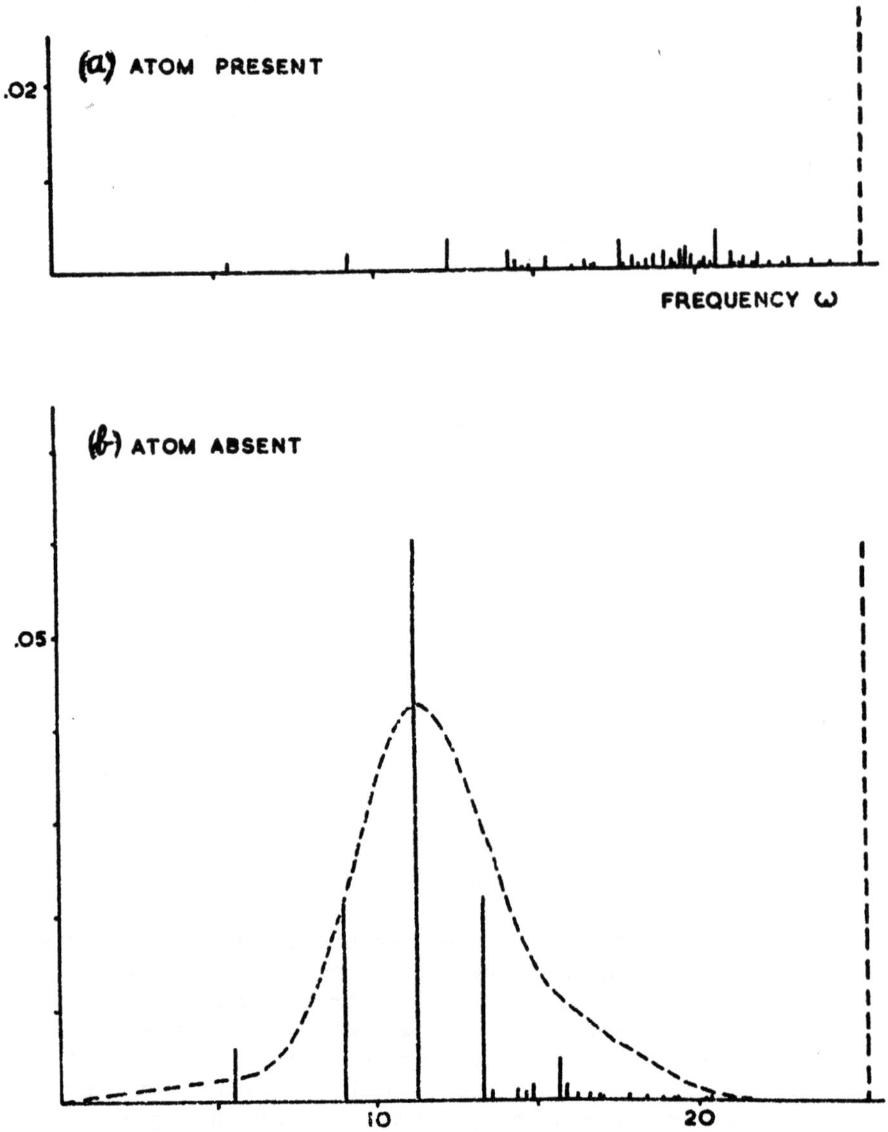

Fig.10. Sixty Shell Diamond model. Square of displacement of the nearest neighbours divided by the cube of the frequency for the spherically symmetric modes (a) with central atom present (b) with vacancy at centre.

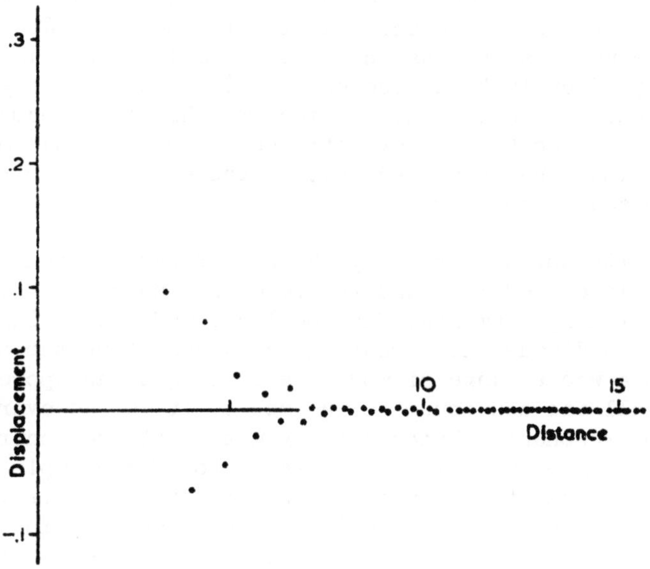

Fig.11. Sixty Shell Diamond Model with central atom replaced by hypothetical hard centre. Localised mode.

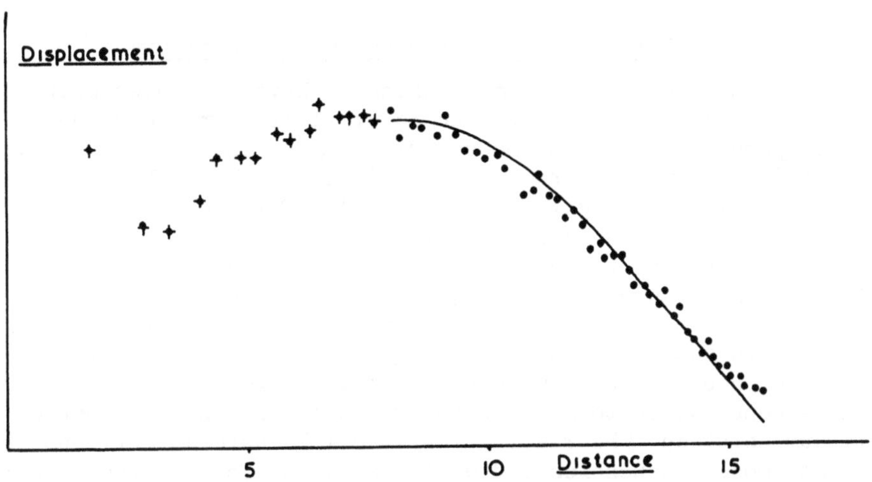

Fig.12. Sixty Shell Diamond Model. Mode obtained by joining to continuum (crosses and solid line) compared with rigid boundary mode (dots) of same frequency.

localised nature of this mode is clearly displayed: this corresponds to the exponential decay of a wave with frequency outside the continuum of perfect lattice frequencies. Just how localised the mode is depends on how far its frequency is from the nearest perfect lattice frequency (which, in turn, depends on the hardness of the centre). In fact the highest frequency perfect lattice mode (of the symmetry being considered) goes continuously into the localised mode as the extra restoring force is increased from zero.

It is worth noting carefully the difference between the genuine localised mode and the resonant enhancement of near neighbour displacements. The localised mode is a single genuine(localised)mode. The resonant enhancement on the other hand affects a whole range of modes, amounting to an appreciable fraction of all the continuum normal modes. It is, however, true that the resonant enhancement may mean that the contribution of modes in a quite small frequency range to, for example, the broadening of an electronic absorption band at the centre, is greatly enhanced and may even dominate other contributions. The effect is essentially the same as if we had a single localised mode, broadened by damping, with frequency at the middle of the range of enhanced modes. In this sense one talks of a quasi-localised mode in the continuum. We shall not have occasion to make use of this idea, the formal nature of which should be emphasised. We shall prefer to talk of resonant modification of the modes or sometimes, more briefly, of "inband resonances."

11. Direct Method: Joining to the Continuum

Even with the 60 shell model which we have been discussing the lowest frequency mode obtained is still at a frequency very large compared with that of modes which may be involved in, say, spin-lattice relaxation processes. To increase the number of shells sufficiently to get a very low value for the lowest frequency mode is clearly not practicable. We can, however, make use of the fact that the very low frequency vibrations of a perfect lattice are essentially just the longitudinal or transverse acoustic waves treated by continuum elasticity theory. This suggests that instead of joining our 60-shells to a rigid lattice we should join them to an infinite surrounding lattice which is constrained to move in the way described by an elastic continuum solution. The resulting system will have a continuum of normal mode frequencies and we can seek the form of the normal mode of the required symmetry for any low frequency which we choose.

The procedure is particularly straightforward for the

"spherically symmetric" modes which we have been considering, if we continue to make the assumption that, to a good approximation, the mode can be taken to be completely spherically symmetric. In that case the motion of the continuum will be a superposition of two spherically symmetric solutions of the continuum wave equations which are both well behaved at infinity: one of these, which varies as $\sin kr/r$ is also well behaved at $r = 0$ (a point which, of course, is outside the region to which we apply the continuum solution) while the other, which behaves at $\cos kr/r$, is not. If we treat the n nearest neighbour shells as able to move in any way they wish and suppose all more remote shells constrained to move with the continuum we have a system with just $n+2$ degrees of freedom. Since we can fix arbitrarily the value of one of the generalised coordinates in the normal displacement (the normalisation can easily be adjusted later to the conventional one) we need $n+1$ equations of motion to determine the other $n+1$ generalised coordinates.

The most obvious $n+1$ equations of motion to take would be those of the n shells which we are treating in detail and of the first shell included with the continuum. This, however, would mean attaching a great deal of weight to the equation of motion of this first shell and it is reasonable to seek a somewhat less arbitrary way of getting $n+1$ equations. The following procedure worked well. For a given choice of displacements of the shells one can calculate for each shell the extra generalised force (the error force) which would have to be applied to it from outside to give the postulated motion. One can consider in this way the error force on $n+r$ shells (the n treated discretely and r from the continuum) and then write down the $n+1$ equations which results from requiring that the $n+1$ generalised coordinates specifying the motion of the system (the other has been fixed arbitrarily) are such as to minimise the sum of the squares of the $n+r$ error forces. The equations which result are of simple form and can easily be handled on a computer. Writing them down is straightforward and as they are a little complicated we shall not give them.

Of course the procedure described is still arbitrary in that we have given no criterion for choosing r. (We may suppose $n+r$ fixed by the size of the computing time available). It is, however, possible to vary r and to check that the results obtained are reasonably independent of the way in which r is chosen. The numbers obtained give satisfactory results for this test.

It is interesting to compare the form obtained for

the lowest frequency mode of the rigid boundary system with the form obtained for a mode with the same frequency by the procedure of joining to a continuum outside the first 15 shells. The results are compared in Fig.12. Even although the frequency involved is not very low the agreement is very satisfactory.

The process of joining to the continuum was applied to determining how the introduction of a vacancy affected the amplitude of vibration of nearest neighbours for extremely low frequency modes. It was found that the ratio of the displacements with and without the vacancy was about two. Since the product matrix elements representing a relaxation process may depend on the fourth power of the nearest neighbour displacement this low frequency enhancement factor may be very significant. It must, of course, be emphasised that the modes involved in the relaxation process may not have spherical symmetry so that we are in effect conjecturing that the enhancement factors for different symmetries have comparable magnitudes.

The main advantage of the direct method is the very clear picture it gives of what is going on. This makes it useful for didactic purposes. As compared with the more powerful methods of investigation which we shall go on to describe it has some advantages if the centre preserves the full cubic symmetry of the perfect crystal and one is concerned only with "spherically symmetric" modes. In that case it can easily deal with changes in force constants extending far from the defect centre and the device of joining to the continuum gives very useful information about the form of very low frequency modes. It may also be useful if the centre has very low symmetry: in this case it is likely that it will be possible to apply the direct method to only a very small portion of crystal, but even this may provide useful information as all methods, including those to be described later, become difficult when the symmetry is very low. For the non-spherical vibrations of a symmetric centre the direct method is unlikely to have a great deal to recommend it compared with the intrinsically more powerful methods which we shall now discuss.

12. Infra-red absorption by a charged isotopic defect.

We turn now to the more powerful methods of dealing with the effects of imperfections on lattice vibrations. All of these exploit the fact that the defect modifies only a very small region of an otherwise perfect crystal. (So long as we are not concerned with interactions between defects it is, of course, sufficient to suppose there is just one defect present). The extreme case of localised modification is that in which an atom of the crystal is replaced by another which has the same interaction as the normal atom with all other atoms in the crystal but which differs from it in mass. The substitution of an isotope for a normal lattice atom fits this description with great precision but one usually has in mind cases in which the substitutional atom is merely chemically similar to the normal lattice atom, so that the interaction with other atoms is only approximately the same. We shall for example treat as an "isotopic impurity" iodine replacing chlorine in KCl.

The problem of the isotopic impurity is the simplest of the type of problem with which we are concerned. It has been used by many authors to illustrate methods of handling modified mode problems and we too shall use it for this purpose. The specific problem we shall take up first is that of the infra-red absorption which arises if the substituted atom carries a charge, all the other atoms being neutral.

As a preliminary let us consider how we can deal with this problem if there is no mass change, so that the lattice, which we shall suppose cubic and monatomic, is perfect apart from the fact that one atom carries a charge e. The result we obtain will lead us rather easily to the solution of the apparently much more difficult problem in which there is a mass change.

Since we are not interested in surface effects we may suppose that the charged atom is at the centre of a large portion of crystal with a boundary which has cubic symmetry. The modes will be chosen to be plane wave modes (we use periodic boundary conditions) and they will further be chosen to have sine and cosine forms. The 3N/2 modes with sine form will be unaffected by a force on the central atom (since they involve no displacement of this atom and so the resulting generalised force on them will be zero). We shall therefore restrict our attention to the 3N/2 cosine modes.

Let us choose as coordinate q'_r for the rth cosine mode the magnitude of the resulting displacement of the central atom: this is, of course, just the maximum displacement in the mode. The displacement of the atoms in the mode will all be along the same direction and denoting the displacement of the atom at the site $\underset{\sim}{R}^0_s$ by y_s we have ($\underset{\sim}{R}^0_s$ being the position vector relative to the central atom)

$$y_s = q'_r \cos \underset{\sim}{k}_r . \underset{\sim}{R}^0_s \qquad (19)$$

The kinetic energy in the mode is given by

$$T = \sum_{s=1}^{N} \tfrac{1}{2} m\dot{y}_s^2 = \tfrac{1}{2} m \dot{q}'^2_r \sum_{s=1}^{N} \cos^2 \underline{k}_r . R^0_s$$

$$= \tfrac{1}{2} \frac{Nm}{2} \dot{q}'^2_r \qquad (20)$$

If we now define q_r by

$$q_r = \sqrt{\frac{Nm}{2}}\, q'_r \qquad (21)$$

the kinetic energy will become

$$T = \tfrac{1}{2} \dot{q}_r^2 \qquad (22)$$

and q_r is clearly the normal coordinate associated with the normalised normal displacement.

If the electric field acts along some definite direction and varies as the real part of $Ee^{i\omega t}$ the force on the central atom is $eEe^{i\omega t}$. If q_r alone varies the resulting component of displacement of the central atom in the direction of the electric field will be $\sqrt{2/Nm}\cos\theta\, dq_r$ where Θ_r is the angle between the displacement in the rth mode and the direction of the electric field. The generalised force on the rth mode will therefore be $\mathfrak{F}_r = \sqrt{2/Nm}\cos\Theta_r\, eEe^{i\omega t}$ and, if the damping force on an atom is $-\kappa \underset{\sim}{v}$ we have

$$\ddot{q}_r + \kappa\dot{q}_r + \omega_r^2 q_r = \sqrt{2/Nm}\cos\Theta_r eEe^{i\omega t} \qquad (23)$$

This leads to

$$q_r = \frac{\sqrt{2/Nm}\cos\Theta_r\; eEe^{i\omega t}}{\omega_r^2 - \omega^2 + i\kappa\omega} \qquad (24)$$

The rate W_r at which work is done by the generalised force on the mode is $\tfrac{1}{2}\omega\,\mathfrak{I}(\mathfrak{F} q_r^*)$ i.e.

$$W_r = \tfrac{1}{2}\frac{2}{Nm}\,\omega\cos^2\Theta_r e^2 |E|^2 \frac{\kappa\omega}{(\omega_r^2 - \omega^2)^2 + \kappa^2\omega^2} \qquad (25)$$

Suppose now that the fraction of normal modes of the crystal with frequency between ω and $\omega + d\omega$ is $\rho(\omega)d\omega$, so that the number of cosine modes in the range is $(3N/2)\rho(\omega)d\omega$. The directions of the displacements in normal modes of the cubic crystal will be distributed isotropically in space and so in summing the W_r over the modes in this frequency range one can replace $\cos^2\theta_r$ by its mean value 1/3. The total absorbed by all the modes of the crystal from the electric field will therefore be given by

$$\frac{1}{2}\,\frac{3N}{2}\cdot\frac{2}{Nm}\cdot\frac{1}{3}\;\omega e^2\;|E|^2\int_0^\infty\frac{\rho(\omega')\kappa\omega d\omega'}{(\omega'^2-\omega^2)^2+\kappa^2\omega^2} \tag{26}$$

Now the factor $\kappa\omega/\{(\omega'^2-\omega^2)^2+\kappa^2\omega^2$ in the integral has, for small κ, a very sharp peak at $\omega' = \omega$ and the integral of this sharply peaked function over all ω' is easily seen to be $\pi/2\omega$ (independent of κ). It follows that in the limit of small κ this factor can be replaced in the integral by $\frac{\pi}{2\omega}\,\delta(\omega'-\omega)$ and we get for the rate of absorption of energy

$$W = \frac{\pi\rho(\omega)}{4\,m}\;e^2\;|E|^2 \tag{27}$$

Note that the only crystal property on which this expression depends is the density of modes at the frequency of the applied electric field. We shall always assume that such properties of the perfect crystal are known and it follows that the expression given solves the problem when there is no mass change. It should be noted that in this case the observed infra-red absorption associated with the defect would lead at once to $\rho(\omega)$, the fractional density of normal modes of the perfect crystal.

It is not difficult to see in very general terms what the effects of changing the mass of the charged particle will be. In the first place if the mass is reduced sufficiently we will get localised modes: if the foreign atom is much lighter than the normal atoms it will clearly be possible for it to vibrate in practically immobile surroundings and a less drastic reduction of mass will allow motions in which the displacement of the surroundings dies away rapidly with distance. The other effect is to modify the continuum modes. The density in frequency of the continuum modes will not be appreciably affected by the single imperfection but the forms of the modes in the vicinity of the imperfection will be changed. We shall not be surprised if we get resonant modifications similar to those discussed earlier for spherically symmetric modes. In so far as these modifications enhance the displacement of the charged atom in the modes concerned they will give an enhanced absorption and so we may anticipate peaks in the absorption curves as a result of the in-band resonances.

It will be convenient for the moment to ignore localised modes which will be considered later, and to fix our attention on absorption at frequencies in the continuum.

We shall now show that we can solve the problem of absorption by the isotopic defect provided that we know the response (after transients) of an atom of the perfect lattice which is acted upon by a sinusoidal force. To be more precise we require the following response function of the perfect lattice. A force $Fe^{i\omega t}$ is applied to an atom of the perfect lattice and the resulting displacement of this atom (after transients have died out so that the only motion is an outgoing wave emanating from the forced atom) is $de^{i\omega t}$ with, of course, F and d complex. The required response function $P(\omega) - i\,Q(\omega)$ is defined by

$$d = \frac{1}{m}\,(P(\omega)-iQ(\omega))F \qquad (28)$$

It is to be emphasised that this refers to the perfect lattice.

The idea of the calculation which follows is to regard the extra mass Δm of the foreign atom as a particle which is rigidly attached to an atom of the perfect lattice which also carries the charge of the foreign atom and to consider explicitly the force exerted by the attached particle on the atom of the perfect lattice to which it is attached. The magnitude of this force depends on the displacement $de^{i\omega t}$ of the extra particle which is, of course, also the displacement of the perfect lattice atom to which it is attached. The acceleration of the extra particle is $-\omega^2 de^{i\omega t}$ and the force exerted by the perfect lattice atoms on it is $-\Delta m\omega^2 de^{i\omega t}$. It follows from Newton's Third Law that the force which it exerts on the atom to which it is attached is $\Delta m\omega^2 d\, e^{i\omega t}$.

The perfect lattice atom which carries the charge and has extra mass is therefore acted upon by the electric force $eEe^{i\omega t}$ and the force exerted by the attached particle which is $\Delta m\omega^2 de^{i\omega t}$. We can therefore write down the relation

$$d = \frac{1}{m}\,(P(\omega)-iQ(\omega))(eE+\Delta m\omega^2 d) \qquad (29)$$

Solving for d we get

$$d = \frac{(P-iQ)\ (e/m)\ E}{(1-\frac{\Delta m}{m}\ \omega^2 P)+i\frac{\Delta m}{m}\ \omega^2 Q} \tag{30}$$

The rate of absorption of energy W will be given by $\frac{1}{2}\omega \mathscr{I}(eEd^*)$, and so

$$W = \frac{\omega}{2m}\ \frac{Q(\omega)\ e^2\ |E|^2}{(1-\frac{\Delta m}{m}\ \omega^2 P)^2 + (\frac{\Delta m}{m}\ \omega^2 Q)^2} \tag{31}$$

This is the required result expressed in terms of $P(\omega)$ and $Q(\omega)$ which (we again emphasise) relate to the response of the perfect lattice. It will be noted that the denominator is of the typical resonance form : we will get a resonance when the first term vanishes.

It will be useful in connection with later applications to put the discussion of the last few paragraphs in a slightly different form. Equation (28) defines a response function $(1/m)(P-iQ)$ for the perfect lattice. We can define also an imperfect lattice response function which gives the relation between force and displacement which we get if we apply an oscillatory force to the atom with the excess mass Δm : we write

$$d = \frac{1}{m}\left\{P_{imp}(\omega)\ -iQ_{imp}\ (\omega)\right\}F \quad \text{(imperfect lattice)} \tag{32}$$

But equation (30), with eE replaced by F, gives the relation between d and F for the imperfect lattice and (32) must agree with this. By writing the complex expression multiplying eE (which plays the part of F) on the right hand side of (30) as the sum of a real and imaginary part and comparing with (32) one finds

$$P_{imp}\ (\omega) = \frac{P - \frac{\Delta m}{m}\ \omega^2\ (P^2+Q^2)}{(1-\frac{\Delta m}{m}\ \omega^2 P)^2 + (\frac{\Delta m}{m}\ \omega^2 Q)^2} \tag{33}$$

$$Q_{imp}\ (\omega) = \frac{Q}{(1-\frac{\Delta m}{m}\ \omega^2 P\ )^2 + (\frac{\Delta m}{m}\ \omega^2 Q)^2} \tag{34}$$

where the dependence of P and Q on ω is understood. In other words one can obtain the imperfect lattice response function in terms of the perfect lattice response function and information about the nature of its imperfection. Some later application of our theory will depend on this fact. The problem under discussion here can be dealt with in this way since combining the expression $\frac{1}{2}\omega \mathcal{J}(eEd^*)$ for W with equation (32) gives

$$W = \frac{\omega}{2m} Q_{imp}(\omega) e^2 |E|^2 \tag{35}$$

and substituting from (34) gives (31).

The problem of determining $P(\omega)$ and $Q(\omega)$ for the perfect lattice remains. We notice, however, that we can immediately determine $Q(\omega)$ by considering the form which the result just derived takes for the case $\Delta m=0$, namely

$$W = \frac{\omega}{2m} Q(\omega) e^2 |E|^2 \tag{36}$$

and comparing it with the result for this case which was obtained earlier

$$W = \frac{\pi}{4m} \rho(\omega) e^2 |E|^2 \tag{27}$$

This gives at once

$$Q(\omega) = \frac{\pi}{2\omega} \rho(\omega) \tag{37}$$

(This result could, of course, have been derived directly, without considering the $\Delta m \neq 0$ case or introducing an electric field, by evaluating the work done by a force $Fe^{i\omega t}$ acting on a perfect lattice atom in exactly the same way as one evaluated the work done by the electric field in the infra-red absorption case. One would simply have derived the result of the earlier argument with $e^2|E|^2$ replaced by $|F|^2$: the result $\frac{1}{m} Q(\omega)|F|^2$ for W follows at once from the definition of the response function and the expression $\frac{1}{2}\omega \mathcal{J}(Fd^*)$ for W).

Knowing $Q(\omega)$ one can obtain $P(\omega)$ because these are related by the Kramers-Kronig relations, which embody the fact that the response function P-iQ must not be such as to imply a response to an impulsive applied force before that force is applied. One has explicitly

$$\begin{aligned} P(\omega) &= \frac{2}{\pi} \mathcal{P} \int_0^\infty \frac{\omega' Q(\omega') d\omega'}{\omega'^2 - \omega^2} \\ &= \mathcal{P} \int_0^\infty \frac{\rho(\omega')\, d\omega'}{\omega'^2 - \omega^2} \end{aligned} \tag{38}$$

Here the symbol $\mathcal{P}$ indicates that one has to take the Cauchy Principal Value of the integral which follows. Explicitly

$$\mathcal{P}\int_0^\infty \frac{\rho(\omega')d\omega'}{\omega'^2-\omega^2} = \lim_{\varepsilon\to 0}\left\{\int_0^{\omega-\varepsilon} \frac{\rho(\omega')d\omega'}{\omega'^2-\omega^2} + \int_{\omega+\varepsilon}^\infty \frac{\rho(\omega')d\omega'}{\omega'^2-\omega^2}\right\} \quad (39)$$

While the main interest of these results lies in their applicability to realistic models of a crystal it will be instructive to see what they give for the Debye model. For this $\rho(\omega)$ is $3\omega^2/\omega_D^3$ for $\omega < \omega_D$ and zero for $\omega > \omega_D$. Thus

$$Q(\omega) = \begin{cases} \dfrac{3\pi}{2\omega_D^3}\,\omega & \omega < \omega_D \\ 0 & \omega > \omega_D \end{cases} \quad (40)$$

The expression for $P(\omega)$ can be written

$$P(\omega) = \frac{3}{\omega_D^3}\mathcal{P}\int_0^{\omega_D} \frac{\omega'^2 d\omega'}{\omega'^2-\omega^2}$$

$$= \frac{3}{2\omega_D^3}\mathcal{P}\int_0^{\omega_D} \left\{2 - \frac{\omega}{\omega'+\omega} + \frac{\omega}{\omega'-\omega}\right\}d\omega' \quad (41)$$

The first two terms contain no singularities and they can be integrated at once to give the contribution

$$\frac{3}{2\omega_D^3}\left\{2\omega_D - \omega\log\frac{\omega+\omega_D}{\omega}\right\}$$

The final term contributes the following

$$\frac{3\omega}{2\omega_D^3}\,\mathcal{P}\int_0^{\omega_D}\frac{d\omega'}{\omega'-\omega} = \frac{3}{2}\frac{\omega}{\omega_D^3}\lim_{\varepsilon\to 0}\left\{\left[\log(\omega'-\omega)\right]_0^{\omega-\varepsilon} + \left[\log(\omega'-\omega)\right]_{\omega+\varepsilon}^{\omega_D}\right\}$$

$$= \frac{3}{2}\frac{\omega}{\omega_D^3}\lim_{\varepsilon\to 0}\{\log\varepsilon - \log\omega + \log|\omega_D-\omega| - \log\varepsilon\}$$

$$= \frac{3}{2}\frac{\omega}{\omega_D^3}\log\frac{|\omega_D-\omega|}{\omega} \tag{42}$$

Thus we get on summing these contributions

$$P(\omega) = \frac{3}{2\omega_D^3}\left(2\omega_D + \omega\log\frac{|\omega_D-\omega|}{\omega_D+\omega}\right) \tag{43}$$

The $P(\omega)$ and $Q(\omega)$ thus obtained for the Debye model are shown in Fig.13. If we regard the triangular $Q(\omega)$ curve as forming a single peak the figure indicates how $P(\omega)$ behaves when $Q(\omega)$ passes through a peak. The behaviour is the same in form as that for the real and imaginary parts of the response of a single damped oscillator, shown in Fig.14 . The curves in this last case also obey the Kramers-Kronig relations, and are indeed just a particularly simple example of $P(\omega)$ and $Q(\omega)$ curves.

We now return to the Debye model and consider the form of the absorption by a charged impurity atom with mass $m + \Delta m$. We had that the absorption as a function of frequency is proportional to

$$\frac{Q(\omega)}{(1-\frac{\Delta m}{m}\omega^2 P)^2+(\frac{\Delta m}{m}\omega^2 Q)^2}$$

As already stated there is a resonant maximum at the frequency which makes $(1 - (\Delta m/m)\omega^2 P)$ zero. For sufficiently large Δm this will occur at low frequencies where P has practically its limiting low frequency value, namely $3/\omega_D^2$ and for the moment we shall suppose this to be the case. The resonant condition may then be written (the resonant frequency being denoted by ω_R)

$$\omega_R^2 = \frac{m}{\Delta m} \frac{\omega_D^2}{3} \tag{44}$$

The absorption at low frequencies is then given by

$$W = \frac{(3\pi/4m)\omega_R^4 \omega^2/\omega_D^3}{(\omega_R^2-\omega^2)^2 + (\frac{\pi}{2}\omega^3/\omega_D)^2} e^2|E|^2 \tag{45}$$

It is easy to see that $\Delta\omega_{1/2}$, the distance from the maximum absorption to the half maximum position is given approximately by

$$\Delta\omega_{1/2}/\omega_R = \frac{\pi}{4}(\omega_R/\omega_D). \tag{46}$$

One sees that for $\omega_R \ll \omega_D$ the fractional change in ω over the absorption band will be negligible as had to be assumed in obtaining this simple expression for $\Delta\omega_{1/2}$.

It should be noted that the reason for the peak getting sharp when it is pushed to low frequencies (by making Δm large) is that $Q(\omega)$ (and that means essentially the density of modes) is small there. It is clearly a general result that the peak will be sharp when it occurs at a frequency for which the density of modes is low.

The position and nature of the resonance peaks can easily be investigated even when we drop the requirement that $\Delta m/m$ should be large and consider the possibility of there being more than one resonance peak for a given value of this mass ratio. One requires, of course, the $P(\omega)$ and $Q(\omega)$ curves (both obtainable, as we have emphasised, from the density of perfect lattice modes) and it will be convenient to illustrate by again referring to the Debye model.

The positions of the resonance are determined by the intersections of the $P(\omega)$ curve with the inverse square curve $(m/\Delta m)(1/\omega^2)$ since it is at these points that $1-(\Delta m/m)\omega^2 P$ vanishes. The form of the inverse square curves is shown on Fig.13. The larger $\Delta m/m$ the lower the corresponding curve lies. For small $\Delta m/m$ the curve will not intersect the $P(\omega)$ curve at all. There may still, of course, be a resonance effect if the curves approach closely. If $\Delta m/m$ is large enough to cause intersection it is clear that the curves must intersect at least twice (this

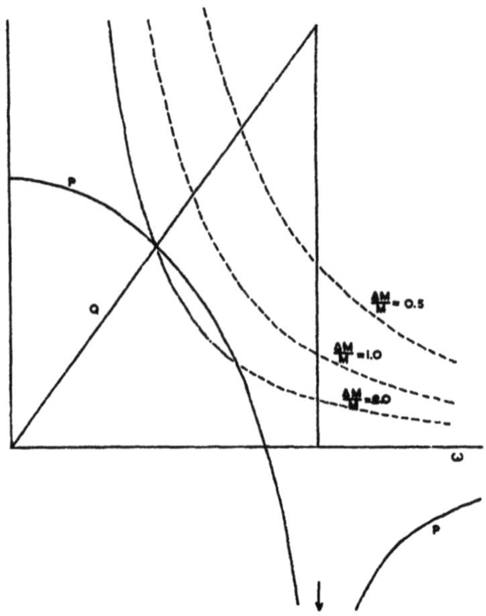

Fig.13. $P(\omega)$ and $Q(\omega)$ curves for Debye Model and $(m/\Delta m)(1/\omega^2)$ for various values of $m/\Delta m$.

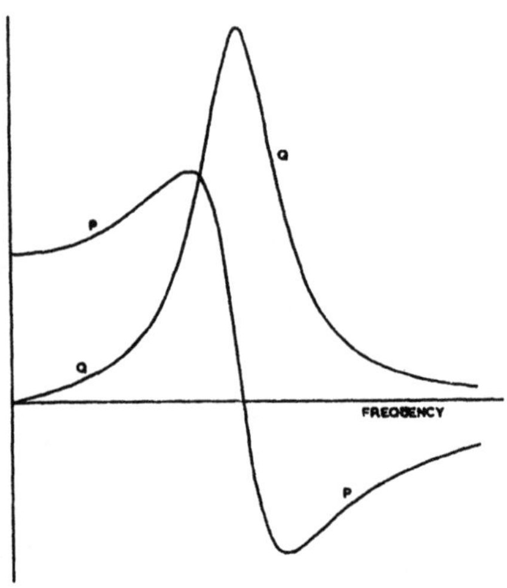

Fig.14. $P(\omega)$ and $Q(\omega)$ curves for harmonic oscillator.

is true in all cases, not only for the Debye model) so there will be two resonances. In the Debye case if these resonances are appreciably separated the lower one will be narrow, in the way we have already discussed, but the density of modes will be large at the higher frequency resonance and the resonance may be very broad. With more complicated densities of modes there may be several resonances and some of them may occur where the density of modes has again become small or, in other words, where we nearly have a gap in the frequency spectrum. The resonances at such positions will be sharp and if the density of modes in the region could be made to drop to zero the behaviour would go continuously into that of the true localised mode which can exist when we have an actual gap in the density of modes.

So far we have considered absorption peaks introduced by the charged isotope which are associated with the behaviour of the resonance denominator in equation. But the $Q(\omega)$ in the numerator may also introduce peaks, since it has all the peaks in the perfect lattice density of modes curve. It is, however, important to notice that two effects in the behaviour of the denominator will tend to suppress such peaks. In the first place $Q(\omega)$ is large at such a peak and the second term in the denominator will become large there : this will reduce the peak very substantially if $(\Delta m/m)\omega^2 Q$ becomes large compared with unity. In the second place there is a tendency for $P(\omega)$ to become large and negative at a peak in $Q(\omega)$ or, in some cases, on the high frequency side of the peak. In the Debye model, for example, we see that $P(\omega)$ tends to minus infinity at the peak in $Q(\omega)$ (at the Debye frequency), but this is, of course, an extreme and exceptional case. When the $P(\omega)$ and $Q(\omega)$ curves are those for a single damped oscillator $P(\omega)$ is zero at the resonance frequency and then goes negative at higher frequencies. In either case the tendency will be either to suppress or to shift the peak in $Q(\omega)$.

The discussion just given shows that one must be very cautious in associating peaks in the infra-red absorption due to a charged isotope with peaks in the density of modes of the perfect crystal. This applies particularly to large peaks which dominate the overall shape of the absorption since the effects in the denominator which have just been considered will operate most strongly on large peaks in $Q(\omega)$. On the other hand the effects on the denominator will be much less important for small peaks in $Q(\omega)$ and these will be expected to be shifted slightly but left relatively unchanged. Similar considerations will apply, as we shall see, when we consider the vibrational structure attendant on sharp zero-phonon absorption lines associated with electronic transitions.

13. Infra-red Absorption by a Defect with Force Constant Change.

The response function technique of the last section deals effectively and simply with the isotope problem. We shall now show how it can be applied to more general cases by considering a slightly more complicated example. We shall suppose that two atoms of the cubic crystal which are nearest neighbours of one another are replaced by foreign atoms 1 and 2, with excess masses Δm_1 and Δm_2. The atom 1 will be supposed to carry a charge e. As to force constant changes we shall make the rather artificial assumption that the two atoms have the same interaction with perfect lattice atoms as does a perfect lattice atom, but that the interaction of atoms 1 and 2 with each other is different from that of two perfect lattice atoms. In the spirit of the approach adopted in the previous section all these changes can be regarded as follows. One atom of the perfect lattice has been given a charge e. This atom and an adjacent one have extra masses Δm_1 and Δm_2 respectively attached to them. Finally an extra "spring", with force constant k connects the two atoms.

Let us now consider an applied electric field $Ee^{i\omega t}$ which is directed along the line joining the two atoms. The resulting displacements of the two atoms will then clearly be along the line joining them and we can denote them by $d_1 \, e^{i\omega t}$ and $d_2 \, e^{i\omega t}$ (the positive directions for each being the same).

We now require two perfect lattice response functions. Suppose a force $F_1 \, e^{i\omega t}$ is applied to a perfect lattice atom the displacement of that atom will be supposed given by

$$d_1 = \frac{1}{m} \left(P_{11}(\omega) - i \, Q_{11} \, (\omega)\right) F_1 \qquad (47)$$

and that of the adjacent atom of the perfect lattice by

$$d_2 = \frac{1}{m} \left(P_{12}(\omega) - i \, Q_{11} \, (\omega)\right) F_1 \qquad (48)$$

It follows from the symmetry of the cubic lattice that if the Force $F_2 \, e^{i\omega t}$ is applied to the adjacent atom the displacements will be given by

$$d_1 = \frac{1}{m} \left(P_{12}(\omega) - i \, Q_{12} \, (\omega)\right) F_2 \qquad (49)$$

$$d_2 = \frac{1}{m} \left(P_{11}(\omega) - i \, Q_{11} \, (\omega)\right) F_2$$

The equations of motion may now be written down in terms of these response functions. The atom 1 (of the perfect lattice) is acted upon by the electric force $eEe^{i\omega t}$, by the force $\omega^2\Delta m_1 d_1 e^{i\omega t}$ exerted on it by the attached mass Δm_1 and by the force $-k(d_1-d_2)$ exerted by the attached spring. The uncharged atom 2 is acted upon similarly by forces $\omega^2\Delta m_2 d_2 e^{i\omega t}$ and $-k(d_2-d_1)$. The motion of atom 1 is composed of a part due to the force applied directly to it and part due to the force on atom 2. Thus

$$d_1 = \frac{1}{m}(P_{11}-iQ_{11})\{eE+\omega^2\Delta m_1 d_1-k(d_1-d_2)\} + \frac{1}{m}(P_{12}-iQ_{12})\{\omega^2\Delta m_2 d_2-k(d_2-d_1\} \tag{50}$$

Similarly

$$d_2 = \frac{1}{m}(P_{11}-iQ_{11})\{\omega^2\Delta m_2 d_2-k(d_2-d_1)\} + \frac{1}{m}(P_{12}-iQ_{12})\{eE+\omega^2\Delta m_1 d_1-k(d_1-d_2\} \tag{51}$$

It is clearly a straightforward matter to eliminate d_2 between these equations and to express d_1, the displacement of the charged particle, in terms of the electric field E. This gives at once the rate of absorption of energy at frequency ω. We could, of course, again express this discussion in terms of an imperfect lattice response function defined as in (32): the analogue of (33) and (34) would be more complicated but could still be obtained explicitly.

The determination of the response functions is only a little more complicated than the determination of the single response function required in the isotope case. Indeed $P_{11}-iQ_{11}$ is exactly the response function P-iQ considered previously. To determine $P_{12}-iQ_{12}$ we need information about the forms of the perfect lattice modes as well as their frequencies. Let us denote by α_{1r} and α_{2r} the components of displacement of atoms 1 and 2 respectively along the line joining them when q_r is unity and all other perfect lattice normal coordinates are zero. If the force on atom 1 is $F_1e^{i\omega t}$, the resulting generalised force on the rth mode will be $\alpha_{1r}F_1e^{i\omega t}$. It follows that the time dependence of the rth generalised coordinate will be

$$q_r = \frac{F_1\alpha_{1r}e^{i\omega t}}{(\omega_r^2-\omega^2)+ik\omega} \tag{52}$$

The resulting displacement of atom 2 will be

$$d_2 = \sum_r \alpha_{2r}\, q_r$$

$$= F_1\, e^{i\omega t} \sum_r \frac{\alpha_{1r}\,\alpha_{2r}}{(\omega_r^2-\omega^2)+i\kappa\omega} \qquad (53)$$

Taking the imaginary part of the sum gives

$$Q_{12}(\omega) = m \sum_r \frac{\alpha_{1r}\,\alpha_{2r}\,\kappa\,\omega}{(\omega_r^2-\omega^2)^2+\kappa^2\omega^2} \qquad (54)$$

To convert this sum to an integral we let $A_{12}(\omega)$ denote the average of $\alpha_{1r}\alpha_{2r}$ for all the modes in a small range of angular frequency about ω. We then have

$$Q_{12}(\omega) = m \int_0^\infty \frac{A_{12}(\omega')\rho(\omega')\kappa\omega d\omega'}{(\omega'^2-\omega^2)^2+\kappa^2\omega^2} \qquad (55)$$

As discussed previously, the factor $\kappa\omega/\{(\omega'^2-\omega^2)^2 + \kappa^2\omega^2\}$ becomes, in the limit of small κ, $(\pi/2\omega)\delta(\omega'-\omega)$ and so we have

$$Q_{12}(\omega) = m\,(\pi/2\omega)\,A_{12}(\omega)\rho(\omega). \qquad (56)$$

This means that $Q_{12}(\omega)$ is related to the density of modes as is $Q_{11}(\omega)$ but the expression for $Q_{12}(\omega)$contains the weighting factor $A_{12}(\omega)$.

The real part $P_{12}(\omega)$ can be obtained from $Q_{12}(\omega)$ by means of the Kramers-Kronig relation. Thus the response function is determined if we know the density function $\rho(\omega)$ and the average displacement product $\alpha_{1r}\,\alpha_{2r}$ for modes in each narrow range of frequencies. When the response functions are known the infra -red absorption can be obtained in the way already indicated.

14. Estimation of Response Functions.

It may be useful at this point to outline briefly the steps in the practical calculation of $\rho(\omega)$ and of $A_{12}(\omega)$.

One starts from the frequencies and forms of the perfect lattice modes associated with a uniform distribution of k-vectors in the first Brillouin zone. The density $\rho(\omega)$ will be estimated by forming a histogram showing the numbers of modes in various frequency ranges and then smoothing it : the division of the frequency range used in the histogram will depend, of course, on the density of the distribution of k-vectors with which one is working. The function $A_{12}(\omega)\rho(\omega)$ can be obtained similarly: for each frequency interval one will find the sum of $\alpha_{1r}\,\alpha_{2r}$ for all the modes with frequency in the range and treat these in exactly the same way as one treated the numbers of modes in the various intervals in arriving at $\rho(\omega)$.

An alternative procedure for estimating response functions is sometimes useful in that it is convenient for machine computation and allows a more physical interpretation of what is being done than does the fairly arbitrary smoothing of a histogram which enters into the procedure outlined above. The idea is to evaluate the response function with a small but finite damping on the lattice. This means that P_{12} and Q_{12} will be approximated by the real and imaginary parts respectively of

$$m \sum_r \frac{\alpha_{1r}\,\alpha_{2r}}{(\omega_r^2-\omega^2)+i\kappa\omega}$$

The arbitrariness in the smoothing of the histogram is replaced by arbitrariness in the choice of κ. In practice one will evaluate the above expression for a range of values of κ. For small κ there will be a good deal of clearly spurious oscillations on the curves obtained while for large κ it will be clear that detail is being lost. If there is a sufficiently larger number of normal modes in the sample of perfect lattice modes employed one will expect to find between these extremes a range of values of κ over which the forms of the curves are reasonably constant. There will then be no difficulty in arriving at reliable estimates of P_{12} and Q_{12} (and similarly of P_{11} and Q_{11}).

It should, however, be pointed out that for some purposes the finite damping method can fail because of the long tails of the Lorentzian curve which it associates with each mode. An example of this may arise if we are interested in more complicated responses to more complicated force systems. In that case the analogue of $\alpha_{1r}\alpha_{2r}$ may depend on the wavelength and so on the angular frequency ω_r of the rth mode. If we suppose this power is positive it will be introduced into the analogue of $A_{12}(\omega)$ and so (cf.52) into $Q_{12}(\omega)$which will therefore vary as some power of ω higher than the first at low frequencies. But each term in the expression (50) which gives the finite damping response varies linearly with ω at low frequencies and so the sum will introduce a spurious linear dependence on ω at low frequencies which will of course dominate the correct higher power of ω behaviour. Since in some cases the response function has to be multiplied by a negative power of ω to get the quantity of interest, spurious infinites at zero frequency may well appear.

15. Further discussion of device of introducing damping.

In the foregoing we introduced the damping κ in order to avoid divergence difficulties. In the main part of the discussion we let κ tend to zero after converting a sum over the modes to an integral. We also introduced the possibility of leaving the damping finite and continuing to work with sums rather than integrals as a matter of computational convenience.

We shall now try to make the nature of these procedures clearer by considering in physical terms what happens when we work with a finite damping in the lattice and let this damping become very small. We first notice that the response function obtained by using a finite sample of modes and introducing finite damping is the exact response function for a simple system. If the modes chosen are those belonging to a uniform distribution of n^3 points in the first Brillouin zone, these are just the modes of an infinite crystal which have periodicity $n\underline{a}_1$, $n\underline{a}_2$, and $n\underline{a}_3$, where $\underline{a}_1$, $\underline{a}_2$, $\underline{a}_3$ are the basic vectors of the lattice. We shall get only these modes of the infinite crystal responding if, instead of applying the force $Fe^{i\omega t}$ to just one atom of the lattice we apply it to each of an infinite sub-array of the atoms chosen to form a lattice with basic vectors $n\underline{a}_1$, $n\underline{a}_2$, $n\underline{a}_3$. Moreover, since the kinetic energy of a mode is normalised within a region of sides $n\underline{a}_1$, $n\underline{a}_2$, $n\underline{a}_3$ (the region enclosed by the boundary on which the periodic boundary conditions are applied) the response of such modes to the array of forces described (which give precisely one force to each normalisation region) will be that assumed in our previous calculations. The resulting displacement of either an atom to which a force is applied or an adjacent atom (as required in evaluating P_{12}, Q_{12}) will therefore contain not only the response of the infinite lattice to the single force just mentioned, but the response to all the other forces on the periodic array of forced atoms. The result will approximate to the response to a single force in an infinite lattice in so far as the damping is sufficient to have caused the waves coming from the other forces to have died away to negligible proportions before reaching the atoms whose response we are considering. The damping introduced should therefore be great enough to ensure this, but should also be so small that the outgoing wave response of an infinite lattice(with this damping)to a single force should, for atoms near the forced atom, differ negligibly from that in the undamped lattice. Clearly if n^3 is large enough it will be possible to satisfy both requirements reasonably well.

Two further points follow from this line of thought. In the first place we can get useful insight into the origin of the rapid oscillations in $P(\omega)$ and $Q(\omega)$ which occur in the finite damping method if κ is chosen too small. We have only to notice that in the model described the displacements of a forced atom and its neighbours are each made up of two parts. The first is due to the (infinite lattice) outgoing wave centred on the particular forced atom considered and the second is the resultant of outgoing waves originating from all the other forced atoms. The wavelengths of the outgoing waves vary with ω and as the number of wavelengths between forced atoms varies the two contributions get in and out of phase. It is only when the damping is big enough for the second part of the displacements to become negligible that the oscillations disappear. As already pointed out it is the first part alone which we wish to determine, but we need it for as near zero damping as possible and so the need for compromise in the choice of κ is again apparent.

The second point concerns the limiting processes involved in determining the exact $P(\omega)$ and $Q(\omega)$ by the procedure described. We have emphasised that we must let κ tend to zero but we must also let the number of modes, and so the size of the periodic volume of crystal, become infinite. The model we have been discussing shows clearly that we must first let the volume become infinite and then let κ go to zero. This is because, for any finite value of the periodic volume, a sufficiently small κ will make the attenuation distance larger compared with the linear dimensions of the periodic volume and so lead to spurious oscillations. Thus there can be no question of letting κ go to zero before the periodic volume has gone to infinity. In our discussion, letting the volume go to infinity took the form of replacing the sum over modes by an integral.

16. Fluctuations in Position of a Foreign Atom

In treating the infra-red absorption by a charged foreign atom we saw that the response function relating the displacement of the foreign atom to a sinusoidal force applied to it could be determined very simply if suitable response functions for the perfect lattice were known. It was immediately clear that the ability to get from one response function to the other enabled us to solve the infra-red absorption problem.

We shall show in this and subsequent sections that this ability also leads in a slightly less obvious but still very simple way to solutions of a variety of other problems. In this section we shall discuss the evaluation of the mean square value of a component of displacement of the foreign atom. This is the quantity which determines the probability of getting a zero-phonon emission (Mossbauer Effect) if the foreign atom is a γ-emitter.

We are leaving the treatment of localised modes till later, so we shall assume for the moment that there are no such modes at the imperfection considered.

To see the method we consider the modes of vibration of the imperfect lattice. We suppose that the rth normal coordinate has coordinate q_r (the corresponding kinetic energy being arranged as usual to be $\frac{1}{2}\dot{q}_r^{\,2}$) and that the contribution from the rth mode to the component of displacement of the foreign atom is $\alpha_r q_r$. If we denote the component of displacement by D we have

$$D = \sum_r \alpha_r q_r \tag{57}$$

and so

$$\overline{D^2} = \sum_r \alpha_r^2 \overline{q_r^2} \tag{58}$$

since the mean values of the cross terms vanish. Now we get an expression for $\overline{q^2_r}$ by equating the mean potential energy in the mode, which is $\frac{1}{2}\omega_r^2 \overline{q^2_r}$, to half the mean energy in the mode. This gives

$$\tfrac{1}{2}\omega_r^2 \overline{q_r^2} = \tfrac{1}{2}\left\{ \frac{\hbar\omega_r}{2} + \frac{\hbar\omega_r}{\exp(h\omega_r/kT)-1} \right\} \tag{59}$$

If we now define $A(\omega)$ by

$$A(\omega)d\omega = \sum_{\omega<\omega_r<\omega+d\omega} \alpha_r^2 \tag{60}$$

we can write

$$\overline{D^2} = \int_0^{\infty} A(\omega)\frac{1}{\omega^2}\left\{\frac{\hbar\omega}{2} + \frac{\hbar\omega}{\exp(\hbar\omega/kT)-1}\right\}d\omega \quad (61)$$

But $Q_{imp}(\omega)$, the imaginary part of the response function relating D to the force applied to the impurity atom, can also be expressed in terms of A (ω). In fact, it is easy to see that the required response function $\frac{1}{m}(P_{imp}-iQ_{imp})$ is given by letting the number of modes go to infinity and then letting κ go to zero in

$$\frac{1}{m}(P_{imp}(\omega)-iQ_{imp}(\omega)) = \lim_{\substack{N\to\infty \\ \kappa\to 0}} \sum_{r=1}^{N} \frac{\alpha_r^2}{(\omega_r^2-\omega^2)+i\kappa\omega} \quad (62)$$

This gives (on taking the imaginary part),

$$Q_{imp}(\omega) = \lim_{\kappa\to 0} m \int_0^{\infty} \frac{A(\omega')\kappa\omega}{(\omega'^2-\omega^2)^2+\kappa^2\omega} d\omega'$$

$$= \frac{\pi m}{2\omega} A(\omega) \quad (63)$$

Thus knowing $Q_{imp}(\omega)$ is equivalent to knowing A (ω) which in turn, as we have just seen, leads us to $\overline{D^2}$. Since $Q_{imp}(\omega)$ can be calculated from perfect lattice response functions it follows that a knowledge of these leads easily to $\overline{D^2}$.

We shall return to a similar indirect application of the response function technique when we consider the vibrational structure in the optical absorption associated with an electronic transition.

17. Scattering of phonons by Imperfections

In considering phenomena such as the conduction of heat by lattice vibrations it is convenient to describe the motion of an imperfect crystal in terms of plane waves. Because of the imperfections these are not normal motions and one has to think of these plane waves as being scattered. To give some indication of how the response function technique can be applied to such scattering problems let us suppose that we have only isotopic defects and that we wish to determine the rate at which energy is scattered from a plane running wave. As before we shall assume for simplicity that the crystal is monatomic and cubic.

To begin with let us suppose that we have one imperfection so that a single atom of the perfect crystal has an extra mass Δm attached to it. The motion of interest will consist of a perfect running wave motion together with an outgoing wave motion produced by the force which the attached mass Δm exerts on the lattice. We must determine the rate at which energy flows away in this outgoing wave. This rate can be found without investigating the asymptotic form of the outgoing wave: it is clearly given by the rate at which the force exerted by Δm would do work if the only displacement in the lattice were the outgoing wave which this force produces. (It is not given by the rate at which the extra mass Δm does work in the total displacement during scattering, including the part due to the plane standing wave. The work done in that case must clearly be zero since the attached mass cannot be feeding net energy into the lattice).

We must determine the part of the displacements, d_f, of the impurity atom which is the displacement of that atom in the forced outgoing wave motion. The displacement which the atom would have if the only motion were the perfect standing wave will be denoted by d_o. The total displacement of the impurity atom will be $d_o + d_f$ and the force F exerted by Δm on the lattice will be given by

$$F = \omega^2 \Delta m(d_o+d_f) \tag{64}$$

But d_f is related to this force by the perfect lattice response function and we can write

$$d_f = \frac{1}{m}(P-iQ)\ \omega^2 \Delta m(d_o+d_f) \tag{65}$$

This gives at once

$$d_f = \frac{(\Delta m/m)\omega^2(P-iQ)}{1-(\Delta m/m)\omega^2(P-iQ)}\, d_o \tag{66}$$

and

$$F = \frac{(\Delta m/m)m\,\omega^2}{1-(\Delta m/m)\omega^2(P-iQ)}\, d_o \tag{67}$$

The rate E_s at which work is done by the force when the

displacement is the forced displacement d_f is $\frac{1}{2}\omega\mathcal{J}(Fd^*_f)$. Thus

$$E_s = \frac{1}{2}\,\frac{(\Delta m/m)^2\ m\ \omega^5 Q}{(1-(\Delta m/m)\omega^2 P)^2+(\Delta m/m)^2\omega^4 Q^2}\,|d_o|^2 \tag{68}$$

As already explained this rate of doing work gives the rate at which energy is scattered and so essentially solves our problem. It is easy to go a little farther and derive an expression for the relaxation time of the mode. If there are N atoms of the crystal per unit volume, and n of these are foreign atoms the energy per unit volume in the running wave will be

$$E = \frac{1}{2}\ N\ m\ \omega^2\ |d_o|^2 \tag{69}$$

and the rate at which energy is scattered, which is $-\,dE/dt$, will be nE_s. This gives

$$\frac{dE}{dt} = -\,\frac{n}{N}\ \frac{(\Delta m/m)^2\ \ \omega^3 Q}{(1-(\Delta m/m)\omega^2 P)^2+(\Delta m/m)^2\omega^4 Q^2}\ E \tag{70}$$

and comparing this with

$$\frac{dE}{dt} = -\,\frac{E}{\tau} \tag{71}$$

leads at once to the expression for τ.

Similar calculations can be done for more complicated centres. It should, however, be noted that for more complicated scattering centres the resultant outgoing wave may have a dependence on direction which is strongly anisotropic. If one is interested in, for example, thermal conductivity one is concerned with any anisotropy in the direction of scatter of energy. This has been emphasised by Krumhansl and Matthew. It is not difficult to see how to modify the response function technique to cope with this problem but we shall not discuss it further here.

17. Optical Absorption Bands

We now turn to considering the effect of lattice vibrations on the optical absorption associated with an electronic transition at an imperfection. Our main objective will again be to see how to take account of the fact that the lattice vibrations will be modified by the presence of the imperfection. The method which we shall give for doing this will be another indirect application of the procedure for getting an imperfect lattice response function from the corresponding response function for the perfect lattice: it is in fact very closely related to the method given earlier for calculating the mean square fluctuation in position of a foreign atom. In what follows we shall content ourselves with treating very simple problems to bring out the main features of the method.

Suppose that the electronic transition is between two non-degenerate electronic states of a substitutional impurity in a cubic crystal. We shall suppose in this first example that the breadth of the absorption band can be regarded as arising from fluctuations in the positions of neighbouring ions which cause the difference in energy between the two electronic states (and so the frequency of the absorbed photon) to vary. In other words we are supposing that the Franck-Condon principle applies. Under these circumstances one expects to get smooth absorption bands of essentially Gaussian shape and so the shape of the absorption band is specified by giving its mean square breadth. Accordingly we shall discuss how to calculate this mean square breadth as a function of temperature.

We shall make two further assumptions. In the first place we shall suppose that, if we expand the energy difference between the two states as a function of the displacements of neighbouring atoms, quadratic and higher powers of the displacements may be neglected. The second assumption is merely to make our problem as simple as possible. It consists in assuming that the energy difference between the two states depends only on the mean radial outward displacement ΔR of the nearest neighbour atoms from their ground state equilibrium position. The first assumption (often described as the assumption of "linear coupling") is very important in some problems but the methods which we shall describe for taking account of the modifications of the normal modes will be applicable when for example, the energy difference is given by

$$E_u - E_g = A - GAR^2 \tag{72}$$

so that

$$\overline{\Delta E^2} = G^2 (\overline{\Delta R^4}) \tag{73}$$

Since the distribution of ΔR is Gaussian

$$\overline{\Delta R^4} = 3(\overline{\Delta R^2})^2 \tag{74}$$

so that any procedure which gives $\overline{\Delta R^2}$ also gives $\overline{\Delta R^4}$. We shall not consider this further here.

The mean radial displacement ΔR, which has a definite value for any given set of displacements $\mathbf{a}_s$ (s=1...n) of the n nearest neighbour ions, may be defined more explicitly as follows. If $\mathbf{e}_s$ is a unit vector directed radially outwards from the equilibrium position of the sth nearest nearest neighbour ion

$$\Delta R = \frac{1}{n} \sum_{s=1}^{n} \mathbf{a}_s \cdot \mathbf{e}_s \qquad (75)$$

The assumption that the energy difference depends only on ΔR would be justified if both electronic wave functions were spherically symmetrical and did not extend appreciably beyond the nearest neighbour ions. The dependence on ΔR must of course be taken to be linear because of the previous assumption and if we denote the energies of the upper and the ground electronic states by E_u and E_g respectively we may write

$$E_u - E_g = A - C\Delta R \qquad (76)$$

with A and C constants.

It is now convenient to introduce the normal coordinates q_r associated with the normal vibrations of the lattice when the imperfection is in its ground electronic state. (The normal vibrations, and so the normal coordinates, will be slightly different when the imperfection is in its excited electronic state: the excited state normal coordinates will not concern us in the present discussion). As usual we shall take it that the scale for q_r has been chosen in such a way that the corresponding kinetic energy is $\frac{1}{2} \dot{q}_r^2$. Let us suppose that when all the normal coordinates except q_r are zero the mean outward displacement of the nearest neighbour ions is $\beta_r q_r$. We may therefore write

$$\Delta R = \sum_r \beta_r q_r \qquad (77)$$

and so

$$E_u - E_g = A - C \sum_r \beta_r q_r \qquad (78)$$

We now calculate the mean value of $E_u - E_g$ which will give us the mean frequency of the absorbed photon (in energy units). The mean will, of course, be calculated by averaging over the positions of neighbouring atoms or, what comes to the same thing, averaging over the q_r. Thus

$$\overline{E_u - E_g} = A - C \sum_r \beta_r \overline{q_r} \tag{79}$$
$$= A.$$

The difference between the frequency of the photon absorbed in a particular absorption process and the mean frequency of the photons absorbed will be given (in energy units) by

$$\Delta E = E_u - E_g - \overline{E_u - E_g} \tag{80}$$
$$= - C \sum_r \beta_r q_r$$

The mean square breadth of the absorption band (in units of energy squared) will therefore be given by

$$\overline{\Delta E^2} = C^2 \sum_{r,s} \beta_r \beta_s \overline{q_r q_s} \tag{81}$$

Now, if $r \neq s$, $\overline{q_r q_s}$ is zero and it follows from (59) that

$$\overline{q_r^2} = \frac{\hbar}{\omega_r} \left\{ \tfrac{1}{2} + \frac{1}{\exp(\hbar\omega_r/kT)-1} \right\}$$
$$= \frac{\hbar}{2\omega_r} \coth (\hbar\omega_r/2kT)$$
$$= \phi(\omega_r, T) \tag{82}$$

where the notation $\phi(\omega,T)$ is introduced for brevity. Thus

$$\overline{\Delta E^2} = C^2 \sum_r \beta_r^2 \phi(\omega_r, T). \tag{83}$$

It will be important to throw this expression into a different form by making use of the fact that the frequencies of the modes constitute a quasi-continuum (we exclude the possibility of there being localised modes). To do this we define a function $\beta(\omega)$ such that

$$\beta(\omega)d\omega = \sum_{\omega<\omega_r<\omega+d\omega} \beta_r^2 \tag{84}$$

In terms of $B(\omega)$ we may rewrite the expression for $\overline{\Delta E^2}$,

$$\overline{\Delta E^2} = C^2 \int_0^\infty B(\omega)\phi(\omega,T)d\omega \ . \tag{85}$$

It will turn out that not only the present problem but more elaborate problems of absorption band structure (including cases to which the Franck-Condon principle does not apply) reduce to the determination of $B(\omega)$ and certain related functions. Our next task is therefore to show how $B(\omega)$ may be determined, account being taken of the fact that the presence of the imperfection will modify the modes of the vibration.

Let us first note that if we neglect the modifications of the modes and have available the eigen frequencies and eigen vectors for a large sample of perfect lattice modes then the determination of $B(\omega)$ denoted in this case by $B_{perf}(\omega)$, is straight forward. A knowledge of the eigen vector for a mode allows one to compute the displacement of each atom in the crystal. It is therefore easy for a computer to determine the displacements of the nearest neighbour atoms in the rth mode and then to evaluate β_r. It can then evaluate the sum of the β_r^2 for all modes in each of a suitably chosen set of frequency intervals. This will then lead (after suitable smoothing) to an estimate of the function $B_{perf}(\omega)$. The choice of the suitable set of frequency intervals will depend on how dense a sample of the normal modes we have to work with: the denser the sample the smaller the intervals and so the greater the detail obtained.

As already suggested we estimate $B_{imp}(\omega)$, the $B(\omega)$ obtained when we take account of the modifications of the modes, by relating $B(\omega)$ to a response function. The appropriate response function is defined as follows. We apply an equal radial force $(F/n)e^{i\omega t}$ to each of the n nearest neighbours and denote the resulting radial displacements by $de^{i\omega t}$ (the resulting displacements of the neighbours will be radial and equal in magnitude because of the cubic symmetry provided there is only one symmetry adapted displacement of the neighbours which belongs to the identity representation of the cubic group and we may assume this to be the case). The required response function is then defined by

$$d = \{P(\omega) - iQ(\omega)\}F \tag{86}$$

Proceeding as in the derivation of (63) one finds easily

$$Q(\omega) = \frac{\pi}{2\omega} B(\omega). \tag{87}$$

This relation applies, of course, both to the perfect and imperfect lattices, so we have

$$Q_{perf}(\omega) = \frac{\pi}{2\omega} B_{perf}(\omega) \qquad (88)$$

and

$$Q_{imp}(\omega) = \frac{\pi}{2\omega} B_{imp}(\omega) \qquad (89)$$

As we have already discussed there is no difficulty in determining $B_{perf}(\omega)$ if we have sufficient information about the normal modes of the perfect lattice. The first of the above two relations then leads to $Q_{perf}(\omega)$ and if we can get from $Q_{perf}(\omega)$ to $Q_{imp}(\omega)$ the second relation will lead us to $B_{imp}(\omega)$ which is what we require.

One gets from $Q_{perf}(\omega)$ to $Q_{imp}(\omega)$ by a process very similar to that used in treating the isotope problem. In the isotope case we thought of the imperfect crystal as being a perfect crystal with an extra mass (possibly negative) attached to one of the atoms and we considered explicitly the force exerted by this extra mass on the perfect lattice. In the present case we regard the imperfect lattice as a perfect lattice with a set of equal springs, with spring constant Δk, attached between the central atom of the crystal (chosen to be the impurity atom) and its neighbours. Here we are, of course, making an assumption, namely that the only effect of the impurity on the lattice dynamics is to change the restoring forces on nearest neighbour ions. An extra mass corresponding to the difference between the mass of the impurity and that of the atom which it replaces can also be regarded as attached to the central atom, but this will not concern us since we can suppose the modes chosen to be either even or odd about the imperfection and only the even modes will contribute to ΔR. Since the central atom will not move in these even modes its changed mass has no effect.

As in the isotope case we can determine the response of the imperfect lattice in terms of the response of the perfect lattice by introducing explicitly the forces exerted on the perfect lattice by the attached masses or springs. In the present case the radial force exerted by the springs will be $-\Delta k d e^{i\omega t}$ where $de^{i\omega t}$ is the radial displacement. This has to be added to the applied force $Fe^{i\omega t}$. The radial displacement $de^{i\omega t}$ and the applied force $Fe^{i\omega t}$ for the imperfect lattice are therefore related by

$$d = \{P_{perf}(\omega) - iQ_{perf}(\omega)\}\{F - \Delta k d\} \qquad (90)$$

which gives

$$d = \frac{P_{perf}(\omega) - iQ_{perf}(\omega)}{\{1+\Delta kP_{perf}(\omega)\}-i\Delta kQ_{perf}(\omega)} F \qquad (91)$$

Comparing this with

$$D = \{P_{imp}(\omega) - iQ_{imp}(\omega)\} F \qquad (92)$$

gives

$$Q_{imp}(\omega) = \frac{Q_{perf}(\omega)}{\{1+\Delta kP_{perf}(\omega)\}^2+\Delta k^2 Q_{perf}(\omega)^2} \qquad (93)$$

We note that the right-hand side contains $P_{perf}(\omega)$ as well as $Q_{perf}(\omega)$, but, as for the response function used in the isotope case, $P_{perf}(\omega)$ can be obtained from $Q_{perf}(\omega)$ by using the Kramers-Kronig relations. We see, therefore, that there is no difficulty in getting from $Q_{perf}(\omega)$ to $Q_{imp}(\omega)$ and so from $B_{perf}(\omega)$ to $B_{imp}(\omega)$. This means that, provided we know the coupling constant C and the eigenvectors and eigenfrequencies for a sufficiently large sample of the modes of the perfect lattice, we can calculate the mean square breadth of the absorption band as a function of temperature for the simple model being considered, taking account of the modification of the modes.

Still considering the transition between non-degenerate electronic states and making the same assumptions about the coupling to the lattice we now turn to the case in which the Franck-Condon approximation breaks down. This happens when the coupling between the electronic states and the lattice vibrations become so weak that a calculation based on the Franck-Condon principle would predict an absorption band with a spread in energy less than the energy of a typical phonon in the lattice. The actual observed absorption structure in this case will be a very sharp line with an attendant structure which lies mainly on the high energy side of the sharp line.

It is not our purpose to give a full discussion of this situation. We merely wish to indicate that, for the problem we are considering, a knowledge of $C(\omega)$ enables us to predict the fraction of the absorption which is in the sharp line and the whole shape of the attendant structure. Thus the possibility of calculating these functions, taking account of the modification of the lattice normal modes by the imperfection, is of great importance in such problems, and indeed, practically leads to their solution.

In order to cover the case in which the Franck-Condon principle breaks down we have to think of the lattice vibrations as quantised and take the following view of the absorption process (we confine ourselves to the case of 0^oK). When the photon is absorbed it is possible that, as well as the electron going to the excited state considered, some of the normal modes of the lattice may go from their ground state to their first excited state. The energy of the photon is equal to the sum of the energy of the electronic transition and the increase in vibrational energy. The spread in the energies of the absorbed photons arises from the spread in the number and frequencies of the modes which are excited. The fraction of the absorption which comes under the sharp line will equal the probability that the electronic transition takes place without any mode being excited. The attendant structure can be regarded as made up by superposing a part coming from transitions in which one mode is excited, a part from transitions in which two modes are excited, and so on.

It is reasonable to expect that we will be able to calculate all these things if we know for each mode the probability that it remains unexcited and so, by subtraction, the probability that it goes to its first excited state. (The observed breadths of absorption bands suggest that only a tiny fraction of the modes of a macroscopic portion of crystal are excited so that we can ignore the possibility of any one mode going to its second excited state).

We can calculate these probabilities as follows. The transition to the excited electronic state changes the potential energy function for the atoms from $\frac{1}{2}\sum_r \omega_r^2 q_r^2$ to

$$A + \frac{1}{2}\sum_r \omega_r^2 q_r^2 - C\sum_r \beta_r q_r$$

Apart from an additive constant which is irrelevant for the dynamics this can be expressed as a sum of squares of the new normal coordinates $q'_r = q_r - C\beta_r/\omega_r^2$, the coefficient of the rth square term still being $\frac{1}{2}\omega_r^2$. This means that in going to the excited electronic state the rth normal coordinate merely has its origin displaced by an amount $C\beta_r/\omega_r^2$. The corresponding harmonic oscillator functions will have their origins displaced with respect to one another by this amount. It is this displacement of the origins which means that harmonic oscillator states associated with the two different electronic states are not orthogonal even if they differ in vibrational quantum number. The probability of the mode being unexcited after the transition is equal to the square of the overlap integral between the two ground state functions displaced relative to one another by the distance $C\beta_r/\omega_r^2$ which turns out to be

$$\exp\left(- C^2\beta_r^2/2\hbar\omega_r^3\right)$$

The probability that the mode is excited to its first excited state will be given to a sufficient approximation by expanding the above expression to first order in the exponent and subtracting from unity. This gives

$$C^2\beta_r^2/2\hbar\,\omega_r^3$$

The probability that the transition will take place without excitation of any mode is the product of the probabilities that each individual mode will be unexcited. It is convenient to write this probability as exp (-S), where S is a quantity for which another significance will emerge later, and we have

$$\exp\{-S\} = \exp\{-\sum_r C^2\beta_r^2/2\hbar\omega_r^3\} \qquad (94)$$

Now S can be expressed in terms of $B(\omega)$ as follows

$$S = \frac{C^2}{2\hbar}\sum_r \frac{\beta_r^2}{\omega_r^3} = \frac{C^2}{2\hbar}\int_0^\infty \frac{B(\omega)}{\omega^3}\,d\omega \qquad (95)$$

This shows that a knowledge of $B(\omega)$ for the imperfect lattice gives at once the fraction of the absorption which is in the sharp no-phonon peak.

The contribution to the absorption from processes in which just one mode is excited can also be expressed in terms of $B(\omega)$. To see this we note that the probability that only the rth mode is excited is the probability that all the other modes remain unexcited multiplied by the probability that the rth mode is excited. To a sufficient approximation this gives

$$e^{-S}\,C^2\beta_r^2/2\hbar\omega_r^3$$

It follows that the probability that only one mode is excited and that this has frequency in the range ω to $\omega + d\omega$ will be

$$e^{-S}(C^2/2\hbar)(B(\omega)/\omega^3)\,d\omega.$$

This gives at once the one-phonon contribution to the absorption curve. The two-phonon contribution also can be expressed in terms of $B(\omega)$ or, more conveniently in terms of the related function $n(\omega)$ where

$$n(\omega) = (C^2/2\hbar)(B(\omega)/\omega^3). \qquad (96)$$

(It may be noted that $n(\omega)d\omega$ gives the mean number of modes with frequencies in ω to $\omega+d\omega$ which are excited: S is just the integral of this overall frequencies so it is just the mean number of modes excited: this is the other significance for S mentioned earlier). The probability that just two phonons are created and that these have total energy in the range $h\omega$ to $h(\omega+d\omega)$ will be given by

$$\frac{1}{2}\, e^{-S} \left\{\int_0^\infty n(\omega')n(\omega-\omega')d\omega'\right\} d\omega.$$

This gives the two-phonon contribution to the absorption curve. The contributions from higher phonon processes may be expressed in terms of further convolutions of $n(\omega)$. It is therefore possible to construct the whole absorption curve from a knowledge of $B(\omega)$ for the imperfect lattice (which, of course, comes to the same thing as a knowledge of $n(\omega)$ for the imperfect lattice).

We shall not attempt to present details of calculations for particular cases. The model we have discussed is too simple to be expected to give agreement with experimental results when the detailed form of the one-phonon curve $n(\omega)$ is important. In cases, however, where an electronic transition is accompanied by the creation of some three or four phonons the convolution process will give the main features of the observed vibrational structure in the absorption, provided the $n(\omega)$ curve used has a dominant peak in roughly the correct position and with roughly the correct breadth. The results of a calculation due to J. Slater (M.Sc. Thesis, Aberdeen, 1962) for a hypothetical transition, with $S = 3.5$, at a vacancy in diamond, based on the $n(\omega)$ curve estimated from Fig. 10 rather than from response functions calculations, is compared with one of the observed diamond emission bands in Fig. 15. The reasonable agreement suggested that the general explanation advanced for the absorption (or mirror image emission) structure was sound.

As an example of the results of a more elaborate calculation we show in Fig. 16 the results of a calculation (M. J. L. Sangster-to be published) of the vibrational structure associated with an electronic transition ($^2E_g \rightarrow {}^4A_{2g}$) at V^{2+} in MgO. Here $S \ll 1$ and the one-phonon curve itself dominates the vibrational structure. In the calculation the assumptions of "spherical symmetry of interaction, with the surroundings" and of non-degenerate electronic states had to be dropped. The charges of force constants at the imperfection and the various coupling constants were treated as adjustable parameters, but a calculation was done to check that the coupling parameters used were reasonable. The agreement with experiment is again very reasonable.

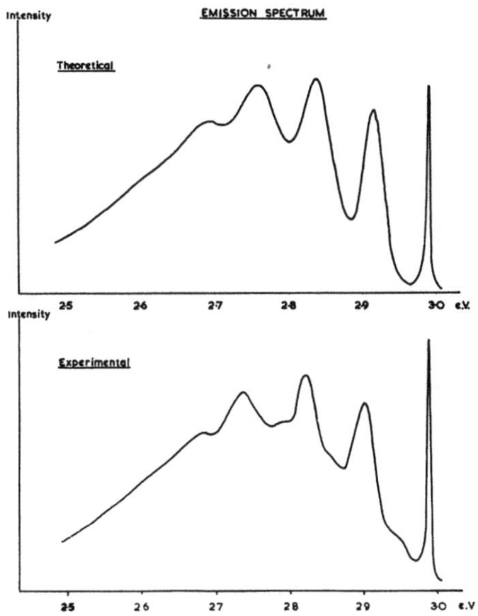

Fig.15. Attendant vibrational structure for emission associated with hypothetical electronic transition at vacancy in diamond calculated from one-phonon curve of Fig.10 and S=3.5. This is compared with one of the defect induced emission structures observed in diamond. (J.Slater. M.Sc.Thesis, Aberdeen 1962).

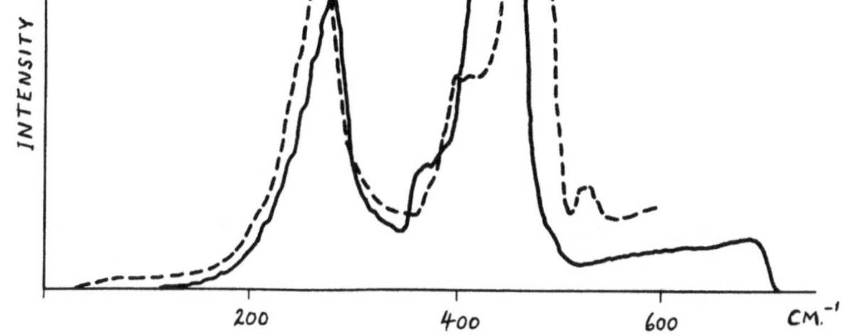

Fig.16. Attendant structure calculated (with some fitted parameters) for the electronic transition $^2E_g \rightarrow {}^4A_{2g}$ at V^{2+} in MgO (solid line) compared with experimental result (dotted line). (M.J.L.Sangster - to be published).

18. Localised Modes

In treating problems by the response function method we have ignored the possibility of the imperfection introducing localised modes. If such modes exist there will be divergences in the response function for the imperfect lattice at the frequencies of these modes and the methods we have given will break down at these points. From the point of view of practical computation a simple device allows us to avoid this difficulty. We simply let the damping on the lattice remain finite instead of letting it go to zero. This will remove the divergences from the imperfect response function and lead to peaks in, for example, the associated infra-red absorption which correspond to infra-red absorption by the local mode. The breadth of the peak will, of course, have no physical significance as it will depend on the precise value of the damping which was arbitrarily introduced. However, the position of the peak and the integrated absorption under it will be essentially independent of κ and these two quantities are all we can expect to get from a theory which does not include a treatment of the anharmonic effects which lead to the true broadening of localised modes.

It is, of course, possible to treat the localised modes, and their contributions to the effects we have been discussing, more directly. It turns out that in simple cases one requires to know two things about the localised mode, namely its frequency, and the scale which must be chosen for the coordinate q which we chose to measure the displacement of the mode if the kinetic energy in the mode is to be given by $\frac{1}{2}\dot{q}^2$. For more elaborate problems one will require to know also the relative displacements in the mode of some of the atoms near the imperfection: the methods we are describing can easily be applied to obtaining such information but we shall not set out the procedures here.

To illustrate let us consider again an electronic transition between two non-degenerate electronic states which are coupled only to the spherically symmetrical displacements ΔR of the nearest neighbours. We now suppose that the centre is hard (i.e. Δk is positive for the extra springs which represent the changes in force constants on nearest neighbours) and that as a result there is a localised mode at a frequency above the maximum perfect lattice mode frequency. We shall suppose that we wish to calculate the probability that when the electronic state changes the localised mode stays in its ground state: this probability will enter as a factor into the probability of getting a no - phonon transition.

We first consider how we can get the frequency of the localised mode from a knowledge of Δk and the appropriate response function for the perfect lattice (i.e. the response function relating a "spherically symmetrical" applied force system and the resulting spherically symmetrical displacement of nearest neighbours). We are interested in frequencies above the maximum lattice mode frequency and at such frequencies $Q(\omega)$ will be zero and $P(\omega)$ will be negative. (That $Q(\omega)$ must be zero follows from the fact that no energy can be absorbed from the force system by the undamped lattice at such frequencies as the lattice cannot sustain an outgoing wave at frequencies outside the range of perfect lattice mode frequencies: that $P(\omega)$ must be negative follows from the fact that all the normal modes of the lattice are now acted upon by a force with frequency greater than their natural frequency and so they will respond in antiphase). The condition for a localised mode is that the force $-\Delta k D e^{i\omega t}$ exerted by the springs should, when acting on the perfect lattice, just be such as to result in the radial displacement $De^{i\omega t}$. This requires

$$D = P_{perf}(\omega)(-\Delta k D) \tag{97}$$

which may be written

$$P_{perf}(\omega) = -\,{}^{1}/\Delta k. \tag{98}$$

Thus the localised mode frequency is determined by the points, above the maximum lattice mode frequency, at which the curve $P_{perf}(\omega)$ is intersected by a straight line at a distance $1/\Delta k$ below the ω-axis. One finds for realistic $P(\omega)$ curves that if Δk is below a certain minimum value there will be no localised mode.

To deal now with the normalisation problem let us suppose that the localised mode is excited and is ocillating in such a way that the radial displacement of the nearest neighbour ions is $D \sin \omega_o t$ with ω_o the localised mode frequency. We start by calculating the kinetic energy in the mode at $t = 0$. This will be the sum of the kinetic energies in the normal modes of the lattice. The rth normal mode is acted upon by the generalised force $-\Delta k D \beta_r \sin \omega_o t$ and so it will satisfy the equation

$$-\omega_o^2 q_r + \omega_r^2 q_r = -\Delta k D \beta_r \sin \omega_o t \tag{99}$$

so that (solving for q_r and then differentiating with respect to time) we have

$$\dot{q}_r = \frac{\Delta k D \beta_r \omega_o \cos \omega_o t}{\omega_o^2 - \omega_r^2} \tag{100}$$

and so at t = 0 the kinetic energy in the mode r, $\frac{1}{2}\dot{q}_r^{\,2}$, is given by

$$\frac{1}{2}\,\frac{(\Delta k)^2 D^2 \beta_r^{\,2} \omega_o^{\,2}}{(\omega_o^{\,2}-\omega_r^{\,2})^2} \tag{101}$$

The total kinetic energy at t = 0 may therefore be written in terms of B(ω) as

$$\frac{1}{2}\int_0^\infty \frac{(\Delta k)^2 D^2 \omega_o^{\,2} B(\omega)}{(\omega_o^{\,2}-\omega)^2}\, d\omega$$

This is the kinetic energy when the rate of change of the displacement of the nearest neighbours is $D\omega_o$. If we denote the radial displacement of the nearest neighbours by q the kinetic energy is therefore of the form

$$\frac{1}{2}\left\{\int_0^\infty \frac{(\Delta k)^2 B(\omega)}{(\omega_o^{\,2}-\omega^2)^2}\, d\omega\right\} \dot{q}^2$$

and if we introduce a coordinate q' defined by

$$q' = \left\{\int_0^\infty \frac{(\Delta k)^2 B(\omega)}{(\omega_o^{\,2}-\omega^2)^2}\, d\omega\right\}^{\frac{1}{2}} q \tag{102}$$

the kinetic energy in the mode will be $\frac{1}{2}\dot{q}'^2$ as required. Moreover the constant of proportionality between the properly normalised coordinate q' and the radial displacement of the nearest neighbour shell will be given by

$$\beta' = \left\{\int_0^\infty \frac{(\Delta k)^2 B(\omega)}{(\omega_o^{\,2}-\omega^2)^2}\, d\omega\right\}^{-\frac{1}{2}} \tag{103}$$

The probability that the mode will not be excited will therefore be (cf. the general expression in terms of β'_r given earlier: β' is, of course, just the β'_r for the localised mode)

$$\exp\,(-C^2\beta'^2/2\hbar\omega_o^{\,3})$$

with β' as given above.

This completes the procedure for handling the particular localised mode problem we have chosen to consider. The problem of calculating the integrated infra-red absorption by the localised mode associated with a light charged isotope can be handled similarly. The solution (arrived at in a somewhat different way) is given by Dawber and Elliott (Proc.Phys.Soc. 81 453 (1963)).

19. Conclusion

We have tried to present a unified, straightforward approach to the treatment of problems involving modified lattice vibrations. We have not attempted to review the many other approaches to these problems which started with the work of I.M.Lifshitz (see Nuovo. Cim. 3 Suppl., 716(1956)) and of Montroll and Potts (see the book by A.A.Maradudin, E.W.Montroll and G.H.Weiss "Theory of Lattice Dynamics in the Harmonic Approximation" Solid State Physics Supplement 3. Ed. Seitz and Turnbull: Academic Press (1963)). Recent reviews of the subject have been given by A.A.Maradudin "Some Effects of Point Defects on the Vibrations of Crystal Lattices" in Reports on Progress in Physics Vol. XXVIII 1965 and by R.J.Elliott "Vibrations of Defects in Lattices" in "Phonons" Ed. R.W.H.Stevenson, Oliver and Boyd 1966.

The approach adopted in these notes and the applications of it which have been quoted have arisen in the course of work done in conjunction with J.A.D.Matthew, A.M.Murray, M.J.L.Sangster and J.Slater.

LATTICE VIBRATIONS OF SOLID SOLUTIONS

J. van der Elsken, R. Metselaar, R. K. Eijnthoven

Laboratory for Physical Chemistry, University of Amsterdam, The Netherlands

It is convenient to make a distinction between low-concentration solid solutions and high-concentration solid solutions. The theoretical approach in the case of low concentrations is rather straight forward and the accordance between theory and the experimental observations is satisfying although many details remain still to be solved. For high concentration solid solutions no firm theoretical approach has yet been given and the interpretation of the experimental behavior of such systems has only been given in an empirical way.

LOW CONCENTRATION SOLID SOLUTIONS

It is known that the normal mode frequencies of a crystal lattice are only slightly affected by the interaction of a single mass effect. Only the frequencies that are associated with the band edges will be displaced out of the bands whereas all other frequencies remain unchanged or shifted by no more than the distance to the next unperturbed frequency[1]. The few displaced frequencies arise from modes that are known as local or gap modes. The experimental behavior of these modes and the theoretical approaches have been treated extensively elsewhere[2,3]. Here we will be concerned only with the great amount of unperturbed frequencies in crystals that have such a low concentration of impurities that the separation of the local and gap modes is justifiable.

Evidence for the existence of a practically unaffected phonon spectrum is given by the spectroscopic observation of transitions between phonon levels or of transitions involving excited vibrational or electronic states plus a phonon transition. To see what the chances are for such observations we have first to investigate the rules of group theory.

Group Theoretical Considerations

In the perfect lattice, the translational invariancy imposes a very strict selection rule, which forbids nearly all phonon transitions. We are, however, interested in the case of dilute solid solutions, which means that one of the ions of the host lattice is replaced by a guest ion. The lattice is now no longer invariant with respect to translations; so the selection rule will be altered too. In general one can say that when the symmetry at the impurity site is lowered further, more phonon transitions will be allowed.

We will investigate the symmetry groups and corresponding character tables of the host crystal and the impurity. If these are known, one should be able to derive the selection rules for the transitions between phonon levels. Apart from these transitions, which lead to absorption of radiation in the far-infrared region, there will be the possibility of absorption in the near infrared, visible and ultraviolet due to a transition in which both a vibration (or electronic state) and a phonon are excited. To begin, some of the results of space group theory will be reviewed, as far as needed here.

The space group is defined as the collection of operations that leave a lattice invariant. In the notation of Seitz (reviewed by Koster[4]) the operators are of the type $\{\alpha|t\}$, which corresponds to a coordinate transformation:

$$\underline{r}' = \underline{R}\,\underline{r} + \underline{t} \tag{1}$$

where r' are the new coordinates expressed in terms of the old ones r. $\underline{R}$ and $\underline{t}$ are real. The real orthogonal matrix $\underline{R}$ stands for a rotation, $\underline{t}$ for a translation. In the operator $\{\overline{\alpha}|t\}$, is the rotational part and t the translational part.

In this notation multiplication of two operators is given by the rule

$$\{\alpha'|t'\}\{\alpha|t\} = \{\alpha'\alpha|\alpha't + t'\} \tag{2}$$

The inverse operator of $\{\alpha|t\}$ is according to (2)

$$\{\alpha|t\}^{-1} = \{\alpha^{-1}|-\alpha^{-1}t\} \tag{3}$$

A pure translation is denoted by $\{\varepsilon|t\}$, a pure rotation by $\{\alpha|o\}$, while $\{\varepsilon|o\}$ is the identity operator. Each space group G possesses an invariant subgroup of translations t of the form $\{\varepsilon|\vec{r}(\ell)\}$, where

$$\vec{r}(\ell) = \ell_1\vec{a}_1 + \ell_2\vec{a}_2 + \ell_3\vec{a}_3 \tag{4}$$

The quantities ℓ_1, ℓ_2, ℓ_3 are integers, and $\vec{a}_1$, $\vec{a}_2$, $\vec{a}_3$ are the three linearly independent primitive translation vectors. The parallelpiped bounded by $\vec{a}_1$, $\vec{a}_2$, $\vec{a}_3$ is called the primitive or unit cell of the crystal. The periodic collection of points generated by the vectors $\vec{a}$ is called the lattice.

In the following discussions we can restrict ourselves to a special type of space groups, called the symmorphic groups. In this type of groups all elements are of the form $\{\alpha|\vec{r}(\ell)\}$, where $\vec{r}(\ell)$ is a primitive translation vector. This is the case when the space group contains no glide planes of screw axes.

The rotational operators α form a point group G_o. Now the operators $\{\alpha|o\}$ in the space group form a subgroup too, though this subgroup is not invariant. For each operator $\{\alpha|o\}$ in this subgroup there is a corresponding element α in the point group G_o. The group $\{\alpha|o\}$ is thus isomorphic with the point group, or in other words, the symmorphic space groups contain the entire point group as a subgroup. Decomposing G in its translational subgroup t and its cosets,

$$G = t + \{\alpha_2|a_2\}t + \dots + \{\alpha_n|a_n\}t \tag{5}$$

The factor group G/t is isomorphic with the point group, as we have seen above. For the actual calculations, we need the traces of the irreducible representations of the space group. These are easily found from the irreducible representations of its subgroups.

Consider first the group t of primitive translations $\{\varepsilon|\vec{r}(\ell)\}$, where $\vec{r}(\ell)$ is of the form (4). Now in fact only an infinite crystal is invariant under all translations or, stated otherwise, the group t is an infinite group. By introducing the Born-Von Kármán periodic boundary conditions, we are confined to a subgroup t_o of the infinite translation group t, which is invariant under the operations

$$t_o = \ell_1 L\,\vec{a}_1 + \ell_2 L\,\vec{a}_2 + \ell_3 L\,\vec{a}_3 \tag{6}$$

where ℓ_1, ℓ_2, ℓ_3 are arbitrary integers, and L is a positive integer. The group t_o is an invariant subgroup of t. Using the fact that irreducible representations of a factor group are also irreducible representations of the full group, it can be stated that "the representations of the infinite translation group t which are invariant under all operations of t_o are the irreducible representations of the group which is isomorphic with the factor group t/t_o". Then

$$\{\epsilon|a_1\}^L = \{\epsilon|a_2\}^L = \{\epsilon|a_3\}^L = \{\epsilon|o\} \tag{7}$$

We see that the group t_o is written here as a direct product of three groups; this is possible because all translations commute; so the three groups themselves commute. Because these groups are Abelian, they have only one-dimensional representations. Now $\{\epsilon|o\}$ must be represented by unity, so the representations of $\{\epsilon|a_1\}$ must be roots of unity

$$\exp\ (2\pi i\, \frac{h_1}{L}) \tag{8}$$

This gives L irreducible representations, one for each value of the integer h_1, from o to L-1. Multiplying the representations of the three one-dimensional translation groups, the representation for $\{\epsilon|\vec{r}(\ell)\}$ becomes

$$\exp\ [2\pi i\vec{k}.\vec{r}(\ell)] \tag{9}$$

where $\vec{r}(\ell)$ is defined by (4) and $\vec{k}$ by

$$\vec{k} = \frac{h_1}{L}\,\vec{b}_1 + \frac{h_2}{L}\,\vec{b}_2 + \frac{h_3}{L}\,\vec{b}_3 \tag{10}$$

with

$$h_1,\ h_2,\ h_3 = 0,\ 1,\ 2,\ \ldots.\ L-1$$

provided the vectors $\vec{b}$ obey the relation

$$\vec{a}_i \cdot \vec{b}_j = \delta_{ij} \qquad\qquad i,\ j = 1,\ 2,\ 3$$

The lattice generated by multiples of the primitive vectors $\vec{b}_1$, $\vec{b}_2$, $\vec{b}_3$ is called the reciprocal lattice. The wavevector $\vec{k}$ is a vector in the reciprocal space, with components $k_i = h_i / L$. We see that $\vec{k}$ defines the irreducible representations of the translation group t_o. There is one irreducible representations of the translation group t_o. There is one irreducible representation for each $\vec{k}$ value, in total L x L x L = N. Speaking in a geometrical way, all allowed values of $\vec{k}$ are found in a unit cell in the reciprocal lattice defined by equation (10). It is, however, not necessary to choose a primitive unit cell to limit the $\vec{k}$-values. Though it gives a very simple definition, there is the disadvantage that the point symmetry of the lattice is not automatically displayed.

In order to obtain a unit cell that transforms into itself under all the operations of the point group of the lattice, one constructs the so called symmetrical unit cell by drawing vectors from the origin of the lattice to all lattice points and then constructing the planes which are the perpendicular bisectors of these vectors. The smallest volume containing the origin which is enclosed by these planes is completely equivalent to the primitive unit cell. The symmetrical unit cell constructed in the reciprocal lattice, is known as the first Brillouin zone. This cell would have been established directly if the following range of h-values had been used:

$$h_1, h_2, h_3 = 0, \pm 1, \pm 2, \ldots \pm \tfrac{1}{2} (L-1) .$$

It is now clear that this concept is very useful for our purpose, because each point in the interior of the Brillouin zone corresponds to an irreducible representation of the group of translations. Any point outside the zone gives rise to a representation identical with some point inside. We only have to keep in mind that each point on the surface is equivalent to at least one other point on the surface, and differs from it by a primitive translation.

Examine now the actual form of the Brillouin zone for an alkali halide crystal with a NaCl type lattice. The two types of ions are placed on two face-centered cubic lattices which are shifted in respect to one another along a half cube diagonal. The crystal as a whole has the same symmetry group as the face-centered lattice.

If the nearest-neighbor distance = r_0, the lattice vectors of the symmetrical unit cell in the direct lattice are

$$\vec{a}_1 = r_o (\vec{j} + \vec{k})$$

$$\vec{a}_2 = r_o (\vec{i} + \vec{k})$$

$$\vec{a}_3 = r_o (\vec{i} + \vec{j}) \qquad (11)$$

where $\vec{i}, \vec{j}, \vec{k}$ are unit vectors along the x, y, z axes. A general lattice vector is given by

$$\vec{r}(\ell) = \ell_1\vec{a}_1 + \ell_2\vec{a}_2 + \ell_3\vec{a}_3 = r_o(\ell_x,\ell_y,\ell_z)$$

$$= r_o \ \{(\ell_2+\ell_3)\vec{i} + (\ell_1+\ell_3)\vec{j} + (\ell_1+\ell_2)\vec{k}\} \qquad (12)$$

with components

$$\ell_x = \ell_2 + \ell_3 \qquad \ell_y = \ell_1 + \ell_3 \qquad \ell_z = \ell_1 + \ell_2$$

The first Brillouin zone is the symmetrical cell in the reciprocal lattice. From the definition $\vec{a}_i.\vec{b}_j = \delta_{ij}$, it follows that

$$\vec{b}_1 = \frac{1}{2r_o}(-\vec{i} + \vec{j} + \vec{k})$$

$$\vec{b}_3 = \frac{1}{2r_o}(\vec{i} - \vec{j} + \vec{k})$$

$$\vec{b}_3 = \frac{1}{2r_o}(\vec{i} + \vec{j} - \vec{k}) \qquad (13)$$

A general vector in the reciprocal space is

$$\vec{b}(h) = h_1\vec{b}_1 + h_2\vec{b}_2 + h_3\vec{b}_3 = \frac{1}{2r_o}(h_x, h_y, h_z) \qquad (14)$$

with components

$$h_x = -h_1 + h_2 + h_3$$

$$h_y = h_1 - h_2 + h_3$$

$$h_z = h_1 + h_2 - h_3$$

The wave vector $\vec{k}$ is a vector in the reciprocal lattice given by

$$\vec{k} = k_1\vec{b}_1 + k_2\vec{b}_2 + k_3\vec{b}_3 = \frac{1}{2r_o}(q_x, q_y, q_z) \qquad (15)$$

with components

$$q_x = -k_1 + k_2 + k_3$$
$$q_y = k_1 - k_2 + k_3$$
$$q_z = k_1 + k_2 - k_3$$

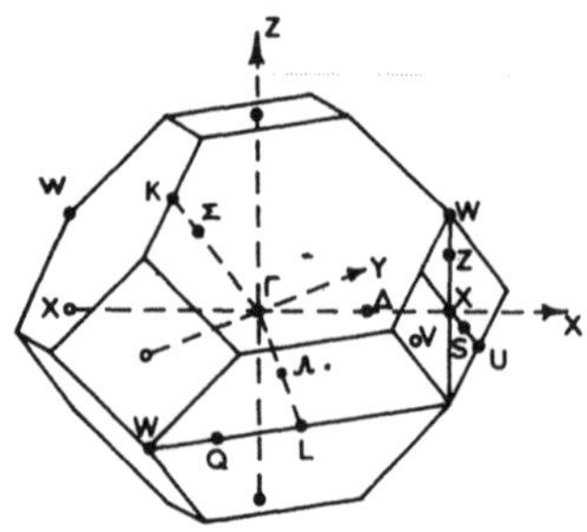

Figure 1. First Brillouin zone of a face-centered cubic lattice.

In Fig. 1 the cell is shown corresponding to the vectors defined by (13). In this cell, which is a truncated octahedron, a number of points with special symmetry properties are indicated. These will be discussed further on in this text.

The presence of the subgroup of translations has led us so far to a factorization of the secular determinant in L x L x L = N determinants of order 3s (s = number of atoms per unit cell). A further factorization is possible if account is taken of the rotational elements of the space group. As noted earlier, the factor group G/t is isomorphic with a point group. The irreducible representations of the factor group give us all irreducible representations of the space group which are invariant under translations. We can now find the irreducible representations of G associated with a given vector $\vec{k}$.

For each wave vector $\vec{k}$ there is a group of operations $\{\beta|b\}$ whose rotational parts β satisfy the condition

$$\beta\vec{k} = \vec{k} \tag{16}$$

The operators $\{\beta|b\}$ form a group n which is called the little group of the second kind. We know that the rotational parts α of the space group $\{\alpha|a\}$ form a point group G_o, isomorphic with G/t. The elements β must form a subgroup of G_o. This is the point group $G_o(\vec{k})$, called the little group of the first kind (or in short "the group of k"). $G_o(\vec{k})$ is isomorphic with n/t, and it contains all elements of this factor group, which leave the given $\vec{k}$ vector invariant. If we now let the $\vec{k}$ vector pass through all possible values within and on the surface of the Brillouin zone, all irreducible representations of G are obtained by finding the appropriate, irreducible representations of $G_o(\vec{k})$ for the given $\vec{k}$ vector.

The operations of G_o which are not contained in $G_o(\vec{k})$, provide (starting from a given $\vec{k}$ vector) a set of vectors $\vec{k}$, $\alpha_1\vec{k}$, $\alpha_2\vec{k}$, ... This set is "the star" of $\vec{k}$. The NaCl lattice corresponds to the space group $O_h^5(F \frac{4}{m} \bar{3} \frac{2}{m})$, a face-centered cubic group. Further details on this group can be found in the work of Bouckaert, et al[5]. As the group contains no glide planes or screw axes, it is a symmorphic group. The group G_o formed by the rotational elements is the point group O_h. The rotational elements α of this group are

$$E,\ 8C_3,\ 3C_4^2,\ 6C_4,\ 6C_2',\ i,\ 8S_6,\ 3\sigma_h,\ 6S_4,\ 6\sigma_d$$

In Fig. 1, a number of special points are indicated. Consider for instance, a point Δ. This point corresponds to a $\vec{k}$ vector along the y-axis, and it is left invariant by the operations

$$E,\ 2C_4(y),\ C_4^2(y),\ \sigma_h(x),\ \sigma_h(z),\ 2\sigma_d(y)$$

This means that the group of Δ, $G_o(\Delta)$ is the point group C_{4v}. The remaining operations of the point group O_h transform this point into five other points; so the star of Δ consists of 6 points. In Table 1 is given the groups $G_o(\vec{k})$ and the number of points in the star of several points in the Brillouin zone. The character tables of these point groups are all well known and can be found in many texts on group theory[4,5,6]. Because for the alkali halides, the space group (symmorphic) is a direct product group of the group of pure translations and the rotational group, the aforementioned character tables together with the star give us all representations of O_h^5.

Phonon Symmetries

At each point $\vec{k}$ in the Brillouin zone, the characters under the operations of the group of $\vec{k}$, $G_o(\vec{k})$, can easily be obtained from an investigation of the transformation properties of the eigen vectors $\vec{w}$. For the points Γ, X, Λ and Δ, the phonons of the six branches can easily be distinguished as acoustic and optical phonons. Furthermore, for these phonons the eigen vectors $\vec{w}$ are always parallel or perpendicular to the wave vector $\vec{k}$; in the first case the phonons are called longitudinal, in the second case transverse. For a general point in the zone however, this characterization is not possible. There are also some points where the phonon symmetries depend on whether the operations of $G_o(\vec{k})$ are carried out with respect to the alkali or the halide lattice. For all these points, the assignment of characters to the several phonon branches has been effectuated here using the actual numerical data for the eigen vectors, as determined from model calculations. The results of this analysis are summarized in Table 2

Table 1

The groups $G_o(k)$ and the number of points in the star of some special points in the first Brillouin zone

	$q_x q_y q_z$	$G_o(k)$	Star
Γ	0,0,0	O_h	1
X	1,0,0	D_{4h}	3
L	½,½,½	D_{3d}	4
W	1,½,0	D_{2d}	6
Δ	q_x,0,0	C_{4v}	6
Λ	q_x, q_x, q_x	C_{3v}	8
Z	1,q_y,0	C_{2v}	12
Σ	q_x, q_x, 0	C_{2v}	12
K	$\frac{3}{4}, \frac{3}{4}$, 0	C_{2v}	12
U	1, $\frac{1}{4}, \frac{1}{4}$	C_{2v}	12
S	1,q_y, q_y	C_{2v}	12
Q	q_x,½,$1-q_x$	C_2	24
α	1,q_y, q_z	C_s	24
β	0,q_y, q_z	C_s	24
γ	q_x, q_x, q_z	C_s	24
	q_x, q_y, q_z	E	48

Table 2

Phonon symmetries for KI at different points in the Brillouin zone, with respect to K^+ and I^- sites

K^+ site	Optical Branches	Acoustical Branches
Γ (O_h)	F_{1u}	F_{1u}
X (D_{4h})	A_{2u} (LO) + E_u (TO)	A_{2u} (LA) + E_u (TA)
L (D_{3d})	A_{2u} (LO) + E_u (TO)	A_{1g} (LA) + E_g (TA)
W (D_{2d})	B_2 + E	A_1 + E
Δ (C_{4v})	A_1 (LO) + E (TO)	A_1 (LA) + E (TA)
Λ (C_{3v})	A_1 (LO) + E (TO)	A_1 (LA) + E (TA)
Z (C_{2v})	A_1 + B_1 + B_2	A_1 + B_1 + B_2
Σ (C_{2v})	A_1 + B_1 + B_2	A_1 + B_1 + B_2
S (C_{2v})	A_1 + B_1 + B_2	A_1 + B_1 + B_2
Q (C_2)	A + 2 B	2 A + B
β (C_s)	2 A' + A"	2 A' + A"
γ (C_s)	2 A' + A"	2 A' + A"

I^- site	Optical Branches	Acoustical Branches
Γ (O_h)	F_{1u}	F_{1u}
X (D_{4h})	A_{2u} (LO) + E_u (TO)	A_{2u} (LA) + E_u (TA)
L (D_{3d})	A_{1g} (LO) + E_g (TO)	A_{2u} (LA) + E_u (TA)
W (D_{2d})	A_1 + E	B_2 + E
Δ (C_{4v})	A_1 (LO) + E (TO)	A_1 (LA) + E (TA)
Λ (C_{3v})	A_1 (LO) + E (TO)	A_1 (LA) + E (TA)
Z (C_{2v})	A_1 + B_1 + B_2	A_1 + B_1 + B_2
Σ (C_{2v})	A_1 + B_1 + B_2	A_1 + B_1 + B_2

(continued)

Table 2 (continued)

I^- site	Optical Branches	Acoustical Branches
S (C_{2v})	$A_1 + B_1 + B_2$	$A_1 + B_1 + B_2$
Q (C_2)	2 A + B	A + 2 B
β (C_s)	2 A' + A"	2 A' + A"
γ (C_s)	2 A' + A"	2 A' + A"

(a part of this table has been published in ref. (7)). Figure 2 gives some dispersion curves for KI, with the phonon symmetries for the six branches pertinent to the I^--site.

Selection Rules

Before we proceed with the derivation of the selection rules of a crystal containing impurities, it will be useful to discuss in short the rules for perfect crystals[8].

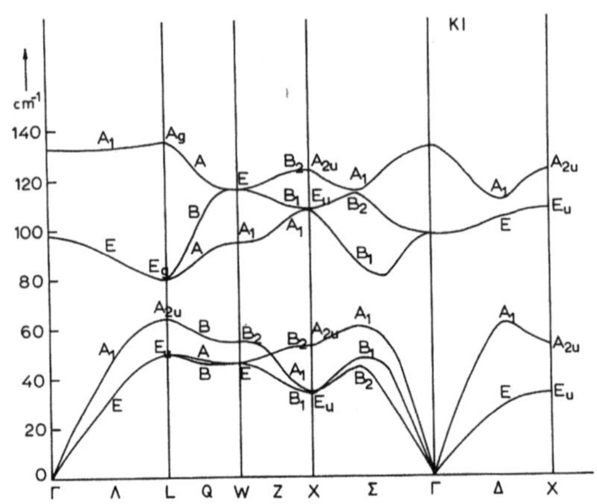

Figure 2. Dispersion curves for KI from Karo and Hardy's deformation dipole model[16]. Irreducible representations of the groups $G_o(\vec{k})$ for the iodine lattice.

The perfect crystal differs mainly from the crystal with impurities or otherwise imperfect crystals in that it is invariant with respect to translations. This invariancy leads to a selection rule as shown below. In general a transition will be allowed if the matrix element

$$< \psi_f | H_q | \psi_i > \tag{17}$$

is not equal to zero. Here ψ_i and ψ_f respectively, denote the initial and final phonon-states, which are coupled by the photon-phonon interaction Hamiltonian. This matrix element can be worked out to

$$< \psi_f | e^{i\vec{k}_q \cdot \vec{r}(\ell)} \text{grad}\,\varepsilon | \psi_i > \tag{18}$$

where grad ε is the gradient in the direction of the polarization vector ε of the incident light wave, and $\vec{k}_q$ the corresponding wave vector. Now (18) will be non-zero only if the totally symmetric representation is contained in the direct product

$$\Gamma_f^* \times \Gamma_q \times \Gamma_i \tag{19}$$

The characters or representations of the translational part of the phonon wave functions were given by Eq. (9). Hence Eq. (18) can only be non-zero if

$$\exp 2\pi i \, (-\vec{k}_f + \vec{k}_q + \vec{k}_i) \cdot \vec{r}(\ell) = 1 \tag{20}$$

for all $\vec{r}(\ell)$, so

$$\vec{k}_f = \vec{k}_i + \vec{k}_q + \vec{b}(h) \tag{21}$$

The general selection rule expressed by Eq. (21) is known as the rule of conservation of crystal momentum. Thus this rule holds only up to an additive constant.

We can simplify this expression by noting that for the infrared spectral region, the photon wave vector $\vec{k}_q$ is of the order 10^3 - 10^4 cm^{-1}. The corresponding wave length 10^4 - 10^5 Å is so large, compared with the cell dimensions (≈10Å), that for all practical purposes this wave vector can be assumed zero. Moreover this means that the reciprocal lattice vector $\vec{b}(h)$ will be zero. In the harmonic approximation, in an electric dipole absorption process, the incident photon creates one phonon. Restricting the discussion to T = 0°K, the initial state ψ_i is totally symmetric, so $\vec{k}_i = 0$. According to Eq. (21), with the above mentioned assumptions it follows that

$$\vec{k}_f = \vec{k}_q \approx 0 \tag{22}$$

Apart from the condition on the wave vector, there is always the law of conservation of energy to be obeyed. For this case it is written as

$$\nu_f = \nu_q \tag{23}$$

where ν_f and ν_q denote the frequency of the phonon and photon respectively. For the incident light waves, the frequency and wave vector are related by the dispersion relation

$$\nu_q = c\ k_q \tag{24}$$

where c = velocity of light.

The frequency approaches zero linearly near $\vec{k} = 0$ for three phonon branches (acoustic branches). The velocity of propagation of these phonons $\left(\frac{\partial \nu_{ac}}{\partial \vec{k}}\right)$ is about 10^{-5} c, being the velocity of sound waves. This means that for the acoustic phonons, Eqs. (22) and (23) cannot be obeyed simultaneously for a non-zero frequency value. Hence in a first order process, no phonons can be excited of the acoustic branches. The three optical branches are threefold degenerate at $\vec{k} = 0$ (cf. Table 2). For wave vectors close to $\vec{k} = 0$, this degeneracy is lifted by the electrostatic forces to form a twofold degenerate transverse mode and a non-degenerate longitudinal mode. The incident light waves are transversely polarized, that is, their polarization vector $\vec{\epsilon}$ is perpendicular to the wave vector $\vec{k}_q$. As the transition matrix element contains the scalar product $\vec{\epsilon}\cdot\vec{w}$, it obviously vanishes for the longitudinal optical (LO) modes, since for these modes the eigen vectors are perpendicular to the wave vector. In group theory, this is expressed by the fact that for the LO modes, which transform as a one-dimensional representation, the direct product with the two-dimensional representation of the light wave does not contain the totally symmetric representation. From similar arguments, one sees that the interaction with the transverse optical phonons (TO) is allowed. Thus for the perfect crystal, the only possible first order absorption is due to the excitation of the transverse optical (TO) phonons near k = 0.

The main difference between the perfect crystal and the crystal with impurities, is that in the second case the lattice is no longer invariant with respect to translations. The consequence is that the wave vector selection rule breaks down, so that many more phonons can be active in first order absorption. Here for this case, we will investigate how the symmetry properties can lead us to selection rules. We now have to look at the group of point operations, that leave the impurity site invariant. This group, which is called the site group S, is necessarily isomorphic with a subgroup of the factor group G/t. In the alkali halides the site group is O_h for the site occupied by an alkali or a halide ion. All possible site groups for a given space group are given by Halford[9]. If the impurity is a polyatomic ion or molecule, the situation is even more restricted. It is now the exact orientation and position of the ion that determines the local symmetry around this ion. In general the ion will be situated in the lattice so that the point group M of the ion and the site group S may have elements in common of which the local symmetry group L is built. The group L consists of the point operations that leave the ion invariant; it is a subgroup of M and S. Thus the symmetry is reduced from that of the space group to that of the local symmetry group. This reduction is easily carried out once the character tables are known.

The calculation will be performed in two steps. The first step will be the reduction of the space group representations into irreducible representations of the site group with highest possible symmetry, which is O_h. For most points in the Brillouin zone, the results are given by Loudon[7]. The next step is the reduction of the representations of the group O_h to those of the local symmetry group. This is easily done by means of correlation tables[6].

The selection rules for a crystal with impurities follow from a consideration of the matrix element:

$$< \psi_f | \vec{r} | \psi_i > \tag{25}$$

where ψ_i and ψ_f denote the initial and final phonon states respectively, coupled by the electric dipole operator $\vec{r}$. This matrix element is only non-zero when the direct product

$$\Gamma_f^* \times \Gamma_{\vec{r}} \times \Gamma_i \tag{26}$$

contains the totally symmetric representation under the local symmetry group L. The representations of the dipole operator are simply given by the transformation properties of the three dimensional vector $\vec{r}$, and are well known for all point groups. The phonon symmetries for most points in the Brillouin zone are given in Table 2. By means of the Loudon tables, we carry out the reduction to irreducible representations of the group O_h. The further reduction to representations of the local symmetry group is done with the aid of correlation tables[6]. For first order absorption processes, Γ_i is totally symmetric. This means that the transition is allowed when Γ_f contains $\Gamma_{\vec{r}}$ under the group L.

During the last two years a number of papers have been published on the far-infrared spectra of alkali halides containing complex ions in solid solution[10,11,12,13]. Without exception, a very strong absorption is observed in the region of the optical phonons. There is however, also a weaker absorption in the region of the acoustic modes, which is not observed in the pure crystals. If the local symmetry group is known, this absorption should be predictable from our theory. In Table 3 the results are given for a number of complex ions. For most cases the local symmetry is rather uncertain and most probable groups are taken as found in literature. Table 3 shows that in nearly all these cases, all phonons can be active in first order absorption.

In connection with the near infrared spectra of nitrate ions in solid solutions, we will also consider the possiblility of combination bands. The discussion holds both for combinations with vibrational transitions, external and internal, or electronic transitions. If the derivation is restricted to transitions from the (totally symmetric) ground state to a singly excited state, we have to investigate the transformation properties of the matrix element[25]. Assume ψ_f is given approximately by $\psi_f = \psi(\text{imp}).\psi(\text{phonon})$, where $\psi(\text{imp})$ denotes the final electronic or vibrational state of the impurity ion and $\psi(\text{phonon})$ the wave function of the phonon that is involved. The transition will be allowed if

$$\Gamma_r \times \Gamma^*(\text{imp}) \times \Gamma^*(\text{phonon})$$

contains the total symmetric representation, or stated otherwise if

$$\Gamma^*(\text{imp}) \times \Gamma^*(\text{phonon}) \text{ contains } \Gamma_r .$$

The Γ's are the representations which are irreducible under the local symmetry group L. Thus, here again one needs to know the position and orientation of the impurity ion in the host lattice. The results for a number of inorganic ions are summarized in Table 3. Clearly in a crystal with an impurity, due to the absorption of electromagnetic radiation

a) both acoustic and optical phonons can be excited
b) phonons can be excited together with the excitation of fundamental vibrations of the impurity ion
c) phonons can be excited together with the excitation of a vibronic state
d) phonons can be excited together with external vibrations (local or gap modes) of the impurity ion.

Of course the group theoretical discussion merely tells whether the transition integrals are zero or non-zero. Only an exact calculation of the matrix elements can provide information about the intensity of the absorption bands. Such a calculation has been carried out for the phonon absorption due to an impurity in the

Table 3

Impurity induced infra-red activity under the local symmetry group

Ion	Mol. group	Loc. group	Fundamentals ν_i	Far-infrared phonons $\nu(\bar{k})$	Near-infrared comb. modes $\nu_i+\nu(\bar{k})$
NO_3^-	D_{3h}	C_{3v}	Active for $i=1, 2, 3, 4$	Active for all **k**	Active for all i and **k**
NO_2^-	C_{2v}	C_{2v}	Active for $i=1, 2, 3$	Active for all **k**	Active for all i and **k**
N_3^-	$D_{\infty h}$	D_{3d}	Active for $i=2, 3$	$\nu(L)$LO and TO forbidden; all other phonons active	$\nu_1+\nu(L)$ LO and TO forbidden; $\nu_2+\nu(\Gamma)$ forbidden; $\nu_2+\nu(X)$ forbidden; $\nu_2+\nu(L)$ LA and TA forbidden; $\nu_3+\nu(\mathbf{k})$ selection rules equal to those for $\nu_2+\nu(\mathbf{k})$; all other combinations active
NCO^-	$C_{\infty v}$	C_{3v}	Active for $i=1, 2, 3$	Active for all **k**	Active for all i and **k**
OH^-	$C_{\infty v}$	C_{4v}	Active	Active for all **k**	Active for all **k**

center of a diatomic linear chain[14]: The extinction coefficients were calculated from an approximation of the dipole moment derivative of the vibrating chain as a function of the mass and the nearest neighbor force constants of the impurity. A very considerable alteration in the intensity distribution over the branches results, dependent on the mass and force constant ratio. This means that the absorption is proportional to the density of states multiplied by a frequency depending weight factor. This weight factor is constant over the whole frequency range only in some special cases. In the case of a combination band (internal vibration-phonon), the transition is forbidden in the harmonic approximation. The nonzero part of the transition matrix elements is due to third and higher order terms in the potential energy function, as well as second and higher order terms in the dipole moment expansion. Neither of these contributions is generally known. If the anharmonic terms are small, perturbation theory can be used for the evaluation of the matrix elements. The denominator of the resulting expression will in that case contain the difference of the frequency of the sideband and the frequency of the main band. This gives a contribution to the weight factor, which causes a decrease in the absorption intensity with increasing frequency difference to the main band (the pure integral transition of the impurity).

Experimental Evidence

In Figs. 3 and 4 there are shown some phonon combination bands that have been observed[15] in the near infrared, superimposed on an internal vibration absorption of the nitrate ion which has been substituted as an impurity in potassium iodide and bromide crystals. Subtraction of the contributions from the pure internal transitions

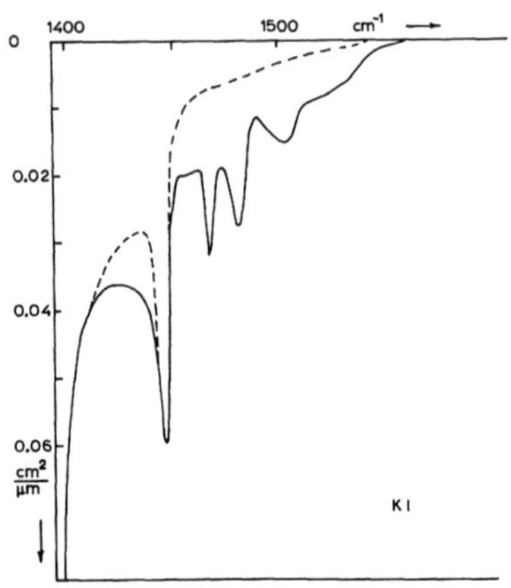

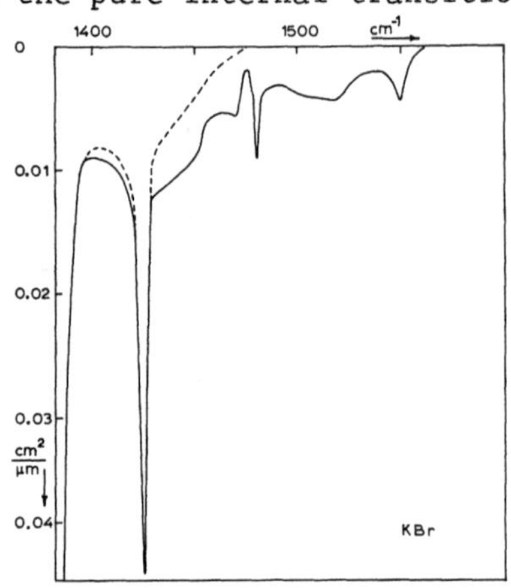

Figures 3 and 4. The high frequency side of the ν_3 internal vibration band of the nitrate ion in solid solution in KI and KBr showing phonon combination bands.

plus the local and gap modes, results in an absorption curve given by the drawn line in Figs. 5 and 6. In these figures are also shown the frequency distribution curves resulting from the deformation dipole model(16) and from the shell model(17) based on neutron scattering experiments. The resemblance of the curves is striking and the implication that there is no need for the introduction of a weight factor in these cases, is rather a surprise.

The accuracy of a frequency distribution (calculated as well as measured) can be checked by calculating the specific heat from the distribution and by comparing this with the available experimental (calorimetric) values. The results for this procedure are shown in Figs. 7 and 8. Another commonly used check is obtained by comparison with Debye temperatures. This is shown in Figs. 9 and 10. One should realize however, that comparison of calculated and calorimetricly measured values for specific heat is not a very sensitive check. Any curve with peaks at about the right place will compare well with calculated distribution curves, and the agreement may here well be fortuitous.

Cundill and Sherman(18) have also measured the phonon combination spectra of a variety of complex anions in the alkali halides. Although phonon combination bands are clearly present in the spectra of CN^- and NCO^- ions in alkali halides, there is no resemblance with the theoretical frequency distribution. The authors have recognized the importance of the directional selection of the modes

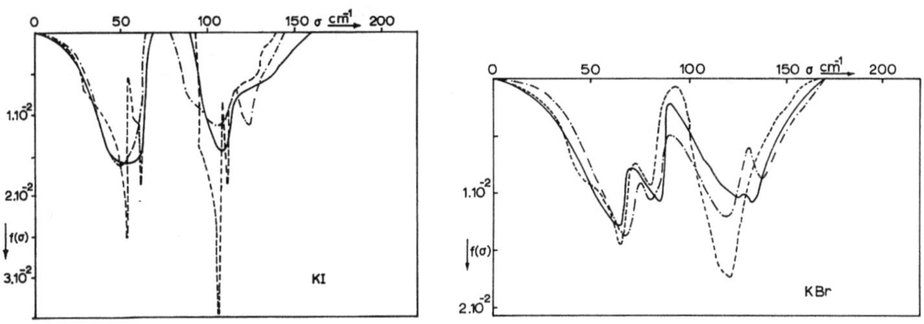

Figures 5 and 6. Experimental frequency distribution for KI and KBr obtained from Figs. 3 and 4 after subtraction of contributions of the internal vibrations and gap modes.

·—·—Distribution curve calculated from the deformation dipole model results(16).

— — —Distribution curve calculated from the shell model results(17).

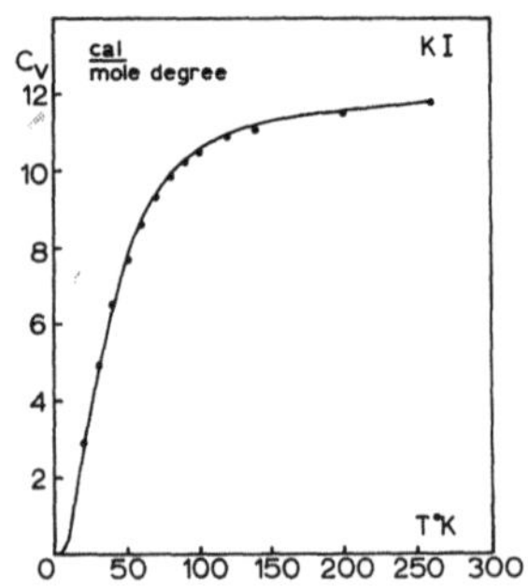

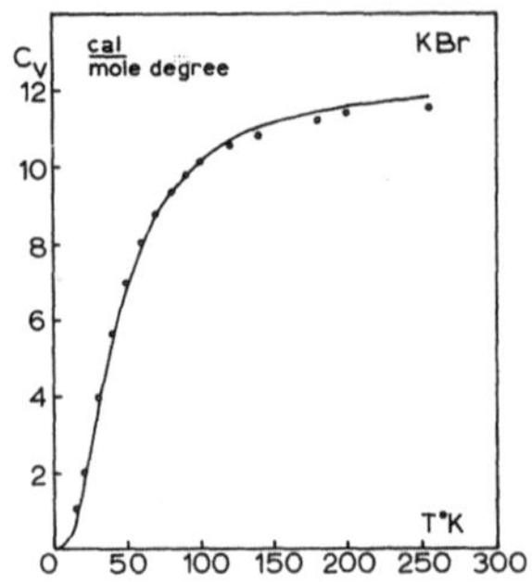

Figures 7 and 8. The specific heat at constant volume of KI and KBr at different temperatures.

Solid line: calorimetric values.

Points: calculated from the experimental frequency distributions in Figs. 5 and 6.

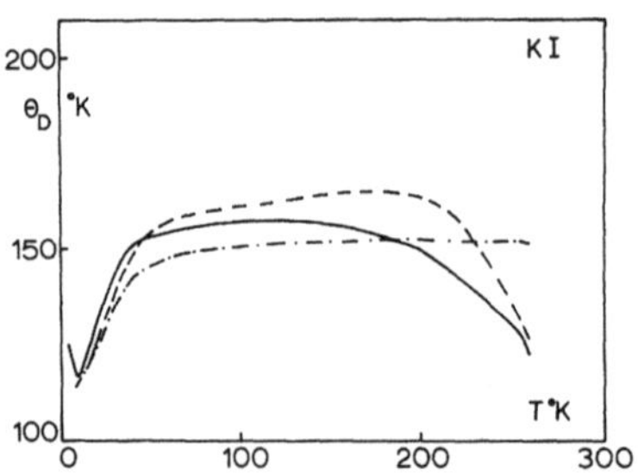

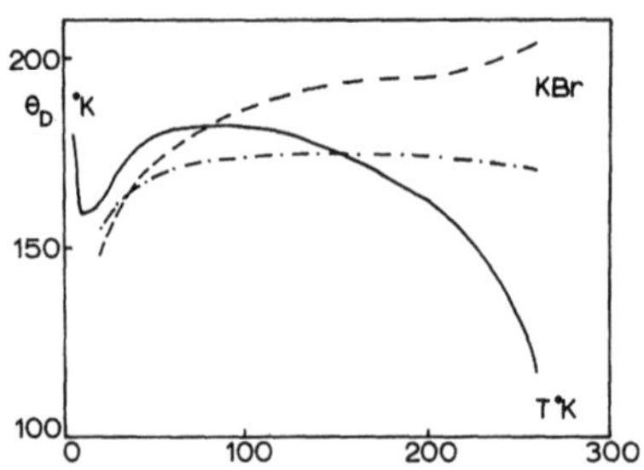

Figures 9 and 10. Debye temperatures as function of the temperature for KI and KBr.

——— calorimetric values.

·—·— calculated from the experimental frequency distribution.

– – – – calculated, deformation dipole model(16).

which appear in strong binary combinations. In NaI crystals where the 154 cm^{-1} peak of the distribution curve does appear in the spectrum of the <001> directed CN^- and not in the <111> directed NCO^- spectrum, it might be suggested that this 154 cm^{-1} peak is connected with one of the phonon branches X, T or L (LA and TA) that are forbidden in this combination under the symmetry group D_{3d}. This symmetry is not quite consistent with the symmetry of the ion, but might well be appropriate as far as the interactions are concerned.

The study of this sort of combination spectra with different ions in known or guessed positions may therefore in principle, serve as a means to identify the various singularities in the phonon spectrum.

Another similar experiment has been published by Timusk and Buchanan(19). These authors have measured the phonon side band of an electronic transition of the Sm^{++} ion in a KBr lattice. The experimentally determined spectrum is here very different from the theoretical frequency spectrum. The introduction of a weight factor that was calculated on the basis of the eigen vectors given by the neutron determined shell model and an interaction mechanism, gave an excellent fit however with only a few exceptions. Remarkable is a difference of about 5 wave numbers between the observed and calculated sharp peak at about 70 cm^{-1}. The authors remark that this is a peak arising from an off symmetry direction saddle point. These off symmetry dispersion curves have not been directly measured in the neutron scattering experiments and therefore it is quite possible that the shell model predictions are in error.

From the few examples quoted here, one may see that by comparison of the phonon combination spectra observed in systems with different symmetries, eventually it might be possible to arrive at a rather detailed knowledge of the composition of the distribution function of solid solutions. It is also clear however, that in general complications will arise due to the effects of finite concentration of the impurities and of disorder in the crystals.

HIGH CONCENTRATION SOLID SOLUTIONS

If the concentration of what in low concentration is called 'impurities' increases, the separation in local and gap modes and unperturbed lattice modes becomes less and less justified. Small but finite concentrations already will give rise to interactions. The frequencies of local modes will split if several defects become aware of one another's presence, and in crystals with such concentrations all defect levels are broadened. Discussion of this effect can in principle be given in terms of the theory we have found applicable for the infinitely diluted solid solutions.

At still higher concentrations the situation can best be described by a disordered lattice and the problem is to find the effect of such disorder on the frequency spectrum of a crystal. In the discussion of this problem there enters an aspect that is more characteristic of liquid and gaseous systems than for the solid state. A disordered lattice can be formed in very many different ways. Each realization of such a lattice gives rise to a frequency distribution. The average frequency spectrum is of physical interest (especially for thermodynamic uses) when averaged over all configurations. To find this averaged distribution function is still quite a problem. The theoretical approaches so far have not been very successful. The existing theories are mainly restricted to a one dimensional system and are very often only valid for impurities with low concentrations.

One disordered lattice problem can be solved exactly. This is the problem of a linear, two component chain in which the mass of one species of atom is infinite[20]. The infinite masses divide the chain into islands of light masses. If p is the probability that a light atom occupies a given lattice site, then the probability of having an island of n light atoms is $p^n (1-p)^2$.

The normal mode frequencies of a chain of n particles of mass m enclosed between rigid walls, are given by $\omega = \omega_n \sin \frac{\pi s}{2(n+1)}$, where $s = 1,2...n$. Therefore larger islands will partially give the same frequencies as the small islands. Taking into account the probability factor which decreases rapidly with increasing n, a frequency spectrum results which consists of a series of discrete lines. If one makes the infinite masses heavy but finite, the degeneracy of the lines are lifted, but the spectrum will show a spiky character even for mass ratios of only 2. These are the results of machine calculations which have been done on chains of finite length[21].

For real solid solutions, there are no really satisfying theories that give phonon spectrum. The theories that do exist are essentially limited to cases of low concentration[1,22]. Since on the other hand very interesting behavior of solid solutions has been found experimentally it is not surprising that frequent use is made of models and approximations to arrive at some interpretation of the experimental results. Many of these approximations and generalizations are not really justified.

The very pronounced optical behavior of mixed crystals for instance has led to a distinction between two types of mixed crystals. There are systems in which the features that represent part of the phonon spectrum (such as absorption, reflection, and Raman spectra), seem to be formed by a superposition of the features of the pure crystals. In other systems these features look on the contrary simply like those observed in pure crystals except for a continuous frequency shift with composition, from that distinguishing one pure crystal to that of the other pure crystal. It has become customary to call this two mode and one mode behavior respectively(23-26). However, this observed simple behavior of optical features alone does not necessarily imply that the crystal modes as such behave in the same simple way. Quite possibly the crystal modes in the mixed crystals are all seriously disturbed, but nonetheless a cooperation of several of these modes might give rise to the observed spectroscopic features.

To illustrate this statement a model calculation is presented below of a one dimensional diatomic chain. Although we are well aware of the limitations of one dimensional model calculations and of the care one must take in applying results of such calculations to actual three dimensional crystals, the results as presented here, provide a believable, clear illustration of the problem of mixed crystal modes, and of the factors that govern the spectroscopic behavior of such mixed crystals.

Our calculations based on the molecular model used by Jaswal(27), include long range as well as short range forces, where interaction of any atom with any other atom of the infinite chain is taken into account. Short range forces are of the form $\lambda \exp(-r/\rho)$, while long range forces have the form e^2/r_{ij}. The values that were taken for the parameters λ, ρ and r, are the values such as found in the actual three dimensional KI crystal; $\lambda = 4.6 \times 10^{-9}$ erg, $\rho = 0.349$ Å and $r = 3.526$ Å. We leave only 39 atoms free to move and keep the rest of the chain fixed. These 39 atoms move in the constant potential field of the rest of the chain. This gives a dynamical matrix of size 39 x 39 which can be diagonalized by standard procedures. To reduce the time needed for the calculations, all the chains are composed with a center of symmetry.

The usual equations of motion are

$$M_\ell u(\ell) = -\sum_{\ell'} x(\ell,\ell')\, u(\ell')$$

where M_ℓ = mass of the ℓth particle

$u(\ell)$ = displacement from equilibrium of ℓth particle

$$x(\ell,\ell') = \left[\partial^2 V / \partial u(\ell)\, \partial u(\ell')\right]_0$$

and

V = potential energy of the complete crystal.

Assume solutions of the form

$$u(\ell) = M_\ell^{-\frac{1}{2}} \, y\,(\ell) \exp\,(i\omega t)$$

$$\omega^2 \, y\,(\ell) = \sum_{\ell'} P\,(\ell,\ell')\; y\,(\ell')$$

$$P(\ell,\ell') = x(\ell,\ell')\,(M_\ell M_{\ell'})^{-\frac{1}{2}}$$

where in matrix notation

$$\omega^2 \, Y = PY$$

Diagonalization of the dynamical matrix P leaves ω^2 at the diagonal positions. Since the work is with a one dimensional chain only longitudinal motions are solutions.

To obtain an impression of the influence of the length of the chain on the results, the frequency distribution for KI chains of different lengths has been calculated. It turns out that this influence is negligibly small for chains of more than about 10 units.

In order to determine a property analogous to the infrared absorption of a real three dimensional crystal, we made a summation of the amplitudes of the positive and negative ions with opposite signs so as to find the value of the changing dipole moments for the different normal modes. The squares of these sums are taken as representative, and are plotted in Fig. 11. As is to be expected for a chain of pure alkali halide, there is one mode with a very large amplitude sum in the pure chains of masses equal to the masses in KI, KBr or KCl. The small values of this sum that are found for a number of other modes, are a consequence of the finite length of the chain and are indicative of the limitations of the model that is used here.

In Fig. 11 there is also plotted the sum of the amplitudes for the modes of some chains of an intermediate composition, that at the left for masses corresponding to KI-KCl, and at the right corresponding to KI-KBr. One should of course for one concentration consider an average over all possible configurations. However, very regular chains or chains with large clusters of the same ion species have a very low probability. Furthermore, it turns out that the distribution of the sum of amplitudes over the modes is very insensitive for rearrangements of small clusters over the chain. Therefore, the result is given for just one configuration for any concentration which is believed to be dominating in a correctly averaged result.

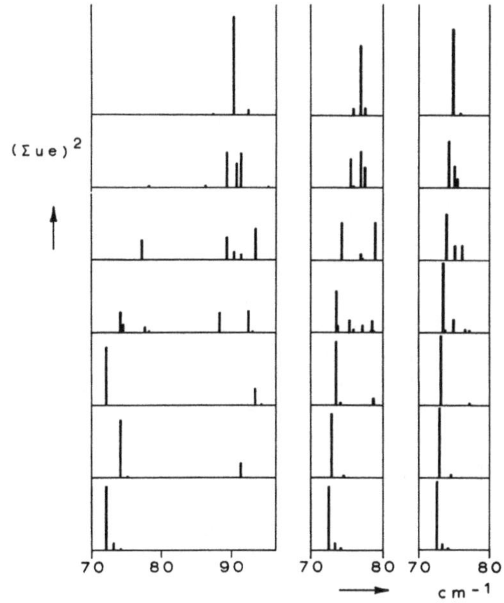

Figure 11. The square of the sum of amplitudes (arbitrary units) of the modes of a linear chain vs. frequency. The series for different concentrations at the left is the result for a chain with masses corresponding to a KI-KCl mixed crystal. At the right for KI-KBr. The series in the middle is for an intermediate ratio of masses KI-K (60). The succession of the halide ions in the chains is schematically indicated.

It is seen from Fig. 11 that the result is very different for different ratios of the substituting masses. If the sum of the amplitudes were associated with optical properties like in a real crystal, the case of the KI-KCl chain would be a clear example of a two mode behavior, whereas KI-KBr would present a clear one mode behavior. Since however, in both cases many more modes participate, it would be more appropriate to speak of one and two band behavior. This result could have been anticipated. According to the group theoretical considerations given earlier, the wave vector selection rule breaks down as soon as the invariancy of the lattice is lifted with respect to translations. For large $\vec{k}$ vectors, the loss of invariancy in a mixed crystal is evident; for small $\vec{k}$ perhaps less so and it is in the region of $\vec{k} \approx 0$ that approximations like the virtual crystal model have been used. However, in a mixed crystal there is a spread in frequency associated with each phonon of wave vector $\vec{k}$ because of the lattice disorder[28]. In a measurement which fixes $\vec{k}$ we will always find the spread in frequencies. Conversely, it is clear that with every frequency there comes a wide choice of $\vec{k}$ vectors.

In our chain model, the eigen vectors become very irregular and there is no way of collecting them according to the $\vec{k}$ value. This gives two reasons why not only the conclusions about one and two mode behavior should be considered with care, but also the experimental results. In the first place there are experiments that never fixed the wave vector at zero, (such as in tunneling spectroscopy[29] and two phonon absorption spectroscopy[30]). Secondly, for the experiments that do fix the wave vector at zero for the pure components, the wave vector selection rule no longer holds for the mixed crystals, and one might expect more active modes because more modes are associated with a varying dipole. In many of the published spectra there is indeed an indication that the features are of a more composite structure.

Finally, it is to note from our model calculation, that different mass ratios alone give rise to extremely different results. This is in agreement with the conclusions drawn from a consideration of the MREI model[31]. General conclusions about the interactions in the systems under consideration can therefore not be drawn simply from the appearance of mixed crystal spectra.

REFERENCES

1. A. A. Maradudin, E. W. Montroll, G. H. Weiss, Solid State Physics Suppl. 3, (1963).
2. C. W. McCombie, this proceedings.
3. A. J. Sievers, this proceedings.
4. G. F. Koster, Solid State Physics 5, 173 (1957).
5. L. P. Bouckaert, R. Smoluchowski, E. Wigner, Physics Rev. 50, 58, (1936).
6. E. B. Wilson, J. C. Decius, P. C. Cross, Molecular Vibrations, McGraw-Hill, New York, (1955).
7. R. Loudon, Proc. Phys. Soc. 84, 379 (1964).

8. R. E. Peirls, Quantum Theory of Solids, Oxford University Press, London, (1964).

9. R. S. Halford, J. Chem. Phys. 14, 8 (1946).

10. A. J. Sievers, C. D. Lytle, Phys. Letters 14, 271 (1965).

11. K. F. Renk, Phys. Letters 14, 281 (1965).

12. H. Dotsch, W. Gebhardt, C. H. Martius, Solid State Comm. 3, 297 (1965).

13. A. Hadni, G. Gorlot, F. Brehat, P. Strimer, C. R. Acad. Sci., Paris, 261, 2605 (1965).

14. L. Genzel, K. F. Renk, R. Weber, Phys. State Sol. 12, 639 (1965).

15. R. Metselaar, J. van der Elsken, Phys. Rev. 165, 359 (1968). R. Metselaar, J. van der Elsken, Phys. Rev. Lett. 16, 349 (1966).

16. A. M. Karo, J. R. Hardy, Phys. Rev. 129, 2024 (1963).

17. G. Dolling, R. A. Cowley, C. Schittenhelm, I. M. Thorson, Phys. Rev. 147, 577 (1966). R. A. Cowley, W. Cochran, B. N. Brockhouse, A. D. B. Woods, Phys. Rev. 131, 1030 (1963).

18. M. A. Cundill, W. F. Sherman, Phys. Rev. 168, 1007 (1968).

19. T. Timusk, M. Buchanan, Phys. Rev. 164, 345 (1967).

20. C. Domb, A. A. Maradudin, E. W. Montroll, J. H. Weiss, Phys. Rev. 115, 18, 24 (1959).

21. P. Dean, J. L. Martin, Proc. Roy. Soc. A259, 409 (1960).

22. A. J. Sievers, A. A. Maradudin, and S. S. Jaswal, Phys. Rev. 138, 272A (1965).

23. F. Kruger, O. Reinkober, E. Koch-Holm, Ann. Physik 85, 110 (1928).

24. P. J. Gielisse, J. N. Plendl, L. C. Mansur, R. Marshall, S. S. Mitra, R. Mykolajewycz and A. Smakula, J. Appl. Phys. 36, 2446 (1965).

25. D. W. Feldmann, M. Ashkin, J. H. Parker, Phys. Rev. Lett. 17, 1209 (1966).

26. H. W. Verleur, A. S. Barker, Phys. Rev. 149, 715 (1966).

27. S. S. Jaswal, Phys. Rev. 140, A 687 (1965).

28. Hin-Chiu Poon, A. Bienenstock, Phys. Rev. 141, 710 (1966).

29. R. A. Logan, J. M. Rowell, F. A. Trumbore, Phys. Rev. 136, A 1751 (1964).

30. R. Braunstein, Phys. Rev. 130, 879 (1963).

31. I. Chang and S. S. Mitra, Phys. Rev. 172, 924 (1968) and references therein.

CHARACTERISTIC ENERGY ABSORPTION SPECTRA OF SOLIDS*

J. N. Plendl

Air Force Cambridge Research Laboratories, L. G. Hanscom Field, Bedford, Massachusetts

ABSTRACT

A method has been developed to derive the fundamental absorption spectra of solids directly from experimental infrared data without need for resorting to computational evaluation. These characteristic lattice absorption spectra represent the true energy absorption spectra of solids and are, through Kirchhoff's Law, directly related to the emission spectra of solids. As opposed to the traditional absorption coefficient (k) versus frequency curve which is regarded as the absorption spectrum for solids in the region of anomalous dispersion, the spectra, as derived by the proposed method, clearly resolve the transverse and optical modes of vibration. They also exhibit the characteristics of anharmonicity and damping. The method uses the combined reflection and transmission data on single crystal and thin film specimens. Its validity has been verified on a wide variety of solids. The most recent data obtained from emission, Laser-Raman or cold neutron scattering techniques, is in complete agreement with the derived values.

A method is advanced also to calculate the various modes of vibration of solids directly from elastic constants. It is simultaneously used to render further support for the developed characteristic energy absorption spectra.

The importance of these spectra is discussed with regard to problems in solid state physics (particularly damping characteristics, temperature effects and non-linear effects of atomic binding in solids).

*The gist of this work was first presented at the ninth European Congress on Molecular Spectroscopy, Madrid (Spain) 12 Sept. 1967.

INTRODUCTION

The characteristic energy absorption spectrum of a solid is defined by Kirchhoff's Law with constant emissive power at anyone specified temperature. As such, it represents the true vibrational energy absorption spectrum which is related to the vibrational emission spectrum of the solid. Such fundamental or characteristic absorption spectra should constitute the frequency distribution function of all modes of oscillation of the crystal, with the maxima in these spectra representing the fundamental or eigenfrequencies of vibration in the major crystallographic directions, as well as their true relative intensity.

In principal, the frequency spectrum of nuclei motions in a solid can be described by the frequency dispersion in k-space of the optical and acoustic branches ($k = 2\pi/\lambda$ wave vector). For the wavelength of infrared radiation in the vibrational region, the wave vector is orders of magnitude below the wave vector at the edge of the Brillouin zone ($k = \pi/2a$). Thus the wave vector associated with the electromagnetic radiation may be considered to be essentially zero. At $k = 0$ the frequencies of the acoustic branches tend to zero and the vibration of the crystal as a whole is represented by the optical branch frequencies which constitute the fundamental vibrational modes (spectrum) of the crystal.

For either intrinsic or experimental reasons, standard techniques of infrared spectroscopy (reflectivity, transmission) and Raman scattering do not, by themselves, readily give all the required information. For instance, the LO modes in an infinite crystal cannot be obtained with the transverse electromagnetic waves at normal incidence. Transmission spectra cannot generally be obtained in the high dispersion area of even the thinnest crystals because they remain opaque. Transmission would require layer thicknesses of the order of a micron, equivalent only to a few thousand atomic layers. Evaporated films or very fine powder layers seem to provide a solution here. Raman data, especially laser-Raman spectra, are rather satisfactory but only produce the Raman active frequencies and again not all fundamental absorption frequencies. They are limited, furthermore, to transparent non-luminescent materials.

It is the purpose of this paper to present a method for developing the fundamental spectra of solids discussed above, directly through evaluation of experimental infrared data without need for recourse to computational evaluation. The proposed method involves the combined data from the infrared reflection spectrum of the single crystal and the infrared transmission spectrum of an evaporated thin film of the same substance.

CHARACTERISTIC ENERGY ABSORPTION SPECTRA FROM INFRARED REFLECTION AND TRANSMISSION DATA

Method

The energy distribution within a very thin film layer is given by

$$\sum R_{Lj}(\nu) + \sum T_{Lj}(\nu) + \sum A_{Lj}(\nu) = 1 \qquad (1)$$

because of the conservation of energy. In equation (1) R stands for reflectivity, T for transmission, A for absorption, L for film layer, ν for frequency and j for multiple paths. If the layer thickness is chosen so that $2\pi nd/\lambda$ remains a very small angle (n = index of refraction, d = thickness of the film layer, λ = wavelength), the interference effects can be neglected between back and front face reflections of a dielectric layer, and permits concentration instead on the characteristics of the infrared energy distribution within the layer. Taking this into consideration, equation (1) reduces to

$$R_L'(\nu) + T_L'(\nu) + A_L'(\nu) = 1 \qquad (2)$$

where the prime stands for single paths.

Since $R_L'(\nu)$ is not generally available experimentally, it is proposed that the thin film be regarded as the top layer of the single crystal, then the $R_L'(\nu)$ term may be substituted for by the reflectivity of the single crystal $R_C(\nu)$ (see Fig. 1). As such $R_L'(\nu)$ is considered representative of the contribution of the top layer of the crystal and proportional to $R_C(\nu)$, but will make less of a contribution to the overall energy balance. Equation (2) may thus be modified in the form

$$R_C(\nu) + T_L'(\nu) + A_L'(\nu) = X \qquad (3)$$

where $X > 1$, and

$$A_L'(\nu) = X - (R_C(\nu) + T_L'(\nu)) \qquad (4)$$

The absolute value of the parameter X is not a constant but changes with d, since $T_L'(\nu)$ increases when d decreases and vice versa. The maximum value of X is obtained as $d \to 0$ when $X_{(d \to 0)} = 1 + R_C(\nu)$ and the minimum as $d \to \infty$ when $X_{(d \to \infty)}$ = unity.

In order to determine $A_L'(\nu)$ from equation (4) the value of X must be ascertained. Since equation (3) is valid for all values which its three terms may assume, a frequency $\underline{\nu}$ may be sought at which $A_L'(\underline{\nu}) \to 0$. At this point $X = R_C(\underline{\nu}) + T_L'(\underline{\nu})$, and the maximum value of X will be the maximum of the sum of the curves $R_C(\nu)$ plus $T_L'(\nu)$. Knowing the value of the parameter X we are now in a position to graphically derive the desired energy absorption spectrum.

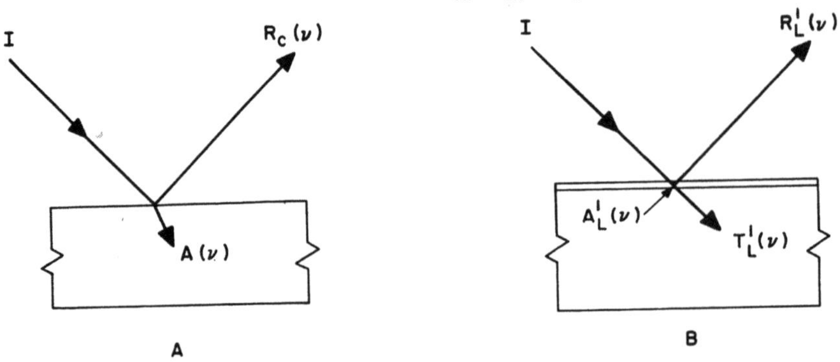

Figure 1. Schematic presentation of energy distribution of incoming radiation for single crystal model (A) and film model (B), as proposed in text.

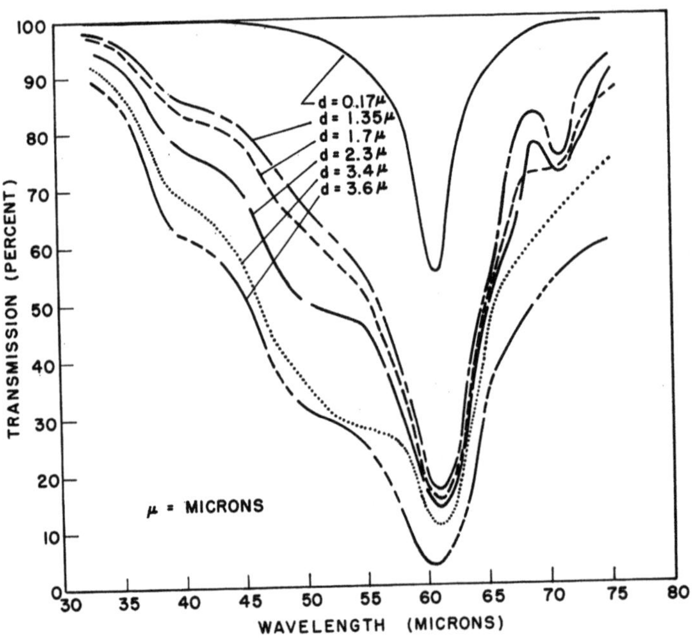

Figure 2. Transmission data of NaCl for various thicknesses, after Barnes and Czerny[1].

Graphical Derivation

Fig. 3a illustrates the case for NaCℓ where $T_L'(\nu)$ is reproduced from Fig. 2 for d = 1.35 microns. In addition, $R_C(\nu)$[1,2] is shown as well as the sum curve $R_C(\nu) + T_L'(\nu)$, of which the maximum limits the value of X. For NaCℓ and d = 1.35, the maximum value of X = 1.57. Fig. 3b shows $A_L'(\nu) = X - (R_C(\nu) + T_L'(\nu))$ which, since it is the inverted $R_C(\nu) + T_L'(\nu)$ curve of Fig. 3a, fixes X as the zero line.

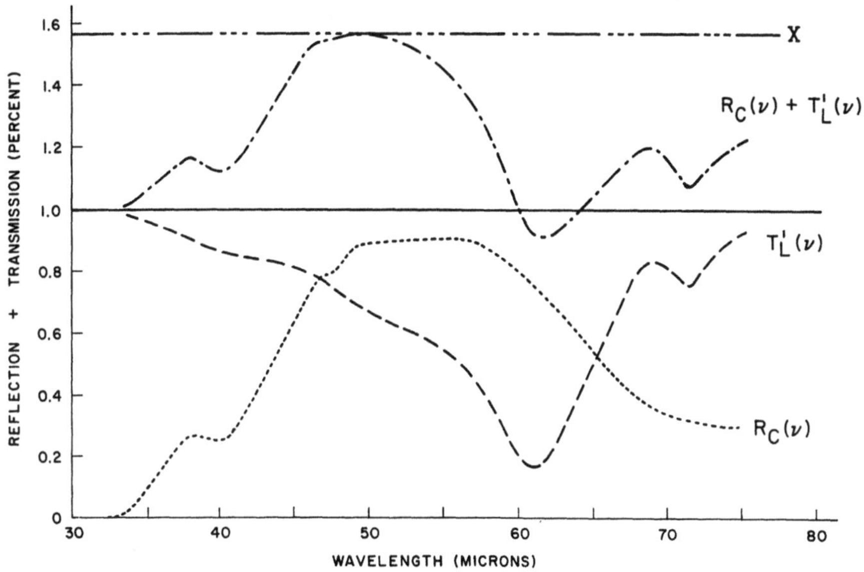

Figure 3a. Reflectivity, transmission and sum value as a function of wavelength for NaCl. Experimental data from reference 1 and 2.

According to the model, the data of $A_L'(\nu)$ may be conceived of as the spectral energy distribution of the absorption within the film layer on top of the single crystal. The term $A_L(\nu)$ refers to the total energy I in Fig. 1. The specific or characteristic absorption may be obtained when $A_L'(\nu)$ is divided by the energy $(1 - R_C(\nu))$ which really enters the crystal. This procedure, however, has two alternatives. $A_L'(\nu)$ may be divided by the energy which enters into the single crystal $[1 - R_C(\nu)]$. In this case the energy absorption spectrum of the single crystal is subsequently approximated and one obtains

$$F[A_C(\nu)] = \left\{ X - [R_C(\nu) + T_L'(\nu)]\right\} \Big/ [1 - R_C(\nu)] \qquad (5)$$

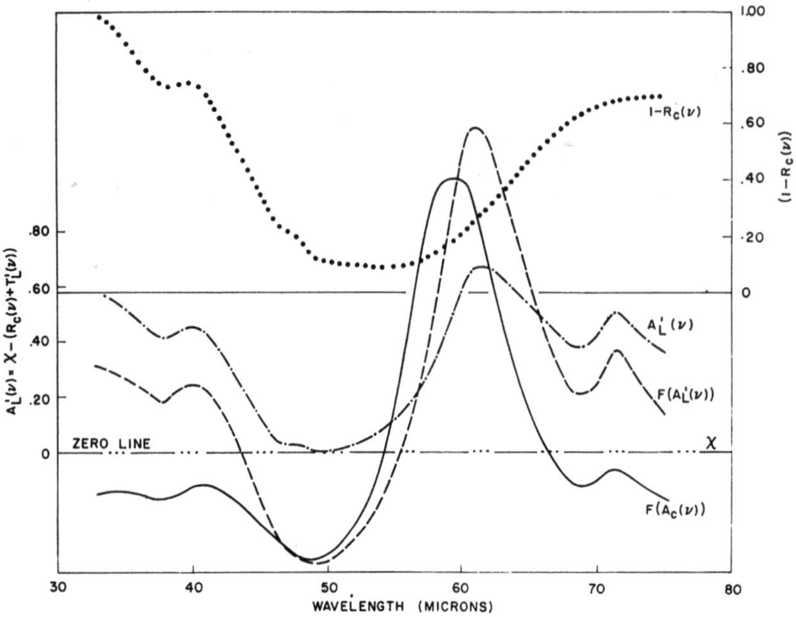

Figure 3b. Evaluation of energy absorption in NaCl from data of Figure 3a using the method of this paper.

$A_L'(\nu)$ may be divided by the energy $[X - R_C(\nu)]$ which enters into the evaporated film layer as such. In this case, the film layer is considered as separated from the single crystal, and thus approximate the energy absorption spectrum of an evaporated film layer by

$$F[A_L'(\nu)] = \left\{X - [R_C(\nu) + T_L'(\nu)]\right\} \Big/ [X - R_C(\nu)] \qquad (6)$$

Both functions have been plotted in Fig. 3b and for various film (crystal) thicknesses in Figs. 4 and 5. Comparing the data of Figs. 4 and 5, a much higher resolution is found for the absorption function $F[A_L(\nu)]$ of the film layer and a much better expression of the anharmonicity (asymmetric absorption curves) for the absorption function $F[A_C(\nu)]$ of the single crystal. Frequency, bandwidth and shape of the main absorption curve in the two sets of data (Figs. 4 and 5) among themselves show virtually no change at all which may be said as well for all other spectral characteristics. Calculations in section III will show that the absorption frequen-

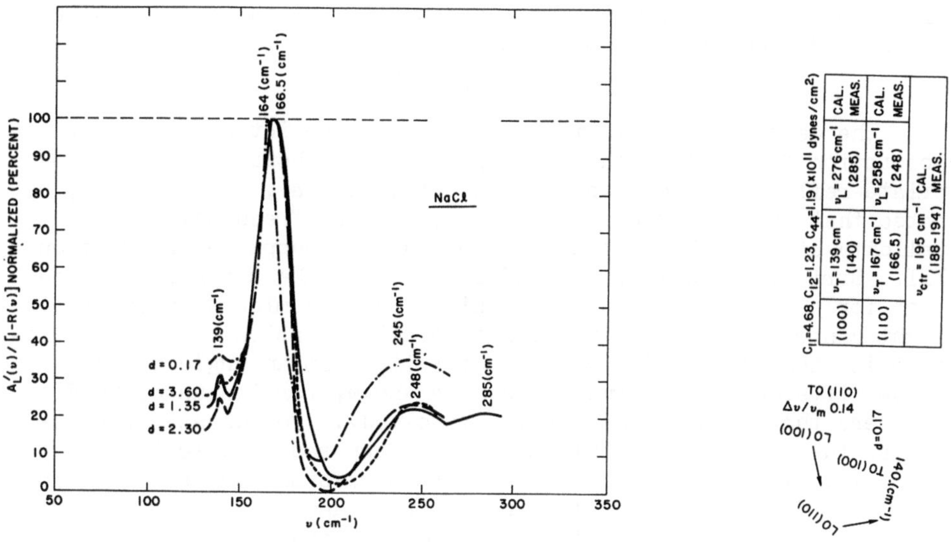

Figure 4. Energy absorption spectrum of NaCl for four thicknesses (microns); Crystal model.

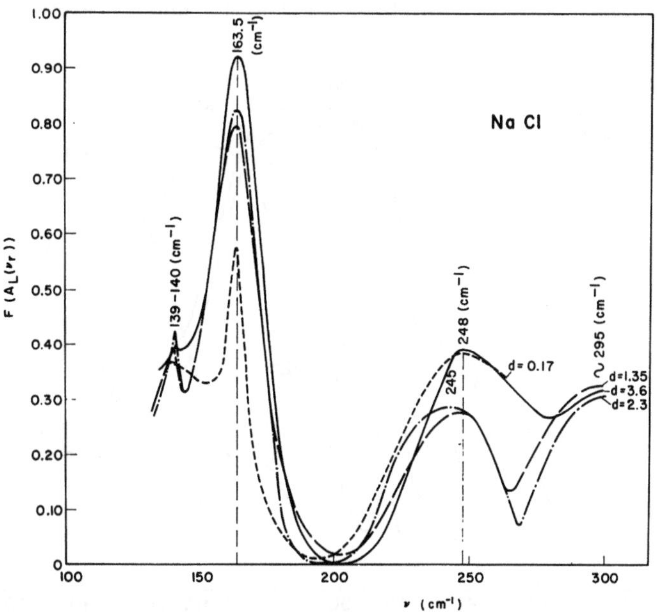

Figure 5. Energy absorption spectrum of NaCl for four layer thicknesses (microns); Film model.

cies from $F[A_C(\nu)]$ show the most exact fit with those frequencies calculated from elastic constants.

Data Background

The degree of resolution and the relative intensities of the absorption frequencies sharply depend on the choice of the appropriate reflection and transmission data. It may, for instance, be noted that $F[A_C(\nu)]$ shows both TO and LO mode frequencies. The validity of this observation will be further supported in section III. Strictly speaking, only TO modes should be shown in reflection for perpendicular incidence since the incoming transverse light cannot couple with the longitudinal vibration. This restriction is, however, no longer valid for light at an angle of incidence different from normal. This is shown, for instance, in Fig. 6, where the LO modes are clearly resolved[3]. Polycrystalline material further enhances this effect, as opposed to oriented single crystals. This is apparent in most transmission spectra which are generally done on polycrystalline material and which show the LO mode region clearly (see Fig. 2 around 40 microns). Thus, to obtain the TO modes and the LO modes with comparable intensities and with relative high resolution, reflection spectra of single crystals should be used which are measured with an angle of incidence in the vicinity of 45°. Hence a combination of 45° incidence, reflection spectra of single crystals and regular transmission spectra of films or powders should give a good approximation for the true ratio between the intensities of the TO and LO modes of vibration.

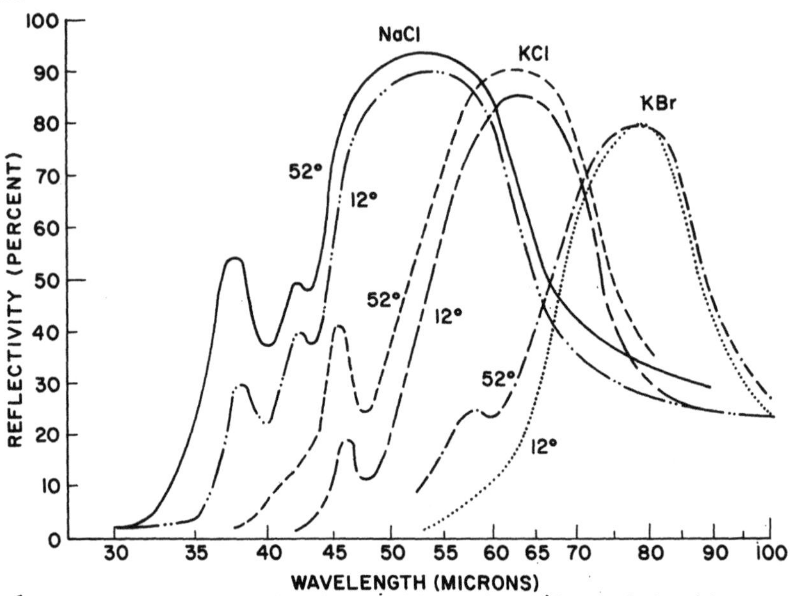

Figure 6. Reflectivity data for different angles of incidence (12 and 52 degrees) of three alkali halides, after Mitsuishi et al[3].

DETERMINATION OF THE CHARACTERISTIC ABSORPTION FREQUENCIES FROM THE ELASTIC CONSTANTS

In order to test the value of the absorption frequencies of the characteristic absorption spectrum, and make a proper assignment, a comparison is needed with similar data from an independent source. The elastic constants have been chosen here because considerable experimental information is available for a great variety of solids.

One successful method for calculating reststrahlen frequencies from elastic data was first proposed by Madelung as early as 1908[4]. In 1940 Madelung's method was simplified by Pohl[5] who defined the eigenfrequency of an atomic layer by the velocity of sound divided by twice the atomic distance (r_o). In applying his formula to the average data from the velocity of sound, Pohl obtained a good agreement with the reststrahlen frequencies for a few solids. In 1957, on modifying Pohl's definition, Plendl[6] calculated the eigenfrequencies of the LO and TO modes from the elastic moduli for a variety of solids and obtained good agreement with the then known infrared frequencies of these solids. Starting from these results, Pohl's definition can now be examined for application to the velocities of sound in the various crystallographic directions (as may be determined from the elastic constants) and extend it to all possible elastic waves within a cubic crystal.

The propagation of energy in a cubic crystal may be resolved along three major directions: (100), (110) and (111). This was shown for example by De Klerk[7]. From the velocities of sound in these directions for both the compressional waves and the shear waves, the fundamental eigenfrequencies of a solid are obtained when we divide the velocities by 2 r_o c, using Pohl's definition (see above). The velocity of light, c, is used only to obtain the frequency in the dimension of cm^{-1}. The basic formulas are given as follows:

$$(100)\begin{cases}\nu_{LO}=\sqrt{c_{11}/\rho}\Big/2\,r_o c \quad (cm^{-1}) & (7)\\ \nu_{TO}=\sqrt{c_{44}/\rho}\Big/2\,r_o c \quad (cm^{-1}) & (8)\end{cases}$$

$$(110)\begin{cases}\nu_{LO}=\sqrt{(c_{11}+c_{12}+2c_{44})/2\rho}\Big/2\,r_o c \ (cm^{-1}) & (9)\\ \nu_{TO}=\sqrt{(c_{11}-c_{12})/2\rho}\Big/2\,r_o c \ (cm^{-1}) & (10)\end{cases}$$

$$(111)\begin{cases}\nu_{LO}=\sqrt{(c_{11}+2c_{12}+4c_{44})/3\rho}\Big/2\,r_o c \ (cm^{-1}) & (11)\\ \nu_{TO}=\sqrt{(c_{11}-c_{12}+c_{44})/3\rho}\Big/2\,r_o c (cm^{-1}) \text{degenerate} & (12)\end{cases}$$

$$\nu_{ctr} = \sqrt{(c_{11} + 2c_{12})/3\rho} \Big/ 2\, r_o c \ (\mathrm{cm}^{-1}) \qquad (13)$$

The TO mode in Eq. (12) for the (111) direction is degenerate and may not be observed in the IR spectrum. The eigenfrequencies calculated on the basis of these equations* may now be compared with the frequencies obtained from the characteristic energy absorption spectrum. This has been done for NaCℓ, KCℓ and LiF in Table I. The supporting spectral data may be found in Figs. 7, 8 and 9. Excellent agreement is found comparing the corresponding data of different origin (infrared and elastic). The average difference is only one percent for the TO modes of $F[A_C(\nu)]$ and somewhat more for the less accurately determined LO modes. It is interesting to note here, as well as in all other cases examined below, that the strongest energy absorption always belongs to the (110) TO mode, whereas the weaker absorption belongs to the (100) TO mode. This result may be explained by the fact that the ions meet less resistance for vibrations in the (110) direction compared to the (100) direction.

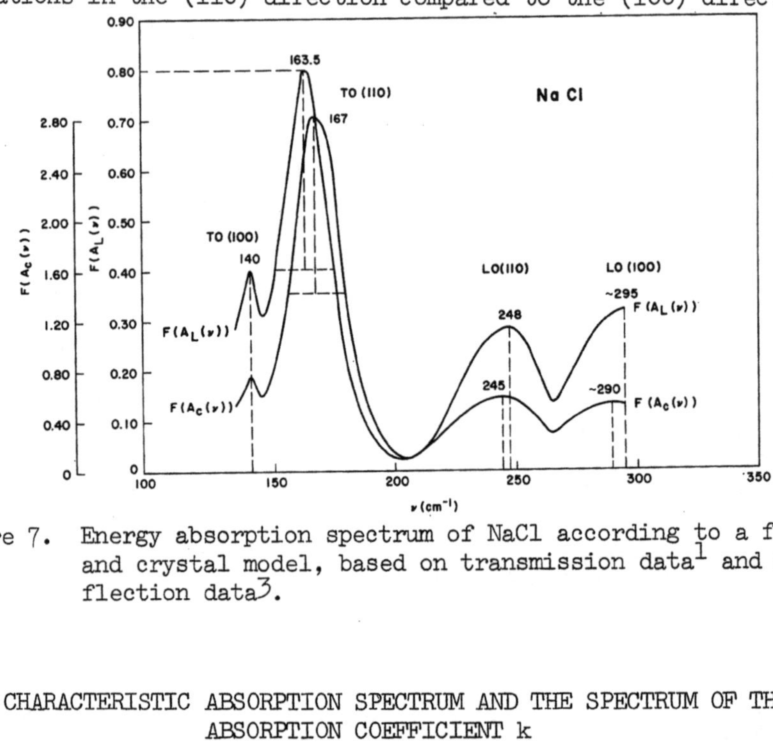

Figure 7. Energy absorption spectrum of NaCl according to a film and crystal model, based on transmission data[1] and reflection data[3].

CHARACTERISTIC ABSORPTION SPECTRUM AND THE SPECTRUM OF THE ABSORPTION COEFFICIENT k

The characteristic energy absorption spectrum may now be compared with the k-spectrum. This is done for LiF and MgO for which

*Compilations of data of elastic constants and pertinent references may be found in References 8 and 9.

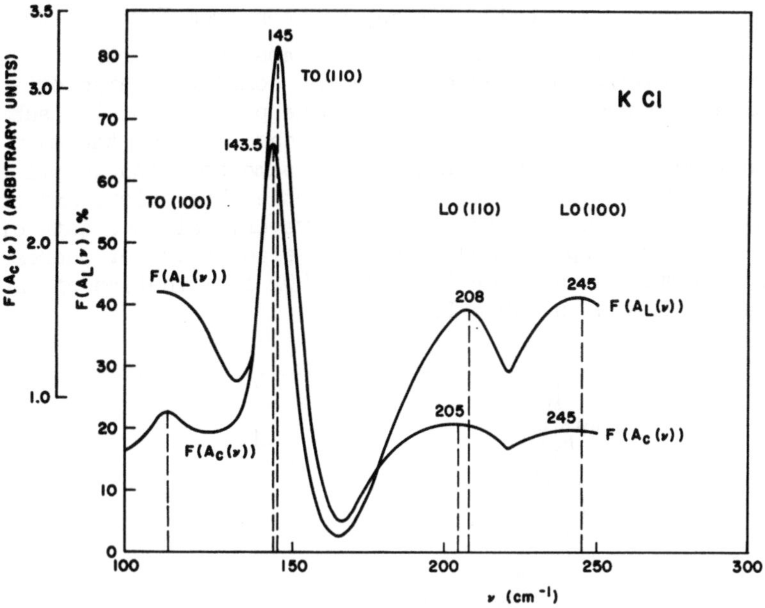

Figure 8. Energy absorption spectrum of KCl according to a film and crystal model, based on transmission data[1,10] and reflection data[3].

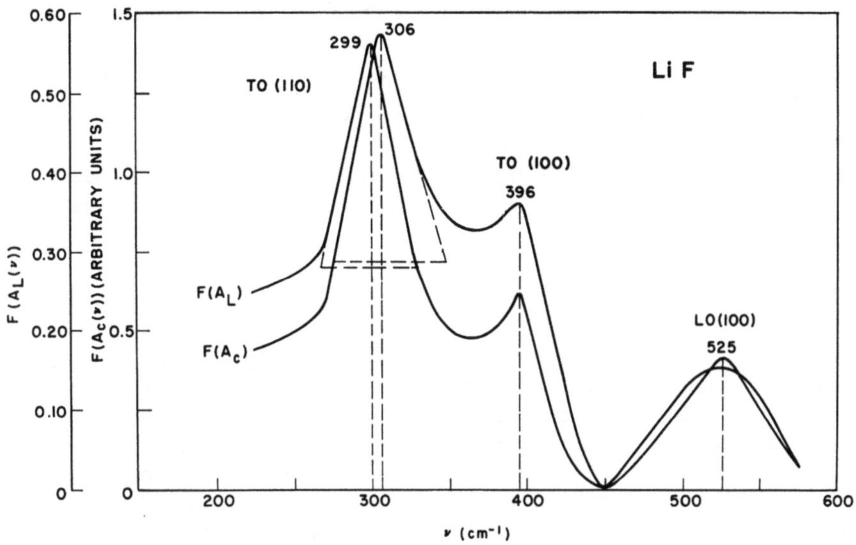

Figure 9. Energy absorption spectrum of LiF according to a film and crystal model, based on transmission data[11] and reflection data[12].

the spectrum of the absorption coefficient k was recently obtained by pole fit approximation[13]. As may be seen from Figs. 10 and 11, the k-analysis which only gives the main absorption frequency (here the (110) TO mode), in a way averages the complete characteristic absorption spectrum without indicating the position of the subsidiary fundamental frequencies. In the case of LiF, the main absorption frequency has approximately the same position in all three types of spectra: that of k, characteristic absorption, and calculated elastic. In the case of MgO, however, only the data of characteristic absorption and calculated elastic agree, whereas the k spectrum gives a lower frequency (402 cm^{-1} instead of 420cm^{-1}). All data are listed in Table I.

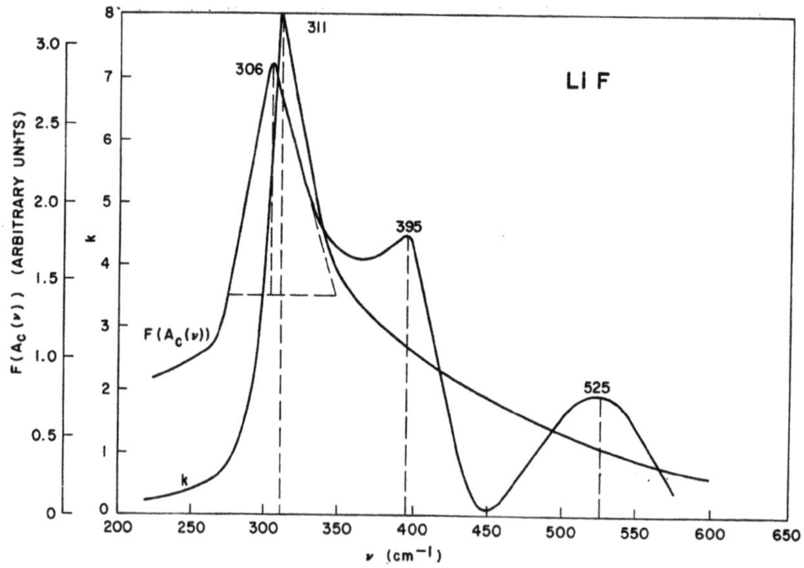

Figure 10. Comparison of energy absorption spectrum and k-spectrum for LiF, based on k-data[13]; references for other data see Fig. 9.

PARTICLE SIZE EFFECT IN POWDER TRANSMISSION SPECTRA

The characteristic absorption spectrum of MgO shown in Fig. 11 indicates insufficient resolution of the various modes of vibration. Although no further data on MgO is available, the main features may be examined, using the very similar characteristic absorption spectrum of MnO shown in Figure 12. Here, two spectra have been evaluated from two transmission curves obtained with different particle sizes. The higher resolution was achieved using a much finer powder than that of the poorer solution. Otherwise, the two absorption spectra are identical, except for the damping parameter.

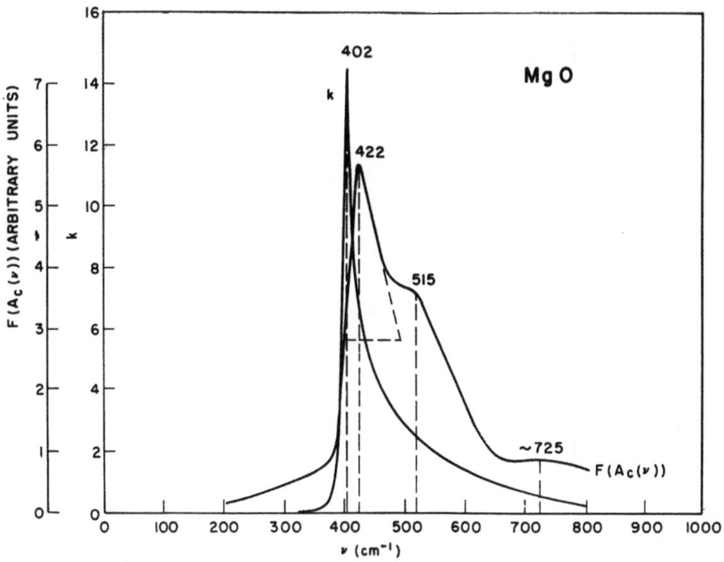

Figure 11. Comparison of energy absorption spectrum and k-spectrum for MgO, based on transmission[14], reflection and k-data[13].

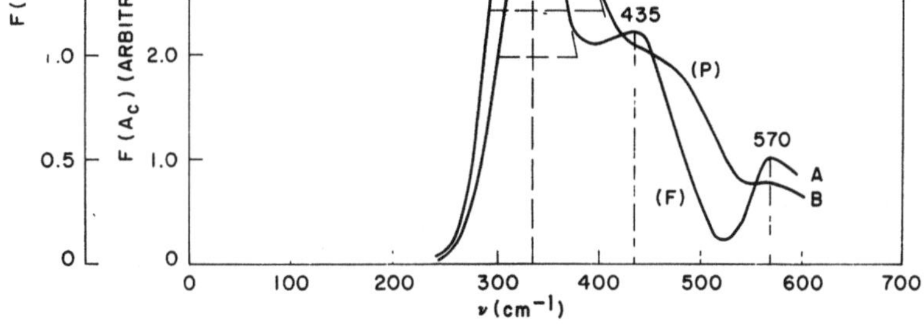

Figure 12. Influence of particle size on characteristic energy absorption spectra for MnO, based on transmission[14,15] and reflection data[16]. Curve A (very fine) and curve B (fine particle) size. Exact size not known.

TABLE I

Characteristic Absorption Frequencies of 15 Solids, as Calculated from Elastic Constants and as Determined from Characteristic Energy Absorption Spectra and/or Cold Neutron Scattering Data

No.	Solid[9]	Structure	ρ	r_o	C_{11}	C_{12}	C_{44}	$\nu_{TO(110)}$	$\nu_{TO(100)}$	$\nu_{LO(100)}$	$\nu_{LO(110)}$	$\nu_{LO(111)}$	ν_{ctr}
			g/cm^3	10^{-8} cm	10^{11} dynes/cm^2			cm^{-1}	cm^{-1}	cm^{-1}	cm^{-1}	cm^{-1}	cm^{-1}
1	MgO	[NaCℓ]	3.59	2.10	28.9	8.80	15.5	420 422	522 515	715 725	765 750	796	520 513
2	LiF	[NaCℓ]	2.64	2.01	11.77	4.33	6.28	310 (303- 307*) 306	405 396	555 525	612	632 640* 655	421 415
3	NaF	[NaCℓ]	2.80	2.31	9.70	2.50	2.80	257 250* 249	229	425 425*	406 400	402 400	302 290
4	NaCℓ	[NaCℓ]	2.16	2.815	4.68	1.23	1.19	167 169* 166.5	139 139	276 290	258 248	253 248	196 190
5	NaBr	[NaCℓ]	3.20	2.98	4.96	1.24	1.29	135 136.5	112 115	220	211	222	156 156
6	NaI	[NaCℓ]	3.67	3.23	3.80	0.70	0.70	112 122* 118	75	175 170*	155 155*	147	119 120
7	KCℓ	[NaCℓ]	1.99	3.14	3.69	0.81	0.79	143 145	106 114	228 245	208 208	200 208	161 162
8	KBr	[NaCℓ]	2.75	3.295	3.49	0.58	0.51	117 121* 118	69 60-70	181 180	155 164* 160	144	120 119
9	KI	[NaCℓ]	3.13	3.525	3.32	0.58	0.62	99 100.5	52 50	154 160	121 125	129 130	103 102

TABLE I (Cont.)

No.	Solid[9]	Structure	ρ	r_o	C_{11}	C_{12}	C_{44}	$\nu_{TO(110)}$	$\nu_{TO(100)}$	$\nu_{LO(100)}$	$\nu_{LO(110)}$	$\nu_{LO(111)}$	ν_{ctr}
			g/cm^3	10^{-8} cm	10^{11} dynes/cm^2			cm^{-1}	cm^{-1}	cm^{-1}	cm^{-1}	cm^{-1}	cm^{-1}
10	RbCℓ	[NaCℓ]	2.80	3.27	3.65	0.61	0.48	121 120.5	66.5 <75	184 >160	156 160	145	123 126
11	RbBr	[NaCℓ]	3.35	3.425	3.10	0.49	0.38	96 92.5	52 <60	148 145	124 125	115 115	98 96
12	RbI	[NaCℓ]	3.55	3.665	2.54	0.41	0.28	78.5 77.5	40 40	121 120	101 100	94 95	81 83
13	CsBr	[CsCℓ]	4.40	3.715	3.07	0.80	0.72	71.5 74.5	58 <60	118 125	110 105	107 105	83.5 84.5
14	CsI	[CsCℓ]	4.50	3.95	2.46	0.65	0.63	60 64.2	50 <50	99 100	93 91	110	70 69
15	TLBr	[CsCℓ]	7.55	3.45	3.73	1.50	0.75	58 53	48 53	107 112	85 85	55 53	84 85.5

Starred values from cold neutron data

Underlined values from I.R. experimental data

SPECTRA OF CHARACTERISTIC ABSORPTION, COLD NEUTRON SCATTERING AND EMISSION

Experimental spectra obtained from cold neutron scattering techniques are available for a number of dielectrics. Hence, a comparison is possible with characteristic absorption spectra, emission spectra and spectral data calculated from elastic constants.

(1) Two different samples of NaI crystal (considered to be of equal orientation) gave two somewhat different cold neutron scattering spectra at 100°K (Fig. 13)[17]. In one sample, the TO mode and the LO mode have almost equal strength, whereas in the other, the TO mode has approximately twice the LO mode intensity. Calculations from elastic constants show that the two TO modes (122 and 124 cm^{-1}) belong to the (110) polarization, whereas the two LO modes belong to two different polarizations: specifically, the weaker mode at 155 cm^{-1} belongs to (110) and the stronger one 170 cm^{-1} to (100), see Table I. The shape of the TO_{110} mode for cold neutron scattering (122 cm^{-1}) is identical to the shape of the TO_{110} mode for IR energy absorption (118 cm^{-1}); the small frequency difference being caused by the temperature difference of the samples.

(2) An experimental cold neutron spectrum of the TO and LO modes of LiF is compared with the complete IR energy absorption spectrum in Fig. 14 using different orientations of the crystal for each mode[19]. The absorption curves of the TO_{110} mode and the LO_{111} mode agree completely in shape, as well as with regard to frequency.

(3) An experimental cold neutron spectrum of the TO and LO modes of KBr (90°K)[17] is compared with the IR energy absorption spectrum in Fig. 15. The neutron spectrum gives only the polarized TO_{110} and LO_{110} modes, which are almost identical to the same modes of the IR energy absorption spectrum, except for a small frequency shift due to the difference in temperatures. In addition, Fig 15 shows an emission spectrum of the TO_{110} mode taken at 339°K[20], which is identical in shape to the TO_{110} absorption. The frequency shift is due to the increase in temperature for the emission spectrum. Fig. 15 also shows the spectrum of $(1 - R_B)$, where R_B is the reflectivity obtained from a KBr film evaporated on a brass plate using polarized light incident at 45°[21]. In this case, the phase relations are such that the TO mode is substantially suppressed. This is due to the metallic reflection of the substratum so that the ratio of intensity between the TO and LO modes appears inverted[18].

(4) The frequency data obtained above (1 to 3) and additional data from cold neutron measurements on NaF[22] and NaCℓ[23], are compared with the frequency data obtained from IR energy absorption

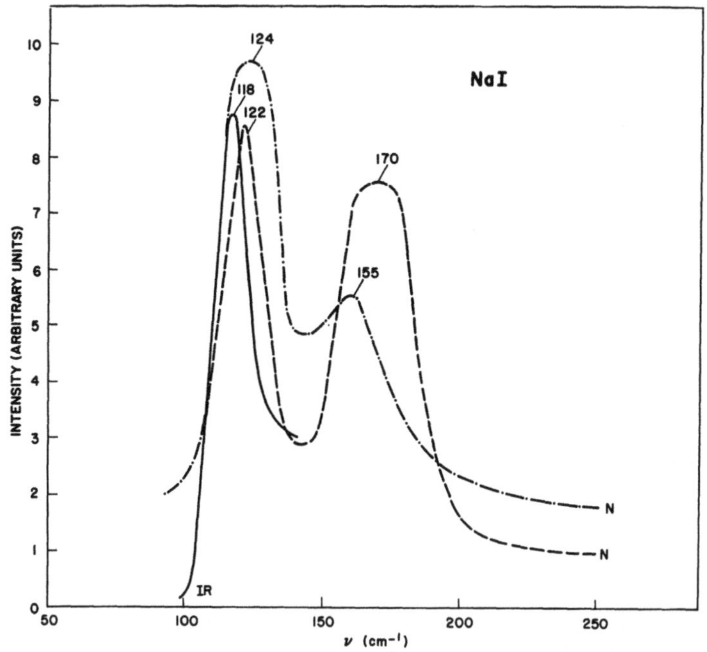

Figure 13. Comparison of cold neutron spectra for two orientations of NaI and its energy absorption spectrum, based on transmission and reflection data[10], and cold neutron data[17].

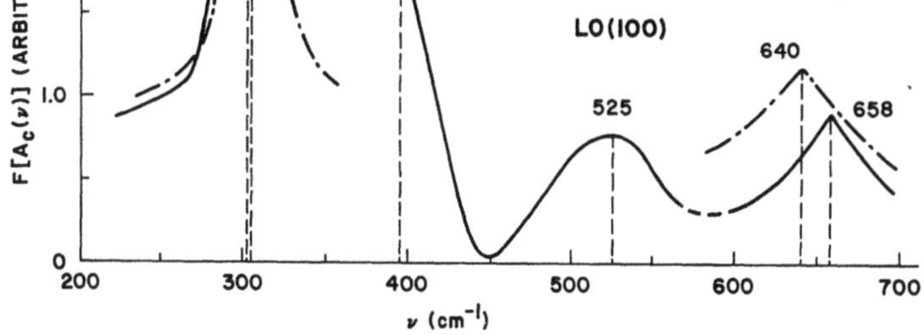

Figure 14. Comparison of cold neutron spectra and the energy absorption spectrum for LiF, the high frequency extension of which is based on transmission[18] and reflection data[12]; cold neutron data[19].

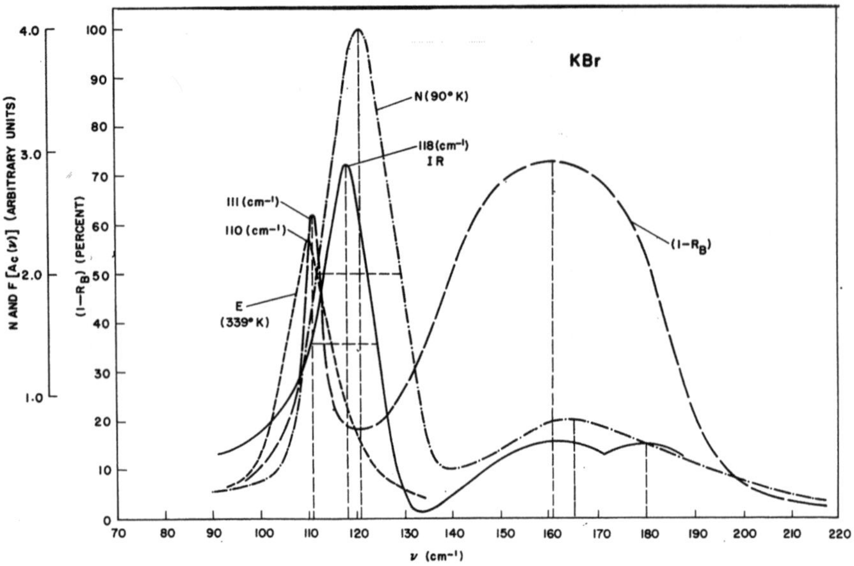

Figure 15. Spectra of emission (E), cold neutron (N) and special reflectivity ($1\text{-}R_B$) of KBr, as compared to the energy absorption spectrum, based on transmission[10], reflection[3], cold neutron[17], emission[20] and special reflectivity data[21].

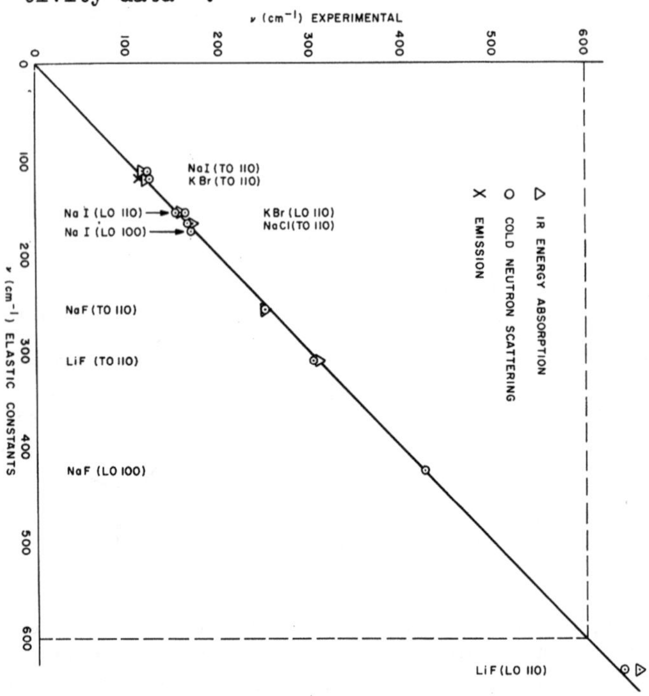

Figure 16. Equality between fundamental frequencies from three types of experimental spectra and those calculated from elastic constants data

and emission, and those calculated from elastic constants in Fig. 16. The 45° slope indicates a virtual identity between the calculated and experimental data.

SPECTRA OF CHARACTERISTIC ABSORPTION AND LASER-RAMAN SCATTERING

Laser-Raman spectra are available for some solids of tetrahedral structure of which characteristic absorption spectra have been evaluated by the methods advanced in this paper. These solids may be predominantly covalent or have both an ionic and a covalent contribution to the total bond strength. In comparing the two types of spectra, the following is found: (1) Solids considered to be almost entirely covalent [like BP (830 cm^{-1}, see Fig. 17) and quartz (466 cm^{-1}, see Fig. 18) for vibrations between the tetrahedrons], show identical absorption curves for the two types of spectra, with the same frequency and shape and the same damping parameter, $\Delta\nu/\nu_0$; (2) Solids considered to have partially ionic and partially covalent binding [like ZnS (350 cm^{-1}, see Fig. 19), or SiC (at 795 cm^{-1}, see Fig. 20), or quartz (close to 1100 cm^{-1}, see Fig. 21) for vibrations within the tetrahedrons], show the same frequency but a considerable difference in the damping parameters of the two types of spectra (see Tables II and III below, where this phenomenon is treated in detail).

CHARACTERISTIC ABSORPTION SPECTRA OF A SERIES OF RELATED SOLIDS

Characteristic energy absorption spectra of a variety of solids have been evaluated from IR data. The characteristic frequencies of many of them have been calculated from elastic constants by the methods advanced in this paper (see Table I). The spectra for some series of related solids are presented in Figs. 22 through 32 in arbitrary units of $F[A_c(\nu)]$. Table I compares the experimental data of the absorption frequencies with the calculated data and gives the proper assignment of the frequencies for 15 solids of cubic structure for which the elastic constants were available. The agreement between experimental and calculated frequencies of the TO and LO modes and of the center frequency may be seen in Figs. 33-35.

An interesting feature common to all of these spectra, is the variation of the asymmetry of the TO mode. The asymmetry is due to the anharmonicity of the binding forces, as was pointed out earlier[45] (see Fig. 36). Tilting of the energy absorption curves towards lower frequencies indicates an anharmonic soft force characteristic, whereas tilting towards higher frequencies indicates an anharmonic hard force characteristic (Fig. 37). As may be readily seen from the spectra (Figs. 22-32), the first case applies to

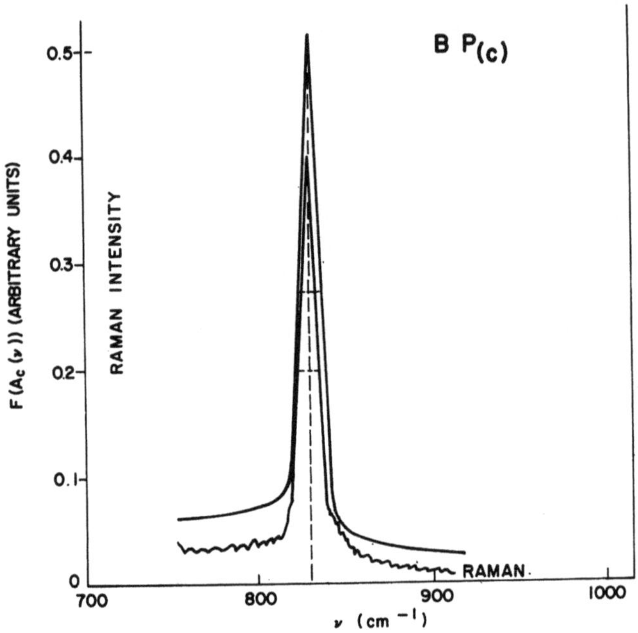

Figure 17. Laser Raman line and corresponding energy absorption line for cubic BP, based on transmission[24], reflection[25] and laser Raman data[26].

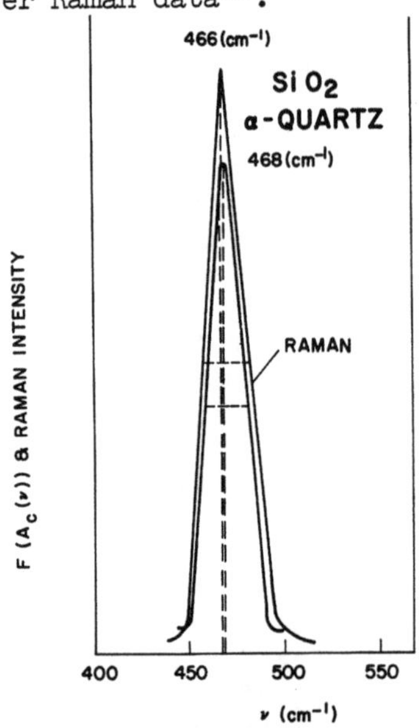

Figure 18. Laser Raman line and corresponding energy absorption line for α-quartz (vibrations between Si-O tetrahedra), based on transmission[15], reflection[27] and laser Raman data[28].

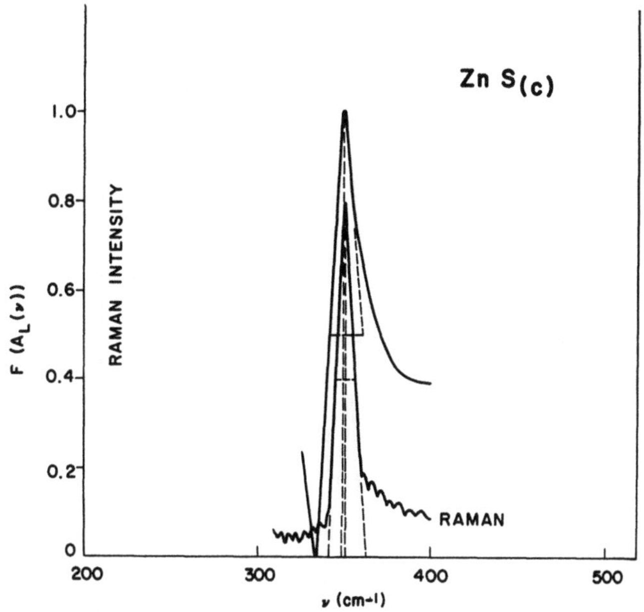

Figure 19. Laser Raman line and corresponding energy absorption line for ZnS (spalerite type), based on transmission[29], reflection[30] and laser Raman data[31].

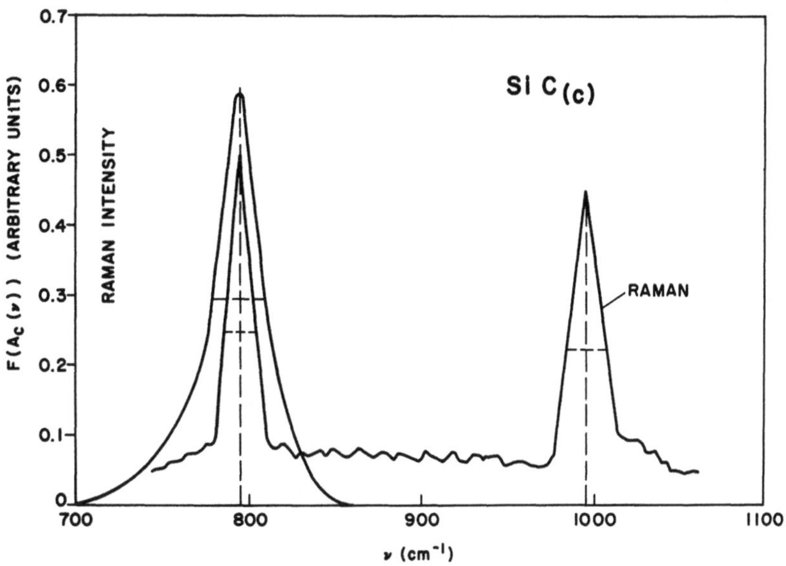

Figure 20. Laser Raman lines and corresponding energy absorption line for cubic SiC, based on transmission and reflection data[32], and laser Raman data.[33]

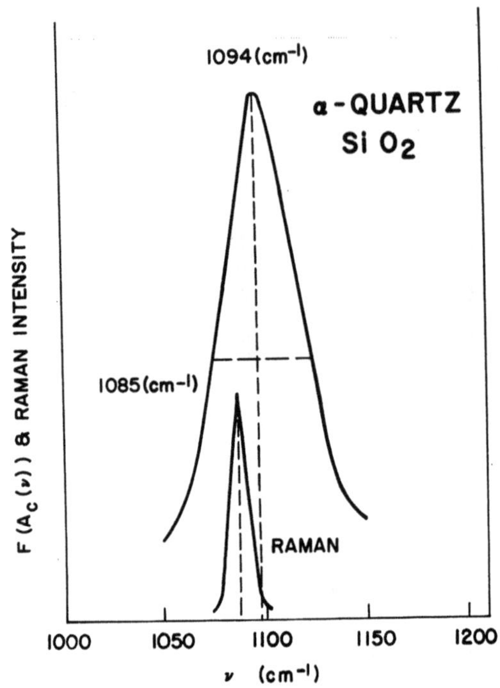

Figure 21. Laser Raman line and corresponding energy absorption line for α-quartz (vibrations within Si-O tetrahedra), based on transmission[15], reflection[27] and laser Raman data[34].

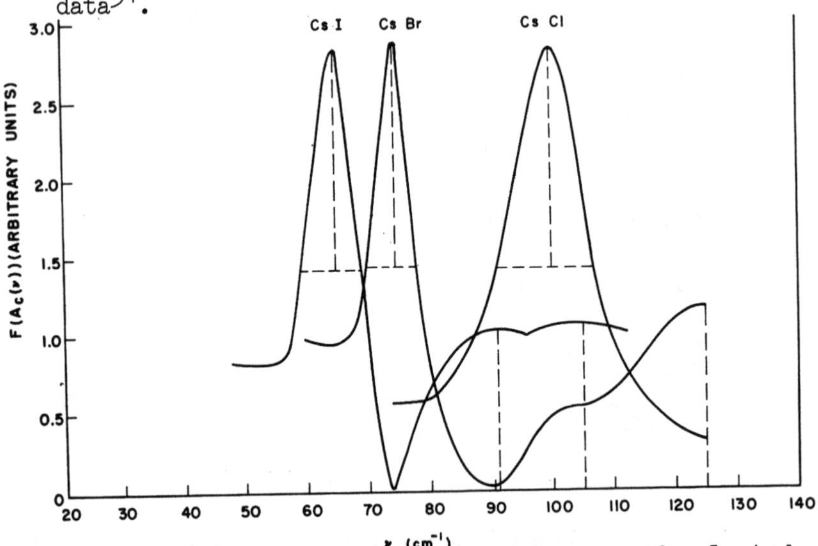

Figure 22. Characteristic energy absorption spectra of selected cesium salts, based on transmission and reflection data of CsI[14,33], CsBr[10], CsCl[10].

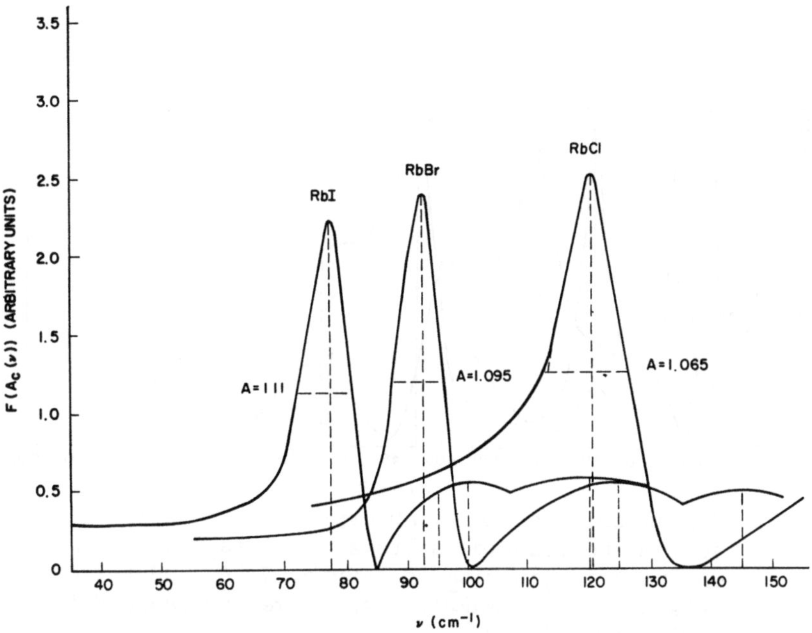

Figure 23. Characteristic energy absorption spectra of selected rubidium salts, based on transmission and reflection data of RbI[10], RbBr[10] and RbCl[10]. A-values refer to anharmonicity.

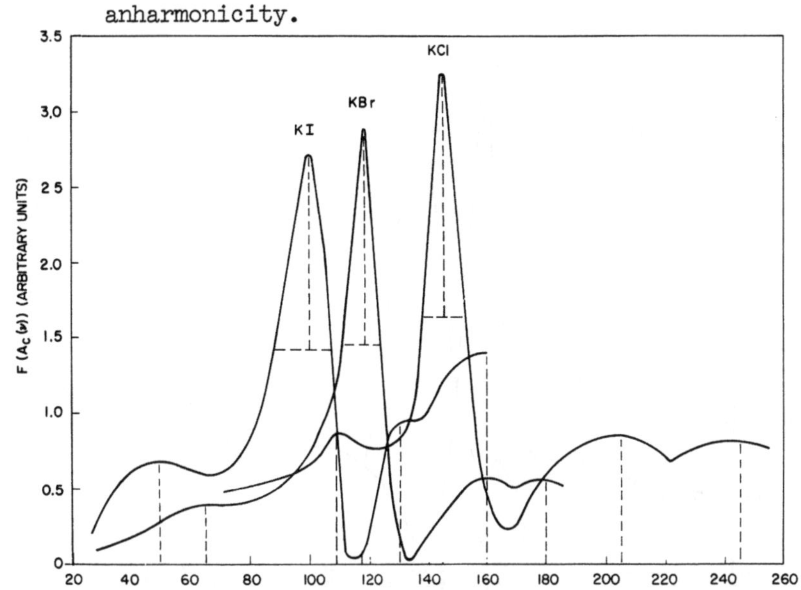

Figure 24. Characteristic energy absorption spectra of selected potassium salts, based on transmission data of KI[10], KBr[10], KCl[1,10] and reflection data of KI[10], KBr[3] and KCl[3]. Low frequency evaluation of the three salts based on special transmission and reflection data[36].

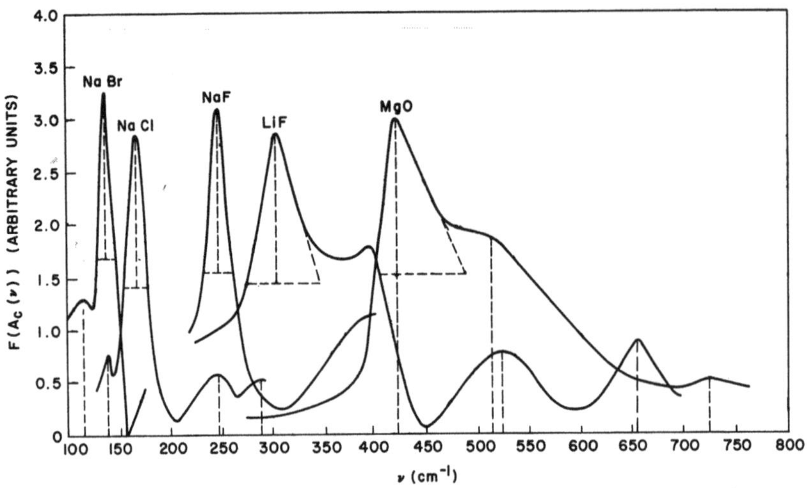

Figure 25. Characteristic energy absorption spectra of selected sodium and fluoride salts as well as that of MgO, based on transmission data of NaBr[10], NaF[11] and reflection data of NaBr[10], NaF[37]. See above for basic data on NaCl, LiF and MgO.

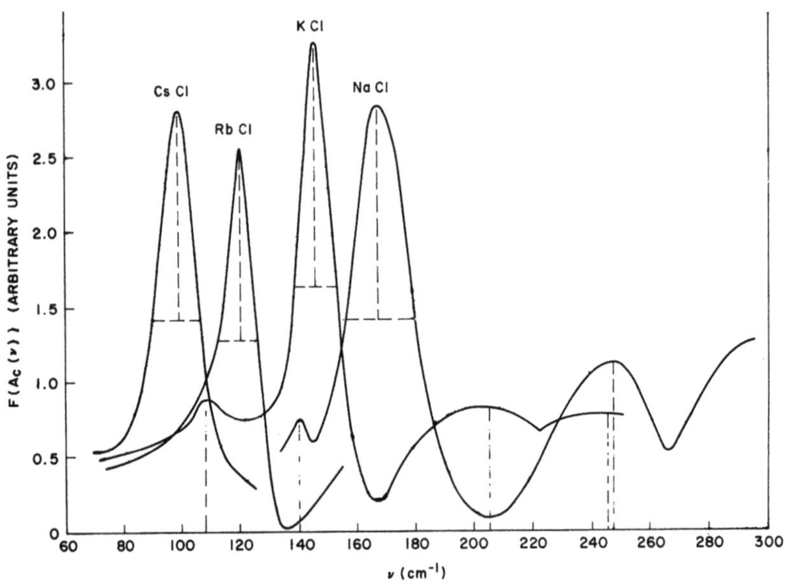

Figure 26. Characteristic energy absorption spectra comparing related chlorides. See above for reference on basic data.

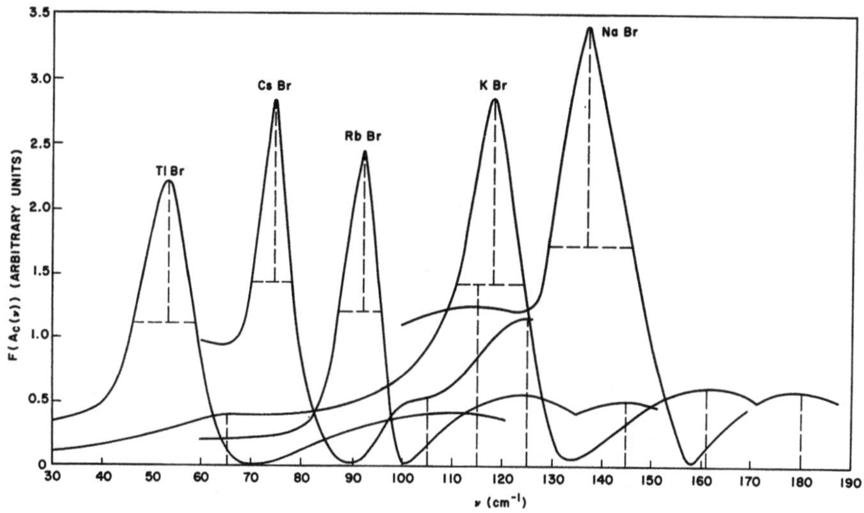

Figure 27. Characteristic energy absorption spectra comparing related bromides, based on transmission and reflection data of TlBr[10]. See above for references on basic data of the other bromides.

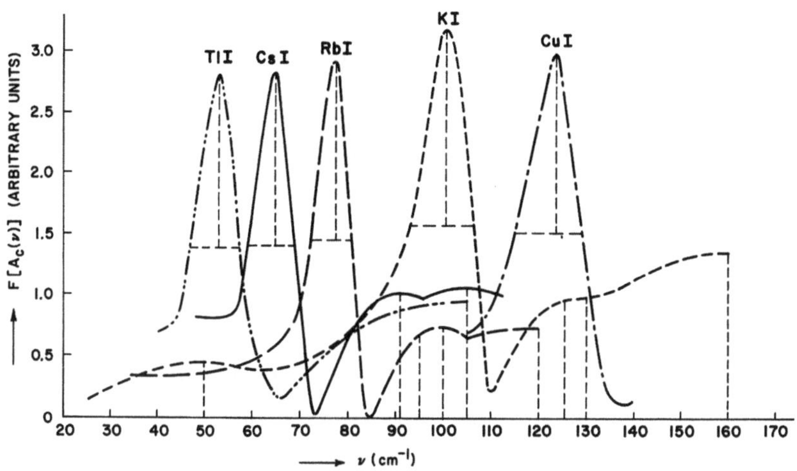

Figure 28. Characteristic energy absorption spectra comparing related iodides, based on transmission and reflection data of TlI[10] and CuI[38]. See above for references on basic data of the other iodides.

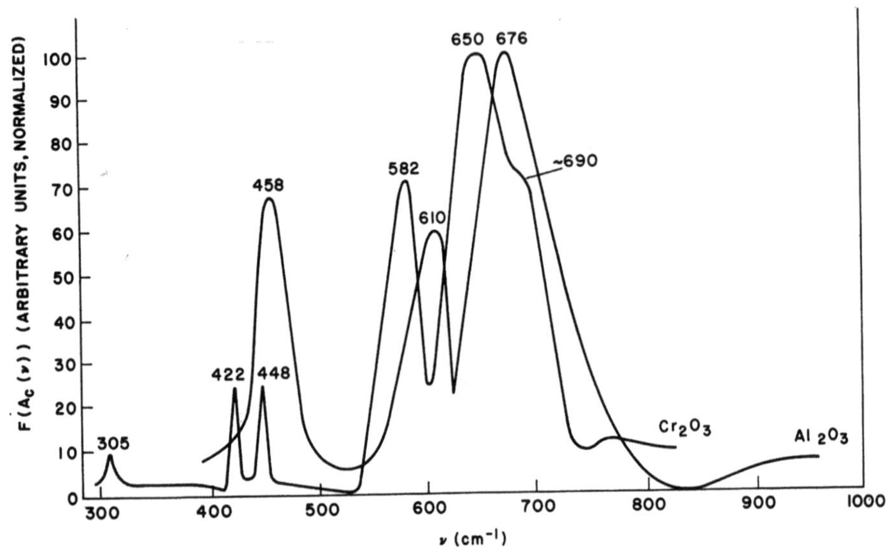

Figure 29. Characteristic energy absorption spectra of Al_2O_3 and Cr_2O_3, based on transmission data of Al_2O_3 and Cr_2O_3[39] and reflection data of Al_2O_3[40] and Cr_2O_3[41]. Both spectra for E⊥C-axis; numbers refer to mode frequencies.

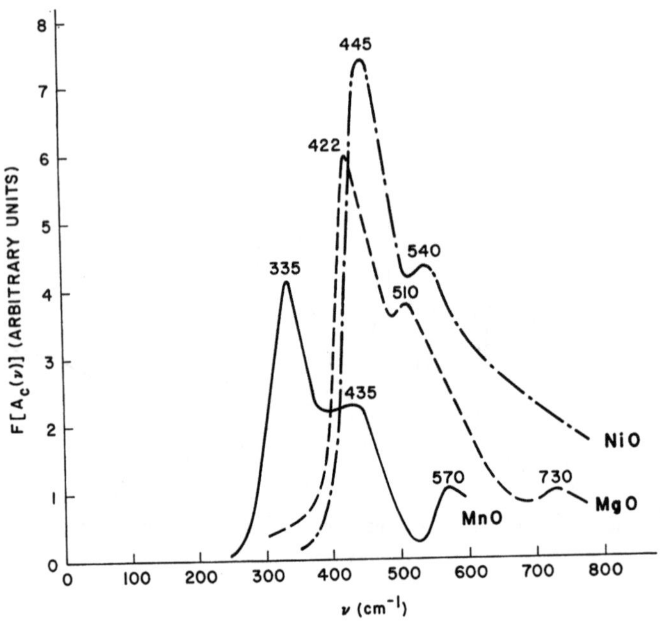

Figure 30. Characteristic energy absorption spectra of selected metal oxides, based on transmission data of NiO[15] and reflection data of NiO;[42] see above for references on basic data of the other oxides.

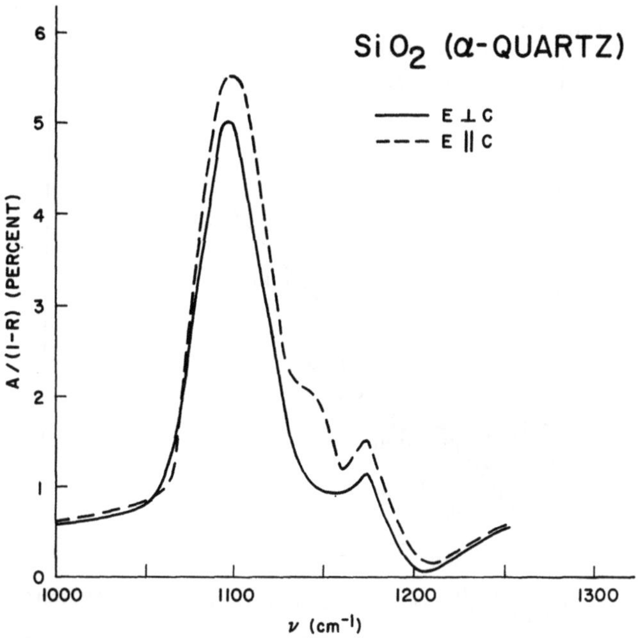

Figure 31. Portion of characteristic energy absorption spectrum of α-quartz for vibrations <u>within</u> the Si-O tetrahedra for E⊥C and E∥C-axis, based on transmission[43] and reflection data[27].

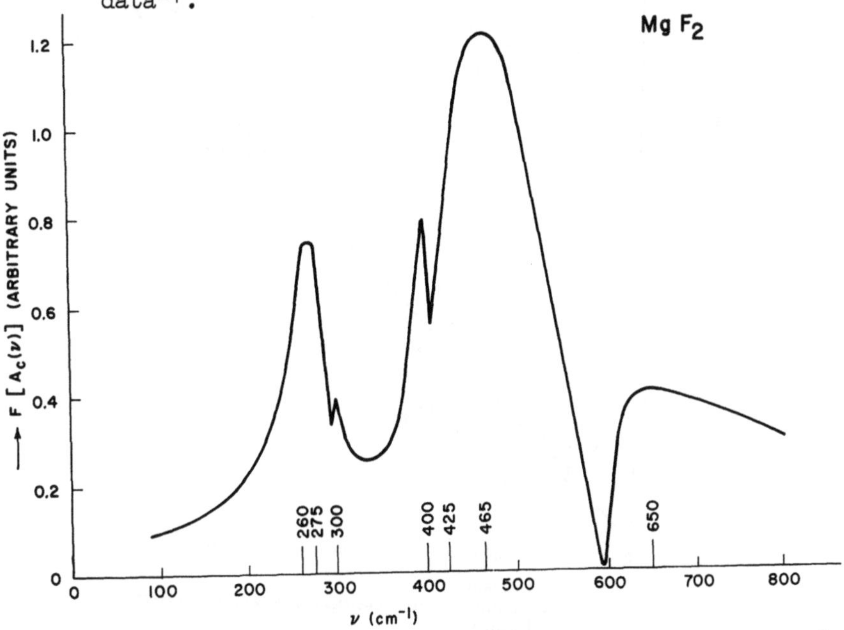

Figure 32. Characteristic energy absorption spectrum of MgF_2 (rutile structure), based on transmission and reflection data[44]. Numerical values refer to the triplets of the two TO modes which result from three different lengths of the deformed octahedral Mg-F edges in the crystalline structure.

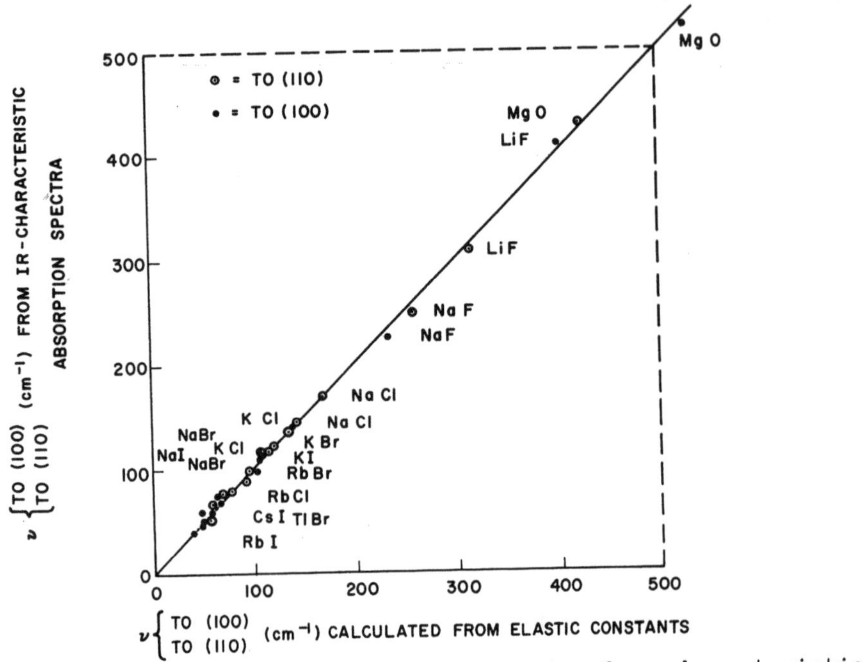

Figure 33. Equality between TO mode frequencies from characteristic energy absorption spectra and those calculated from respective elastic constants.

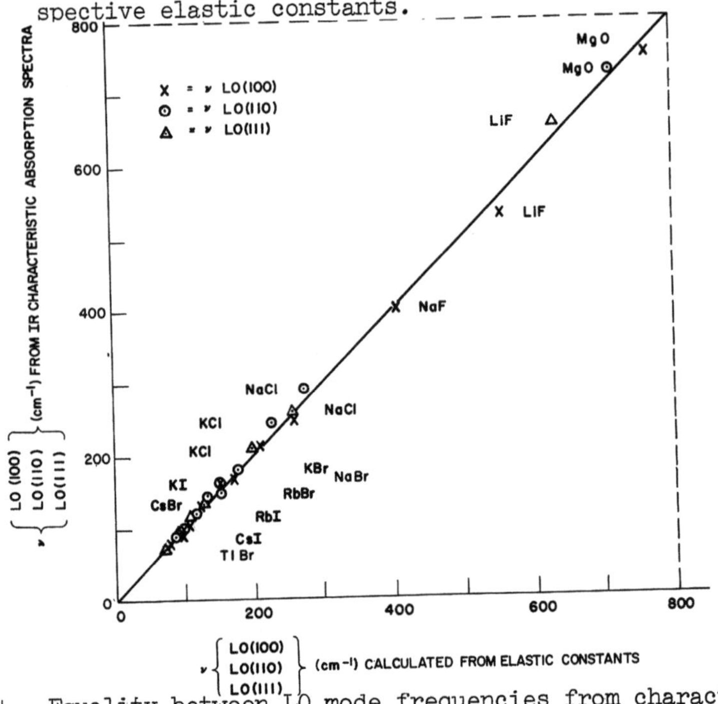

Figure 34. Equality between LO mode frequencies from characteristic energy absorption spectra and those calculated from respective elastic constants.

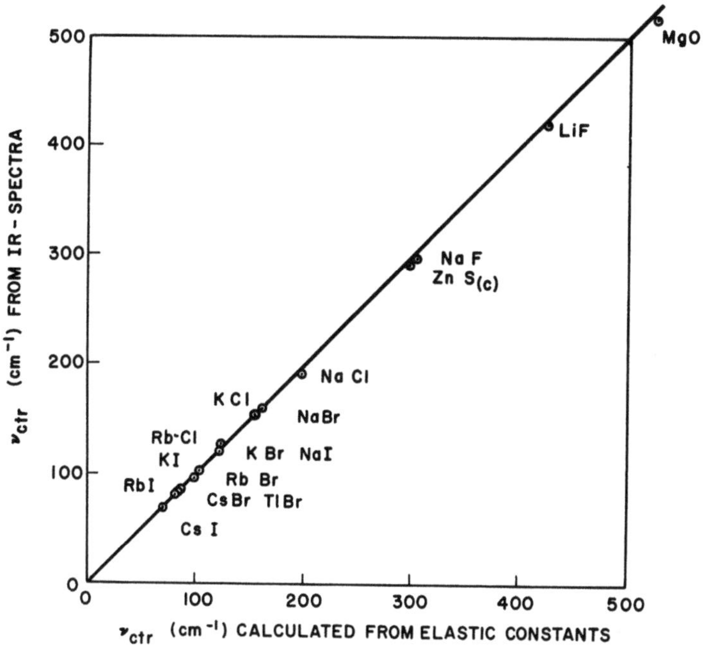

Figure 35. Equality between center-frequencies from IR-spectra (reflectivity and absorption)[45] and those calculated from respective elastic constants.

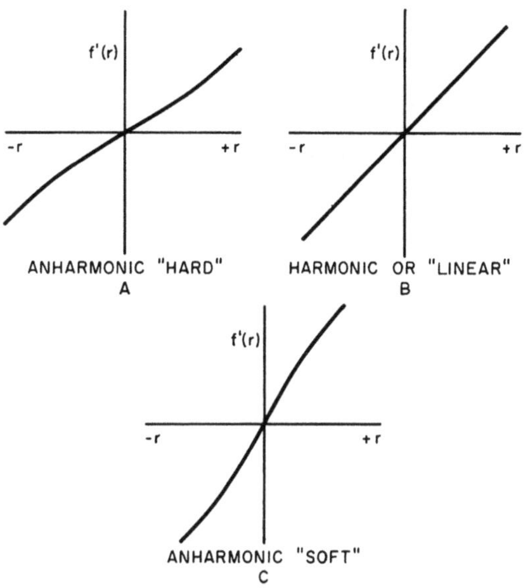

Figure 36. Schematic presentation of different types of anharmonic force characteristics.[45]

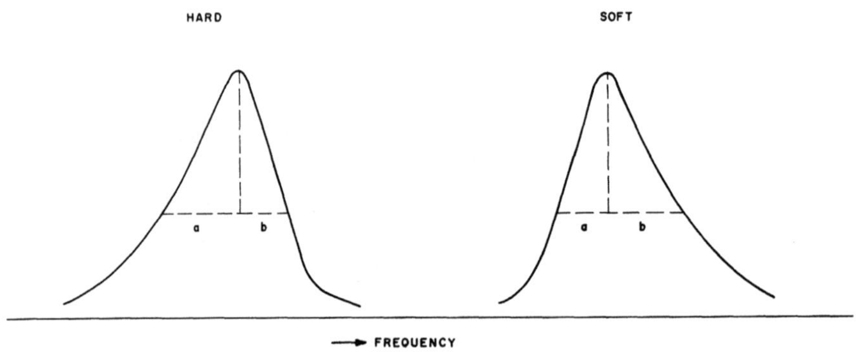

Figure 37. Differentiation between anharmonic soft and hard force characteristic in the energy absorption curve.[45] For a and b values see text.

all solids having an electron density smaller than that of the argon configuration of an ion pair, whereas the second case applies to solids having an electron density larger than that of the argon configuration. The transition between the two is a quasi-harmonic (linear) force characteristic, which is present in the $KC\ell$ absorption curve, since it represents the argon configuration of an ion pair (see Figs. 36 and 40). From the asymmetry of the absorption curves, an empirical formula may be derived for a factor of anharmonicity[45]

$$\Psi(A) = 1 + \frac{1}{\pi}\left(1 - \frac{b}{a}\right) \qquad (14)$$

where a and b are defined by Fig. 37 and $\Psi(A)$ is the anharmonicity factor.

When the maximum value of the entire vibrational energy is equated to the total cohesive energy, it follows that

$$\tfrac{1}{2} m_r \Sigma v_{ij}^2 = U_o/N \qquad (15)$$

where $v_{ij} = \sqrt{C_{ij}/\rho}$. This relation should be strictly satisfied in the case of harmonic vibrations (for example; for $KC\ell$), but deviates from equality for solids having anharmonic binding forces[45] (Plendl 1961). Using here the maximum values of the elastic constants of $KC\ell$ (measured at very low temperature, e.g. $C_{11} = 4.83$, $C_{12} = 0.63$, $C_{44} = 0.66$ in units of (10^{11} dynes/cm^2), and $\rho = 2.04$ (g/cm^3)[9], the maximum energy of the six vibrational modes defined by Eqs. (7) through (12) is obtained as

$$\tfrac{1}{2} m_r \Sigma v_{ij}^2 = 11.75 \times 10^{-12} \text{ ergs/molecule}$$

This may be compared to the total cohesive energy

$$U_o/N = (1/N)\ 169\ (\text{kcal/mole}) = 11.7 \times 10^{-12} \text{ ergs/molecule.}$$

The equality of the two data indicates that the frequencies of the six vibrational modes fully account for the entirety of the active vibrational spectrum of the solids of $NaC\ell$ structure.

Determination of the characteristic frequencies of the various modes of vibration permits an examination of their relationship to the interatomic distance. An example is shown in Fig. 38 where the frequencies of the two TO modes, (110) and (100), of 22 solids are plotted versus the intgratomic distance. Considering distances greater than 2.2×10^{-8}cm, it is clear that the frequency of the TO_{110} mode is higher and much more intense than the frequency of the TO_{100} mode. However, a rapid change occurs near the atomic distance of NaF. Here the TO_{100} mode assumes higher frequency values: It also shows a substantial increase in intensity and rate of growth. This is also indicated in Fig. 39 which shows the corresponding behavior of the two shear moduli c_{44} and $\frac{c_{11}-c_{12}}{2}$. The same phenomenon has been found with the two LO modes (100) and (110),

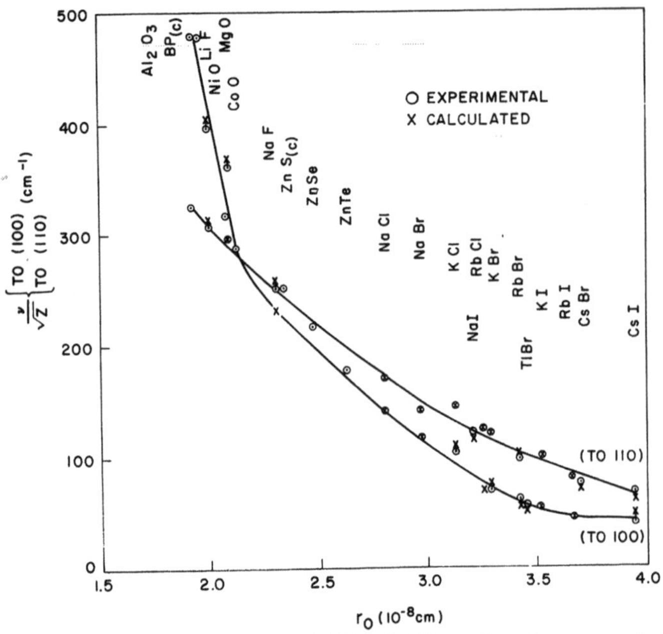

Figure 38. Relation between modified TO mode frequencies and interatomic distances of solids. The Z parameter refers to the valency of the compounds.

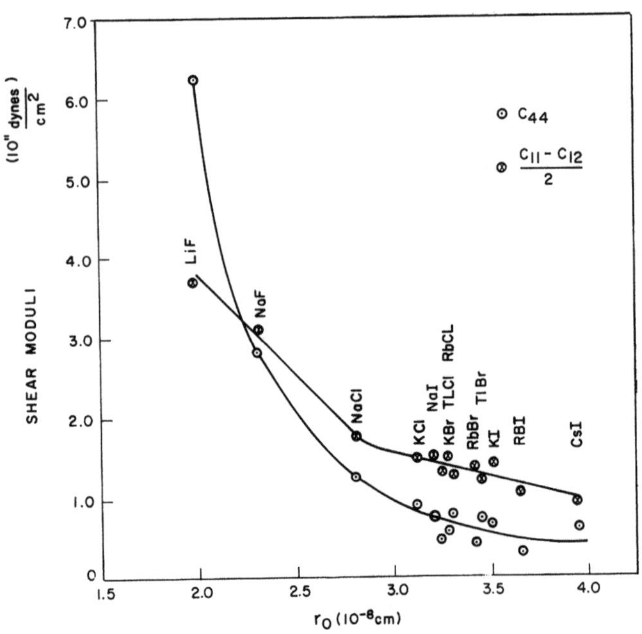

Figure 39. Relationship between the shear moduli and interatomic distances of solids, corresponding to the relation of Fig. 38.

where the change occurs in the opposite direction and the LO_{110} mode becomes stronger than the LO_{100} mode when the atomic distance becomes smaller than 2.2×10^{-8} cm. The change-over in strength occurs at the same atomic distance (neon configuration) as the cross-over from "soft-core" to "hard-core" solids in the hardness of crystalline materials[46,47]. As was pointed out in this study on hardness, solids having electron densities lower than the neon configuration display superior strength.

THE DAMPING OF LATTICE VIBRATIONS*

The damping of lattice vibrations is substantially caused by collisions between the atoms during vibrations. Hence it may be treated like damping in the kinetic theory of gases.

In any binary compound of simple structure, each atom pair is surrounded by a pair of coordination polyhedrons, each of which consists of the respective component atoms. Each atom represents the center of such a polyhedron and at the same time, one corner of the next neighbor polyhedron. Hence the number of similar atom pair collisions and those with next neighbor atoms are given by

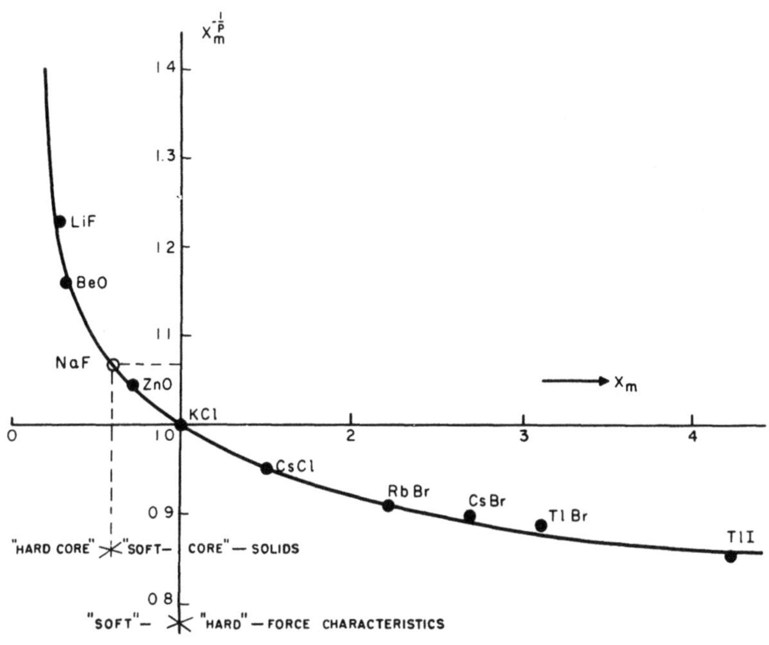

Figure 40. Differentiation between hard and soft core solids based on anharmonicity values ($X_m^{-1/p}$). Change-over takes place at NaF (neon configuration).[46,47]

*A review of pertinent literature may be found in Ref. 13.

$CN_a + CN_b - 1$, where CN_a and CN_b are the coordination numbers. For example: CN = 4, = 6, = 8, for the ZnS, NaCl and CsCl structures, respectively; but $CN_a = 4$ and $CN_b = 2$ for quartz. Proceeding from one atom pair to the next in any direction along which the propagation of energy takes place (as given by the vibrational mode which are taken into consideration), each atom successively belongs to two atom pairs, which it constitutes with its next neighbor atoms at each side. Hence, it is necessary to divide by 2 and so obtain 1/2 (2 CN-1) collision possibilities per atom pair.

The molecular volume of an atom pair is given by $M/\rho N$, where M = molecular weight, ρ = density and N = Avogadro's number. It is practical to refer to the interatomic distance r_o and put

$$M/\rho N = C_{str}\, r_o^{3} \quad (cm^3) \tag{15}$$

where C_{str} represents a characteristic constant of the crystalline structure.

The effective cross section of collision may simply be defined as

$$\pi \left(\frac{r_a + r_b}{2} \right)^2 = \frac{\pi}{4} r_o^{2} \quad (cm^2) \tag{16}$$

where r_a and r_b represent either the covalent or the ionic radii of the two component atoms of a binary compound, depending on which damping mode is to be examined.

The average velocity of the colliding atoms during vibration is defined by the frequency of the vibrational mode, according to Eqs. (7) through (12). For example,

$$V_{TO} = 2\, r_o\, \nu_{TO} \quad (cm/sec) \tag{17}$$

On the basis of these parameters, the collision frequency of an atom pair may now be determined for a particular mode of vibration as

$$Z = \frac{(\frac{1}{2})(2\ CN - 1)}{C_{str}\, r_o^{3}} \left(\frac{\pi}{4} r_o^{2} \right) (2\, r_o\, \nu_{TO}) = \frac{\pi}{4} \frac{2\ CN-1}{C_{str}} \nu_{TO} (1/sec) \tag{18}$$

where $\frac{2\ CN-1}{C_{str}}$ may be conceived as a single parameter which is characteristic for each crystalline structure (see below).

The damping besides being a function of the collision frequency, is also a function of two additional parameters, namely the atomic size factor[48] and the anharmonicity of the binding forces. First, the atomic size factor r_b/r_a may be defined in such a way that r_a represents the radius of the smaller, electro-positive metal ion and r_b the radius of the electro-negative. In the case of covalent binding, the metal atom generally has the larger radius. However, the distinction between metal and non-metal is not always clearly defined. Hence, in the case of covalent binding, we consider r_b as the smaller, radius and r_a the larger radius in order to remain consistent.

As to the anharmonicity of the force constant, the energy term of the factor $\Psi(A)$ defined above may be used (see Eq. (14)). Here one must distinguish between the two types of anharmonicity, corresponding to either a hard or a soft anharmonic force characteristic.[45] The latter stands for a mutual interpenetration of the outer electron shells (overlapping of the ionic radii) corresponding to an inelastic collision. With the hard, anharmonic force characteristic, no overlapping of the ionic radii takes place. Hence we may assume that the anharmonicity of the force constant only plays a major role in damping with a soft anharmonic force characteristic defined by $\Psi(A) \leq$ unity. For this reason, we may consider only values of $\Psi(A) <$ unity, and may set $\Psi(A) =$ unity, when $\Psi(A) \geq$ unity.

Introducing these two additional parameters into the equation for the damping function, together with the collision frequency Z, leads to

$$\gamma = Z\,(r_b/r_a)\,(\pi\,\Psi(A))^{-2} \tag{19}$$

Combining Eqs. (18) and (19) and introducing $\Delta\nu = \gamma/2\pi$, the damping parameter may be calculated as

$$\Delta\,\nu/\nu_{TO} = (1/8\,\pi^2)\,[(2\text{ CN-1})/C_{str}]\,(r_b/r_a)\,\Psi(A)^{-2} \tag{20}$$

which may be experimentally obtained as the resonance width at half intensity of the energy absorption curves. This damping equation of lattice vibrations applies to binary solids of simple structures, whether they have the ionic or the covalent binding, using the ionic or the covalent radii, respectively. In the case of covalency $\Psi(A)$ equals unity.

The characteristic constant C_{str} of the crystalline structure is a function of the coordination number, as is the characteristic parameter $(2\text{ CN-1})/C_{str}$. They assume the following values:

CN = 2:4 (quartz),	C_{str} = 4, and	$(2\ CN-1)/C_{str}$ = 1.20
= 4:4 (ZnS)	" = π	= 2.23
= 6:6 (NaCℓ)	" = 2	= 5.50
= 8:8 (CsCℓ)	" = $\pi/2$	= 9.55
= 3:6	= 2.6	= 3.10

The characteristic structure parameter, $(2\ CN-1)/C_{str}$, within limits, is shown to be proportional to CN^2.

The large variety of infrared energy absorption spectra of solids, so far evaluated by the methods advanced in this paper, permits a testing of the damping equation with respect to both the ionic and covalent binding. When the total damping is examined (measured as the band width at half intensity, of either the TO_{110} mode of the IR energy absorption spectra or of corresponding laser Raman lines), comparison reveals that; (1) the IR damping data of 24 solids agree with the damping data calculated by Eq. (20), using ionic radii (see Table II), and (2) that the Raman damping data of 15 solids agree with the damping data calculated by Eq. (20), using covalent atomic radii (see Table III).* The coordination polyhedrons of these solids are either tetrahedrons (zincblende, wurtzite or quartz-structure), or octahedrons (NaCℓ and rutile structure), or cubes (CsCℓ structure). The tetrahedral solids are generally considered to have mixed ionic and covalent binding components. Hence they may exhibit two different damping data, with respect to their respective characteristic frequencies. This is shown for 6 solids for which the two damping data are experimentally available: BeO, ZnS, ZnSe, ZnTe, CdS, SiO_2 (Table II and III). They are in good agreement with the damping data obtained by calculation based on either the ionic or the covalent radii, respectively.

The damping data of Tables II and III have been plotted in Figures 41 and 42, where the 45° slopes stand for equality of data. As may be seen, experimental and calculated damping data agree excellently, considering the fact that the experimental damping data, as obtained from different measurements of the same kind of solids, may sometimes differ by ± 10 percent in the case of large damping values and ± 20 percent in the case of small damping values.

*Tables II and III cover many solids of which characteristic absorption spectra have not been shown here. They will be treated separately in future publications.

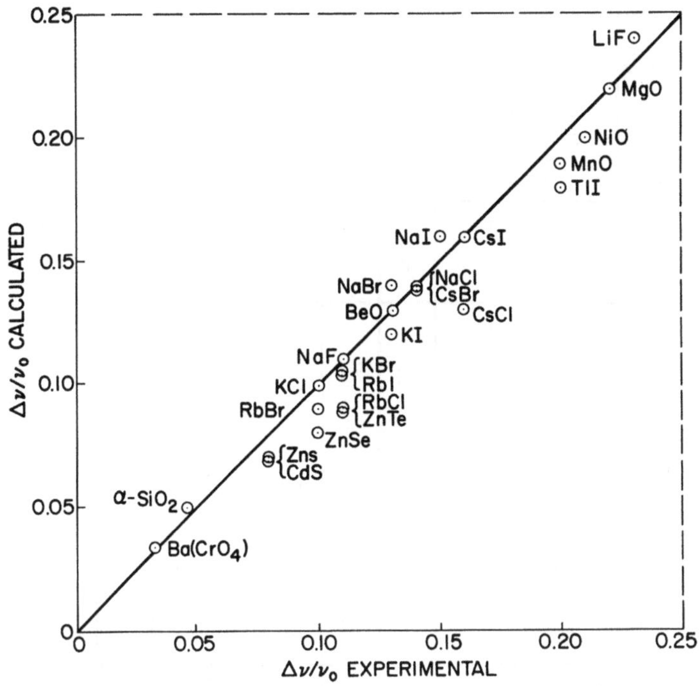

Figure 41. Equality between damping values obtained from IR energy absorption spectra and calculated from ionic radii.

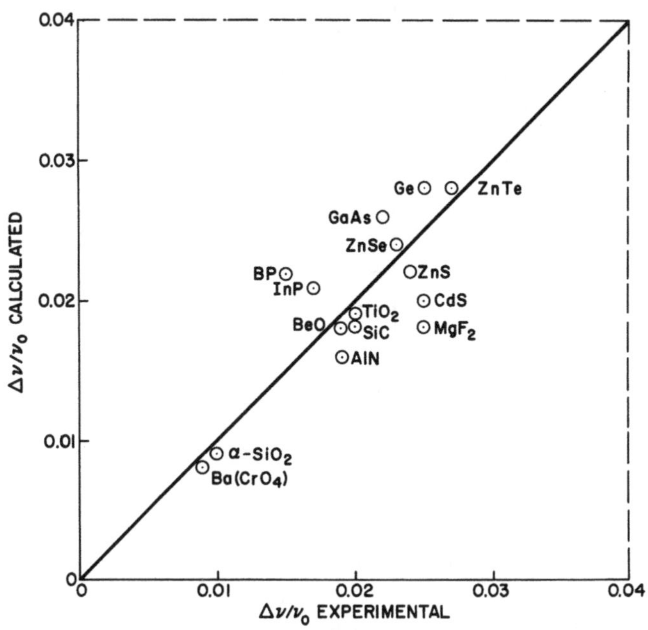

Figure 42. Equality between damping values obtained from laser Raman lines and calculated from covalent radii.

TABLE II

Damping Data of the Main TO Modes of 24 Solids as Determined from Characteristic Energy Absorption Spectra and Calculated for Ionic Binding (see Text).

No.	Substance	Structure	CN	C_{str}	r_o	r_a	r_b	r_b/r_a	Ψ(A)	$\Delta\nu/\nu_{TO}$ Calculated	ν_{TO}	$\Delta\nu/\nu_{TO}$ Experimental
					10^{-8} cm							
1	MgO	[NaCℓ]	6:6	2	2.10	0.66	1.40	2.12	0.83	0.22	422	0.22
2	NiO	[NaCℓ]	6:6	2	2.085	0.69	1.40	2.02	0.85	0.20	448	0.21
3	MnO	[NaCℓ]	6:6	2	2.215	0.70	1.40	2.0	0.87	0.19	335	0.20
4	LiF	[NaCℓ]	6:6	2	2.01	0.60	1.36	2.27	0.81	0.24	306	0.23
5	NaF	[NaCℓ]	6:6	2	2.31	0.95	1.36	1.43	0.95	0.11	249	0.11
6	NaCℓ	[NaCℓ]	6:6	2	2.815	0.95	1.81	1.90	0.97	0.14	166.5	0.14
7	NaBr	[NaCℓ]	6:6	2	2.98	0.95	1.95	2.04	0.99	0.14	137	0.13
8	NaI	[NaCℓ]	6:6	2	3.23	0.95	2.20	2.32	>1.0	0.16	116	0.15
9	KCℓ	[NaCℓ]	6:6	2	3.14	1.33	1.81	1.36	1.0	0.10	145	0.10
10	KBr	[NaCℓ]	6:6	2	3.295	1.33	1.95	1.47	>1.0	0.105	118	0.11
11	KI	[NaCℓ]	6:6	2	3.525	1.33	2.20	1.66	>1.0	0.12	100	0.13
12	RbCℓ	[NaCℓ]	6:6	2	3.27	1.48	1.81	1.23	>1.0	0.09	120.5	0.11

TABLE II (Cont.)

No.	Substance	Structure	CN	C_{str}	r_o (10^{-8} cm)	r_a (10^{-8} cm)	r_b (10^{-8} cm)	r_b/r_a	Ψ(A)	$\Delta\nu/\nu_{TO}$ Calculated	ν_{TO}	$\Delta\nu/\nu_{TO}$ Experimental
13	RbBr	[NaCℓ]	6:6	2	3.425	1.48	1.95	1.32	>1.0	0.09	92.5	0.10
14	RbI	[NaCℓ]	6:6	2	3.665	1.48	2.20	1.49	>1.0	0.105	77.5	0.11
15	CsCℓ	[CsCℓ]	8:8	π/2	3.56	1.69	1.81	1.08	>1.0	0.13	100	0.16
16	CsBr	[CsCℓ]	8:8	π/2	3.715	1.69	1.95	1.15	>1.0	0.14	74.5	0.14
17	CsI	[CsCℓ]	8:8	π/2	3.95	1.69	2.20	1.30	>1.0	0.16	64.2	0.16
18	TℓI	[CsCℓ]	8:8	π/2	3.62	1.44	2.20	1.53	>1.0	0.18	53	0.20
19	BeO	wurtzite	4:4	π	1.645	0.31	1.40	4.5	1.0	0.13	759	0.13
20	ZnS	[ZnS]	4:4	π	2.345	0.74	1.84	2.50	1.0	0.07	350	0.08
21	ZnSe	[ZnS]	4:4	π	2.45	0.74	1.98	2.67	1.0	0.08	215	0.10
22	ZnTe	[ZnS]	4:4	π	2.64	0.74	2.21	2.98	1.0	0.09	180	0.11
23	CdS	[ZnS]	4:4	π	2.52	0.97	1.84	1.90	0.88	0.07	242	0.08
24	SiO_2	SiO_4 - tetra-hedron	2:4	4	1.60	0.41	1.40	3.40	0.93	0.05	1095	0.046

TABLE III

Damping Data of 15 Solids as Determined from Laser Raman Spectra and Calculated for Covalent Binding

No.	Substance	Structure	$CN_1:CN_2$	r_o	r_a	r_b	r_b/r_a	$\Delta\nu/\nu_o$	ν_o	$\Delta\nu/\nu_o$
				10^{-8} cm	10^{-8} cm	10^{-8} cm		Calculated	cm^{-1}	Experimental
1	BeO (49)	Wurtzite	4:4	1.645	1.07	0.66	0.62	0.018	684	0.019
2	BP (26)	Zinc-blende	4:4	1.96	1.10	0.89	0.80	0.022	830	0.015
3	AℓN (26)	Wurtzite	4:4	1.87	1.26	0.70	0.55	0.016	610	0.019
4	SiC (33)	Zinc-blende	4:4	1.89	1.17	0.77	0.65	0.018	795	0.020
5	SiO_2 (34)	SiO_4-tetrahedron	2:4	1.60	1.17	0.66	0.56	0.009	1085	0.010
6	ZnS (31)	Zinc-blende	4:4	2.345	1.31	1.04	0.80	0.022	350	0.024
7	ZnSe (50)	Zinc-blende	4:4	2.45	1.31	1.14	0.87	0.024	252	0.023
8	ZnTe (50)	Zinc-blende	4:4	2.64	1.31	1.31	1.0	0.028	207	0.027
9	CdS (51)	Zinc-blende	4:4	2.52	1.48	1.04	0.71	0.020	302	0.025
10	GaAs (52)	Zinc-blende	4:4	2.435	1.26	1.18	0.93	0.026	273	0.022

TABLE III (Cont.)

No.	Substance	Structure	$CN_1:CN_2$	r_o	r_a	r_b	r_b/r_a	$\Delta\nu/\nu_o$	ν_o	$\Delta\nu/\nu_o$
				10^{-8} cm	10^{-8} cm	10^{-8} cm		Calculated	cm^{-1}	Experimental
11	InP(52)	Zinc-blende	4:4	2.54	1.44	1.10	0.76	0.021	304	0.017
12	$Ba^-(CrO_4)$(15)	CrO_4-tetrahedron	2:4	1.9	~1.30	0.66	0.51	0.008	863	0.009
13	TiO_2(53)	Rutile	3:6	1.96	1.36	0.66	0.49	0.019	143	0.020
14	MgF_2(53)	Rutile	3:6	1.96	1.40	0.64	0.46	0.018	341	0.025
15	Ge(54)	Diamond	4:4	2.44	1.22	1.22	1.00	0.028	301	0.025

The atomistic formulation of damping given in Eq. (20) and confirmed by experimental data on a large variety of solids, includes atomic parameters which determine the stability of a crystalline structure[55]. Hence, a detailed study of the change of damping with temperatures (see below) and pressure may shed some new light on both structure stability and polymorphic modifications.

TEMPERATURE DEPENDENCE OF THE DAMPING OF LATTICE VIBRATIONS*

(Interrelation Between Damping and Thermal Expansion)

Damping data calculated by Eq. (20) refer to room temperature only. The damping changes substantially for temperatures very different from room temperature, as may be seen from Figures 43 and 44. The figures show an exponential increase of energy absorption and a corresponding decrease of line-width of the main TO mode of RbI and CuI, with temperatures decreasing from room temperature to liquid He temperature. Note the corresponding frequency

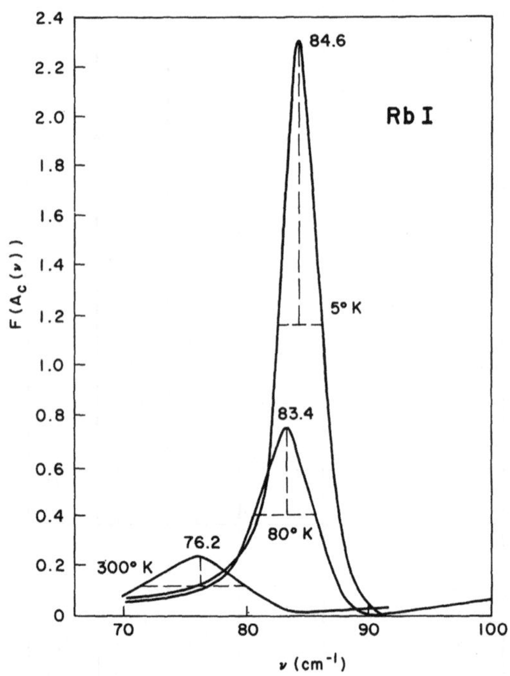

Figure 43. Variation of energy absorption, absorption frequency and damping of RbI for temperatures below 300° K.

*See, for example, references 13, 20 and 56, which give reviews of pertinent literature.

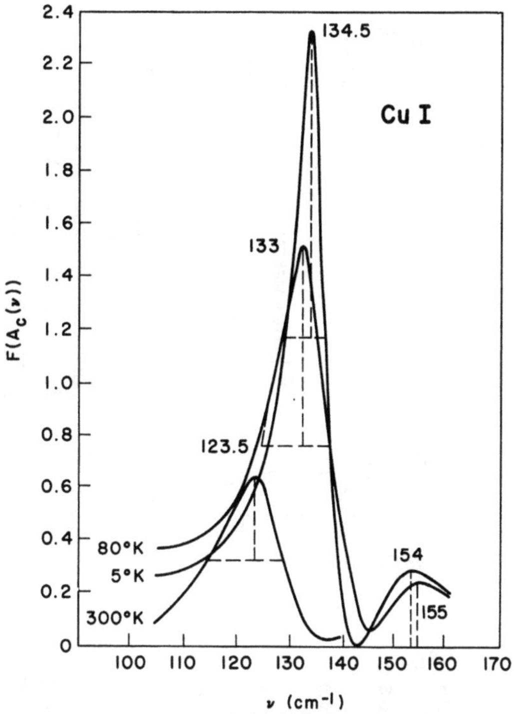

Figure 44. Variation of energy absorption, absorption frequency and damping of CuI for temperatures below 300° K.

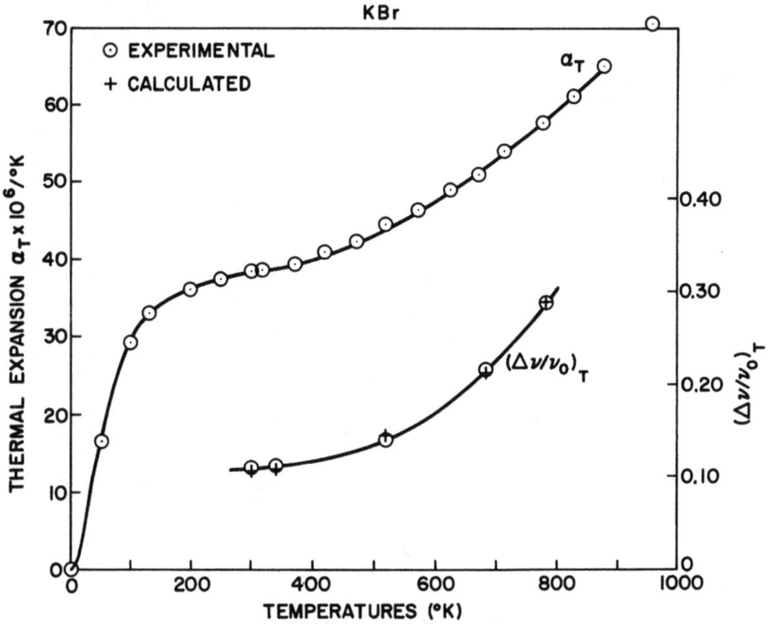

Figure 45. Damping values derived from IR emission spectra of KBr[20] (300°-800°K, Table IV) and data of linear thermal expansion[56,57]. o, experimental damping values $(\Delta\nu/\nu_o)_T$, +, damping values $(\Delta\nu/\nu_o)_T$ calculated according to Eq.(22).

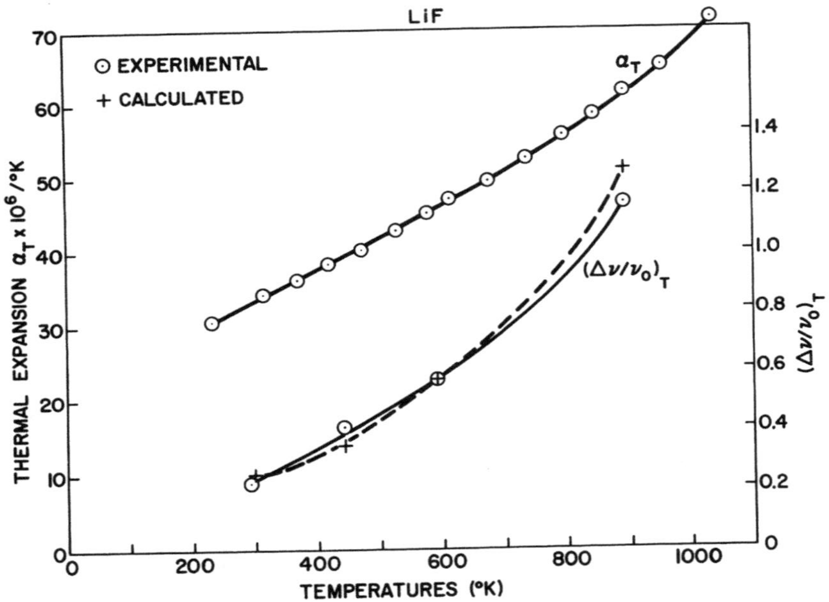

Figure 46. Damping values derived from IR reflection spectra of LiF[58] (300°-900° K, Table IV) and data of linear thermal expansion.[57] o, experimental damping values, modified as $\pi \frac{\gamma}{\omega_0}$ to equal, at room temperature, $(\Delta\nu/\nu_o)$ of energy absorption spectrum; +, damping values $(\Delta\nu/\nu_o)_T$ calculated according to Eq. (22).

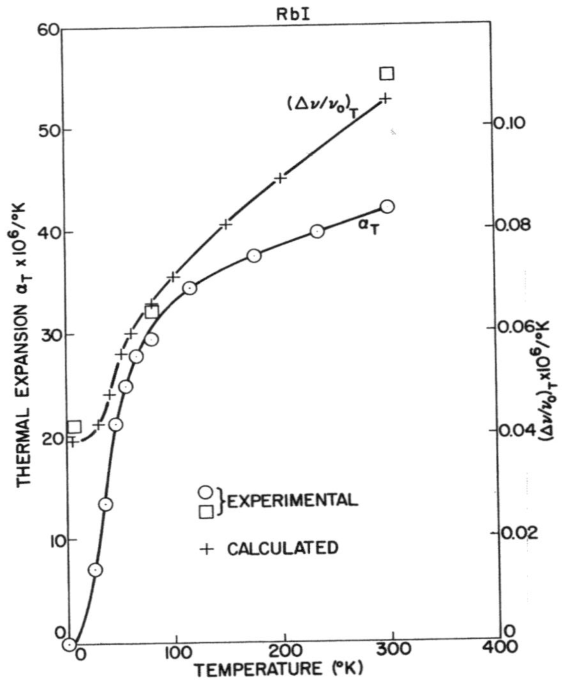

Figure 47. Damping values derived from energy absorption spectra of RbI (300°-5° K, Fig. 43 and Table IV) and data of linear thermal expansion.[59] o, experimental damping values $(\Delta\nu/\nu_o)_T$; +, damping values $(\Delta\nu/\nu_o)$ calculated according to Eq. (22).

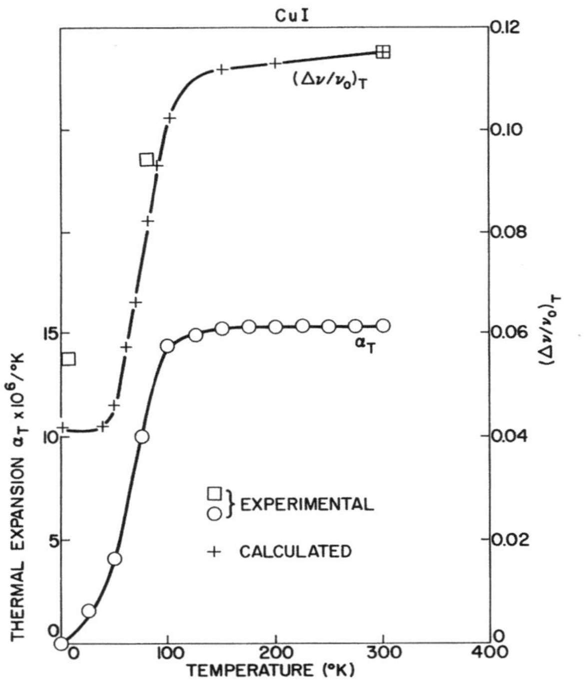

Figure 48. Damping values derived from energy absorption spectra of CuI (300°-5° K, Fig. 44 and Table IV) and data of linear thermal expansion[60]. o, experimental damping values $(\Delta\nu/\nu_o)_T$; +, damping values $(\Delta\nu/\nu_o)_T$ calculated according to Eq. (21).

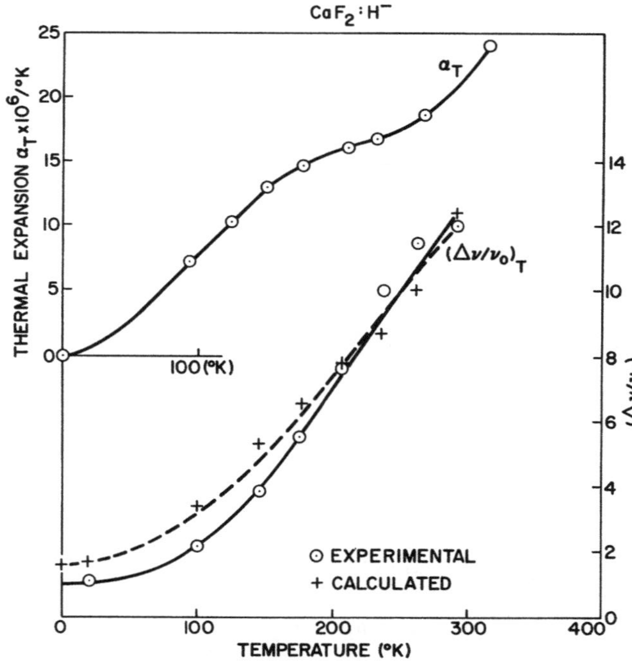

Figure 49. Damping values derived from absorption lines of localized modes of (H^-) in CaF_2[60], (290°-20° K, Table IV), and data of linear thermal expansion of CaF_2[62]. o, experimental damping values $(\Delta\nu/\nu_o)_T$, second harmonic. +, damping values $(\Delta\nu/\nu_o)_T$ calculated according to Eq. (21). In this case: b=2=constant; no θ-T/θ correction for (H^-) ions in CaF_2 as host lattice.

shift. To take these phenomena into consideration, a temperature dependent factor may be introduced. It may be determined from experimental data, either by analytically expressing observed damping curves or, a search may be made for a relationship between damping curves and one of the well known temperature dependent characteristics, such as specific heat or thermal expansion. The specific heat as a function of temperature, exhibits a saturation curve which differs substantially from the damping curves, whereas, the curves of thermal expansion bear a definite relationship to the damping curves. This may be seen in Figures 45 through 49, where the data of linear thermal expansion α, is compared with damping data for various solids of which both data are available; KBr, LiF, RbI, CuI, and CaF_2: (H^-), in that order.

In each case, a striking similarity is observed between the two sets of curves, damping and thermal expansion vs temperatures, which extends from close to absolute zero to close to the melting point. To formulate the apparent relationship, introduce the following expression,

$$(\Delta\nu/\nu_o)_T = (\Delta\nu/\nu_o)_{RT} \cdot \exp\left\{ b\,\frac{\alpha_T - \alpha_{RT}}{\alpha_{RT}} \right\} \tag{21}$$

where $b = 2$ for $\theta < T < T_m$, and $b = 2 - \frac{\theta - T}{\theta}$ for $0 < T < \theta$.

Applying this expression to Eq. (20), the temperature dependent damping is obtained as,

$$(\Delta\nu/\nu_o)_T = \frac{1}{8\pi^2}\,\frac{2\,CN - 1}{C_{str}}\,\frac{r_b}{r_a}\,\Psi\,(A)^{-2} \cdot \exp\left\{ b\,\frac{\alpha_T - \alpha_{RT}}{\alpha_{RT}} \right\} \tag{22}$$

Data calculated by Eqs. (21) or (22) (depending on available data) are presented in Table IV and graphically in Figures 45 through 59. The available experimental data, including those derived from the localized modes of CaF_2: (H^-)[61] (see below), generally agree with the calculated data.

Eventually, it may be of interest to test Eq. (22) by calculating the damping of the localized modes of the H^- vibration in CaF_2. Introducing in this case for Ca: $r_a = 0.99 \times 10^{-8}$ cm; and for H^-: $r_b = 2.08 \times 10^{-8}$ cm (according to Pauling), furthermore (one H^- between two C_a^{++}) CN = 1:2, $C_{str} = 4\pi/3$, and $\Psi(A) = 1$, then calculation gives $\Delta\nu/\nu_o = 0.0125$, which compares to 0.012 from experimental data, both for room temperature.

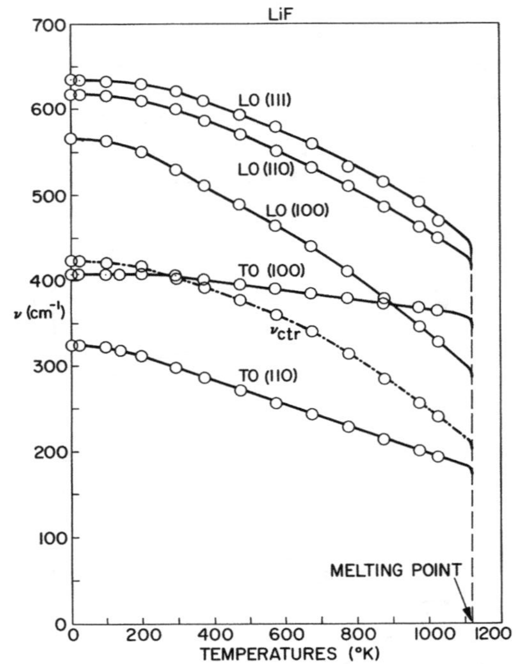

Figure 50. Frequency spectrum of the fundamental modes and center frequency of LiF as a function of temperatures from the absolute zero to the melting point. Note the cross-over between TO_{100} and LO_{100} at 900° K.

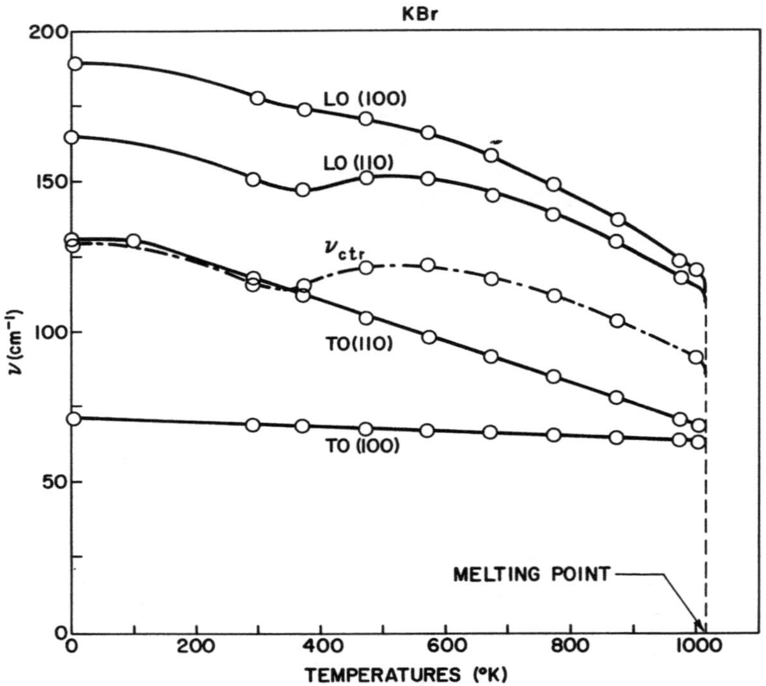

Figure 51. Frequency spectrum of the fundamental modes and center frequency of KBr as a function of temperatures from the absolute zero to the melting point. Note the anomaly of the LO_{110} and center frequency at 350° K which corresponds to a point of inflection in the curve of thermal expansion, see Fig. 45.

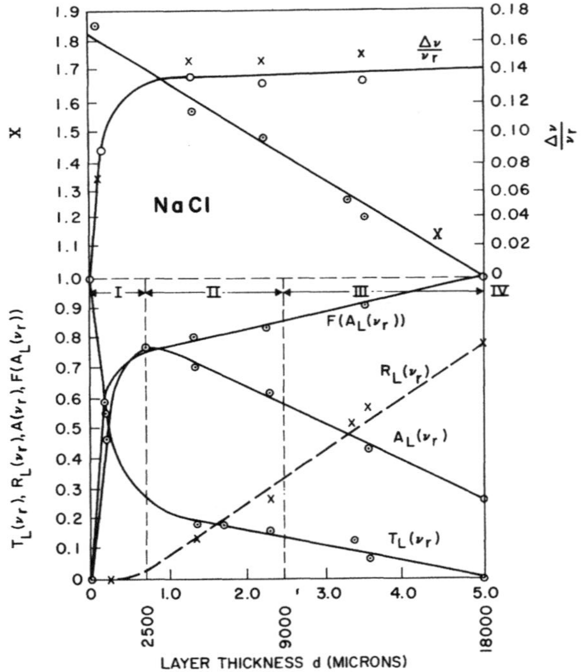

Figure 52. Transmission $T_L(\nu_r)$, Reflection $R_L(\nu_r)$, Absorption $A_L(\nu_r)$, Absorption Function $F[A_L(\nu_r)$, Damping $\Delta\nu/\nu_r$, and Parameter X as a function of layer thickness d for an evaporated film of NaCl.

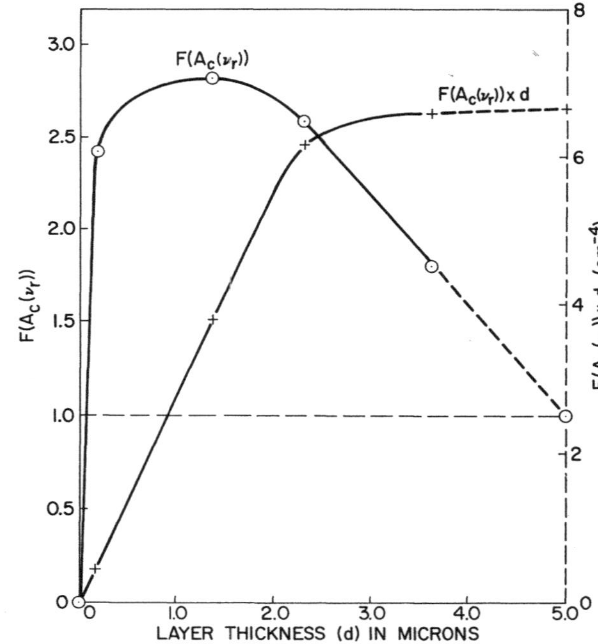

Figure 53. Absorption Function $F[A_c(\nu_r)$ and specific absorption function $d\cdot F[A_c(\nu_r)]$ versus layer thickness for NaCl (single crystal model).

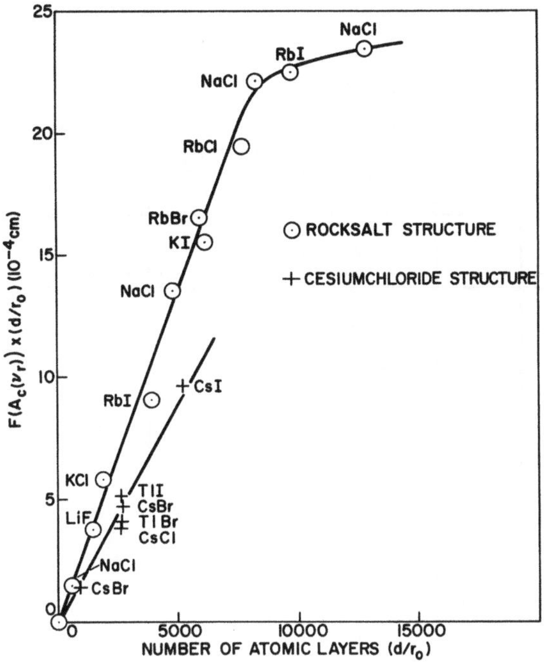

Figure 54. Influence of structure on the specific absorption function (see Text).

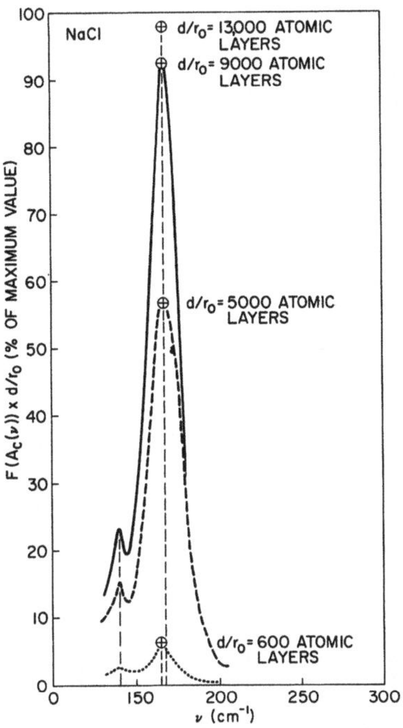

Figure 55. Increase of specific energy absorption in percent with penetration depth for the two TO modes of NaCl.

TABLE IV

Temperature Dependence of the Damping Data of 5 Solids as Determined from IR Spectral Data and Calculated for Ionic Binding (see text)

No.	Substance	Θ	T	α	$b\,\frac{\alpha_T-\alpha_{RT}}{\alpha_{RT}}$	$(\Delta\nu/\nu_o)_T$	$(\Delta\nu/\nu_o)_T$
		^oK	^oK	$x10^6/^oK$		Experim.	Calcul.
1	KBr	170	300	38.5	1	0.11	0.105
			339	39.0	1.03	0.113	0.108
			521	44.5	1.36	0.14	0.143
			680	52.0	2.02	0.215	0.213
			780	58.0	2.72	0.285	0.285
2	LiF	620	300	33.0	1	0.23	0.24
			443	39.5	1.42	0.41	0.34
			593	47.5	2.36	0.56	0.56
			893	61.0	5.35	1.16	1.28
3	RbI	120	300	42.0	1	0.11	0.105
			200	38.5	1/1.18		0.090
			150	36.5	1/1.30		0.081
			100	33.0	1/1.49		0.071
			80	30.5	1/1.59	0.064	0.066
			60	26.5	1/1.75		0.060
			50	23.5	1/1.88		0.056
			40	17.0	1/2.20		0.048
			30	11.0	1/2.50		0.042
			5	~ 0	1/2.72	0.042	0.039
4	CuI	180	300	15.3	1	0.115	0.115
			200	15.2	1/1.015		0.113
			150	15.1	1/1.025		0.112
			100	14.2	1/1.12		0.102
			90	13.0	1/1.24		0.093
			80	11.5	1/1.40	0.094	0.082

TABLE IV
(cont)

No	Substance	Θ	T	α	$b\,\frac{\alpha_T - \alpha_{RT}}{\alpha_{RT}}$	$(\Delta\nu/\nu_o)_T$	$(\Delta\nu/\nu_o)_T$
		^oK	^oK	$x10^6/^oK$		Experim.	Calcul.
4	CuI (cont.)		70	9.0	1/1.75		0.066
			60	6.5	1/2.0		0.057
			50	4.0	1/2.5		0.046
			40	2.5	1/2.7		0.042
			5	~0	1/2.72	0.055	0.042
5	$CaF_2:(H^-)$		290	21.0	1	12.0($x10^3$)	12.5($x10^3$)
			260	18.5	1/1.25	11.5	10.0
			235	17.0	1/1.45	10.4	8.7
			205	15.9	1/1.60	7.6	7.8
			175	14.5	1/1.9	5.5	6.6
			145	12.5	1/2.3	3.8	5.4
			100	7.7	1/3.6	2.1	3.4
			20	0.5	1/7.1	1.0	1.7
			0	0	1/7.4		1.6

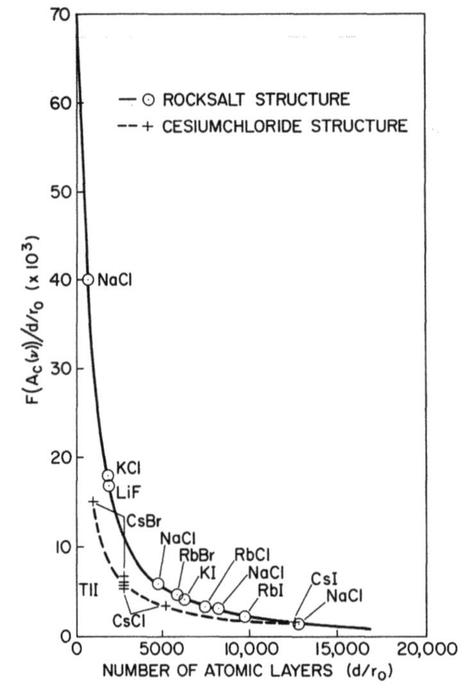

Figure 56. Influence of structure on the incline of the specific energy absorption as the number of atomic layers increases.

TEMPERATURE EFFECTS ON CHARACTERISTIC ABSORPTION SPECTRA

These effects can be summarized as follows: First calculating the characteristic absorption frequencies on the basis of elastic constants permits an examination of the various modes of vibration and of center frequency. This can be done as a function of temperatures from absolute zero to the melting point, since the elastic constants of some solids are available[9] for the entire temperature ranges of the solid state. An example is shown in Figures 50 and 51 for LiF and KBr. The pertinent data is compiled in Tables V and VI. The frequencies of the various modes of vibration decrease in a different manner with increasing temperatures, showing a cross-over between the LO_{100} mode and the TO_{100} mode at 900° K for LiF. The frequency curves of the various modes start with a flat slope at zero temperature, similar to specific heat and thermal expansion curves. They steadily decline towards the melting point, the TO_{100} mode being affected less by temperature than the other modes. An anomaly of the LO_{110} mode of KBr at 350° K is also expressed in the curve for the center frequency. It corresponds to a point of inflection in the curve of thermal expansion, as may be seen in Figure 45. The center frequency averages the frequencies of the various modes, considering that they have very different intensities. Second, the frequency shift observed in IR absorption spectra due to a change of temperatures (see Figures 43 and 44 for RbI and CuI) may also be calculated from the elastic constants. For example, the frequencies of the TO_{110} mode of CsBr and CsI have been calculated for room temperature and liquid He temperature. The calculated frequency shifts compare to the experimental data from the characteristic absorption spectra as follows: CsBr, 3.5 cm^{-1} to 5.5 cm^{-1}; CsI, 2.5 cm^{-1} to 2.3 cm^{-1}, respectively.

INFRARED PROPERTIES OF EVAPORATED ATOMIC LAYERS

The terms introduced in Eqs. (1) through (6) have been calculated for NaCℓ at the frequency of main resonance ν_r, based on data measured by Barnes and Czerny (1932).[1] The calculated data are plotted in Figures 52 and 53, as a function of the thickness d = zero to d = 5 microns (extrapolated). Some remarks may be in order. Figure 52 shows the experimental data of the transmissivity, $T_L(\nu_r)$, determined from Figure 2. It decreases exponentially from unity to zero. Using this data and the data of the reflectivity of the single crystal $R_C(\nu_r)$, the following data has been derived from Eqs. (1) through (6): (1) The parameter X decreases linearly from a maximum value of $[1 + R_C(\nu_r)]$ to unity. (2) The reflectivity of the film layer $R_L(\nu_r)$, indicates complete transparency at very small values of d, followed by a linear increase to the value of the single crystal $R_C(\nu_r)$, which is achieved at d = 5 microns. (3) The absorption function $A_L(\nu_r)$ reaches a maximum value at d = 0.8 micron and afterwards decreases to $[1-R_C(\nu_r)]$, which is reached at d = 5 microns. (4) The characteristic absorption function of the film layer $F[A_L(\nu_r)]$ exhibits a saturation curve which reaches unity at d = 5 microns. (5) The damping function $\Delta\nu/\nu_r$ (experimentally determined), represents a saturation curve which reaches the constant value of 0.14 for NaCℓ at d = 1 micron. Figure 53, on the other hand, shows the characteristic absorption function of the single crystal $F[A_C(\nu_r)]$. The trend of this curve is very similar to that of the absorption curve $A_L(\nu_r)$ in Figure 52. A saturation curve may be obtained here by multiplying $F[A_C(\nu_r)]$ by d, as may be seen in Figure 53. The meaning of the products curve may best be shown, on introducing the parameter

TABLE V

Temperature Dependence of the Characteristic Absorption Frequencies of LiF for the Temperature Range Absolute Zero to the Melting Point

No.	Solids[9]	Temperature	ρ	r_o	C_{11}	C_{12}	C_{44}	$\nu_{TO(110)}$	$\nu_{TO(100)}$	$\nu_{LO(100)}$	$\nu_{LO(110)}$	$\nu_{LO(111)}$	ν_{ctr}
		(^oK)	(g/cm^3)	(10^{-8} cm)	10^{11} dynes/cm^2			cm^{-1}	cm^{-1}	cm^{-1}	cm^{-1}	cm^{-1}	cm^{-1}
1	LiF	4	2.69	1.995	12.26	4.20	6.375	324	408	566	618	635	424
2		20	2.69	1.995	12.26	4.20	6.375	324	408	566	618	635	424
3		100	2.68	1.996	12.12	4.15	6.375	322	408	563	616	633	421
4		140	2.675	1.998	11.94	4.17	6.375	318	408	558	615	632	420
5		200	2.66	2.001	11.65	4.22	6.365	312	407	552	610	630	418
6		280	2.63	2.008	11.22	4.22	6.31	304	406	542	606	628	415
7		373	2.61	2.015	10.01	3.81	6.19	285	402	512	586	610	392
8		473	2.58	2.023	9.13	3.60	5.98	270	396	490	570	594	378
9		573	2.55	2.031	8.18	3.25	5.77	256	391	465	550	580	360
10		673	2.515	2.039	7.22	2.86	5.57	242	386	440	531	560	340
11		773	2.48	2.047	6.23	2.38	5.36	228	380	410	510	531	314
12		873	2.45	2.055	5.23	1.86	5.14	213	373	377	485	515	283

TABLE V (Cont.)

Temperature Dependence of the Characteristic Absorption Frequencies of LiF for the Temperature Range Absolute Zero to the Melting Point

No.	Solids[9]	Temperature	ρ	r_o	C_{11}	C_{12}	C_{44}	$\nu_{TO(110)}$	$\nu_{TO(100)}$	$\nu_{LO(100)}$	$\nu_{LO(110)}$	$\nu_{LO(111)}$	ν_{ctr}
		(^oK)	(g/cm^3)	(10^{-8}cm)	10^{11} dynes/cm^2			cm^{-1}	cm^{-1}	cm^{-1}	cm^{-1}	cm^{-1}	cm^{-1}
13	LiF	973	2.415	2.063	4.38	1.42	4.97	200	368	345	462	492	256
14		1023	2.40	2.069	3.97	1.21	4.87	193	364	328	450	474	240

TABLE VI

Temperature Dependence of the Characteristic Absorption Frequencies of KBr for the Temperature Range Absolute Zero to the Melting Point

No.	Solids(9)	Temperature	ρ	r_o	C_{11}	C_{12}	C_{44}	$\nu_{TO(110)}$	$\nu_{TO(100)}$	$\nu_{LO(100)}$	$\nu_{LO(110)}$	ν_{ctr}
		(^oK)	(g/cm^3)	(10^{-8}cm)	10^{11} dynes/cm^2			cm^{-1}	cm^{-1}	cm^{-1}	cm^{-1}	cm^{-1}
1	KBr	4	2.82	3.285	4.30	0.57	0.56	130.5	71.5	190	165.5	129
2		293	2.75	3.295	3.42	0.44	0.51	118	69	178.5	151	116
3		373	2.73	3.305	3.23	0.54	0.50	112	68.5	174	147.5	115
4		473	2.69	3.32	3.11	0.81	0.48	104	67.5	171	151.5	122
5		573	2.65	3.33	2.94	0.94	0.47	97.5	66.5	167	151	123
6		673	2.60	3.34	2.60	0.84	0.46	92	66	158	145	117
7		773	2.56	3.36	2.28	0.80	0.44	85	65.5	149	138.5	112
8		873	2.51	3.38	1.93	0.69	0.43	78	64.5	137	130	104
9		973	2.47	3.395	1.54	0.46	0.42	70	63.5	123	118	90
10		1003	2.45	3.405	1.51	0.53	0.41	69	63	121	117.5	90

d/r_o (which stands for the number of atomic layers), instead of the thickness d of the film layer. Plotting the product $\Phi F(A_C(\nu_r))$ by (d/r_o) as a function of d/r_o, we obtain the saturation curve of Figure 54, which is now applicable to all solids having the NaC structure. This may be seen when we plot in Figure 54 the data of $F[A_C(\nu_r)] < (d/r_o)$ for all solids of the NaCℓ structure on which the data of d were measured, and transmission measurements were available. These data are compiled in Table VII for NaCℓ , LiF, KCl, KI, RbCl, RbBr and RbI. All these data fit a curve derived for NaCℓ in Figure 54. Another curve having a different slope, is obtained for similar data for the solids of CsCl structure, and is shown in Figure 54 for CsCl, CsBr, CsI, TlBr and TlI (see Table VII). The increase of slope is approximately inversely proportional to the coordination number CN of the solids. A substantially higher energy absorption is obtained for the same number of atomic layers, for solids with the NaCℓ-structure as compared to solids with the CsCl-structure. It is inversely proportional to the damping reported above (see Eq. (20)). Plotting the product $F[A_C(\nu_r)] <$ (d/r_o) as a function of (d/r_o), the energy absorption of the single crystal is defined as a function of the penetration depth. This may be seen in Figure 55 which shows the absorption curves of NaCℓ in percent of the total absorption, having d/r_o as a parameter. For example, observe that about ninety percent of the total energy (which enters the crystal) is absorbed after a penetration of about 8,000 atomic layers. By contrast, the ratio $F[A_C(\nu_r)]/(d/r_o)$ may also be plotted as a function of (d/r_o), as is shown by curve (a) of Figure 56, which again is applicable to all solids having the NaCℓ-structure, whereas, curve (b) is applicable to all solids of the CsCl-structure. The curves exhibit a strong exponential decline of the energy absorbed per atomic layer with increasing penetration depth. For example, observe for curve (a) that the energy absorption per atomic layer decreases to about ten percent of its initial value after a penetration of 4,000 atomic layers by IR energy (and to one percent at 18,000).

The curves (a) of Figure 54 and 56, derived from NaCℓ data of various layer thicknesses, exactly fit the data of all solids of the NaCℓ-structure (of which we have both the data of transmissivity and of layer thickness available, see Table VI). This fact actually confirms that the energy absorption of these solids is sufficiently characterized by the number of atomic layers penetrated, the electronic charges of the atoms and the coordination number of the crystalline structure, besides being a function of temperature (see Figs. 43 and 44). However, in the vicinity of room temperature this is only a small effect.

From Figure 52, the following conclusions may be drawn with regard to transmission measurements for NaCℓ:

(1) An evaporated film layer does not act in a crystal-like manner when the film thickness is less than 0.7 microns (2500 atomic layers),

TABLE VII

Dependence of the Specific Absorption on the Number of Atomic Layers for Rocksalt and Cesium Chloride Structures

No.	Sub-stance	Struc-ture	CN	r_o	d	d/r_o	$F[A_c(\nu)]$	$F[A_c(\nu)]\ (\frac{d}{r_o})$	$\frac{F[A_c(\nu)]}{d/r_o}$
				10^{-8} cm	10^{-4} cm	Number of Atomic Layers		Depth Absorption	Atomic Layer Absorption $\times 10^{-4}$
1	NaCl	NaCl	6:6	2.815	0.17	600	2.4	1450	40
2	NaCl	NaCl	6:6	2.815	1.35	4800	2.8	13500	5.9
3	NaCl	NaCl	6:6	2.815	2.3	8200	2.7	22200	3.3
4	NaCl	NaCl	6:6	2.815	3.6	12800	1.85	23400	1.45
5	LiF	NaCl	6:6	2.01	0.3	1500	2.5	3750	16.7
6	KCl	NaCl	6:6	3.14	0.56	1800	3.2	5700	17.6
7	KI	NaCl	6:6	3.52	2.20	6200	2.5	15500	4.0
8	RbCl	NaCl	6:6	3.27	2.52	7700	2.5	19500	3.3
9	RbBr	NaCl	6:6	3.42	2.03	5900	2.8	16500	4.7
10	RbI	NaCl	6:6	3.66	1.50	4100	2.2	9000	5.3
11	RbI	NaCl	6:6	3.66	3.54	9700	2.3	22500	2.4

TABLE VII (Cont.)

No.	Sub-stance	Struc-ture	CN	r_o 10^{-8} cm	d 10^{-4} cm	d/r_o Number of Atomic Layers	$F[A_c(\nu)]$	$F[A_c(\nu)]$ $(\frac{d}{r_o})$ Depth Absorption	$\frac{F[A_c(\nu)]}{d/r_o}$ Atomic Layer Absorption x 10^{-4}
1	CsCl	CsCl	8:8	3.56	0.97	2720	1.4	3800	5.2
2	CsBr	CsCl	8:8	3.72	0.35	940	1.4	1300	15.0
3	CsBr	CsCl	8:8	3.72	1.02	2750	1.7	4600	6.1
4	CsI	CsCl	8:8	3.95	2.06	5200	1.8	9400	3.4
5	CsI	CsCl	8:8	3.95	5.0	12600	2.3	2900	1.8
6	TlBr	CsCl	8:8	3.41	0.92	2700	1.5	4000	5.5
7	TlI	CsCl	8:8	3.62	0.98	2700	1.9	5100	7.0

region I.
(2) A film layer emulates the property of a single crystal when the thickness of the film layer exceeds 5 microns (18,000 atomic layers), region IV.
(3) Interference effects become noticeable* in transmission data at a layer thickness > 2.5 microns (> 9000 atomic layers), region III.
(4) Optimal conditions, with regard to both crystal-like behavior and avoidance of distortion by interference effects, prevail when film deposits have a thickness between 0.7 and about 2.5 microns (2500 to 9000 atomic layers), region II. This corresponds to a transmission minimum somewhere between 25 and 12 percent. Film deposits of less than 1500 atomic layers i.e. having a minimum value of larger than 35 percent, deviate noticeably from the crystal-like behavior of such a substance.

*Interference effects in the case of NaCℓ made distortions visible only at the minima of absorption.

CONCLUDING REMARKS

The concepts introduced and developed in this work may form the basis for the study of problems in solid state research. Among these may be included the abnormal behavior in thermal expansion of solids at low temperatures; the variation of mechanical properties in solid solution crystals as a function of composition; and the stability of a crystal structure as related to polymorphism.

ACKNOWLEDGMENTS

For obtaining the wealth of basic data used in this work, the author gratefully acknowledges the excellent collaboration he was fortunate to have over many years with Dr. P. J. Gielisse, Dr. S. S. Mitra, Dr. A. Hadni, Dr. A. Smakula, Dr. C. H. Perry, Dr. P. Tarte, Dr. J. R. Jasperse, Dr. J. R. Aronson, Dr. W. G. Spitzer, Dr. K. D. Moeller, Dr. H. M. Otte, Dr. H. F. Schaake, Dr. R. A. Kromhout and Dr. H. S. Plendl, from universities and industries in U.S.A. and from universities abroad.

The author is particularly grateful to Dr. P. J. Gielisse, Dr. S. S. Mitra, Dr. A. Hadni and L. C. Mansur for many helpful discussions and suggestions with regard to this work.

The author wants to thank especially Dr. P. J. Gielisse for his assiduous assistance which has greatly improved this work for the final publication.

BIBLIOGRAPHY

1. R. B. Barnes and M. Czerny, Z. Physik 72, 447 (1931).
2. M. Czerny, Z. Physik 56, 600 (1930).
3. A. Mitsuishi, H. Yoshinaga and S. Fujita, J. Phys. Soc. Japan 14, 110 (1959).
4. E. Madelung, Nachr. Ges. Wiss. Goettingen (Math. Phys. Kl.) 100, (1908) and 43 (1910).
5. R. W. Pohl, "Optik", Springer Verlag Berlin 1940, 43.
6. J. N. Plendl, "Resonance Vibrations of the Ionic Lattice from the Mechanical Aspect", presented at the "European Conference of Molecular Spectroscopists" at Freiburg (Br.), Germany, 1957.
7. J. de Klerk, Rev. Scientific Instr. 36, 1540 (1965).
8. W. P. Mason, "Physical Acoustics", Vol. III/B of "Lattice Dynamics", Academic Press, New York, 1965.
9. G. Simmons, "Single Crystal Elastic Constants and Calculated Aggregate Properties", J. Graduate Res. Center, Dallas, Texas, Vol. XXXIV, No. 1 and 2, March 1965.
10. C. H. Perry, Diss. Queen Mary College, London 1960; and spectral measurements, M.I.T. Reports on AFCRL Contract No. AF 19 (628) 6066 (1966/68).
11. R. B. Barnes, Z. Physik 75, 723 (1932).
12. J. R. Aronson and H. G. McLinden, spectral measurements, A. D. Little Co. Reports on AFCRL Contract No. AF 19 (604) 8504 (1961/65).
13. J. R. Jasperse et al., Phys. Rev. 146, 526 (1966).
14. A. Hadni, spectral measurements, Univ. of Nancy Reports on AFCRL Contract No. AF 61 (052) 518 and AF 61 (052) 957 (1961/69).
15. P. Tarte, spectral measurements, Univ. of Liege Reports on AFCRL Contract No. AF 61 (052) 917 (1966/69).
16. J. N. Plendl, S. S. Mitra et al., Solid State Com. 7, 109, (1969).

17. A. D. B. Woods et al., Phys. Rev. 131, 1025 (1963).

18. D. W. Berreman, Phys. Rev. 130, 2193 (1963).

19. G. Dolling et al., Phys. Rev. 168, 972 (1968).

20. J. E. Mooij, Phys. Letters 24A, 249 (1967).

21. A. Hadni et al., Phys. Rev. 163, 836 (1967).

22. R. E. Schmunk, Bulletin Am. Phys. Soc. March 1967, 281.

23. W. J. L. Buyers, Phys. Rev. 153, 923 (1967).

24. B. Stone and D. Hill, Phys. Rev. Letters 4, 282 (1960).

25. P. J. Gielisse et al., Phys. Rev. 155, 1039 (1967).

26. O. Brafman et al., Sol. State Com. 6, 523 (1968).

27. J. R. Jasperse, spectral measurements, A. D. Little Co. Reports on AFCRL Contract No. AF 19 (628) 4351 (1964/65).

28. S. M. Shapiro et al., Phys. Rev. Letters 19, 361 (1967).

29. H. Yoshinaga and R. A. Oetjen, J. Opt. Soc. Am. 45, 1085 (1955).

30. G. Picus, unpublished (E. Burstein and P. E. Egli, Adv. Electronics and Electron Phys., VII, 24 (1955).

31. O. Brafman and S. S. Mitra, Phys. Rev. 171, 931 (1968).

32. W. G. Spitzer et al., article in "Silicon Carbide", 347, Pergamon Press, Oxford 1960.

33. J. S. Ziomek and P. B. Pickar, Phys. Stat. Sol. 21, 271 (1967).

34. J. F. Scott and S. P. S. Porto, Phys. Rev. 161, 903 (1967).

35. A. Hadni et al., C. R. de 3e Conf. Int. d'Electronique Quantique, Paris.

36. K. D. Moeller and S. T. Varma, spectral measurements, Fairleigh Dickinson University. Reports on AFCRL Contract No. AF 19 (628) 4084 (1964/69).

37. H. W. Hohls, Ann. Phys. 5. Folge, 29, 433 (1937).

38. J. N. Plendl, A. Hadni et al., Appl. Optics 5, 397 (1966).

39. P. Tarte, Rev. Silicates Industriels, 7-8, 1 (1963).

40. A. S. Barker Jr., Phys. Rev. 132, 1474 (1963).

41. D. R. Renneke and D. W. Lynch, Phys. Rev. 138, A530 (1965).

42. P. J. Gielisse, J. N. Plendl et al., J. Appl. Phys. 36, 2446 (1965).

43. P. Tarte, Acad. Roy. Belgique, Memoirs XXXV (1965).

44. G. R. Hunt, C. H. Perry and J. Ferguson, Phys. Rev. 134, A688 (1964)

45. J. N. Plendl, Phys. Rev. 123, 1172 (1961).

46. J. N. Plendl and P. J. Gielisse, Z. Kristal. 118, 404 (1963).

47. J. N. Plendl, "New Concepts in the Physics of Solids", Monograph, Spec. AFCRL Report No. 52 (1966).

48. W. Hume-Rothery et al., Phil. Trans. Roy. Soc. A 233, 1 (1934).

49. E. Loh, Phys. Rev. 166, 673 (1968).

50. S. S. Mitra, spectral measurements, Univ. of Rhode Island Reports on Contract No. AF 19 (628) 6042 (1963/69).

51. B. Tell et al., Phys. Rev. 144, 771 (1965).

52. A. Mooradian and G. B. Wright, Sol. State Com. 4, 431 (1966).

53. S. P. S. Porto et al., Phys. Rev. 154, 522 (1967).

54. J. H. Parker et al., Phys. Rev. 155, 712 (1967).

55. J. N. Plendl and P. J. Gielisse, "Compressibility and Polymorphism of Solids", submitted for publication.

56. R. A. Cowley, Advances in Physics 12, 42 (1963).

57. A. Eucken and W. Dannoehl, Z. Elektrochem. 40, 814 (1934).

58. G. Heilmann, Z. Physik 152 368 (1958).

59. D. E. Schuele and C. S. Smith, J. Phys. Chem. Solids 25, 801 (1964).

60. H. F. Schaake, thermal expansion measurements Martin Marietta Corp. Reports on AFCRL Contract No. AF 19 (628) 5663 (1965/69).

61. R. J. Elliott et al., Proc. Roy. Soc. (London) 289, 1 (1965).

62. Critical Tables Vol. III, 43, Natl. Acd. Sci. 1928; and Report 2389 - 11 - S, Willow Run Labs., U. Michigan.

HIGH PRESSURE RESEARCH IN THE FAR INFRARED REGION*+

John R. Ferraro

Chemistry Division, Argonne National Laboratory

Argonne, Illinois USA

ABSTRACT

The capability of obtaining vibrational spectra in the low frequency region below 300 cm^{-1} under non-ambient pressures to 80 kbars, is a comparatively new technique. The instrumentation necessary to do this will be reviewed. Studies made with ionic solids and coordination compounds will be discussed. Problems and limitations in the method will be presented.

INTRODUCTION

The first spectroscopic measurements at non-ambient pressures were made by Drickamer. His experiments covered the regions of ultraviolet, visible and near infrared.[1-2] It is the infrared region that is of concern to us in this paper. In 1956, Drickamer studied the effect of pressure on the vibrational frequency of bonds containing hydrogen to 5.0 microns (2000 cm^{-1}). Extension to 35.0 microns (285 cm^{-1}) was made by Weir, Van Valkenburg, and Lippincott.[3-5] In 1966, the capabilities of making low frequency measurements at high pressures up to 200 microns (50 cm^{-1}) were developed by Ferraro, Postmus and Mitra[6-7] using a grating spectrophotometer. Using an interferometer, Gebbie[8] has reported low

*Based on work performed under the auspices of the U. S. Atomic Energy Commission.

+Presented at the Nato Advanced Study Institute on Far Infrared Properties of Solids, Delft, Netherlands, August 5-23, 1968.

frequency spectra to 1000 microns (10 cm^{-1}), and McDevitt et al.[9] to 40 cm^{-1}. It thus appears that adequate techniques are now available to conduct far infrared studies at high pressure. This is not to say that the ultimate in instrumentation has been reached, however. Many instrumental problems exist and experiments are tedious and time-consuming.

This paper will discuss the present instrumentation and applications of the technique, and in conclusion, will discuss recent high pressure-Raman techniques.

AVAILABLE INSTRUMENTATION FOR FAR INFRARED-HIGH PRESSURE STUDIES

Optical High Pressure Cells

It would be advisable to briefly describe the available instrumentation capable of making low frequency measurements under pressure. A summary of the optical high pressure cells available with their advantages and disadvantages is shown in Table I.

The shock wave techniques can obtain pressures up to 1000 kbars. The shock wave is generated by explosives. The technique would be extremely difficult to use with a scanning spectrophotometer, since the duration of the shock is only of a short time.

The piston-cylinder developed by Drickamer and Balchan[10] for use from 0.2 to 5.0 microns used sapphire windows of 1/2 inch diameter and 1/2 inch thickness. These windows proved fragile and Drickamer[11-13] converted to sodium chloride windows thereafter.

Perhaps the most useful cell is the opposed anvil cell developed in recent years by Weir, Van Valkenburg, and Lippincott,[3-5] using diamond anvils. They used type II diamonds which are transparent in the regions of the ultraviolet, visible, and infrared, except for a portion at 4-6 microns where absorption occurs. Pressures up to 200 kbars have been claimed with this cell.[14]

Link of Optical Cell with Infrared Spectrophotometer

Table II lists the apparatus in current use for low frequency high pressure studies. The instrumentation using a grating spectrophotometer were developed by Ferraro, Mitra, and Postmus.[6-7] The technique is capable of reaching 200 μ, and involves the use of a Perkin-Elmer Model No. 301 double-beam grating spectrophotometer, used in double beam operation, a 6x beam condenser and the high

TABLE I

Summary of Available Optical High Pressure Cells

Type of Cell	Pressures	Advantages	Disadvantages
Shock Wave	1000 kbar	1) Highest obtainable pressures	1) Pressure exerted over short time
Piston and Cylinder	200 kbar	1) Largest specimen volumes	1) Not enough optical clarity to permit optical observation or photography
		2) Considered to give hydrostatic pressures	2) Specimen may interact with salt matrix
			3) Must be calibrated with respect to pressures measured in a different type of cell
Opposed Anvils (Diamonds)	200 kbar	1) Micro-quantities of material necessary	1) Pressure gradient exists
		2) Compact--can be used with spectrophotometers	2) Absorption of diamonds from 4-6 μ

TABLE II

High Pressure Apparatus Currently Used for Low Frequency Studies

Workers	Spectrophotometer or Interferometer	Wavelength Range μ	Optical Cell
Weir, Van Valkenburg, and Lippincott[3-5]	Commercial double-beam spectrophotometer with beam condenser	2-35	Diamond Anvil
Jacobsen and Brasch[15-16]	Perkin-Elmer No. 521*	2-35	Diamond Anvil
Ferraro, Mitra and Postmus[6-7]	Perkin-Elmer No. 301* Beckman IR-11** Beckman IR-12**	16-200 16-200 2-40	Diamond Anvil " " " "
McDevitt, Witkowski, and Fateley[9]	FS-520 interferometer	to 250	Diamond Anvil
Bradley, Gebbie *et al.*[8]	Michelson interferometer	50-1000	Anvil, quartz window

*With 6x beam condenser.

**With 8x beam condenser, (see L. Basile, et al. Spec. Letters, 1(5), 189 (1968)).

NOTE: For operation to 200 μ with a grating spectrophotometer, a cost of about $12,000 is necessary for a beam condenser and the diamond cell.

pressure diamond cell of High Pressure Diamond Optics, Inc., McLean, Virginia. Certain modifications were necessary. The size of the ellipsoid mirrors on the beam condenser had to be reduced, to accept the diamond cell. To allow for easy alignment of the cell, a machine lathe micrometer attachment was added to give movement in the x,y,z directions. To allow for more energy for operation below 100 μ, larger diamonds (~0.8 sq. mm. area) were used, although above 100 μ diamonds of 0.25 sq. mm. can be used. An expansion scale was also built into the No. 301 spectrophotometer, to amplify weak vibrations. This allows for less compensation in the reference beam and the use of narrower slits. For all of the pressure work a microscope is absolutely essential, to determine if one has a good solid load between the diamonds, and if a phase transition is occurring.

In the interferometer tie-up with the diamond cell, a brass cone light pipe is used (angle 9°37').[9] The infrared radiation comes out of the instrument and enters this cone. The cone bends the radiation toward the small opening in the other end (.060" opening). The cone is machined to accept the diamond cell piston. The energy then passes through the diamonds and enters the detector by means of another light pipe.

Calibration of Diamond Cell

It has already been indicated that a pressure gradient exists in the multiple anvil diamond cell. Duecker and Lippincott[17] have demonstrated that the pressure gradient across the diamond face is parabolic, with pressures in the center reaching 1 1/2 times those on the edges. Thus, any measure of pressure in the contact area of the diamond gives only an average pressure.

There are several ways to make a pressure calibration of the cell. These are as follows:

1. One measures the compression of the spring by means of a Dillon force gauge. The contact area of the diamond is determined by means of micro photographs. One then knows the force per unit area or pressure.

2. The use of solids which undergo phase transformations at known pressures (e.g., KBr - 18 kbars; KCl - 20 kbars; $NaNO_2$ - 14 kbars).

3. Calibration can be made by using nickel dimethylglyoxime, which shows a change in spectral properties with pressure.[17-18]

APPLICATIONS

This paper will attempt to present some of the applications of the far infrared-high pressure technique that have thus far been investigated. It will indicate the newest developments in this area, and further point out other applications that are contemplated.

Lattice Vibration Studies

Table III lists some ionic lattice frequencies. Most of the frequencies for the transverse optical lattice modes (ν_{TO}) are found below 300 cm^{-1}. The capability of measuring vibrational spectra in the far infrared region under pressure allows one to study these vibrations for the first time in this manner. The longitudinal mode (ν_{LO}) is more difficult to study in the infrared. At 90° incident radiation it is not normally observed. Berreman[19] observed the longitudinal mode for a thin film of LiF with an oblique incident radiation. Longitudinal optical modes of the silver halides have recently been studied with similar techniques.[20] With the diamond cell and the highly converging oblique radiation coming from the beam condenser, the longitudinal modes can be observed. They appear as shoulders on the main intense transverse vibrational bands, and are not easily studied in the infrared, for they are less pressure sensitive than the ν_{TO}, and because of the high frequency shift of the ν_{TO} band, can no longer be observed. In more covalent solids the ν_{TO} is very close to the ν_{LO}. In a homopolar covalent crystal, ν_{TO} may be equal to ν_{LO}. In this instance a very broad absorption is observed, which shows very little frequency shift with pressure.

Figure 1 shows a comparison of several ν_{TO} frequencies with pressure. The ν_{LO} pressure dependence for NaF is shown in Fig. 2. In this system the separation between the ν_{LO} and the ν_{TO} is sufficiently large to make possible a determination of the pressure dependency of both optical modes.

In all cases studied, ionic lattice vibrations shifted toward higher frequency with increasing pressure. The shifts at pressures up to 50 kbars can be considerable. However, not all ionic lattice vibrations will show dramatic shifts, since the compressibility of the solid is involved. The relationship between the change in frequency with pressure is given in equation 1,

$$\gamma \chi \nu = \left(\frac{\partial \nu}{\partial p}\right)_T \qquad (1)$$

where γ is the Grüneisen parameter, χ is the isothermal compressi-

TABLE III

Lattice Vibrations for Alkali Halides[1] cm^{-1}

	ν_{TO}
LiF	307
LiCl	191
LiBr	159
NaF	246
NaCl	164
NaBr	134
NaI	117
KF	190
KCl	141
KBr	113
KI	98
RbF	156
RbCl	118
RbBr	88
RbI	77
CsCl[2]	99
CsBr[2]	74
CsI[2]	62

ν_{TO} = Frequency of transverse lattice vibration

1. S. S. Mitra and P. J. Giellisse, "Infrared Spectra of Crystals," AFCRL-69-395, June (1965).

2. C. M. Randall, R. M. Fuller and D. J. Montgomery, Solid State Comm., 2, 273 (1964).

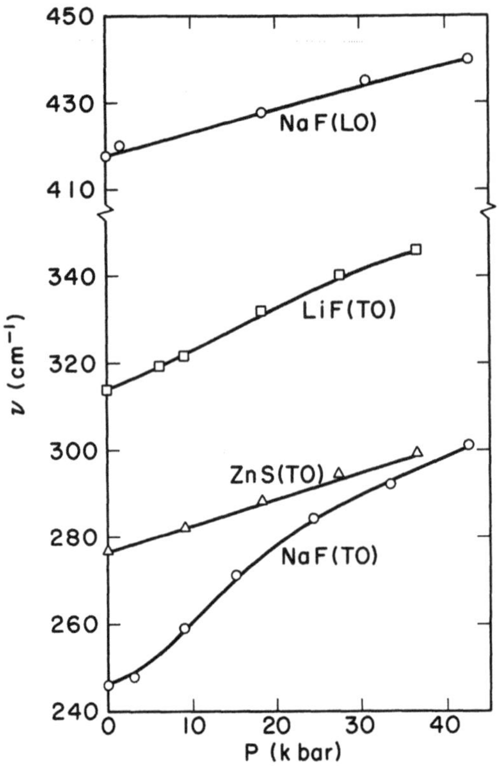

Fig. 1. Pressure Dependencies of Several Lattice Vibrations.

bility of the solid and ν is the frequency of the particular lattice mode. For non-compressible solids it is possible that only small shifts will occur. For example, the lattice modes of zirconia and hafnia failed to show significant shifts at 40 kbars.[9]

The data obtained from studies of the pressure dependency of the k∼0 lattice vibrations of ionic crystals[21] when combined with data from studies of these modes with temperature,[22] can give a better understanding of the anharmonic interactions in these solids. It is possible to distinguish between the purely volume-dependent contribution and the contribution from various anharmonic terms in the crystal Hamiltonian.[23] Figure 3 shows a plot of $\ln\nu/\nu_0$ vs. $\ln v/v_0$ for several optic modes. The data has been obtained from pressure and high temperature studies,[21-23] P-V data from Drickamer[24] and from Cline and Stephens.[25] It can be observed that the straight line extrapolated from the pressure domain does not coincide with the line obtained from the temperature data. The difference may be attributed to the anharmonic contribution to the frequency shift (known as the "self-energy" shift), and increases steadily with increasing temperature for

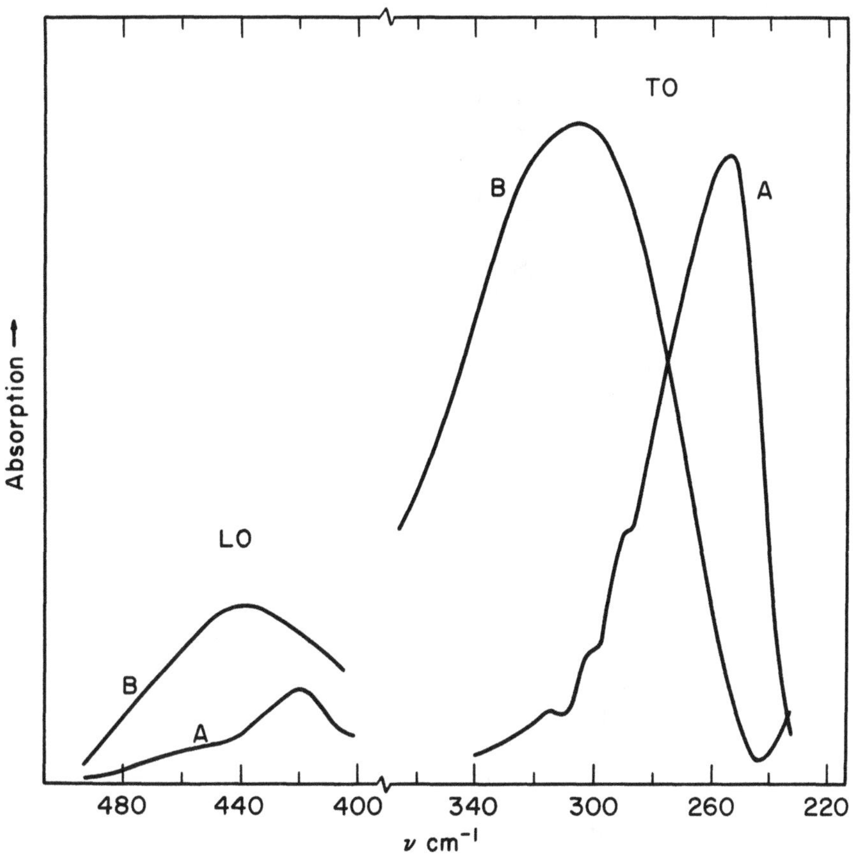

Fig. 2. Comparison of the Transverse and Longitudinal Modes in NaF with and without Pressure.

LiF. A similar analysis was made for KBr and in this crystal the "self-energy" shift is negligible. A similar result was obtained for RbI using other techniques.[22,26]

The results obtained for the Grüneisen parameters for the long-wavelength optical modes from equation 1 and equation 2,

$$\gamma_j(k) = \frac{\partial \ln \nu_j(k)}{\partial \ln V} \qquad (2)$$

are tabulated in Table IV. The agreement with the calculations made from those assuming a rigid-ion model with central forces incorporating repulsion terms of the Born-Mayer [exp(-r/p)] and inverse-power (r^{-n}) type agree well. The results using Cowley's theory give somewhat larger values of γ.

TABLE IV
Gruneisen Parameters for the Long Wavelength Optical Modes

	Experimental from		Calculated			Cowley and Cowley[30]	
	Eq. 1	Eq. 2	Born-Mayer[27]	r^{-n} repulsion[28]	Cowley[29]	Model I	Model II
LiF	2.15	2.59	2.44	3.46			
NaF	2.80	2.95	2.43	3.00			
KBr		2.83	2.52	2.95	3.27	~ 3.0	~ 2.6
KCl		2.46	2.52	2.92			

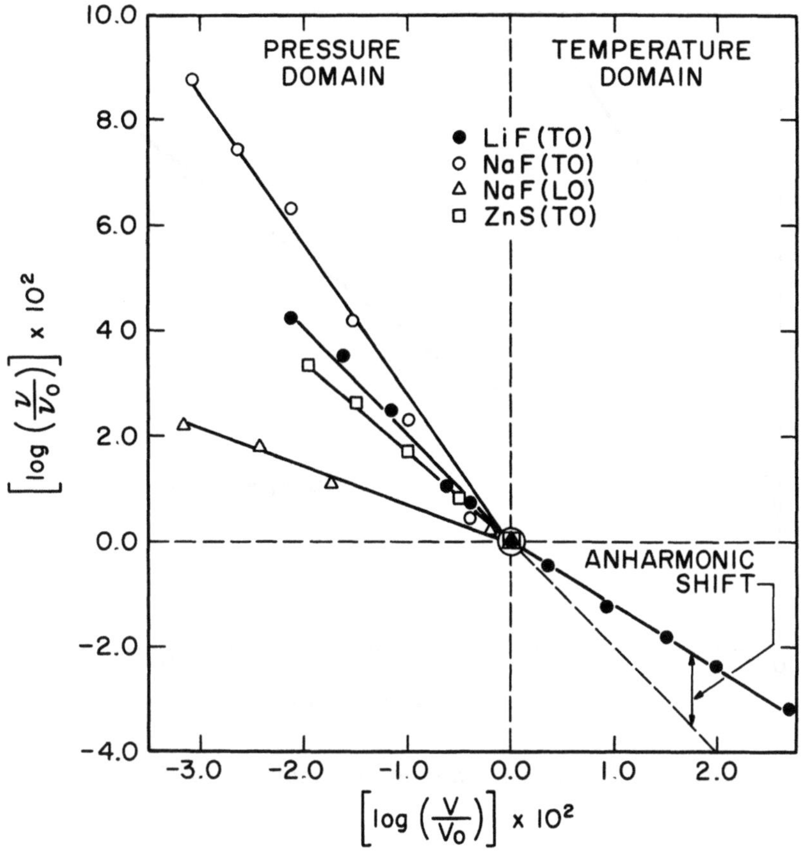

Fig. 3. Plot of $\ln(\nu/\nu_0)$ vs. $\ln(V/V_0)$ for Several Optic Modes.

Pressure dependencies of two phases of a solid can be determined by these techniques.[23] Such studies have been made with KBr[23] and KCl.[9,23] Figures 4 and 5 show the TO mode of KBr and KCl as a function of pressure. The difficulty of studying phase transitions is illustrated by the figures. The pressure gradient across the diamond anvils prevents the detection of a sharp transition pressure and both phases appear over a range of pressure. The high pressure phase (CsCl structure) appears at 16 kbars for KBr and at 24 kbar for KCl. The conversion to the high pressure phase is complete at 26 kbars and 30 kbars for KBr and KCl respectively. For a limited average pressure range the TO mode of both the low and high pressure phases can be detected, with a gradual decrease in the intensity of the low pressure phase and the in-

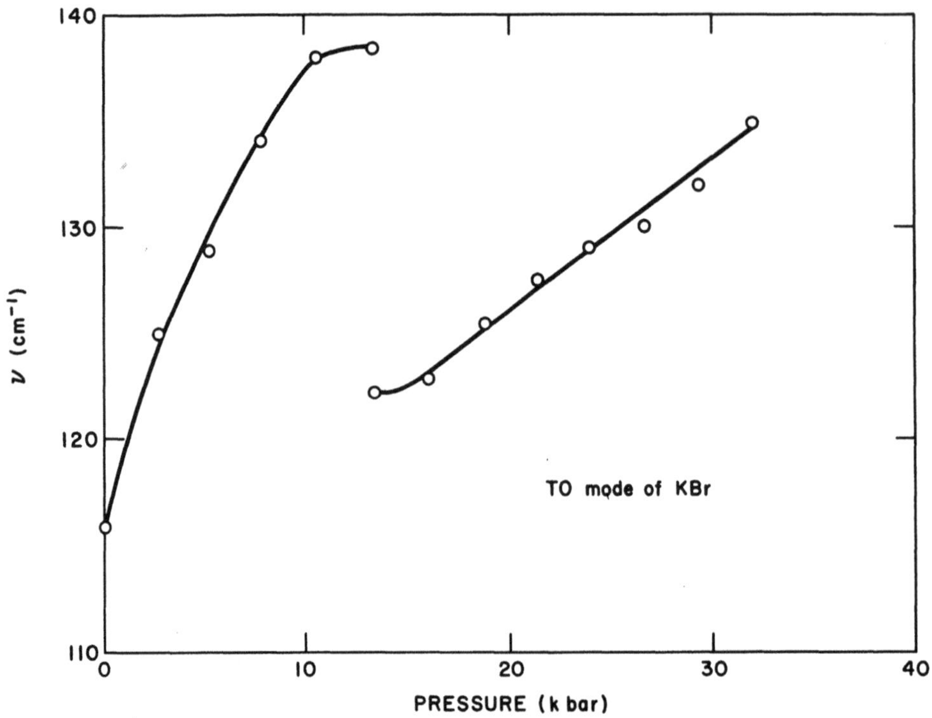

Fig. 4. TO Mode of KBr as a Function of Pressure.

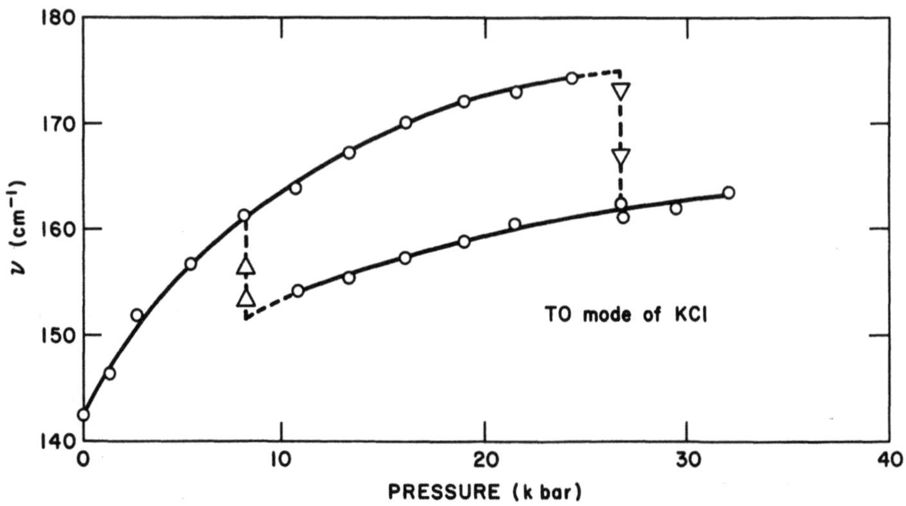

Fig. 5. TO Mode of KCl as a Function of Pressure.

crease in that of the high pressure phase. With a reduction of the pressure, the frequencies of the CsCl phase do not coincide with those determined from increasing pressures (maximum experimental error $\pm$ 2 cm^{-1}). The low pressure phase does not reappear until 11 and 16 kbars for KBr and KCl respectively. Part of these discrepancies can be attributed to the sluggishness of the system and friction within the cell.

The vibrational frequency decreases at the transition pressure by about 10-12%. The ratio of TO frequency of the CsCl phase to that of the TO frequency in the NaCl phase should equal the square root of the coordination number of each phase; e.g., $\sqrt{6/8}$ or 0.87. The observed ratio for KBr is 0.88 and that of KCl is 0.92.

It has been found[31] that the k ~ 0 TO frequency of the alkali halides of NaCl structure is proportional to $(a/\chi\mu)^{1/2}$, where a is the lattice constant and μ is the reduced mass per unit cell, and χ is the compressibility. The same is true for the CsCl structure. Figure 6 illustrates this. Thus, it may be possible to relate the frequency-pressure data to the structure of the ionic crystal.

Pressure studies with mixed ionic crystals are possible using the diamond cell technique and are now under way.

Certain molecular lattice modes were investigated by McDevitt, Witkowski and Fateley.[9] The experiments are more difficult to perform, since a thicker sample is needed, and spacers must be used to accomplish this. Molecular lattice vibrations have been observed to also shift toward higher frequency with increasing pressure. It would be expected that their pressure dependencies would be greater than ionic lattice modes due to their greater compressibilities.

Coordination Compounds (Effect of Pressure on Molecular Vibrations)

Whereas certain ionic solids when submitted to pressures show shifts of the optical modes toward higher frequencies, internal modes in other compounds show very little pressure shifts. Nevertheless, certain intensity changes are possible under pressure. For example, a number of coordination complexes having two halides or nitrogens per metal atom would be expected to show an asymmetric and a symmetric stretching vibration. Wherever these two vibrations are resolved into two bands, pressure effects one mode more than an other.[32] Figures 7 and 8 illustrate the effect of pressure on the asymmetric and symmetric stretching vibrations of $[\phi_4As][GeCl_3]$ and $[\phi_4As][SnCl_3]$. The verification as to which is

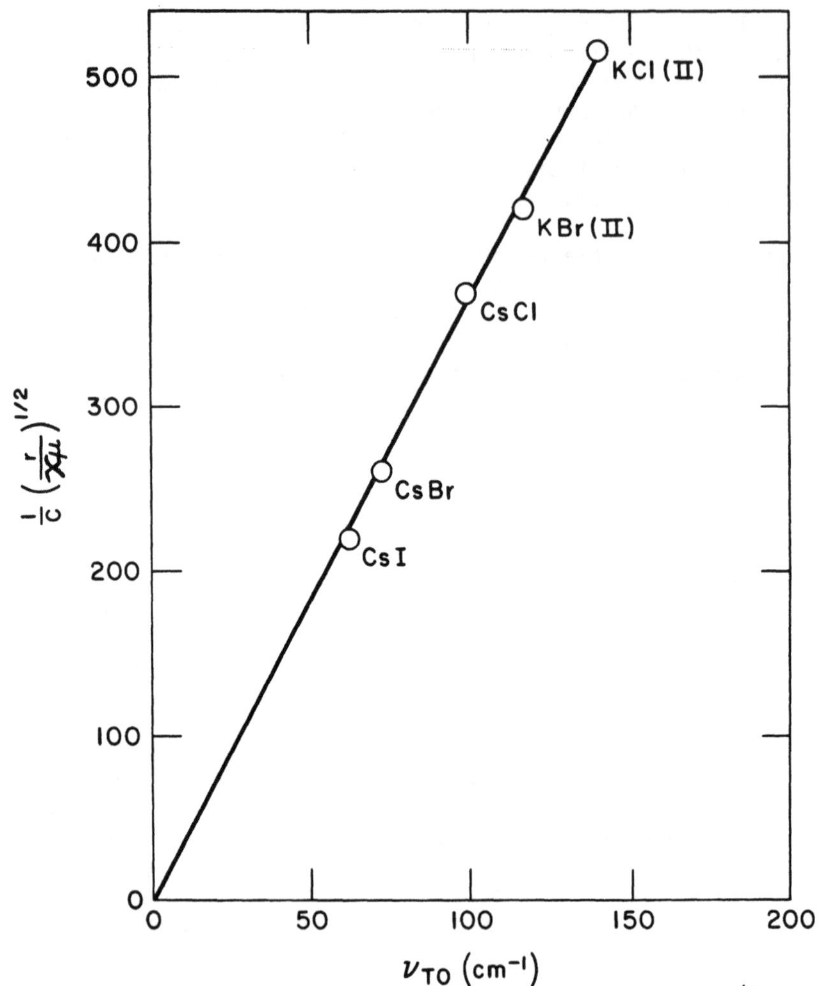

Fig. 6. Plot of TO Mode Frequency vs. $1/c(\frac{r}{\chi\mu})^{1/2}$.

low frequency band splits into a doublet, and that the high frequency band decreases in intensity. Under pressure the degeneracy of the F vibration (ν_4) is completely removed. It is concluded that in the spectrum of $KMnO_4$ at ambient pressure the low frequency band is the E species and the high frequency band is the A_1 specie. The results may be attributed to a further lowering of the symmetry of the solid under pressure to C_{2v} and illustrate that pressure can induce symmetry lowering. However, verification of this interpretation can only result when the F vibration (ν_3) is also studied under pressure. It is possible that not all solids will lend themselves to a lowering of the symmetry with the same ease, and one must consider the crystal packing involved for each solid.

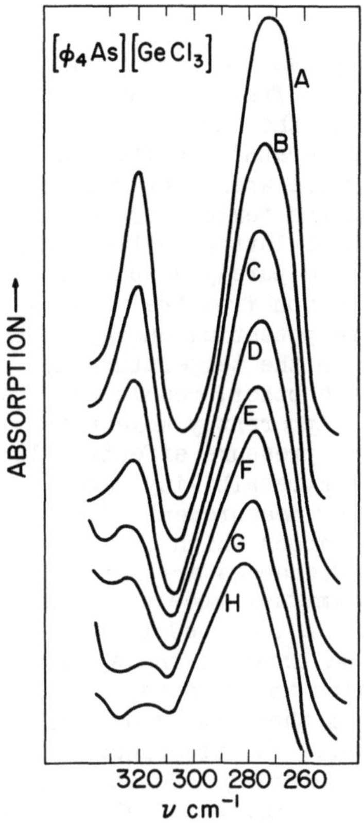

Fig. 7. Low Frequency Spectra of $(\phi_4As)(GeCl_3)$ with and without Pressure.

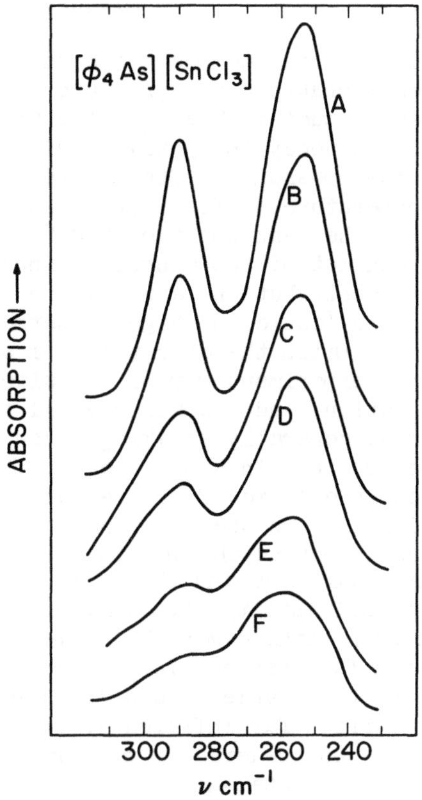

Fig. 8. Low Frequency Spectra of $(\phi_4As)(SnCl_3)$ with and without Pressure.

The pressure gradient existing in the multiple anvil diamond cell has already been discussed. In an attempt to eliminate as much of the gradient as possible, experiments with molybdenum steel gaskets were made. It has been our experience that applying pressure on a solid with a gasket does not give hydrostatic pressures. However, using a gasket with the solid (HgI), and adding a drop of Nujol and then applying pressure gave what appeared under the microscope to be a single pressure.[36] Similar results were obtained recently by Davies[18] with nickel dimethylglyoxime with other organic solvents. Further studies are now under way in our laboratory with other solids and other liquids. The use of a liquid under these conditions, with a freezing point at a known pressure can also give a pressure calibration point besides a hydrostatic pressure.

Discussion

It has been demonstrated that in compressible ionic solids lattice vibrations are pressure sensitive to frequency shift. In all cases studied thus far, the shift is in the direction of higher frequency in agreement with low temperature shifts. However, the pressure shifts may be considerable when compared with the temperature shifts. It is known that the temperature dependence of thé peak position and half-width of lattice vibrational modes consist of two contributions: (1) the purely volume dependent contribution; and (2) the contribution from the various anharmonic (cubic and higher) terms in the potential energy of the lattice. Often the directions of shifts in the temperature dependence experiment may be opposite to each other, resulting in a cancellation, and a small overall shift. Apparently molecular lattices, generally speaking, show similar pressure effects. It may be possible in some cases to obtain compressibility data in compressible solids from pressure-frequency measurements (see Equation 1). On the other hand, internal modes which consist of motions involving changes in interatomic distances are not
the symmetric mode and which is the antisymmetric mode was made by Raman polarizability data in solution. It is observed that although both peaks diminish in intensity and broaden, the symmetrical vibration (A_1 species) is the most sensitive to pressure. It is important to be able to distinguish between these two modes in making correct far infrared interpretations. The technique is able to do this on powdered or polycrystalline material. Heretofore, only single crystal-dichroism infrared measurements, single crystal or solution-Raman polarizability studies would make this differentiation.

Attempts to induce structural changes in coordination compounds upon the application of pressure to the solids have all failed[33] Of course, only pressures to about 60 kbars have been used. At these pressures only color changes were observed, similar to those found for nickel dimethylglyoxime.[17-18] These spectral changes with pressure can be observed under the microscope or in a visible spectrophotometer. These changes are not to be confused with any structural changes (e.g., Td $\rightarrow D_{4h}$ etc). It becomes apparent that to determine if any structural changes occur, one should go to solution studies. Such studies are now being initiated.

Newer Developments

Our interest in coordination compounds has motivated us to make a basic study of effects of pressure on the uncomplexed ligands as compared to those in the complexes. Although only preliminary results are available at this time, it appears that some molecular vibrations involving out-of-plane motions in ligands containing heterocyclic rings (e.g., pyrazine, dipyridyl etc.) are very sensitive to pressure.[34] Some in-plane motions are also sensitive. Thus, motions, generally involving large volume changes, may become pressure sensitive.

Although Lippincott et al.[4,5] examined some inorganic compounds with pressure, no systematic studies have been made in the mid-infrared region and certainly none made in the low frequency region. A systematic basic study of inorganic compounds from 1600-100 cm^{-1} has been commenced. Preliminary results[35] with salts such as K_2SO_4, K_2CrO_4, $KMnO_4$ etc. in the ν_4 region (MO_2 deformation vibration) indicate some pressure sensitivity. For example, the ν_4 (a triply degenerate vibration) vibration in $KMnO_4$ at ~400 cm^{-1} is seen as a doublet at ambient pressure. The doublet is attributed to a lowered symmetry (probably to C_{3v}). In this symmetry the two vibrations are the A_1 and E species. Differentiation be-

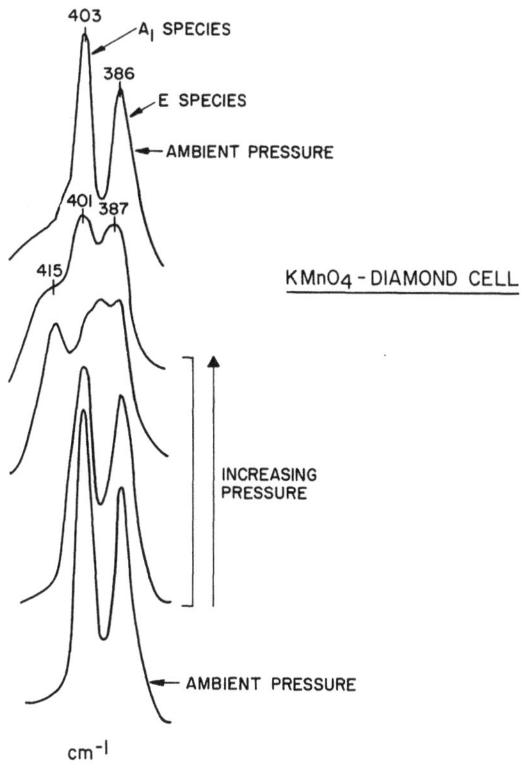

Fig. 9. Low Frequency Spectra of $KMnO_4$ with and without pressure.

tween the two species is normally difficult. Figure 9 illustrates the effect of pressure on the doublet. It is observed that the similarly effected. No noticeable shifts, or at the most, only slight shifts are observed. The direction of the slight shift may be either a blue or red shift. In hydrogen bonded compounds, studies of the OH stretching vibration has resulted in a red shift.[5,37] Red shifts were also observed for $HgCl_2$ and its dioxane complex by Brasch and Jacobsen.[38] In most of our studies with internal modes, blue shifts have been observed. The attempts at converting a structural isomer to another having failed in the solid state, indicates that in the pressures used (up to 60 kbars) little effect on the interatomic distances or bond angles occur. However, the results do point to the fact that intermolecular distances are changed (e.g., distance between ions or molecules) under these pressures. It is conceivable that higher pressures are needed to successfully effect the interatomic distances. It should be mentioned, however, that pressure effects appear to be sufficiently large to cause changes in symmetry in a molecule.

The pressure effects on the ionic or molecular lattice modes may be accounted for by the contraction of the solid under pressure, placing ions or molecules closer to each other. Obviously, repulsive and attraction forces in the solid are involved. For internal modes which shift only slightly toward higher frequency the repulsive force must be greater. The shifts of out-of-plane and in-plane vibrations in ring compounds toward higher frequency under pressure, may be attributed to the more energy required to cause the vibration, since the molecules are in closer proximity to each other. Several reasons can be cited for the splitting of bands. The discussion concerns bands with no change in phase occurring with an increase in pressure. One involves the failure to resolve a band into two components at room temperature and ambient pressure. With high pressure this is possible because of the pressure shifts occurring, and one component being shifted a greater amount than the other. Another reason may be due to the removal of degeneracy in degenerate modes. Broadening of all bands occurs and this may be due, at least in part, to the pressure gradient existing in the diamond cell.

The changes in intensities that occur under pressure probably involves large changes in dipole moments for certain modes of vibration. It has been definitely established that other physical effects occurring simultaneously with the electromagnetic effect, can cause changes in the dipole moment for a particular vibration. For example, electrical effects cause dipole moment changes to be non-zero ($\frac{\partial\mu}{\partial Q} \neq 0$) and forbidden frequencies can occur. Pressure effects may cause similar changes, although tending toward $\frac{\partial\mu}{\partial Q} \to 0$. It has been observed that the A_1 species is particularly sensitive in this regard.

RAMAN-HIGH PRESSURE STUDIES

Raman spectra of gases at high pressures have been obtained.[39-40] The first Raman spectra of micro quantities of solid using the diamond-anvil cell was reported in 1968.[41-42] The method complements the high pressure-infrared technique and allows a complete vibrational analysis of a molecule to be made at non-ambient pressures.

The instrumentation consisted of a Spectra-Physics Model 125 He-Ne laser (90 mW) with a Spex double monochromator at a slit width of 2 cm^{-1} resolution. Figure 10 shows the schematic diagram of 0° laser excitation. Figure 11 shows the actual apparatus used. Figure 12 shows the Raman spectra for HgI_2 at various pressures. The sample must be contained in a molybdenum steel gasket to build up the thickness of the sample. Best results were obtained for 0° radiation, although 180° radiation may also be used.

The Raman results are important, since the LO mode of crystals showing a first-order Raman spectrum are more intense in the Raman spectra, and thus are more easily studied. In addition, various infrared inactive or infrared weak vibrations may be studied with this technique.

LIMITATIONS OF TECHNIQUES

The above applications have been presented to attempt to illustrate the potential use of the high pressure technique for study of molecules in the low frequency range. The technique is new and much work remains to be done. Further, a better understanding is necessary as to what mechanisms are involved when a material is under pressure stress. More work with gaskets and with hydrostatic pressures is needed. Certainly precautions are necessary when using the average pressures obtained in the diamond anvil cell, and relating these to other pressure data. This is particularly dangerous in attempts to obtain quantitative data such as anharmonic contributions in ionic crystals. For qualitative work, the use of an average pressure appears quite satisfactory. The technique has warranted the attention it has received in the past two years. Although it may never realize all of its potentials ultimately, it has already demonstrated that it is a valuable tool for the chemist, physicist, spectroscopist, and solid-state scientist.

ACKNOWLEDGMENT

The author wishes to acknowledge the valuable contribution of Dr. Clarence Postmus in the aforementioned work.

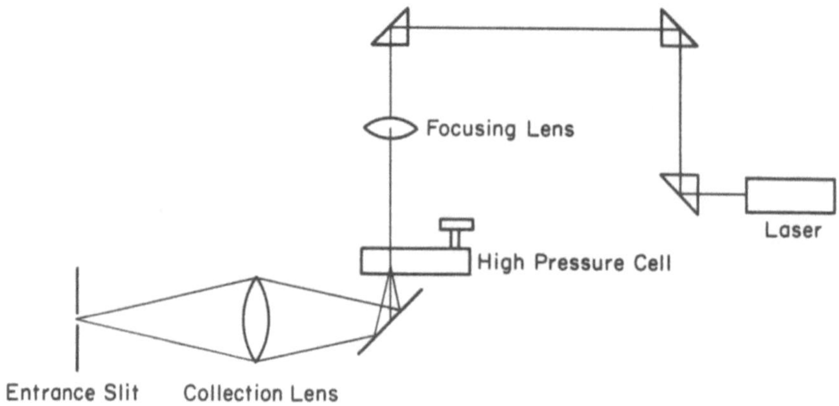

Fig. 10. Schematic Diagram for Laser Raman-High Pressure Experiment.

Fig. 11. Apparatus Used in Laser Raman-High Pressure Experiment.

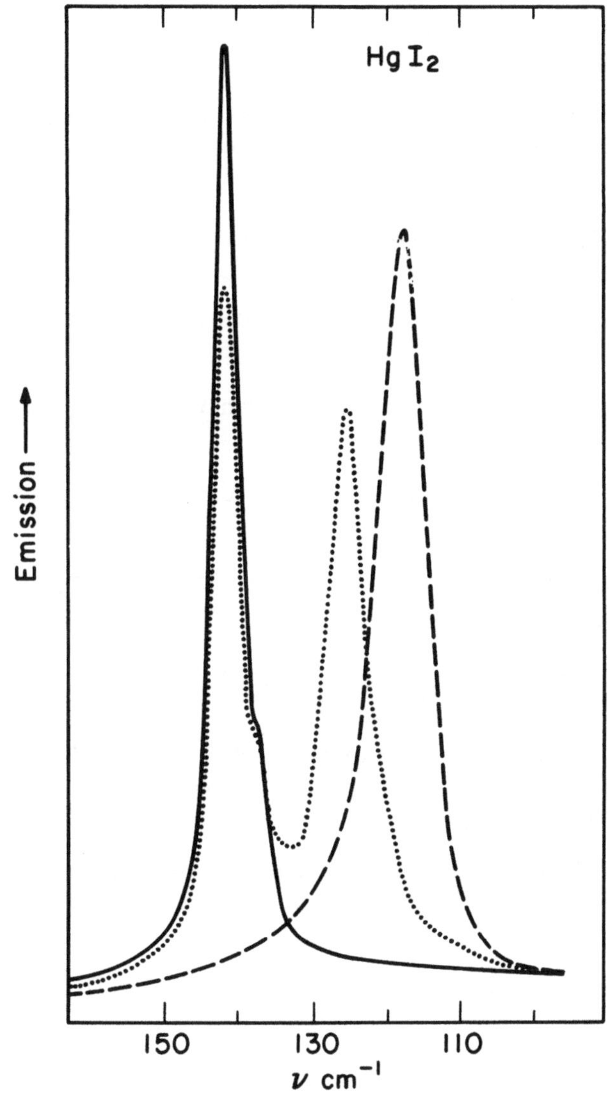

Fig. 12. The Spectra of HgI_2 at Various Pressures.

---- Atmospheric pressure (red phase only). Peak at 119 cm^{-1}.

oooo Intermediate pressure (red and yellow phases present). Peaks at 126 cm^{-1} and 143 cm^{-1} and a weak shoulder at 138 cm^{-1}.

—— High pressure (yellow phase only). Peak at 143 cm^{-1} with a weak shoulder at 138 cm^{-1}.

BIBLIOGRAPHY

1. E. Fishman and H. G. Drickamer, Anal. Chem. 28, 804 (1956).

2. H. G. Drickamer, "Optical Studies at High Pressure," Progress in Very High Pressure Research, J. Wiley and Sons, New York, 1961, pp. 16-27.

3. C. E. Weir, A. Van Valkenburg and E. R. Lippincott, J. Res. Nat. Bur. Stds. 63A, 55 (1959).

4. E. R. Lippincott, F. E. Welsh and C. E. Weir, Anal. Chem. 33, 137 (1961).

5. E. R. Lippincott, C. E. Weir, A. Van Valkenburg and E. N. Bunting, Spec. Acta 16, 58 (1960).

6. J. R. Ferraro, S. S. Mitra and C. Postmus, Inorg. Nucl. Chem. Letters, 2, 269 (1966).

7. C. Postmus, J. R. Ferraro and S. S. Mitra, Inorg. Nucl. Chem. Letters, 4, 55 (1968).

8. C. C. Bradley, H. A. Gebbie, A. C. Gilby, V. V. Kechin and J. H. King, Nature, 211, 839 (1966).

9. N. T. McDevitt, R. E. Witkowski and W. G. Fateley, Abstract, 13th Colloquium Spectroscopium Internationale, June 18-24, 1967, Ottawa, Canada.

10. H. G. Drickamer and A. S. Balchan, "Modern Very High Pressure Techniques," R. H. Wentorf, Jr., Editor, Butterworths, London 1962.

11. R. W. Parsons and H. G. Drickamer, J. Chem. Phys. 29, 930 (1958).

12. R. A. Fitch, T. E. Slykhouse and H. G. Drickamer, J. Opt. Soc. Am. 47, 1015 (1957).

13. A. S. Balchan and H. G. Drickamer, Rev. Sci. Instr. 31, 511 (1960).

14. T. Takahashi and W. A. Bassett, Science 145, 483 (1964).

15. J. W. Brasch and R. Jakobsen, Spec. Acta, 21, 1183 (1965).

16. J. W. Brasch, J. Chem. Phys. 43, 3473 (1965).

17. H. C. Duecker and E. Lippincott, Doctoral Thesis, Univ. of Maryland, (1964).

18. H. W. Davies, to be published.

19. D. W. Berreman, Phys. Rev. 130, 2193 (1963).

20. C. L. Bottger and A. L. Geddes, J. Chem. Phys. 46, 3000 (1967).

21. S. S. Mitra, C. Postmus and J. R. Ferraro, Phys. Rev. Letters, 18, 455 (1967).

22. J. R. Jasperse, A. Kahan, J. N. Plendl and S. S. Mitra, Phys. Rev. 146, 526 (1966).

23. C. Postmus, J. R. Ferraro, and S. S. Mitra, Phys. Rev. 174(3), 983 (1968).

24. M. Pagannone and H. G. Drickamer, J. Chem. Phys. 43, 2266 (1965).

25. C. F. Cline and D. R. Stephens, J. Appl. Phys. 36, 2869 (1965).

26. S. S. Mitra, Phys. Stat. Solids, 9, 519 (1965).

27. M. Born and K. Huang, "Dynamical Theory of Crystal Lattices," (Oxford Univ. Press, London, 1954), p. 52.

28. F. Seitz, "Modern Theory of Solids" (McGraw-Hill Book Co., New York, 1940), p. 80.

29. R. A. Cowley, Advan. Phys. 12, 421 (1963).

30. E. R. Cowley and R. A. Cowley, Proc. Roy. Soc. A287, 259 (1965).

31. S. S. Mitra and R. Marshall, J. Chem. Phys. 41, 3158 (1964).

32. C. Postmus, K. Nakamoto and J. R. Ferraro, Inorg. Chem. 6, 2194 (1967).

33. C. Postmus, J. R. Ferraro and K. Nakamoto, unpublished results.

34. R. Bayer and J. R. Ferraro, unpublished results.

35. J. R. Ferraro and C. Postmus, unpublished results.

36. C. Postmus and J. R. Ferraro, unpublished results.

37. A. M. Benson and H. G. Drickamer, Far. Soc. Disc. 39 (1956).

38. Y. Mikawa, R. J. Jakobsen and J. W. Brasch, J. Chem. Phys. 45, 4528 (1966).

39. H. L. Welsh, P. E. Baschler and B. P. Stoicheff, Can. J. Phys. 30, 99 (1952).

40. A. D. May, V. Degen, J. C. Stryland and H. L. Welsh, Can. J. Phys. 39, 1769 (1951).

41. C. Postmus, V. A. Maroni, J. R. Ferraro and S. S. Mitra, Inorg. Nucl. Chem. Letters, 4, 269 (1968).

42. J. W. Brasch and E. Lippincott, Chem. Phys. Letters, 2, 99 (1968).

RECENT PROGRESS ON THE LOW FREQUENCY SPECTRA OF COORDINATION COMPOUNDS*+

John R. Ferraro

Chemistry Division, Argonne National Laboratory

Argonne, Illinois USA

ABSTRACT

A review of the present status of the coordinated metal-ligand atom stretching vibration (where coordination involves oxygen, nitrogen, or halide of ligand) will be presented. The usefulness of the low frequency region of the infrared in elucidating structural information for coordination compounds will be discussed. Limitations of the technique will be reviewed.

INTRODUCTION

The proliferation of research in the area of far infrared spectroscopy of inorganic and coordination compounds has been phenomenal. As a result, no attempt will be made to provide a complete, comprehensive review of the subject. Instead, the discussion will bring the topic to its present state, and will concern only coordination compounds. It is understood that much of the discussion will also apply to inorganic molecules.

Of interest to us will be the region of 650 cm^{-1} and lower, as this range is exceedingly rich for coordination compounds.

---

*Based on work performed under the auspices of the U. S. Atomic Energy Commission.

+Presented at the NATO Advanced Study Institute on Far Infrared Properties of Solids, Delft, Netherlands, August 5-23, 1968.

However, it should be mentioned that the "far infrared" is considered to be the region beyond 200 cm^{-1} to 10 cm^{-1}.[1] Since we are discussing the region from 650 cm^{-1} and lower, we will call this the low frequency region rather than the "far infrared." The present low frequency instrumentation available, the usefulness of the region together with limitations, the present status of metal-ligand vibrations, and future direction of the technique will be covered.

Several reviews, which bear on the present discussion, are published. These are reviews by Stewart,[2] Wilkinson et al.,[3] Brasch et al.,[4] and Ferraro.[5]

Present Low Frequency Instrumentation

The tremendous surge of interest in the low frequency region began with the development of commercial far infrared instruments. First came the extension of the mid-infrared region by means of prisms such as KBr, CsBr or CsI. These could be inserted into the instruments in place of the NaCl prism. This caused the first ripple of interest. Then in 1962, the first far infrared grating instrument, specifically built for the "far infrared" was introduced. This was the Perkin-Elmer No. 301 spectrophotometer. It was followed in 1964 by the Beckman IR-11. This caused a major explosion of interest in this region. Gratings then were used to replace prisms in the conventional instruments, to allow one to go beyond 650 cm^{-1}. Today there are at least 20 commercial prism and/or grating instruments capable of going beyond 650 cm^{-1}. Exclusive of the interferometers there are 4 commercial grating instruments able to go lower than 100 cm^{-1}. At least 5 interferometers are now being commercially manufactured. Thus, it appears that adequate instrumentation is presently available. Far infrared spectroscopy is no longer the exclusive tool of the pioneer instrumentalists. The relative ease in which spectra can be obtained in the low frequency range has made the region accessible to chemists, analysts, technicians and any other curious individual.

Usefulness of Low Frequency Infrared Region

The type of information obtainable from this region for inorganic and coordination compounds is listed in Table I. The multitude of uses explains why the region has become important in the past 5-6 years. Of particular interest to the coordination chemist are items 3,4,5,6 and 12, with the main concern being the metal-ligand vibrations and the determination of stereochemistry. The mid-infrared region has not proven to be especially helpful for the study of coordination compounds, because it concerns itself only with ligand vibrations, as modified by complexation. In the low-frequency region, the vibrations which are observed involve metal-ligand motions, and are of the utmost importance in determining the stereochemistry of the complex. The factors determining the position of metal-ligand vibrations, and cited by Clark,[6] are shown in Table II. It becomes quite clear from the

TABLE I

Information Obtainable from the Far Infrared Region for Inorganic and Coordination Compounds

1. Ionic Lattice Vibrations.
2. Molecular Lattice Vibrations.
3. Coordinated Water Modes of Vibration.
4. Librations in Coordinated Polyatomic Anions (hindered rotatory motions).
5. Chain Vibrations in Inorganic Polymers.
6. Metal-Ligand Vibrations:

   e.g., M-O, M-N, M-S, M-X, M-C, where X = halogen.

7. Potential Barriers to Internal Rotation.
8. Potential Barriers to Inversion.
9. Rotational Spectra of Gases.
10. Low-lying Vibrations in Hydrogen Bonded Compounds.
11. Semiconductor Studies (electronic transition).
12. Determination of Stereochemistry.
13. Translation and Hindered Rotations of Gas Molecules in Clathrates.
14. Ion-pair Vibrations.
15. Thermodynamic Functions.

TABLE II

Factors Determining Position of Metal-Ligand Vibrations

Oxidation Number of Metal	→	Higher Oxidation Number - Higher Frequency
Mass of Metal and Ligand	→	Larger Mass - Lower Frequency
Coordination Number of Metal	→	Higher C.N. - Lower Frequency
Stereochemistry of Complexes	→	Frequency decreases from a D_{4h} to T_d to O_h Structure
Basicity of Ligand	→	Higher the Basicity - Higher the Frequency(1)
Counter Ion Effect	→	Increased Size of Counter-ion - Decreased Frequency
Bridging or Non-Bridging Ligand	→	Non-Bridging Ligands - Vibration at Higher Frequency.
High Ligand Field Stabilization Energy	→	Higher Energy - Higher Frequency.

(1) Only when sigma bonding is involved

Table that some idea as to the coordination number and oxidation state of the metal, stereochemistry of the complex and basicity of ligand may be possible. Furthermore, some notions as to type of bonding existing in the complex can be obtained. For example, the normal order of trend listed in Table II is followed in regards to the basicity of the ligand if sigma bonding predominates. Namely, the higher the basicity of the ligand the higher the frequency of the metal-ligand vibration. However, where π bonding predominates the trend reverses. This is quite possible in many of the transition metal elements with incomplete d shells. These elements have electrons available for back bonding (metal → ligand) and π bonding increases. In addition, it should be possible to distinguish between a bridging and non-bridging ligand. Some examples of this are available in the literature, although more studies of established bridged complexes are necessary.

Table III shows the number of modes of vibration theoretically possible for each stereochemistry.[6] Knowing how many vibrations are possible and how many are actually found experimentally, will help in determining the stereochemistry, at least in theory. This procedure is based on being able to make a correct interpretation of the spectrum in regards to the metal-ligand vibrations. However, it is not always possible to find all the vibrations needed to decide upon a particular stereochemistry. This is due to a multiple of problems occurring when one works in the low frequency region. Sampling of inorganic or coordination compounds is difficult, because some of these materials are hard to grind and a good dispersion in a matrix is difficult. This may result in a poorly resolved spectrum. Coupling of vibrations is very important in the low energy region and broad bands, not easily resolvable, ensue. These broad bands may hide weak vibrations. Coupling of the stretching vibration may occur with the bending, ligand and lattice modes. Further, many bands may be weak due to a significant lack of dipole moment change between the atoms involved in a vibration. The metal-nitrogen stretching vibration is one such vibration that is generally weak. High symmetry in a molecule may also contribute toward lowering the intensity of a vibration. On the other hand, more than the expected number of bands may be found due to the activation of some forbidden ligand vibrations upon complexation. This makes the interpretation of the spectra very difficult. These problems must be kept in mind when using the low frequency region to determine the stereochemistry of a complex. Despite these limitations, the method is particularly attractive to the coordination chemist since some inference as to the stereochemistry of the molecule may be made in a very short time and with little effort. In addition, for insoluble, intractible solids, it may not be possible to use any of the other structural tools. Whenever possible, other structural tools should be used in complementation with the low frequency infrared method. At best, the infrared method regardless of the region, is only an inference method, and in the final analysis, one of the diffraction techniques must be used for an answer with finality.

Certain measures are possible to aid interpretation in the low frequency region and these are listed in Table IV. The measures are not without difficulties. Deuteration of some hydrogen-bonded compounds fails to shift the low-lying $\nu_{OH...O}$ vibration. However, deuteration studies of the hydrates of transition metal salts and of rare earth sulfates have been successful in assigning the low-frequency coordinated water vibrations. The conversion of a chloride complex to a heavier halide complex (e.g., bromide, iodide), and the expectation of observing the ν_{MX} shift toward lower frequency doesn't always materialize. Examples are now known where the chloride complex may involve octahedral symmetry with chloride bridging, while the bromide involves a tetrahedral

TABLE III

Distribution of Normal Modes of Vibration in Compounds of Stoichiometry MX_mL_n*

Type	Stereochemistry	Symmetry Group	ν_{MX}	Number Infrared Active	ν_{ML}	Number Infrared Active
MX_4	Tetrahedral	T_d	$a_1 + f_2$	1	---	---
MX_3L	Tetrahedral	C_{3v}	$a_1 + e$	2	a_1	1
MX_2L_3	Tetrahedral	C_{2v}	$a_1 + b_1$	2	$a_1 + b_1$	2
MX_4	Planar	D_{4h}	$a_{1g} + b_{1g} + c_{1g}$	1	---	---
MX_3L	Planar	C_{2v}	$2a_1 + b_1$	3	a_1	1
MX_2L_2	trans-Planar	D_{3h}	$a_g + b_{3u}$	1	$a_g + b_{2u}$	1
MX_2L_2	cis-Planar	C_{3v}	$a_1 + b_1$	2	$a_1 + b_1$	2
MX_2L_4	trans-Octahedral	D_{4h}	$a_{1g} + a_{2u}$	1	$a_{1g} + b_{1g} + e_u$	1
MX_2L_4	cis-Octahedral	C_{2v}	$a_1 + b_1$	2	$2a_1 + b_1 + b_2$	4
MX_3L_3	trans-Octahedral	C_{2v}	$2a_1 + b_1$	3	$2a_1 + b_2$	3
MX_3L_3	cis-Octahedral	C_{3v}	$a_1 + e$	2	$a_1 + e$	2
MX_2L_2	Polymeric Octahedral	C_i	$2a_g + 2a_u$	2	$a_g + a_u$	1

*R. J. H. Clark and C. S. Williams, Inorg. Chem. 4, 351 (1965).

TABLE IV

Measures Used to Aid Low Frequency Infrared Interpretation

1. Deuteration Studies in Hydrogen-Containing Compounds.
2. Converting a Chloride Complex to a Bromide and/or Iodide Complex.
3. Low Temperature Studies.
4. Observation of Breadth and Intensity of Ligand and Metal-Ligand Absorptions.
5. Slow Neutron Scattering Studies of Hydrated Materials.
6. High Pressure Studies.
7. Complementation with Raman Spectroscopy.

TABLE V

Comparison of Infrared and Raman Intensities of Several Complexes (cm^{-1})

Complex	IR		Raman
$Zn(bipy)Cl_2$	323(vs)	ν_{ZnCl}	314(m)
	241(w)	ν_{ZnN}	237(m)
$Zn(bipy)Br_2$	261(s)	ν_{ZnBr}	260(m)
	250(s)	ν_{ZnN}	242(s)
$Zn(bipy)I_2$	250(vs)	ν_{ZnN}	243(s)
	217(vs)	ν_{ZnI}	not observed
$Mn(pyrazine)Cl_2$	246(s)	ν_{MnCl}	250(w,sh)
	220(s), 202(s)	ν_{MnN}	215(s)

TABLE VI

The Metal-Halogen Stretching Vibration (ν_{MX}) in Various M(II) Complexes (cm^{-1})

	$MX_2 \cdot 2L$ or $MX_2 \cdot LL$ Complexes			$MX_2 \cdot 2L$ Complexes			$MX_2 \cdot 4L$ Complexes
MX_2	Tetrahedral	Distorted Polymeric Octahedral	Polymeric Octahedral	Square Planar Trans	Square Planar cis	Square Planar	Monomeric Octahedral
$CrCl_2$		328,303					
$MnCl_2$	272-311		233				
$MnBr_2$	230-248		< 200				
$FeCl_2$	288-315		(222)				
$FeBr_2$	235		< 200				
$CoCl_2$	274-351		(224-242)				230
$CoBr_2$	237-274		< 200				< 200
CoI_2	237						
$NiCl_2$	282-343		238				246
$NiBr_2$	234-258		< 200				< 200
NiI_2	< 200		< 200			218	< 200
$CuCl_2$	305-319	294,235					
$CuBr_2$	219-229	255,202					
$ZnCl_2$	277-345		218-225				
$ZnBr_2$	198-265		188				

TABLE VI (Continued)

MX_2	$MX_2 \cdot 2L$ or $MX_2 \cdot LL$ Complexes			$MX_2 \cdot 2L$ Complexes			$MX_2 \cdot 4L$ Complexes
	Tetrahedral	Distorted Polymeric Octahedral	Polymeric Octahedral	Square Planar Trans	Square Planar Cis	Square Planar	Monomeric Octahedral
$CdCl_2$		< 200(?)	205-230				
$CdBr_2$			186				
$HgCl_2$	254-342	292					
$HgBr_2$	198-214	215					
$RhCl_2$							
$PdCl_2$				332-362	306-335		
$IrCl_2$							
$PtCl_2$				328-344	277-344		
$PtBr_2$				227-259	211-254		

( ) = Unassigned

L = pyridine and related ligands

LL = bidentate ligand

symmetry with no halide bridging. Thus, the ν_{MBr} vibration will be found at the same or higher frequency that the ν_{MCl} vibration. The use of low temperature to distinguish between internal modes and lattice modes proves to be ineffectual in some cases, since the shifts of lattice modes due to temperature effects may be very minimal. Observation of the breadth and the intensities of the low-frequency vibrations can be of some help at least qualitatively. Ligand bands are usually sharp, whereas metal-ligand bands are broad. Generally, ν_{MO} vibrations are more intense than ν_{MN} vibrations, and ν_{MX} vibrations are usually equal to or more intense than ν_{MN}. The effect of pressure will be discussed elsewhere. The use of laser Raman instrumentation hasits limitations, too. If the material is finely powdered it may still be difficult to obtain a good Raman spectrum. In theory, the use of Raman should be helpful in compounds manifesting high symmetry (e.g., O_h, D_{4h}, T_d). It is also of use in differentiation between a metal-nitrogen and a metal-halide vibration. The metal-halide vibration will be normally more intense in the infrared than the metal-nitrogen. The reverse should be true in the Raman, and thus if a good Raman spectrum can be obtained, differentiation of the two bands may be easily made. Table V lists some recent unpublished Raman data[7] for some dipyridyl complexes.

Metal-Ligand Vibrations in Coordination Complexes

Stretching vibrations involving metal-halide, metal-nitrogen and metal-oxygen in coordination complexes have been assigned in the region 650-200 cm^{-1}. Less well defined are the metal-sulfur, metal-phosphorus, metal-selenium, etc. stretching vibrations. It must be noted that the stretching vibrations are the best characterized and then only the ν_{MCl} vibrations are well established. The ν_{MBr} vibration is less established, and few ν_{MI} vibrations assigned. The bending vibrations are less well defined. They are located, in most cases, below 200 cm^{-1} in a region involving the low frequency modes (e.g., molecular lattice modes), which complicate the interpretation. Only the metal-ligand stretching vibrations in the complexes of the transition metals of the first row and of the lower oxidation state, have been studied extensively.

Metal-Halide Stretching Vibration. Table VI presents the metal-halide stretching data for various metal(II) complexes. These data illustrate some of the trends cited by Clark.[6] For example, as the structure changes from $T_d \rightarrow O_h$ and the coordination number (CN) from $4 \rightarrow 6$, the frequency lowers. These data also illustrate the lower frequency of a bridged halide vs. a terminal halide. Although only a few data are available, it is strange to note that little difference is found between the ν_{MX} in a polymeric octahedral complex with bridging halides, and a monomeric octahedral complex with terminal halides.

TABLE VII

The Metal-Halogen Stretching Vibration (ν_{MX}) in Various M(III) Complexes (cm^{-1})

	Monomeric Octahedral $[ML_4X_2]^+$ or $[ML_2L_2X_2]^+$	$MX_3 \cdot 3L^3$	$[ML_3^2X]^+$ Sq. Planar
$[Rh\ L_2 \cdot L_2^1Cl_2]^+$	343		
$[Rh\ L_2 \cdot L_2^1Br_2]^+$	223		
$[Rh\ L_2 \cdot L_2^1I_2]^+$	178		
$[Pd\ L_3^2Cl]^+$			333
$[Pd\ L_3^2Br]^+$			241
$[Pd\ L_3^2I]^+$			<200
Ti,V,Cr (Chlorides)		300-364	
$[Ir\ Cl_2L_4]^+$	335(trans)		
$[Ir\ Cl_2L_4]^+$	333,327(cis)		
$CrCl_3$		364,341,307	
$RhCl_3$		355,332,295(trans)	
$RhCl_3$		341,325(cis)	
$IrCl_3$		329,318,307(trans)	
$IrCl_3$		325,217,303(cis)	
QuH $FeCl_4$	369		
QuH $FeBr_4$	291,280		
$InCl_3(bipy)_{1.5}$	251-300		
$In\ Br_3(bipy)_{1.5}$	191-200		

TABLE VII (Continued)

	Monomeric Octahedral $[ML_4X_2]^+$ or $[ML_2L_2X_2]^+$	$MX_3 \cdot 3L^3$	$[ML_3^2X]^+$ Sq. Planar
$InI_3(bipy)_{1.5}$	117-134		

1. ethylenediamine ligand

2. diethylenediamine ligand

3. pyridine

QuH = quinolinium

It is also possible to differentiate between the cis and trans square planar of complexes of the type $MX_2 \cdot 2L$, since the ν_{MX} vibration is at lower frequency for the cis isomer, and the cis compound will show two ν_{MX} stretches while the trans shows only one.

Table VII presents metal-halide stretching data for transition metal(III) complexes. Fewer data are available for the complexes of the higher oxidation state of the transition metals. It can be observed that all ν_{MCl} vibrations are in the region of 300-350 cm^{-1}, whereas the ν_{MCl} vibrations in M(II) complexes are centered at 200-350 cm^{-1}. In contrast to the M(II) complexes, little difference is noted for the M(III) complexes in cis and trans square planar isomers.

Table VIIIA shows a few coordination complexes of Ti(IV) and Sn(IV) fluorides. Both the stretching and bending vibrations have been assigned.[8] It is observed that the ν_{MF} stretching vibration is located at much higher frequency than the other halides, as expected. Table VIIIB shows the metal-halogen vibrations in several complexes of Ti(IV), V(IV) and Sn(IV). For a review of metal-halogen frequencies see Clark[8] and Adams.[9]

TABLE VIII A

The Metal-Fluoride Vibrations in Several Complexes (cm^{-1})

	ν_{MF}	δ_{FMF}
$TiF_4 \cdot bipy$[1]	634,562,450(?)	311,272,254
$TiF_4 \cdot phen.$	645,570,462	308,276,256
$Ti(acac)_2F_2$	629	306(?)
$Sn(acac)_2F_2$	581	---

TABLE VIII B

The Metal-Halide Vibrations in Several Complexes of Ti(IV), V(IV) and Sn(IV) (cm^{-1})

	ν_{MX}
$TiCl_4 \cdot bipy$[2]	384,366
$TiCl_4 \cdot phen.$	379,356
$V\ Cl_4 \cdot bipy$	366,356
$V\ Cl_4 \cdot phen.$	380,361
$MeSnCl_3 \cdot bipy$[3]	289,267
$MeSnCl_3 \cdot phen.$	296,272
$SnCl_4 \cdot bipy$[4]	282-337
$SnBr_4 \cdot bipy$	220-236
$SnI_4 \cdot bipy$	176-189

Metal-Nitrogen Stretching Vibration. Table IX lists the data for the ν_{MN} vibrations in various complexes involving the M(II) transition metals. The metal-nitrogen vibrations are generally weak in the infrared, and there has been much difficulty in making good interpretations. In the early work several of these bands were missed. The various trends cited for the ν_{MX} vibrations also are existent for these data.

Table X presents similar data for the M(III) transition metals.

For a review on the metal-nitrogen vibrations see Adams[9] and Durig and Wertz.[10]

Figure 1 lists several correlations made for metal-ligand vibrations. These correlations should be used with great caution, since some of them are based on a limited number of data.

Metal-Oxygen Stretching Vibration. Figure 2 lists several correlations made for metal-oxygen vibrations. These involve several coordination complexes and in addition, several inorganic compounds of polyatomic salts, in which coordination between metal and oxygen exists. The metal-oxygen stretching band can occur anywhere between 1100-200 cm^{-1}. The highest frequencies are associated with double bonds, while those below 650 cm^{-1} with single bonds.

Summary

It appears that, as of the present data, the following ranges exist for metal-ligand stretching vibrations:

$\nu_{MX}(cm^{-1})$

	ν_{MF}	ν_{MCl}	ν_{MBr}	ν_{MI}
M(II)		200-350	275 and lower	< 200
M(III)		300-350	223-241	< 200
M(IV)	450-650			

$\nu_{MN}(cm^{-1})$

M(II)	220-270	pyridine complexes
	430-520	ethylenediamine complexes
	298-334	amine complexes
M(III)	230-290	pyridine complexes
	400-540	amine complexes
	475-610	ethylenediamine complexes

TABLE IX

The Metal-Nitrogen Stretching Vibration (ν_{MN}) in Various M(II) Complexes (cm^{-1})

	$MX_2 \cdot 2L$ or MX_2LL Complexes			$MX_2 \cdot 4L$
	Tetrahedral	Distorted Polymeric Octahedral	Polymeric Octahedral	Monomeric Octahedral
$CrCl_2$		219		
$MnCl_2$			212	
$MnBr_2$			212	
$FeCl_2$				
$FeBr_2$				
$CoCl_2$	252			217 (trans)
$CoBr_2$	250			
CoI_2	246			
$NiCl_2$	239,227		<200	236(trans)
$NiBr_2$	239		<200	
NiI_2	240		<200	
$CuCl_2$		268		
$CuBr_2$		269		
$ZnCl_2$	220			
$ZnBr_2$	220			
ZnI_2	220			
$CdCl_2$		<200		
$CdBr_2$			<200	
CdI_2			<200	
$HgCl_2$		<200		
$HgBr_2$		<200		
$PdCl_2$				516,446[a]
$PtCl_2$				430-520[a]

L = pyridine and related ligands except where noted.

(a) = ethylenediamine

TABLE X

The Metal-Nitrogen in Stretching Vibration (ν_{MN}) in Various M(III) Complexes (cm^{-1})

	$[MX_4L_2]^-$	Monomeric Octahedral ML_3X_3	$[ML_4X_2]^+$ or $[ML_2L_2X_2]^+$
Cr		221	
Ir	265(trans) 262,255(sh)(cis)		
Rh		265,245,230(trans) 266,245 (cis)	246(trans)
Ir		272,264,255(trans) 270,266(sh)(cis)	260,255(sh)(trans) 287,273(cis)
Cr			560,520[(a)]
Co			503-610[(a)]
Rh			475-590[(b)]

L = pyridine unless noted

a = ethylenediamine

b = N-methylethylenediamine

For M(III) amines the range is 400-540 cm^{-1}

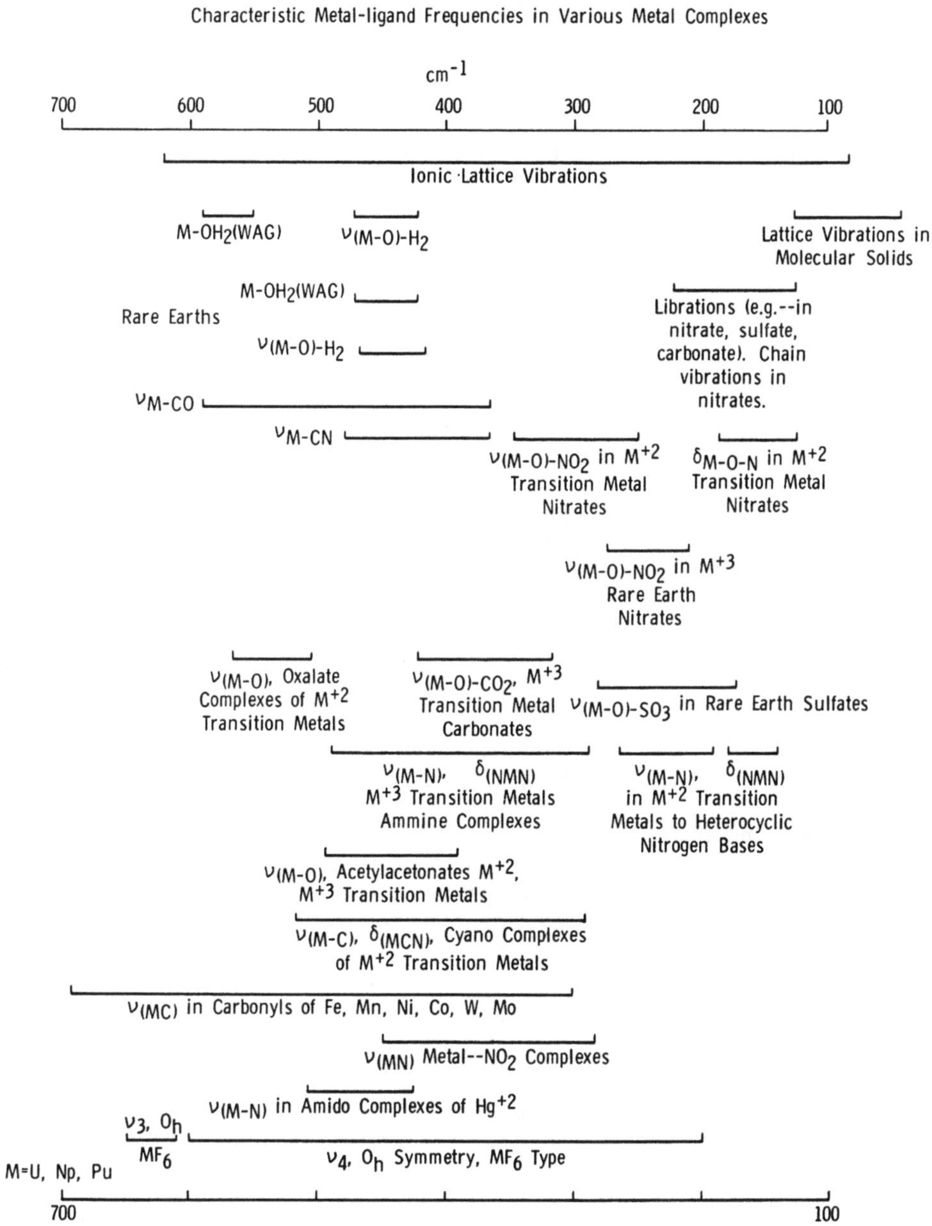

Fig. 1. Low Frequency Correlations for Metal-Ligand Vibrations.

$\nu_{MO}(cm^{-1})$

200-650 cm^{-1} with M(III), always higher than M(II).

$\nu_{M\text{-}CO}(cm^{-1})$[9]

For binary metal carbonyls - $M(CO)_n$, where n = 4, 5 or 6.

ν (M-CO) 357-556
δ (MCO) 300-785

Neutral binary carbonyls and hydrocarbonyls

ν (M-CO) 357-477
δ (MCO) 460-752

Substituted Octahedral Carbonyls

ν (M-CO) 366-428
δ MCO 468-787

$\nu_{M\text{-}CN}$[9]

Linear Dicyanides

ν (M-CN) 360-452
δ (MCM) 250-354

The assignments for ν_{MS} stretching vibration are less definite. However, recent work[11] indicates that for a sulfur atom acting on a monatomic ligand, $\nu_{M\text{-}S}$ is in the range 352-504 cm^{-1}. In cases where it acts as a bidentate atom, ν_{MS} may be in the range 278-376 cm^{-1}. Flint and Goodgame[12] report it at 200-300 cm^{-1} in thiourea complexes.

Future Direction

It has been demonstrated that the low frequency region is a very useful one for the coordination chemist. Much information can be acquired with a minimum of time and energy expended. Certain limitations and pitfalls have been mentioned. As more and more information is accumulated on metal-ligand vibrations in the ensuing years, and as more comparisons are made with other structural tools, a better understanding as to the reliability of the structural inferences obtained from low frequency data will be possible. There is much to be done. More work must be done with M(III) transition metals (of the other rows). More studies are necessary with the bromides and iodides. The region from 200-50 cm^{-1} must be assigned if possible. At present little bending vibrations in coordination compounds have been assigned. In summary, the use of low frequency infrared appears to hold much promise for the coordination chemist. Much useful information can be obtained despite the problems. However, the limitations and dangers present in the technique must not be overlooked or minimized. As more reliable research develops, it is hopeful that the technique will be found to live up to its present promise.

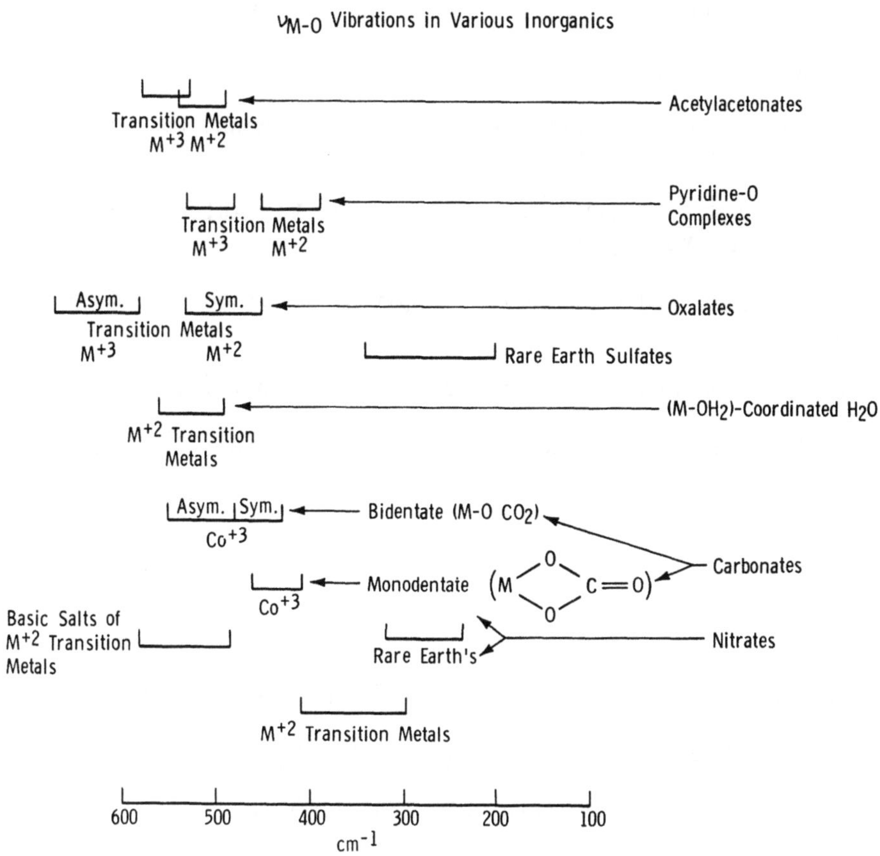

Fig. 2. ν_{MO} Vibrations in Various Inorganics.

Bibliography

1. Triple Commission on Spectroscopy, as reported in J. Opt. Soc. Am., 52, 476 (1962).

2. J. Stewart, "Interpretative Spectroscopy," J. K. Freeman, Ed., Reinhold Publishing Corp., New York (1965).

3. G. K. Wilkinson, S. A. Inglis and C. Smart, "Spectroscopy," M. S. Wells, Ed., Institute of Petroleum, London, 1966, pp. 157-80.

4. J. W. Brasch, Y. Mikawa, and R. J. Jakobsen, "Applied Spectroscopy Reviews," Vol. I, M. Dekker, New York, 1968.

5. J. R. Ferraro, Anal. Chem. 10, 24A (1968).

6. R.J.H. Clark and C. S. Williams, Inorg. Chem., 4, 351 (1965).

7. J. R. Ferraro, unpublished data.

8. R.J.H. Clark, "Halogen Chemistry," V. Gutmann, Ed., Academic Press, New York (1967).

9. D. M. Adams, "Metal-Ligand and Related Vibrations," St. Martins Press, New York (1968).

10. J. R. Durig and D. W. Wertz, Applied Spectroscopy, 22, 627 (1968).

11. D. M. Adams and J. B. Correll, J. Chem. Soc. (A), 1300 (1968).

12. C. D. Flint and M. Goodgame, J. Chem. Soc.(A), 144 (1966).

References for Table VI

1. R.J.H. Clark and C.S. Williams, Inorg. Chem. 4, 351 (1965).

2. R.J.H. Clark, Spec. Acta, 21, 955 (1965).

3. R. A. Walton, Spec. Acta, 21, 1795 (1965).

4. W. J. Eilbeck, F. Holmes, C. E. Taylor and A. E. Underhill, J. Chem. Soc. (A), 6 (1968); 128 (1968).

5. J. Burgess, Spec. Acta, 24A, 277 (1968).

6. R.J.H. Clark and C. S. Williams, Chem. and Ind. 1317 (1964).

7. W. R. McWhinnie, J. Inorg. Nucl. Chem., 27, 2573 (1965).

8. M. Goldstein and E. F. Mooney, Spec. Acta, 21, 105 (1965).

9. R. G. Inskeep, J. Inorg. Nucl. Chem., 24, 763 (1962).

10. J. R. Allan, D. H. Brown, R. H. Nuttall and D.W.A. Sharp, J. Inorg. Nucl. Chem., 27, 1305 (1965).

11. C. W. Frank and L. B. Rogers, Inorg. Chem., 5, 615 (1966).

12. D. H. Brown, R. H. Nuttall, J. McAvey and D.W.A. Sharp, J. Chem. Soc. (A), 892 (1966).

13. W. J. Eilbeck, F. Holmes, and A. E. Underhill, J. Chem. Soc. (A), 5 (1967).

14. N. S. Gill and H. J. Kingdon, Aust. J. Chem., 19, 2197 (1966).

15. R.J.H. Clark and C. S. Williams, Spec. Acta, 23A, 1055 (1967).

16. P. J. Hendra and N. Sadisivan, Spec. Acta, 21, 1271 (1965).

17. M. Goldstein and E. F. Mooney, J. Inorg. Nucl. Chem., 27, 1601 (1965).

18. G. E. Coates and D. Ridley, J. Chem. Soc., 166 (1964).

19. C. Postmus, J. R. Ferraro, and W. Wozniak, Inorg. Chem. 6, 2030 (1967).

20. D.M. Adams, "Metal-Ligand and Related Vibrations," St. Martins Press, New York (1968).

21. R.J.H. Clark, "Halogen Chemistry," V. Gutmann, Editor, Academic Press, New York (1967).

References for Table VII

1. R.J.H. Clark and F. B. Taylor, J. Chem. Soc. (A), 693 (1967).

2. A. J. Carty, S. J. Patel and D. G. Tuck, Can. J. Chem., accepted for publication.

3. C. H. Perry, D. D. Athans, E. F. Young, J. R. Durig, and B. R. Mitchell, Spec. Acta, 23A, 137 (1968).

4. R.J.H. Clark and C. S. Williams, Inorg. Chem., 4, 351 (1965).

5. R.J.H. Clark, "Halogen Chemistry," V. Gutmann, Ed., Academic Press, New York (1967).

6. D. M. Adams, "Metal-Ligand and Related Vibrations," St. Martins Press, New York (1968).

References for Tables VIII A and B

1. R.J.H. Clark, "Halogen Chemistry," V. Gutmann, Ed., Academic Press, New York (1967).

2. R.J.H. Clark, J. Chem. Soc., 1377 (1963).

3. R.J.H. Clark and C. S. Williams, Spec. Acta, 21, 1861 (1965).

4. M. F. Farona and J. G. Grasselli, AFML-TR-67-272, October, 1967.

References for Table IX

1. D. M. Adams, "Metal-Ligand and Related Vibrations," St. Martins Press, New York (1968).

2. J. Durig and D. W. Wertz, Applied Spectroscopy, 22, 627 (1968).

3. Most all references appearing in Table VI.

4. R.J.H. Clark and C.S. Williams, Inorg. Chem., 4,351 (1965).

References for Table X

1. D. H. Adams, "Metal-Ligand and Related Vibrations," St. Martins Press, New York (1968).

2. L. Sacconi, A. Sabatini, and P. Gans, Inorg. Chem. 2, 1772 (1964).

3. I. Nakagawa and T. Shimanouchi, Spec. Acta, 22, 759 (1966).

4. R.J.H. Clark and C. J. Williams, Inorg. Chem., 4, 351 (1965).

LOW FREQUENCY VIBRATIONS OF POLYMERS

Giuseppe Zerbi

Istituto di Chimica Industriale Del Politecnico, Milano

Italy

I. MOLECULAR DYNAMICS

The branch of science which deals with the study of the atomic vibrations of finite molecules and crystals is generally represented by two groups of workers who classify themselves as physicists if they work in a Cartesian space and as physical chemists if they work in an internal coordinate space. Since most of the theories described in the books of solid state physics are expressed in terms of a Cartesian space, we wish first to discuss briefly the principles, usefulness and limitations of the so-called Wilson-Eliashevitch algorithm widely used by chemists and neglected by most of the physicists. We believe this method gives a deep insight into the problems of molecular and crystal dynamics.

In order to describe the motion of the nuclei of an N atomic molecule, a set of 3N independent coordinates is required. For a non-linear molecule, the vibrational motion of the nuclei can be described by 3N-6 coordinates when they are suitably chosen as to satisfy the Eckart-Sayvetz conditions[3,4]. Of the six remaining coordinates three are required to describe the translational motion of the system as a whole and three are required to describe the rotational motion of the system as a whole about its center of mass.

If the quadratic approximation is used (springs and balls model) the vibrational potential energy V can be written, in matrix notation, in a bilinear quadratic form:

$$2V = \tilde{q} f q \qquad (1)$$

where $\mathbf{q}$ is the column vector of the 3N working coordinates and f is a square matrix whose elements are

$$f_{ij} = \left(\frac{\partial^2 V}{\partial q_i \partial q_j}\right)_{\text{equilibrium}} \quad (2)$$

$\sim$ indicates the transpose of a matrix. Classical mechanics provide the way of finding a linear transformation from the working coordinates to a set of 3N generalized coordinates Q_i, which allows one to write the potential energy V and kinetic energy T as

$$2V = \tilde{\mathbf{Q}} \mathbf{\Lambda} \mathbf{Q} \quad (3a)$$

$$2T = \tilde{\dot{\mathbf{Q}}} \dot{\mathbf{Q}} \quad (3b)$$

where $\mathbf{Q}$ is the column vector of the so-called normal coordinates and $\dot{\mathbf{Q}}$ is the column vector of the corresponding velocities. $\mathbf{\Lambda}$ is a diagonal matrix which contains the roots of a determinantal equation of order 3N. The vibrations described by the Q's are called normal vibrations. During a normal vibration all atoms move with the same frequency $\nu_i(\text{cm}^{-1}) = \frac{\sqrt{\lambda_i}}{2\pi c}$ in such a way that the cartesian components of the displacements are sinusoidal. The actual vibrational motion of the molecular system is obtained by a superposition of the 3N-6 genuine normal vibrations which correspond to the 3N-6 non-zero λ's.

The chemical feeling comes into play with the suggestions (independently proposed in 1939 by E. B. Wilson[1] and by Eliashevitch[2]) of setting up the vibrational problem in terms of 3N-6 "Internal Displacement Coordinates". It is indeed possible to find 3N-6 linear combinations of the cartesian displacement coordinates for defining the following basic types of internal displacement coordinates (hereafter indicated by R): stretching of the chemical bond, bending of the valence angle, change of the dihedral angle between the two planes defined by four atoms bonded in sequence, change of the angle between a bond and a plane defined by two bonds all having one atom in common. Let the linear transformation be written in matrix form as

$$\mathbf{R} = \mathbf{B}\mathbf{X} \quad (4)$$

where the elements of the $\mathbf{B}$ matrix are constants[5] determined by the geometry of the molecule (which is generally known). The vibrational potential energy V can then be written as

$$2V = \tilde{\mathbf{R}} \mathbf{F} \mathbf{R} \quad (5)$$

where $\mathbf{F}$ is the matrix of the coefficients in R space of the quadratic form analogous to those of eq. 2. Let us point out that f_{ii} is the force constant of the ideal spring associated with the i-th internal coordinate and f_{ij} is the interaction constant between the i-th and j-th internal coordinates. $\mathbf{F}$ describes the more General

Valence Force Field (GVFF). The vibrational kinetic energy is written as

$$2T = \tilde{\dot{\mathbf{R}}}\mathbf{G}^{-1}\dot{\mathbf{R}} \qquad (6)$$

where

$$\mathbf{G} = \mathbf{B}\mathbf{M}^{-1}\tilde{\mathbf{B}} \qquad (7)$$

$\mathbf{M}^{-1}$ is a diagonal matrix determined by the inverse of the atomic masses. Tables for the construction of the $\mathbf{G}$ matrix elements are available in the literature [6-8] for those who do not wish to construct the $\mathbf{B}$ matrix and perform the necessary matrix multiplications.

We then look for a linear transformation which relates the working coordinates R to a set of 3N-6 normal coordinates. Let the linear transformation be written as:

$$\mathbf{R} = \mathbf{L}\mathbf{Q} \qquad (8)$$

Substitution in eqs. (5) and (6) and taking into account eq. (3a) and (3b) one can write:

$$2V = \tilde{\mathbf{Q}}\tilde{\mathbf{L}}\mathbf{F}\mathbf{L}\mathbf{Q} = \tilde{\mathbf{Q}}\boldsymbol{\Lambda}\mathbf{Q} \qquad (9a)$$

$$2T = \tilde{\dot{\mathbf{Q}}}\tilde{\mathbf{L}}\mathbf{G}^{-1}\mathbf{L}\dot{\mathbf{Q}} = \dot{\mathbf{Q}}\mathbf{E}\dot{\mathbf{Q}} \qquad (9b)$$

where $\mathbf{E}$ is the unit matrix. One can then write:

$$\tilde{\mathbf{L}}\mathbf{F}\mathbf{L} = \boldsymbol{\Lambda} \qquad (10a)$$

$$\tilde{\mathbf{L}}\mathbf{G}^{-1}\mathbf{L} = \mathbf{E} \qquad (10b)$$

From eq. (10b), $\tilde{\mathbf{L}} = \mathbf{L}^{-1}\mathbf{G}$; upon substitution in eq. (10a) and premultiplication by $\mathbf{L}$ one obtains the well known eigenvalue equation

$$\mathbf{G}\mathbf{F}\mathbf{L} = \mathbf{L}\boldsymbol{\Lambda} \qquad (11)$$

This is a set of simultaneous equations which provides the way to find the desired transformation matrix $\mathbf{L}$ introduced in eq. (8). Non-trivial solutions of eq. (11) are obtained by finding the characteristic roots of the secular determinant:

$$\left|\mathbf{G}\mathbf{F} - \mathbf{E}\lambda_{\kappa}\right| = 0 \qquad (12)$$

Once the matrix $\boldsymbol{\Lambda}$ is obtained, substitution in eq. (11) gives the desired matrix $\mathbf{L}$.

A few details need to be pointed out:

1) In eqs. (11) and (12) masses and geometry enter through the $\mathbf{G}$ matrix defined in eq. (7).
2) A decision must be made as to the values of the elements of the matrix $\mathbf{F}$ introduced in eq. (5). This matrix has (3N-6) (3N-5)/2 independent elements for a completely asymmetrical molecule. Even if symmetry drastically decreases the number of independent force constants, they are still,

in general, too many to be determined from the experimental data. In order to make the problem solvable, approximations are introduced in the following way:

i) arbitrarily introduce in the GVFF many zeroes or some relationship between force constants.

ii) Set up a particular model of force field and apply it to several molecules.

The complete freedom left to the worker by the former method is removed by the second method since it forces zeroes and numbers at least in a systematic way. The reader is referred to the enormous amount of work reported in the literature for making up his own judgment on the validity of the various force fields proposed[9]. It can be stated, however, that at the present time the validity of a given force field is judged in terms of its capability of reproducing the fundamental frequencies of many similar molecules (i.e. it must be transferable from molecule to molecule), its capability of reproducing the other spectroscopic or physical quantities which are related to the fundamental frequencies and to the corresponding calculated amplitudes of vibration.

3) The values of the force constants for any given force field adopted are obtained by solving eq. (12) in a somewhat different way. If the experimental frequencies are known, the elements of the **F** matrix are taken as unknown quantities. Mathematical techniques have been developed for this purpose and a large amount of papers is available in the literature for the interested reader(9, 10-14).

4) Electronic computers are generally used for performing these calculations since the size of the molecules (i.e. the number of degrees of freedom) is very large. Computer programs are available in several specialized laboratories.

A few more theoretical details need to be pointed out for a better understanding of the problems which will follow.

5) When the number of R's is larger than the number of degrees of freedom (unfortunately this is the most common case!) it is said that the R's form a redundant set of coordinates (1) and that one or more linear (5,15) relations exist among them. Some matrices which are derived carry the annoying property of being singular(16). This should be avoided generally. The removal of the redundant coordinates (hereafter called local redundancies) require the exact knowledge of the explicit form of the geometrical relationship. Methods have been proposed for finding these relations. They can be classified as: a) explicit derivation of the geometrical relation (17-19), b) brute force numerical procedure by electronic computers (20).

6) In order to take full advantage of the symmetry of the molecule by application of group theory, it is possible to find linear combinations of the R's which factor the secular determinant (Eq. 12) into blocks of smaller size, thus reducing the computational labor[1,21,22]. Another advantage of the collection of the normal frequencies and normal modes into symmetry species is that they give immediate information to be matched with the experimental data.

The same concepts can be extended when infinite chain molecules (polymers) must be studied. Let us define a polymer chain as an infinite sequence of molecules linked to each other by some type of forces. Let the chain be linear and ideally perfect. If the concepts just discussed are applied to an infinite chain one must deal with infinite **F** and **G** matrices. We wish to go from a problem of infinite size to a finite one[23,25].

Let p be the number of atoms in each chemical repeating unit. Then let the index n label each chemical unit which repeats itself along the chain and let R_i^n describe the i-th internal displacement coordinate which belongs to the n-th chemical unit. In other words each chemical unit has its own set of R's which interact among themselves and with those of the neighboring units. The vibrational potential energy in the harmonic approximation (25) can be written as:

$$2V = \sum_{n,n'} \sum_{i,k} F_{ik}^{nn'} R_i^n R_k^{n'} \tag{13}$$

where

$$F_{ik}^{nn'} = \left(\frac{\partial^2 V}{\partial R_i^n \, \partial R_k^n} \right)_{eq.} \tag{14}$$

hence

$$F_{ik}^{nn'} = F_{ki}^{nn'} \tag{15}$$

The periodicity of the chain requires that $F_{ik}^{nn'}$ must depend on the difference $|n-n'| = s$. Equation (13) becomes

$$2V = \sum_{n} \sum_{i,k} F_{ik}^{o} R_i^n R_k^n + \sum_{n,s} \sum_{i,k} \left(F_{ik}^{s} R_i^n R_k^{n+s} + F_{ki}^{s} R_i^n R_k^{n-s} \right) \tag{16}$$

The kinetic energy can be written in an analogous way, in terms of the momenta P_i^n conjugated to the coordinate R_i^n and in terms of the **G** (defined in eq. (7), as

$$2T = \sum_{n} \sum_{i,k} G_{ik}^{o} P_i^n P_k^n + \sum_{n,s} \sum_{i,k} \left(G_{ik}^{s} P_i^n P_{k+}^{n+s} G_{ki}^{s} P_i^n P_k^{n-s} \right) \tag{17}$$

The equation of motion can be written using eqs. (16) and (17) thus leading to a system of an infinite number of second order differential equations in the unknowns R_i^{n+s} whose solution can be assumed to be a plane wave of the type

$$R_i^{n+s} = A_i \cdot e^{-i(\omega t + s\phi)} \tag{18}$$

In eq. (18) the amplitude A_i is independent of n; ϕ is the phase shift between two equivalent adjacent internal displacement coordinates and ω is the circular frequency. Substitution of eq. (18) in the system of differential equations yields a set of 3p simultaneous homogeneous linear equations in the unknowns A_i whose nontrivial solutions are given by the determinantal equation

$$|\mathbf{G}(\phi)\mathbf{F}(\phi) - \lambda(\phi)\mathbf{E}| = 0 \tag{19}$$

where

$$\mathbf{G}(\phi) = \mathbf{G}^{0} + \sum_{s} (\mathbf{G}^{s} e^{is\phi} + \mathbf{G}^{s} e^{-is\phi}) \tag{20a}$$

$$\mathbf{F}(\phi) = \mathbf{F}^{0} + \sum_{s} (\mathbf{F}^{s} e^{is\phi} + \mathbf{F}^{s} e^{-is\phi}) \tag{20b}$$

Eq. (19) is quite similar to eq. (12), but each of the matrices appearing in eq. (19) is a function of the phase shift ϕ. Eq. (19) is of the 3n-th degree in λ. There are 3p characteristic roots for each value of the phase ϕ. The 3p functions $\lambda(\phi) \rightarrow \nu(\phi)$ can be interpreted as branches of a multiple valued function $\nu(\phi)$ which is the well known dispersion relation for a polyatomic chain. The shape of the 3p frequency branches depends on the extent of coupling with neighboring units through both **G** and **F** matrices. Methods have been proposed[24,25] for the solution of the secular equation which contains complex numbers. Computer programs are becoming available for the calculations of the quantities appearing in the equations discussed above. A few details need to be pointed out:

a) If p is the number of atoms of the chemical repeating unit, then 3p is the size of the $\mathbf{G}(\phi)$ and $\mathbf{F}(\phi)$ matrices appearing in eq. (19) when local redundancies are removed.

b) Two frequency branches always reach zero for $\phi = 0$. These branches correspond to the acoustical branches. It may happen that one or both acoustical branches reach zero again for values of $\phi \neq 0$. This can be predicted by geometrical considerations. The other 3p-2 frequency branches for which $\nu(\phi) \neq 0$ for any value of ϕ correspond to the optical branches since for some particular ϕ the vibrational motion will interact with the electromagnetic radiation, thus producing an absorption of light.

c) From actual numerical calculations it turns out that some of the frequency branches may cross at particular ϕ values, while some others repel and never cross throughout the first Brillouin zone. This fact is particularly important when the density of vibrational states is studied. The chemical structure, the geometry of the polymer chain and the basic linear approximation on which the whole theory of normal vibrations is based, account for the crossing properties of the dispersion curves. Sometimes almost accidental crossings can occur and calculations must be performed at very close interval of ϕ in order to have the precise details of the dispersion curves (see for instance ref.25). Group theory techniques may provide a priori information on the crossing properties of the branches.

d) The theory of vibrations of polymers could well be set up in terms of the wave vector $\vec{k}$ instead of the phase shift ϕ as it is usually done in all textbooks of solid state physics. The GF approach as in eq. (19) gives a more direct description of the normal modes and a more intuitive feeling of what the atoms in the polymer chain are doing.

e) In the case of polymers the use of the Internal Displacement Coordinates introduces new redundancies in addition to the local ones, which are found also in finite molecules. It is possible to find four geometrical relations between the internal coordinates of the backbone chain which express the conditions that: i) the length of the crystallographic repeating unit must be constant during the vibration, ii) there must be no internal twist about the helix axis, iii) the crystallographic repeating unit cannot move away from the helix axis in either of the two directions perpendicular to the helix axis. Explicit geometrical relations have been worked out[26,27], but most of the time brute force numerical methods have been used. Because of these relations, the rank of the secular equation is $3p-4$, hence 4 zero roots are always found when eq. (19) is solved. These four zero roots can also be interpreted as arising from the 4 non-genuine vibrations of an infinite polymer chain corresponding to the rigid translations and rotations of the whole chain. Indeed if a chain is infinite it can have only three free translations along the three cartesian axes and one free rotation about its helix axis. Rotation about the two other axes would imply a deformation of the whole chain.

II. POLYMERS AS ONE DIMENSIONAL CRYSTALS AND THEIR VIBRATIONAL SPECTRUM

A few remarks on the geometry of the polymer chains which are found in nature are necessary for a better understanding of what will follow (see ref. 9). When a given monomeric unit suitable for polymerization finds itself in a favorable energetic condition, a chemical reaction takes place which produces a chain formed by a regular repetition of the starting monomeric unit. We generally deal with polymers obtained by addition or condensation reaction in which chemical units are linked to each other by covalent bonds. When the monomer unit contains an asymmetric carbon atom, the selective factors affecting the mechanism of the polarization reaction must supervise not only the regular linking of the units, but also must supervise the orientation of each monomer in entering the chain. A very large branch of chemistry studies the problems of steroregular polymerization. For our purposes, we assume a working model consisting of an ideal steroregular polymer.

After the chain has been formed, intermolecular interactions (whose precise origin is not yet known) will force the molecule to take up the most stable molecular conformation such as to minimize its energy content (i.e. the molecule finds the most comfortable conformation). The obtained geometry (bond lengths, bond angles, internal rotational angles) is the result of a balancing of several phenomena which are intimately related to the size of the atoms, their electronic structure and the state of hybridization of the molecular orbitals. Since in some cases several minima may exist in the conformational potential energy, the molecule will take up one or another conformation depending on the environmental conditions prevailing.

Once the polymer chains have reached a high degree of configurational and conformational regularity, they can pack into a three-dimensional lattice whose parameters depend on the extent of the inter-molecular interactions. The origin of these interactions is again not yet known, although several explanations have been proposed.

In the previous section, we discussed the vibrational problem of an infinite isolated polymer chain. We have neglected, and still neglect, the existence of a three-dimensional lattice. We are then working on a single chain model, which in the ideal case must be: i) chemically perfect, i.e., perfect and regular type of linking between monomeric units; ii) stereoregularly perfect, i.e. 100°/o isotactic or syndiotactic; iii) conformationally perfect, i.e., straight chain with no bends or folds with a regular sequence of internal rotational angles along the chain backbone. Such a chain can be geometrically generated by applying a screw symmetry operation $\Psi(\theta, \ell)$ to the starting monomer unit (rotation of an angle θ and translation of a distance ℓ along the screw axis). The crystallographic repeat unit of length d is generated after performing a given number of such operations.

A large number of works by crystallographers have shown that in nature polymers exist as helices of different types. Sometimes helices degenerate into a planar zig-zag chain. Each of the helices is characterized by an operator $\Psi(\theta, \ell)$. The physical properties of these polymers strongly depend upon their structure. It is quite obvious that the quantities θ, ℓ and d are connected by geometrical relationships[26,27]. Once the proper geometry of the infinite chain as described above has been chosen and a reliable force field is adopted, the $\mathbf{G}(\phi)$ and $\mathbf{F}(\phi)$ matrices of eq. (19) can be constructed and the calculations performed.

It is worthwhile briefly to mention what kinds of information can be obtained and how it can be used.

1) Solution of eq. (19) at constant ϕ intervals gives the dispersion curves. Several frequency branches come out flat or almost flat, some others are strongly curved.

2) For each frequency ν_ϕ^j (j refers to the frequency branch), we can calculate the amplitudes of vibration $L_{\phi\alpha}^j$ (α refers to the atoms), i.e., the atomic displacements. Thus a detailed quantitative description of the corresponding normal modes can be provided.

3) It turns out that several normal modes are well localized within the chemical repeating unit such that they can be described by curious names as bending, wagging, twisting, rocking, etc. of a particular group of atoms. Other motions instead turn out to involve most of the atoms of the chemical repeating units or even more than one unit. Such motions cannot be qualitatively described. The former types of vibrations are called "group frequencies" and are of particular value and will be discussed later. A mathematical definition of "group vibration" can be found in the literature by the interested reader[10,28]. The existence of group vibrations is the reason for the enormous success of infrared spectroscopy in chemical studies[29,30].

4) From the dispersion curves, it can be deduced that some normal modes are phase independent. This means that atoms vibrate while ignoring what the equivalent neighboring atoms are doing. Thus intra-molecular coupling does not exist between them. Strongly curved frequency branches show that the nature of the vibrations depend strongly upon a cooperative effect of several equivalent atoms along the chain for a given distance. It has to be pointed out that the intra-molecular coupling in organic polymers seems not to extend further than the second or third neighbor.

5) It is possible to apply group theory techniques to polymer chains in a way quite similar (even if somewhat more involved) to what is done in the case of finite molecules. We cannot delve into group theory, but wish only to point out that the symmetry operations which leave the chain invariant, form a space group (or line group) which is isomorphous to one of the well known point groups[31,33]. A consideration of the momentum or wave vector conservation predicts[34] which of the infinite normal vibrations distributed in the 3p frequency branches can interact with the electromagnetic wave, i.e. they can be observed with a spectrometer (infrared or Raman). Group theory states that motions with $\vec{k} \simeq 0$ can be spectroscopically active. At $\vec{k} \simeq 0$, all atoms in neighboring equivalent crystallographic units vibrate in phase. In terms of the phase shift, the conditions for $\vec{k} \simeq 0$ are reached for values of $\phi = 0$, θ and 2θ, etc. In particular infrared active motions are those for which $\phi=0$ and $\phi=\theta$; Raman active motions are those for which $\phi=0$, $\phi=\theta$ and $\phi = 2\theta$. Not all the $\vec{k} \simeq 0$ motions however, can appear in the

Infrared or Raman spectra. Only those motions which belong to an irreducible representation which contains one of the translations will occur in the infrared. Only those motions which belong to an irreducible representation which contains one of the components of the polarizability tensor will occur in the Raman spectrum*. Preferential absorption by an oriented sample will occur if the linearly polarized light is placed alternatively parallel to the three transition moments. Dichroic measurements on oriented specimens are very useful for the detection of the symmetry species of the normal modes which absorb in the infrared[35,37]. The higher the symmetry of the polymer, the smaller is the expected number of infrared bands.

6) It is then quite apparent that spectroscopic techniques are very selective methods which experimentally visualize only a few points of the calculated dispersion curves. Thus they give only a partial check to the reliability and validity of the calculations, which all depend upon the geometry and upon for the force field chosen.

We can now examine in more detail the origin of the bands in the vibrational spectrum of molecules in general and consequently of polymers. The polymers most commonly studied are always organic polymers, i.e. they are generally made up by H,C,O and N atoms. Sometimes F and Cl are found. Very rarely S is found. Carbon atoms are mainly found in sp^3 or sp^2 states of hydridization, i.e. they are linked to each other by single bond or by double bonds. The sp state is only found for the C ≡ N group in polyacrylonitrile. The most common groups of atoms which are found in the helix chain are: CH, CH_2, CH_3, CC, CCC, CCCC, CN, CO, C=O, C=C, C=N, C-F, C-Cl, C-Br, etc. We can then divide the vibrational spectrum in sections which are characteristic of some "group vibrations" as shown below:

3400-3200 cm^{-1}	N-H, O-H	stretchings
3200-2800 cm^{-1}	CH, CH_2, CH_3	stretchings
2200-2000 cm^{-1}	C ≡ N	stretching
1800-1650 cm^{-1}	C = O	stretching
1650-1600 cm^{-1}	C = C	stretching
1470-1200 cm^{-1}	CH_3, CH_2, CH, NH, NH_2	deformations
1100-950 cm^{-1}	C-C, C-N	stretchings

In the lower frequency region several deformation vibrations occur which cannot be considered any more as group vibrations, since they

*The discussion is restricted here to $\vec{k} \simeq 0$ fundamental transitions. For a more comprehensive discussion see refs. 33 and 39.

are strongly coupled and therefore depend on the geometry and structure of the molecule considered.

Let us arbitrarily set the frequency of 650 cm^{-1} as the upper limit of the far infrared spectrum. Toward lower frequencies we can generally find the stretching vibrations of halogen atoms (C-Cl,C-Br,C-I), whose extent of coupling with other motions is not yet completely clarified because of the lack of a good force field. We also find absorptions due to the deformation modes of the skeleton (C-C-C, $^{C}_{C}{>}$C-C) and the torsional modes (CH_3 torsion, CCCC torsion, etc.). All these motions are not group frequencies. It should be stressed here that we are still working on a single chain model of polymers and no intermolecular interactions due to lattice forces are considered.

The main types of information which can be obtained from the analysis of the vibrational spectrum (infrared and Raman) of solid substances or solutions are the following:

a) The near infrared (3500-650 cm^{-1})
   1. is the clue for identification of the chemical nature of an unknown sample.
   2. gives considerable detailed information on the structure of the unknown sample (i.e. types of bonds, mutual position in the molecule of different atomic groups, etc.).
   3. provides information on symmetry, hence on the spatial arrangement of the atoms in the molecule.
   4. provides the value of the fundamental frequencies. This allows one to determine the principal and interaction force constants of a few atomic groups.

b) The far infrared (650-50 cm^{-1})
   1. shows absorptions which are very characteristic of certain chemical structures or spatial arrangements of atoms which do not have any correspondence in the near infrared.
   2. provides data for the calculations of force constants of skeletal deformations.
   3. is the unique source of data for torsional oscillations from which the shape of the potential hindering the oscillation can be inferred.

c) The analysis of band shapes for solids or solutions which provides some information on the motions of the whole molecules in the condensed states.

III. INTER-MOLECULAR INTERACTIONS AND LOW FREQUENCY DENSITY DISTRIBUTION

In the first part of this section, we wish to analyze the spectroscopic phenomena which occur when the polymer chain is organized in a three-dimensional lattice. In the second part, we shall briefly investigate some other ways of obtaining information on the low frequency motions of polymers.

A. Let us assume that when polymer chains have reached a high degree of structural regularity, they pack into a three-dimensional lattice. In principle, the vibrational problem of polymer crystals can be treated in a way analogous to those used for inorganic crystals and treated in the books of solid state physics. Practical calculations become however, very cumbersome. We satisfy ourselves again with the "springs and balls" model for which intermolecular forces are assumed to be of an elastic type. Since these binding forces are generally very weak relative to the intramolecular forces, crystalline organic polymers are to be considered as molecular crystals. The ordered inter-atomic forces will, in general, produce slight frequency shifts in the spectrum of the free molecule. A multiplication of the number of bands is also expected due either to the lowering of the symmetry of the molecular environment or due to the mechanical coupling with neighboring chains. Moreover, low frequency lattice modes will appear in the vibrational spectrum (for a discussion on organic crystals, see refs. 38 and 39).

All calculations start by assuming an ideal crystal lattice whose vibrations must be accounted for by a suitably constructed dynamical matrix. For this purpose the traditional GF technique in terms of internal displacement coordinates is practically abandoned and the problem is treated in terms of cartesian displacement coordinates[40]. Because of its more direct physical meaning, however, the concept of internal coordinates is retained as much as possible in the calculations. Since the linear transformation from cartesian to internal coordinates can be easily constructed by geometrical considerations (see eq. 4), the dynamical matrix in cartesian coordinates is obtained after expressing the potential energy matrix in terms of R's. The problem becomes more tractable, however, if all the quantities are expressed as functions of the wave vector $\vec{k}$, instead of the phase shift ϕ as discussed in the case of the single chain model.

Let us consider a crystal made up of an infinite number of unit cells[40]. Each unit is labeled by a triplet of integers ℓ_1, ℓ_2 and ℓ_3 which can be either zero, positive or negative. Let ℓ indicate any given triplet. Let the cartesian and internal displacement coordinates of the ℓ-th unit be collected in the matrix vectors x_ℓ and r_ℓ respectively. The kinetic energy of the whole crystal in terms of x_ℓ becomes:

$$2T = \sum_{\ell\ell'} \tilde{\dot{\mathbf{X}}}_\ell \mathbf{M} \dot{\mathbf{X}}_{\ell'} \qquad (\ell = \ell') \tag{21}$$

The quadratic potential energy is written in terms of $\mathbf{r}_\ell$ as

$$2V = \sum_{\ell\ell'} \tilde{\mathbf{r}}_\ell \mathbf{F}_{\ell\ell'} \mathbf{r}_{\ell'} \tag{22}$$

In eq. (21) $\mathbf{M}$ is a diagonal matrix of order 3q (q = number of atoms per unit cell) whose elements are the masses of the atoms contained in the units cell and suitably arranged. In eq. (22) $\mathbf{F}_{\ell\ell'}$ is the square matrix of order 3q of the harmonic force constants which describe the interactions between the internal coordinates of unit ℓ with those of unit ℓ'.

Let us define the cartesian and internal phonon coordinates:

$$\mathbf{X}(\vec{k}) = \frac{1}{\sqrt{2\pi}} \sum_{-\infty}^{\infty} \mathbf{X}_\ell \; e^{-i\vec{k}\cdot\vec{t}\,(\ell)} \tag{23}$$

$$\mathbf{R}(\vec{k}) = \frac{1}{\sqrt{2\pi}} \sum_{-\infty}^{\infty} \mathbf{r}_\ell \, e^{-i\vec{k}\cdot\vec{t}(\ell)} \tag{24}$$

where $\vec{k}$ is the wave vector. We then wish to find the transformation matrix $\mathbf{B}(\vec{k})$ from cartesian to internal phonon coordinates; analogous to that of eq. (4). After some algebra it turns out that:

$$\mathbf{R}(\vec{k}) = \sum_{-m}^{m}{}_j \; \mathbf{B}_j \; e^{-i\vec{k}\cdot\vec{t}(j)} \, \mathbf{X}(\vec{k}) \tag{25}$$

The desired transformation matrix is then

$$\mathbf{B}(\vec{k}) = \sum_{-m}^{m}{}_j \; \mathbf{B}_j \; e^{-i\vec{k}\cdot\vec{t}(j)} \tag{26}$$

Let us rewrite the potential energy matrix which appears in eq.(22) in a way analogous to that discussed in eq. (20).

$$\mathbf{F}_R(\vec{k}) = \mathbf{F}^{\circ} + \sum_s \tilde{\mathbf{F}}^s \; e^{-i\vec{k}\cdot\vec{t}(s)} + \mathbf{F}^s \; e^{i\vec{k}\cdot\vec{t}(s)} \tag{27}$$

Using eq. (26) the potential energy matrix of eq. (27) can be rewritten in terms of phonon cartesian coordinates as:

$$\mathbf{F}_x(\vec{k}) = \tilde{\mathbf{B}}(\vec{k}) \; \mathbf{F}_R(\vec{k}) \, \mathbf{B}(\vec{k}) \tag{28}$$

A further transformation to mass weighted phonon cartesian coordinates

$$\mathbf{X}(\vec{k}) = \mathbf{M}^{\frac{1}{2}} \mathbf{X}(\vec{k}) \tag{29}$$

gives the dynamical matrix in the hermitian form

$$\mathbf{D}(\vec{k}) = \mathbf{M}^{-\frac{1}{2}} \, \mathbf{F}_x(\vec{k}) \, \mathbf{M}^{-\frac{1}{2}} \tag{30}$$

suitable for machine computations. The dispersion curves for a given crystal can be easily obtained when a proper set of $\vec{k}$ values is inserted in the following eigenvalue equation:

$$[\mathbf{D}(\vec{k}) \;\; -\omega^2 \; (\vec{k}) \, \mathbf{E}] \, \mathbf{L}_x(\vec{k}) = 0 \tag{31}$$

A few interesting comments should be made here:

a) According to eq. (27) the whole problem can be first set up in terms of internal coordinates. One has to define a set of internal coordinates which belong to the molecular chain (the same as those of the single chain model) and a set of internal coordinates which can reasonably represent the more probable intermolecular interactions. Transformation to cartesian coordinates is then performed according to eqs. (29) and (30).

b) While numerical values of the intramolecular force constants are available, very little is known for the intermolecular potentials. This is particularly true for molecular crystals in general and even more for polymers. The GF approach to crystal vibrations allows a more direct way of verifying the innumerable semi-empirical inter-atomic potentials spread everywhere in the literature. Viceversa, the use of experimental data observed in the vibrational spectrum allows obtaining an estimate of the intermolecular forces in the harmonic approximation.

c) Let us examine the order of magnitude of the force constants associated with each type of internal coordinates, which are generally defined when a crystalline polymer is studied. The intramolecular force constants (all in mdyne/A) are as follows: Stretching 5.00, angle bending 0.5, torsion 0.02, intermolecular forces 0.02. With these constants, the following comments apply:

Splitting of a few wavenumbers (a sort of small perturbation[38]) will be observed for the stretching and bending modes in the near infrared, in going from the single chain model to the three-dimensional model.

Remarkable perturbations can occur in the far infrared for the low frequency deformation modes and for the torsional oscillations. Inter-and intra-molecular interactions are of the same magnitude.

Absorptions in the low frequency region (infrared and Raman) due to lattice motions can occur and have been observed (see ref. 9 for the case of polymers). The analysis of lattice motions can give useful suggestions as to the symmetry of the space group, i.e. to the type of unit cell of the material.

Because of points discussed above, torsional and lattice motions can be coupled to a remarkable extent. While the shape of the normal vibrations of the modes occurring in the near infrared region is practically the same if obtained from a single chain model or from a three dimensional one, a more reliable description of the torsional motions can be obtained only from a three dimensional calculation.

d) Very little is still known as to the low frequency vibrations of crystalline polymers. Many polymers must be experimentally studied and their vibrations theoretically analyzed before we can obtain a comprehensive view of the problem of the intermolecular forces of these important materials.

e) Due to the lack of knowledge of a good inter-molecular force field, the shape of the dispersion curves, especially in the low frequency region, is quite dubious since they depend strongly on the values of the force constants adopted.

B. The problem of the uncertainty of the dispersion relations for crystalline polymers just pointed out is very important and requires a more thorough discussion. As a matter of fact, many physical properties of the crystals depend on the dispersion curves mainly in the lower frequency region. It has already been pointed out that spectroscopic techniques are very selective methods of investigation of the molecular and crystal vibrations, since only very few normal modes of the infinite number possible in a large crystal can interact with the electromagnetic wave. It has been shown that different vibrational potentials do reproduce the $\vec{k}=0$ observed fundamentals but give rise to drastically different dispersion curves[41].

Additional information on the vibrational properties of crystals in general and particularly of organic polymers, can be obtained from the study of the Inelastic Neutron Scattering Spectrum (hereafter indicated by INS). This technique has been only recently applied to the study of molecular crystals[42-45]. The INS spectrum is determined mainly by the density of vibrational states or frequency distribution $g(\nu)$. Since $g(\nu)$ can be obtained from the dispersion curves, the comparison between observed and calculated frequency distributions provides some information on the validity of the calculations performed. Let $g(\nu)d\nu$ be the fraction of normal modes with frequencies in the interval between ν and $\nu + d\nu$ where in the limit $d\nu \to 0$. From graphically determined dispersion curves, the normalized frequency distribution is defined as:

$$g(\nu) = C \sum_j \frac{\phi_{j+1} - \phi_j}{\Delta\nu} \qquad (32)$$

where ϕ_i are the roots of $\nu = \nu(\phi)$ for a fixed value of ν and C is a normalizing factor. In eq. (32) the total density is obtained as the sum over 3p branches. An exact analytical derivation of the function $g(\nu)$ from the dynamical matrix is impossible for these large systems. There are several numerical methods which provide an hystogram of $g(\nu)$[40], whose accuracy can be improved by reducing the interval $\Delta\nu$ in eq. (32).

For the ideal model treated in this lecture, some of the peaks of the frequency distribution correspond to the high concentration of vibrational states in the neighborhood of the limiting spectroscopically active $\vec{K} = 0$ states. These peaks must coincide (in frequency, nothing can be yet said about intensities) with the absorption bands in the infrared or Raman spectra. Other peaks of $g(\nu)$ do not find any coincidence in the vibrational spectrum. They correspond to flat sections of the dispersion curves for intermediate $\vec{k}$ values. They can be revealed, in some particular cases, by the INS spectrum.

The basic principles of neutron scattering experiments are the following. Since the energy of the thermal neutrons is of the same order as that of the lattice vibrations, when neutrons pass through a crystal they may gain or lose energy. The study of the energy gain or loss provides direct information on the distribution of the vibrational energy inside the crystal, hence on the frequency spectrum of the crystal, especially in the lower frequency region.

For these studies a beam of cold neutrons is made monoenergetic by suitable experimental devices. The neutrons impinge on to a crystal in a given direction and the scattered beam is observed at a given angle. The geometry of the experiment defines the direction of the wave vectors of the ingoing and outgoing neutrons $\vec{K}_o$ and $\vec{K}_f$ respectively. Neutrons can be scattered coherently and incoherently, depending on whether interference occurs or does not occur. The energy distribution of neutrons inelastically and incoherently scattered is then analyzed. It has been shown that the inelastic neutron scattering cross section of a crystal can be expressed in terms of a few quantities which are accessible to the vibrational spectroscopist[46]. It is also known that the scattering cross section of hydrogen is much larger than that of other atoms (H=82.5 barns, C=5.5 barns). INS studies are then particularly suitable for hydrogenic compounds like most of the common hydrogenic polymers. The direct determination of the frequency spectrum from INS experiments is possible for cubic lattices or for polycrystalline materials which have been shown to behave approximately like cubic lattices[47]. For non-cubic symmetries a more general expression of the one-phonon incoherent differential cross section has been derived in terms of the eigenfrequencies, eigenvalues and other quantities[46].

A few comments can be made here:

a) Peaks of $g(\nu)$ correspond to peaks or shoulders in the experimental $g(\nu)$ obtained from INS experiments. No comparison can be made between the calculated intensities of the peaks of $g(\nu)$ and the experimental ones from the INS spectrum, since 1) most of the time crystals do not behave like isotropic scatterers, 2) the temperature factor is assumed to be equal to unity in deriving the experimental $g(\nu)$, 3) multiphonon scattering may occur even at room temperature, 4) the larger the vibrational amplitude the more important is the contribution of the temperature factors to the scattering process.

b) Flat regions of the dispersion curves do however find nice coincidences with the experimental density distribution. A few polymers have been studied experimentally and theoretically (see ref. 9).

c) A priori theoretical calculations have been attempted for polyethylene, isotactic polypropylene and hexagonal polyoxymethylene (see ref. 9). Calculations have been carried out for single chain model or three-dimensional models of polymers[48,41]. It must be pointed out that lattice motions strongly affect the shape of $g(\nu)$ only for light polymer chains like polyethylene and polyoxymethylene. For heavier polymer chains, the single chain model is enough for the interpretation of the observed features of the experimental $g(\nu)$ (see ref. 49).

d) The experimentally observed points of $g(\nu)$ may provide additional data to be used for the determination of those force constants which mainly determine the low frequency dispersion curves.

REFERENCES

1. E. B. Wilson, J. Chem. Phys. 7, 1047 (1939); E. B. Wilson, J.C. Decius and P.C. Cross, "Molecular Vibrations", McGraw-Hill, New York (1955).

2. M. Elyashevitch, Comp. Rend. Accd. Sci. U.R.S.S. 28, 605 (1940); M. V. Volkenstein, M. A. Elyashevitch and B. I. Stepanov, Kolebanya Molekul, II, Moscow (1949).

3. C Eckart, Phys. Rev. 47, 552 (1935).

4. A. Saivetz, J. Chem. Phys. 6, 383 (1939).

5. M. Gussoni and G. Zerbi, Chemical Physics Letters 2, 145 (1968).

6. J. C. Decius, J. Chem. Phys., 16, 1025 (1948).

7. S. M. Ferigle and A. G. Meister, J. Chem. Phys. 19, 982 (1951).

8. R. Lohmann, unpublished tables.

9. For a recent discussion of the problem and for a list of up to date references, see: G. Zerbi "Molecular Vibrations of High Polymers", Advances in Applied Spectroscopy", edited by E.G. Brame, Marcel Dekker Co., Vol. 2 (1969).

10. W. T. King, Thesis, University of Minnesota, Minneapolis, 1958, Dissertation Abstract 18, 2172 (1958).

11. E. Curtis, Thesis, University of Minnesota, Minneapolis, 1956.

12. J. Overend and J. R. Scherer, J. Chem. Phys. 32, 1289 (1960).

13. I. M. Mills, Spectrochim. Acta, 16, 35 (1960).

14. J. H. Schachtschneider and R. G. Snyder, Spectrochim. Acta 19, 17 (1963).

15. M. Gussoni and G. Zerbi, Rend. Acad. Naz. Lincei, Rend. Classe Sci. Fis. Mat. Nat, 40, 843 (1966).

16. R. Gold, J. M. Dowling and A. G. Meister, J. Mol. Spectry, 2 9 (1958).

17. B. L. Crawford, Unpublished results.

18. S. Califano and B. L. Crawford, Z. Elektrochem. 64, 571 (1960).

19. M. Gussoni and G. Zerbi, Atti Acad. Nazl. Lincei, Rend. Classe Sci. Fis. Mat. Nat. 40, 1032 (1966).

20. J. H. Schachtschneider, Shell Report, 57/65.

21. E. P. Wigner, "Group Theory", Academic Press, New York (1959).

22. M. Gussoni and G. Zerbi, J. Mol. Spectry, 26, 485 (1968).

23. P. W. Higgs, Proc. Roy. Soc. (London) A220, 472 (1953).

24. T. Miyazawa, J. Chem. Phys., 35, 693 (1961).

25. L. Piseri and G. Zerbi, J. Chem. Phys., 48, 3561 (1968).

26. T. Miyazawa, J. Polymer Sci., 55, 215 (1961).

27. D. Lee, European Research Office, United States Army, Technical Report 1965; Contract No. DA-91-691-EUC-3606.

28. W. T. King and B. L. Crawford, J. Mol. Spectry, 5, 421 (1960).

29. R. N. Jones in "Chemical Applications of Spectroscopy" in Physical Methods of Organic Chemistry, Technique of Organic Chemistry, vol. IX, Editor A. Weissberger; Interscience, New York (1956).

30. L. J. Bellamy "The Infrared Spectra of Complex Molecules, Methuen, London (1958).

31. V. Heine "Group Theory in Quantum Mechanics" International series of Monographs on Pure and Applied Mathematics; Editors I. N. Sneddon and S. Ulam, Pergamon Press, London (1960).

32. R. S. Halford, J. Chem. Phys. 14, 8 (1946).

33. H. Winston and R. S. Halford, J. Chem. Phys. 17, 607 (1949).

34. R. Zbinden, "Infrared Spectroscopy of High Polymers", Academic Press, New York (1964).

35. S. Krimm, Fortschr. Hochpolymer, Forsch., 2, 51 (1960).

36. A. Elliott, J. Polymer Sci., C7, 37 (1964).

37. S. Krimm "Infrared Spectroscopy and Molecular Structure", Chapter 8, Editor M. Davies, Elsevier Publishing Company, Amsterdam (1963).

38. W. Vedder and D. F. Hornig, Infrared Spectra of Crystals in "Advances in Molecular Spectroscopy", Vol. 2, Editor. H. W. Thompson; Interscience, New York (1961). S.S. Mitra, Infrared Spectra of Crystals in"Progress in Infrared Spectroscopy" Vol. 2, Editor, S. Szymanski, Plenum Press, New York (1961).

39. D. Dows in "Physics and Chemistry of the Organic Solid State" Vol. I, Editors: D. Fox, M. M. Labes and A. Weissberger, Interscience, New York (1963).

40. L. Piseri and G. Zerbi, J. Mol. Spectry 26, 259 (1968).

41. G. Zerbi, 8th European Congress on Molecular Spectroscopy, Madrid, 1967.

42. G. J. Safford and A. W. Naumann "The Technique of Neutron Scattering and its Application to Polymers" in Macromolecular Reviews, Vol. 2, Editors A. Peterlin, M. Goodman, S. Okamura, B. H. Zimm and H. F. Mark, Interscience, New York (1967).

43. S. Trevino and H. Boutin, J. Chem. Phys. 45, 2700 (1966).

44. W. Myers, G.C. Summerfield and J. S. King, J. Chem. Phys., 44, 184 (1966).

45. S. Trevino and H. Boutin, J. Macromol. Sci. Chemistry A1, 723 (1967).

46. A. Sjölander, Arkiv. fur Fyisik, 14, 315 (1958).

47. L. V. Tarasov, Soviet Physics Solid State, 3, 1039 (1961).

48. T. Miyazawa and K. Kitagawa, J. Polymer Sci., B2, 395 (1964). K. Kitagawa and T. Miyazawa, J. Chem. Physics. 47, 317 (1967).

49. G. Zerbi and L. Piseri, J. Chem. Physics., 49, 3840 (1968).

ELECTRONIC RAMAN SPECTRA OF TRIVALENT LANTHANIDES*

J. A. Koningstein

E.W.R. Steacie Bldg. of Chemistry, Carleton University

Ottawa, Ontario, Canada

ABSTRACT

The theoretical background of electronic Raman transitions of trivalent rare earth ions in crystals is discussed. A survey is given of the experimental data which have now become available.

INTRODUCTION

Since the discovery of the Raman effect the greater part of the spectra has dealt with the excitation of transitions of rotational or vibrational levels of molecules. Only in a few cases in the years since the advent of the laser have transitions between electronic levels of molecules or ions in crystals been observed. Consequently the importance of this effect, or rather, the ultimate interpretation of the properties of the electronic Raman transitions and also the differences between the vibrational and the electronic Raman effect were not well known.

It is perhaps interesting to remark here that during the last decade many crystal field calculations have been performed on the spectra of the trivalent rare earth ions which in turn are playing such an important role in laser technology. The optically active 4f electrons of the Lanthanides are screened from external forces by filled 5s and 5p orbitals and the effect of the crystalline field is weak. This has permitted rather detailed crystal field studies and the results have been promising.

*Research supported by the National Research Council and Defense Research Board of Canada

The levels originating from the 4f electronic configurations are split into different crystal field components if the rare earth ions are placed in a crystal and these levels are situated anywhere from 0- 70 000 cm^{-1} above the ground state[1]. Pure electric dipole transitions between the electronic levels are, however, not allowed because the states have the same parity. The Raman operator $\alpha_{\rho\sigma}$ (where σ and ρ stand for x,y or z)is, however, an even operator and electronic Raman transitions are allowed. The intensity of the Raman lines are related to the elements of a tensor of three rows and three columns. This tensor for the vibrational Raman effect is a symmetric tensor but for the electronic Raman transitions pure anti-symmetric scattering tensors are possible[2]. The occurrence of such tensors have rather profound effects with regard to the polarization properties and the selection rules.

The positions of the electronic levels is dependent on the strength and the symmetry of the crystal field but the position of ground and electronic states may also change under the effect of an external applied magnetic field which permits Zeeman-Raman spectroscopy.

THEORY

In the Raman effect we are concerned with the interaction of light with matter. Contrary to absorption spectroscopy, the frequency ν_o of the incident radiation is in most cases such that it is far removed from the frequency of an absorption band of the molecule or atom. Let us assume that this molecule is in a certain quantum state k. As soon as the external perturbation is applied, that is if the bombardment of the photons sets in, the wavefunction in the perturbed case will become different from the wavefunction ψ_k^o. The former can be calculated from first order time dependent perturbation theory and the result is:

$$\psi_k = \psi_k^o + \frac{1}{h}\sum_r \psi_r^o \left\{ \frac{A\langle k|M|r\rangle}{\nu_{rk} - \nu_o} e^{-i/h(E_k + h\nu_o)t} + \frac{A^*\langle k|M|r\rangle}{\nu_{rk} + \nu_o} e^{-i/h(E_k - h\nu_o)t} \right\} \quad (1)$$

A and A* are related to the electric vector of the incident electric vector in the following way; $E = A^* e^{-1/h\cdot\nu_o t} + A\, e^{=i/h\cdot\nu_o t}$. The matrix element of the electric dipole operator $\langle k|M|r\rangle$ stands for the integral $\int \psi_k^o \sum_i e_i r_i \psi_r^o\, d\tau$ where e_i is the charge and r_i is a position vector (describing the position of the ith electron) and $\nu_{rk} - \frac{1}{h}(E_r - E_k)$.

There are now some interesting points to be made. Firstly, the question arises whether there exists a direct moment of the perturbed state. This moment, represented by the integral

$\int \psi_k^* \Sigma e_i r_i \psi_k \, d\tau$ can be worked out by substituting in eq. 1. The result shows that there does indeed exist a moment and the time dependence of this moment is exactly identical to that of the incident radiation. Also the frequency of the dipole irradiated light is identical to the frequency of the incident radiation. This type of scattering is called Rayleigh scattering and the main features are thus that it is colinear and copolarized and in phase with the incident photon. The next question is to ask for the details of the moment between the state k and another state n of the molecule or atom. It is found that the matrix element $\langle k'|M|n'\rangle$ can be written as:

$$\langle k'|M|n'\rangle = \langle k|M|n\rangle e^{-2\pi i \,\nu_{kn} t} + \frac{1}{h}\sum_r \left\{ \frac{A\langle k|M|r\rangle\langle r|M|n\rangle}{\nu_{rk} - \nu_o} + \frac{\langle k|M|r\rangle\, A.\langle r|M|n\rangle}{\nu_{rn} - \nu_o} \right\} \times e^{-2\pi i(\nu_{kn} + \nu_o)t} + \frac{1}{h}\sum_r \left\{ \frac{A^*\langle k|M|r\rangle\langle r|M|n\rangle}{\nu_{rk} + \nu_o} + \frac{\langle k|M|r\rangle\, A^*\langle r|M|n\rangle}{\nu_{rn} + \nu_o} \right\} \times e^{-2\pi i(\nu_{kn} - \nu_o)t} \quad \ldots\ldots(2)$$

The first term on the R. H. S. of eq. (2) represents radiation of frequency $\nu_k - \nu_n$, however in the next terms radiation occurs of frequency $\nu_{kn} + \nu_o$ and $\nu_{kn} - \nu_o$ and the latter represents the Raman scattered radiation. In fact the third member of eq. (2) can be represented by a vector R_{kr} so that:

$$[A.\langle k|M|r\rangle]\langle r|M|n\rangle = A\langle k|M|r\rangle \cdot \{ \langle r|M_x|n\rangle i + \langle r|M_y|n\rangle j + \langle r|M_z|n\rangle k\}$$

where i, j and k are orthonormal vectors. Similarly the inner product $A\langle k|M|r\rangle = A_x\langle k|M_x|r\rangle + A_y\langle k|M_y|r\rangle + A_z\langle k|M_z|r\rangle$ and by defining $R_{kn} = \alpha_{kn}\cdot A$ the following expression can be written for the Raman scattering tensor;

$$(\alpha_{\rho\sigma})_{kn} = \frac{1}{h}\sum_r \left\{ \frac{\langle k|M_\rho|r\rangle\ \langle r|M_\sigma|n\rangle}{\nu_{rk} - \nu_o} + \frac{[\rho \leftrightarrow \sigma]}{\nu_{rn} + \nu_o} \right\} \qquad (3)$$

here ρ, σ stand for x,y or z.

Up to now the quantum state k and also n have not been defined sharply. In most cases k shall be the ground state of the system while n can then be a vibrational excited or rotational state of the molecule associated with the ground electronic state of the molecule. It is, of course, also possible that n refers to another electronic state of the particle and it might even be an excited rotational or vibrational state of an excited electronic state! It is clear that in the latter two cases changes occur in the electronic quantum number as opposed to the normal vibrational or rotational Raman effect.

Our interest here lies with the electronic type Raman effect and in the following we shall try to explain the main differences with the normal vibrational Raman scattering.

The wavefunction of the ground state of the particle is a product of wavefunctions. We shall ignore here the rotations of the system and by denoting the wavefunction of the electrons with $\psi(\vec{r},\vec{Q})$ and the vibrational wavefunctions with $\phi(\vec{Q})$ the total wavefunction in the zero order Born-Oppenheimer approximation can then be written as: $\psi_{tot} = \psi_k(\vec{r},\vec{Q})\cdot\phi(\vec{Q})$. The coordinates of the electrons, however, are dependent to a certain extent on the movements of the nuclei of the atoms of the molecule and, if it is assumed that this effect is small, then the Hamiltonian H of the molecule can be expanded in terms of one (or more) of the normal modes of vibration of the molecule Q_j. Thus

$$\mathcal{H} = \mathcal{H}^o + \left(\frac{\partial\mathcal{H}}{\partial Q_j}\right)_{Q_j=o} Q_j + \cdots$$

From first order perturbation theory follows:

$$\psi_k(\vec{r},\vec{Q}) = \psi_k^o(\vec{r}) + \sum_{\ell\neq k} h_{\ell k}^{Q_j} \cdot \psi_\ell^o(\vec{r})\, Q_j \qquad \text{where}$$

$$h_{\ell k}^{Q_j} = \int \psi_k^o(\vec{r}) \left(\frac{\partial\mathcal{H}}{\partial Q_j}\right)_{Q_j=o} \psi_\ell^o(\vec{r})\, d\tau / E_\ell - E_k$$

Substitution of the expanded wavefunction in the matrix elements of the elastic dipole operator leads to the following expression for an element of the scattering tensor between the ground state k and the u-th vibrational state of the normal coordinate Q_j and the electronic state n and the w-th vibrational state of the same normal mode Q_j.

$$(\alpha_{\rho\cdot\sigma})_{ku;nw} = \frac{1}{h}\langle w|u\rangle \sum_r \left\{ \frac{\langle n|M_\rho|r\rangle\langle r|M_\sigma|k\rangle}{\nu_{rk}-\nu_o} + \frac{[\rho\longleftrightarrow\sigma]}{\nu_{rn}+\nu_o} \right\} +$$
$$\frac{1}{h}\langle w|Q_j|u\rangle \sum_r \frac{\langle n|M_\rho|r\rangle\{\langle r|M_\sigma|\ell\rangle + \langle\ell|M_\sigma|r\rangle\} h_{\ell k}^{*\,Q_j}}{\nu_{rk}-\nu_o} +$$
$$\frac{1}{h}\langle w|Q_j|u\rangle \sum_r \frac{[\rho\longleftrightarrow\sigma]}{\nu_{rn}+\nu_o} \qquad (4)$$

The main conclusion which can be drawn here is that the constants $h_{\ell k}^{Q_j}$ are all small and thus the main contribution to the intensity of the electronic transition comes from the scattering tensor between the electronic states k and n. The vibronic coupling effect on the intensity of electronic Raman transitions is apparently a small effect.

One notes also that $(\alpha_{\rho\sigma})_{kn} \neq (\alpha_{\sigma\rho})_{kn}$ or, in other words, the scattering tensor is not symmetric for the electronic Raman effect. It is a rather simple matter to write down the tensor for a normal vibrational Raman effect in the ground state of the molecule. One only has to replace n by k and the following expression results:

$$(\alpha_{\rho\sigma})_{ku;ku'} = \frac{1}{h}\Big[\sum_r \sum_{\ell\neq k}\Big\{h^{Q_j}_{\ell k}\,(\langle k|M_\rho|r\rangle\langle r|M_\sigma|\ell\rangle + \langle \ell|M_\rho|r\rangle\langle r|M_\sigma|k\rangle)$$

$$+ h^{Q_j}_{\ell r}(\langle k|M_\rho|r\rangle\langle \ell|M_\sigma|k\rangle + \langle k|M_\rho|\ell\rangle\langle \ell|M_r|k\rangle)\Big\}(\nu_o - \nu_{kn})^{-1}$$

$$+ \Big\{(\rho \longleftrightarrow \sigma)\cdot(\nu_o + \nu_{rk})^{-1}\Big\} \times \langle u|Q_j|u'\rangle$$

now $(\alpha_{\rho\sigma})_{ku;ku'} = (\alpha_{\sigma\rho})_{ku;ku'}$ and the scattering tensor is symmetric. In order to derive Raman selection rules, one mostly employs the powerful method of group theory. If we define the operator $\hat{\alpha}_{\rho\sigma}$ as:

$$\hat{\alpha}_{\rho\sigma} = \frac{|\sum_i e_{i\rho i}|r\rangle\langle r|\sum_i e_i\sigma_i|}{\nu_{rk} - \nu_o} + \frac{[\rho \longleftrightarrow \sigma]}{\nu_{rn} + \nu_o}$$

then a Raman transition between the states k and n is allowed if the integral $\int \psi_k^* \hat{\alpha}_{\rho\sigma} \psi_n d\tau \neq 0$. This is the case if the totally symmetric representation of the point group of the molecule is contained in the product of the representations of the states n and k and the operator $\hat{\alpha}_{\rho\sigma}$. The transformation properties of wavefunctions and operators are in general easy to obtain and of special interest here is the impact of the antisymmetric tensor on the Raman selection rules.

TENSOR SYMMETRY AND SELECTION RULES

When discussing symmetry of tensor components, it is customary to describe a second order cartesian tensor by products of the type α_{xy}, etc. It is easily shown that a tensor component α_{xy} transforms as the product xy, if the latter product indeed transforms as an irreducible representation. This is, however, rarely the case; more often products of the type xy will transform into products of the type yx. Most physical cartesian tensor and in particular the ordinary rotational and vibrational Raman effect are symmetric so that $\alpha_{xy} = \alpha_{yx}$, so the correlation with cartesian product of the type xy = yx is immediately clear. This correlation is much less clear when the tensor is non-symmetric, that is, expression of the type $\alpha_{xy} - \alpha_{yx}$ describe nonvanishing quantities. the similar expression xy-yx will in general be vanishing quantities and in order to stretch the analogy between these products and the tensor components one must introduce the fiction that quantities like x and y are non-commutative so that the quantities xy-yx will be non-vanishing. In such a case, which is in fact the appropriate one when discussing the electronic Raman effect, it seems profitable to skip this somewhat artificial analogy with coordinate products and instead introduce some quantities that are more closely related to the whole concept of symmetry groups and symmetry operations.

Under general rotations of the molecular framework, the nine cartesian tensor components will in general transform into each other in a very complicated way and the symmetry transformation must be described by a 9 x 9 matrix. It is, however, possible to choose certain linear combinations of the cartesian tensor components that under an arbitrarily rotation around a particular axis transform into themselves and to collect the components in three sets so that members of each set under an arbitrarily rotation transform into only members of the same set. These linear combinations often given the name irreducible tensor components are connected to the usual cartesian components in the way shown in Table I.

To get a picture of what these irreducible tensor components stand for, we may remark that there exists a close analogy between the three sets of irreducible tensor components and the well known s , p and d orbitals. Thus α_0^0 is completely invariant under all rotations while the three members of the α_Q^1 set generally transform into combinations of themselves under arbitrarily rotations; so, of course, do the five components of the α_Q^2 set in complete analogy with the five d orbitals. There is, however, a slight difference between the three α_Q^1 components and the three p orbitals p_x, p_y and p_z. The latter change sign under inversion of the coordinate system while all the tensor components are invariant under this operation.

As stated in the preceding section, electronic Raman transitions are allowed if the matrix element $\langle \psi_n | \hat{\alpha}_{\rho\sigma} | \psi k \rangle$ is different from zero. Since the $\hat{\alpha}_{\rho\sigma}$ operator can be expressed as a linear combination of α_Q^k's by the relations given in Table I, the problem reduces to the investigation of matrix element $\langle \psi_n | \hat{\alpha}_Q^k | \psi_k \rangle$. As stated earlier, the condition for being non-vanishing is that the product of the three functions making up the matrix element remain invariant under the symmetry operations of the scattering system. The ψ_n and ψ_k will already, being eigenfunctions of the Hamiltonian, transform as irreducible representations and all that is left, therefore, is to determine how the operator components α_Q^k transform under the operation of the symmetry group. We have investigated these transformation properties in some detail and the results for the most common point groups are collected in Table II

From Table II it is seen that the tensor components α_Q^1 representing the anti-symmetric part of the scattering tensor in general transforms completely different from the α_Q^2's that represent the

Table I

Relation between the cartesian and irreducible components of the Raman scattering tensor

$$\alpha^{o}_{o} = -(1/\sqrt{3})\,(\alpha_{xx} + \alpha_{yy} + \alpha_{zz})$$

$$\alpha^{1}_{1} = \frac{1}{2}\,(-\alpha_{xz} + \alpha_{zx} - i\,\alpha_{yz} + i\alpha_{zy})$$

$$\alpha^{1}_{o} = (1/\sqrt{2})\,(i\alpha_{xy} - i\alpha_{yx})$$

$$\alpha^{1}_{-1} = -\frac{1}{2}\,(\alpha_{xz} - \alpha_{zx} - i\alpha_{yz} + i\alpha_{zy})$$

$$\alpha^{2}_{2} = \frac{1}{2}\,(\alpha_{xx} - \alpha_{yy} + i\alpha_{xy} + i\alpha_{yx})$$

$$\alpha^{2}_{1} = -\frac{1}{2}\,(\alpha_{xz} + \alpha_{zx} + i\alpha_{zy} + i\alpha_{yz})$$

$$\alpha^{2}_{o} = (1/\sqrt{6})\,(2\alpha_{zz} - \alpha_{xx} - \alpha_{yy})$$

$$\alpha^{2}_{-1} = \frac{1}{2}\,(\alpha_{xz} + \alpha_{zx} - i\alpha_{yz} - i\alpha_{zy})$$

$$\alpha^{2}_{-2} = \frac{1}{2}\,(\alpha_{xx} - \alpha_{yy} - i\alpha_{xy} - i\alpha_{yx})$$

Table II. Transformation Properties of the Irreducible Tensor Components

	C_s	C_i	C_2	C_3	C_4	C_6	D_2	D_3	D_4	D_6	C_{2v}	C_{3v}	C_{4v}	C_{6v}	C_{2h}	C_{3h}	C_{4h}	C_{6h}
α^0_0	A'	A_g	A	A	A	A	A	A_1	A_1	A_1	A_1	A_1	A_1	A_1	A_g	A'	A_g	A_g
α^1_1	A''	A_g	B	E	E	E_1		E	E	E_1		E	E	E_1	B_g	E''	E_g	E_{1g}
α^1_0	A'	A_g	A	A	A	A	B_1	A_2	A_2	A_2	A_2	A_2	A_2	A_2	A_g	A'	A_g	A_g
α^1_{-1}	A''	A_g	B	E	E	E_1		E	E	E_1		E	E	E_1	B_g	E''	E_g	E_{1g}
α^2_2	A'	A_g	A	E	B	E_2		E		E_2		E		E_2	A_g	E'	B_g	E_{2g}
α^2_1	A''	A_g	B	E	E	E_1		E	E	E_1		E	E	E_1	B_g	E''	E_g	E_{1g}
α^2_0	A'	A_g	A	A	A	A	A	A_1	A_1	A_1	A_1	A_1	A_1	A_1	A_g	A'	A_g	A_g
α^2_1	A''	A_g	B	E	E	E_1		E	E	E_1		E_g	E	E_1	B_g	E''	E_g	E_{1g}
α^2_2	A'	A_g	A	E	B	E_2		E		E_2		E_g		E_2	A_g	E'	B_g	E_{2g}
$\alpha^1_1 + \alpha^1_{-1}$							B_2				B_1							
$\alpha^1_1 - \alpha^1_{-1}$							B_3				B_2							
$\alpha^2_1 + \alpha^2_{-1}$							B_3				B_2							
$\alpha^2_1 - \alpha^2_{-1}$							B_2				B_1							
$\alpha^2_2 + \alpha^2_{-2}$							A		B_1		B_1		B_1					
$\alpha^2_2 - \alpha^2_{-2}$							B_1		B_2		B_2		B_2					

Table II (Continued)

	D_{2h}	D_{3h}	D_{4h}	D_{6h}	D_{2d}	D_{3d}	D_{4d}	D_{6d}	S_4	S_6	T_d	O_h	$C_{\infty v}$	$D_{\infty h}$
α_0^0	A_g	A_1'	A_{1g}	A_{1g}	A_1	A_{1g}	A_1	A_1	A	A_g	A_1	A_{1g}	A_1	g^+
α_1^1		E''	E_g	E_{1g}		E_g	E_3	E_5	E	E_g	T_1	E_{1g}	E_1	$\bar{u}$
α_0^1	B_{1g}	A_2'	A_{2g}	A_{2g}	A_2	A_{2g}	A_2	A_2	A	A_g	T_1	T_{1g}	A_2	u^+
α_{-1}^1		E''	E_g	E_{1g}	E	E_g	E_3	E_5	E	E_g	T_1	T_{1g}	E_2	$\bar{u}$
α_2^2		E'		E_{2g}		E_g	E_2	E_2		E_g			E_2	g
α_1^2		E''	E_g	E_{1g}	E	E_g	E_3	E_5	E	E_g	T_2	T_{2g}	E_1	g
α_0^2	A_g	A_1'	A_{1g}	A_{1g}	A_1	A_{1g}	A_1	A_1	A	A_g	E	E_g	A_1	g^+
α_1^2		E''	E_g	E_{1g}	E	E_g	E_3	E_5	E	E_g	T_2	T_{2g}	E_1	g
α_2^2		E'		E_{2g}		E_g	E_2	E_2		E_g			E_2	g
$\alpha_1^1 + \alpha_{-1}^1$	B_{2g}													
$\alpha_1^1 - \alpha_{-1}^1$	B_{3g}													
$\alpha_1^2 + \alpha_{-1}^2$	B_{3g}													
$\alpha_1^2 - \alpha_{-1}^2$	B_{2g}													
$\alpha_2^2 + \alpha_{-2}^2$	A_g		B_{1g}		B_1				B		E	E_g		
$\alpha_2^2 - \alpha_{-2}^2$	B_{1g}		B_{2g}		B_2				B		T_2	T_{2g}		

symmetric parts of the scattering tensor. This means, in general, that the selection rules for electronic Raman transitions are different from those of vibrational or rotational Raman transitions since for the latter only the selection rules corresponding to α^2_Q components should be considered while for the electronic scattering α^1_Q and α^2_Q components must be considered. In this way we have one way to distinguish electronic and vibrational or rotational Raman transitions.

It is clear from the relation of Table I that the polarization properties for electronic Raman transitions in a crystal, say, will be generally different from vibrational transitions in the same crystal. It is less directly seen, although a very interesting fact, that for systems with a random orientation in space, like powder or freely rotating molecules, the depolarization ratios of the two types of transitions will be completely different. The evaluation of these properties is fairly direct but somewhat outside the scope of this chapter so we will just give the results of the detailed derivation which has been given elsewhere.

For vibrational Raman transitions the depolarization ratios for linearly polarized light will always be $\leq 3/4$, but for electronic transitions this depolarization ratio may be zero and ∞. Thus, again here we have found a very sensitive probe for the separation of electronic and ordinary rotational or vibrational Raman spectra. Let us illustrate the usefulness of the above mentioned procedures by discussing briefly a specific example. The trivalent Europium ion has the electronic groundstate $^7F_0$ and low lying excited states $^7F_1$ and $^7F_2$. When this ion is placed in a crystal, the symmetry of its environment will be lower than that of free space and the degenerate levels $^7F_1$ and $^7F_2$ will in general split. As an example, in an environment of symmetry D_{2d} the $^7F_1$ state will split into two levels of symmetry A_2 and E, while the $^7F_2$ state will split into levels with symmetry A_1, B_1, B_2 and E. The groundstate $^7F_0$ has the same symmetry as an S-orbital and will, of course, not be split in any environment. Using the multiplicating tables of point groups, we then find that the $^7F_1(A_2) \leftarrow {}^7F_0$ transition is allowed as an electronic Raman transition with tensor component α^1_0 and that the transition $^7F_1(E) \leftarrow {}^7F_0$ is allowed with tensor components α^1_1, α^1_{-1} and α^2_1, α^2_{-1}. The last result should, however, be taken with a grain of salt for the following reason. The symmetry characterization scheme used in the derivation here does only concern the point group D_{2d} and does not take into account the fact that the $^7F_0$ and $^7F_1$ wavefunctions are almost completely free ion wavefunctions. As said previously, the α^2_Q components transform under rotations as d-orbitals, that is, they have angular momentum $\mathcal{J} = 2$. The $^7F_0$ and $^7F_1$ states have, of course, angular momenta 0 and 1 respectively and we see from the vector model for angular momenta that these states can never be connected by an operator having angular momentum equal to 2. Thus, if the tensor components of the type α^2_Q should be observed in a $^7F_1 \leftarrow {}^7F_0$ transition, it would be due solely to a deviation of the wavefunctions of the ion from that of a free ion and such a deviation, we may remark, is known from other sources to be quite small. Thus, we may safely say that the transition $^7F_1 \leftarrow {}^7F_0$ of the Eu^{3+} ion is completely governed by the anti-symmetric part of the scattering tensor. By the same argument it may be seen that the $^7F_2 \leftarrow {}^7F_0$ transitions will be determined solely by the symmetric part of the scattering tensor and we thus have in this situation a very unique and very interesting separation of these two types of scattering symmetries. It is perhaps worthwhile to point out that the $^7F_0 \leftarrow {}^7F_1$ transition and thus a completely anti-symmetric scattering tensor has indeed been observed experimentally recently.

EXPERIMENTAL OBSERVATION OF ELECTRONIC RAMAN EFFECT OF TRIVALENT RARE EARTH IONS

The energy levels of trivalent rare earth ions below $70{,}000 cm^{-1}$ are nearly all associated with the $(4f)^n$ electronic configurations of these ions. The wavefunctions of these levels have all the same parity and the Raman operator is an even operator. This suggests that electronic Raman transitions between the levels are possible and this is contrary to the case for the direct (electric dipole) transitions. Such transitions are, however, nearly always observed in the far infrared, infrared and visible regions of the electromagnetic spectrum because the parity of the wavefunctions can change if the ions are imbedded in crystals. The change in parity can be caused by either vibronic coupling effects or by the effect of the crystal fields but it is perhaps worthwhile to point out here that these effects are small, which result in weak absorptions in the spectra of the trivalent Lanthanides. Our knowledge of energy levels and wavefunctions of the 4f electrons is rather extensive and it has in fact been possible to calculate the Raman processes for some of the trivalent ions[3-5]. It turns out that the cross sections of the process is small and in the most favorable case we can expect the intensity of the electronic Raman lines to be equal to intensities of the weaker phonon modes of such crystals as garnets[6] and chlorides[7]. The amount of anti-symmetry of the scattering tensor has also been calculated[4]. For rare earth ions with less than seven 4f electrons, the anti-symmetric contribution can be considerable compared to the contribution of symmetric tensors to the intensity. This situation is completely reversed for the ions with more than seven 4f electrons.

The energy levels of the Lanthanides are such that in only some instances the radiation of the He-Ne or Argon ion gas laser lines is directly absorbed[8]. Only electronic transitions of $PrCl_3$

were excited with conventional light sources, all other electronic Raman spectra of trivalent rare earth ions were excited with either the 6328 Å radiation of a He-Ne laser or the 4880 Å and 5140 Å radiation of an Argon ion gas laser. Contrary to the spectrum excited with the mercury arc which was photographed, all other electronic Raman spectra were photoelectrically recorded.

In this section a discussion is given of the experimental data which have now become available. There is no doubt that the advent of the laser, coupled with the construction of double monochromators and better electronic detection systems has made possible the detection of electronic Raman transitions of moderate intensity and with small shifts from the exciting line.

a. Ytterbium ion

The electronic configuration of Yb^{3+} is $Xe(4f)^{13}$, which is equivalent to one hole in the 4f shell. The electronic ground state is $^2F_{7/2}$ and the $^2F_{5/2}$ state is situated some 10,000 cm^{-1} above the ground state. Electronic transitions of Yb ion have been observed for Ytterbium Gallium Garnet[9] and Ytterbium Aluminum Garnet[10]. In Fig. 1 the low temperature Raman spectrum of Yb;YGaG, YbGaG and YGaG are compared. From a comparison of the phonon spectrum of the crystals, it could be concluded that the electronic transitions occur at 552 cm^{-1} and 539 cm^{-1} of YbGaG and at slightly different frequencies of Yb;YGaG. By applying an external magnetic field of up to 50,000 Gauss, it is observed that the strong electronic transition at 552 cm^{-1} decreases in intensity and becomes larger in width whilst at strong field the lines are considerably broadened. The Zeeman-Raman experiment[11] is shown in Fig. 2.

Electronic transitions have also been observed in YbAlG. They are slightly weaker than the transitions of the Galium Garnet crystals but the intensity was strong enough to measure the polarization properties of the Raman lines. The observation of electronic Raman transitions of the ytterbium ion in garnet crystals is of particular interest because due to presence of strong vibronic coupling effects in the absorption and fluorescence spectra in these crystals the presence of the pure electronic transitions was sufficiently obscured to make a definite assignment possible. All the above mentioned transitions take place between the crystal field levels of the $^2F_{5/2}$ manifold, and the symmetry labelling of the crystal field states and the elements of the scattering tensor should occur within the framework of the double group D_2 which describes the proper symmetry of the ytterbium ion in Garnet crystals[9]. The occurrence of double groups is so far unique that it is only associated with electronic Raman transitions. The full Raman selection rules for the ytterbium garnet crystals can only be derived if the wavefunctions of ground and excited crystal field levels are known and these data are not yet available. From theoretical arguments, it follows, however, that the tensor for the case of double groups is not different from the tensors for the normal point groups and the determination of the α^k_Q's is straightforward.

b. Thulium Ion

The electronic groundstate of the Thulium ion is $^3H_6$ and this state is split in 13 components of TmGaG. Only one electronic transition has been observed[12] employing He-Ne excitation. It appears that the scattering tensor for this transition does not contain asymmetry.

c. Erbium Ion

The groundstate of this ion is $^4I_{15/2}$ and this state is split in 8 components for ErAlG. Only one electronic level was observed[13] in the Raman spectrum of this compound. Using He-Ne excitation fluorescence was excited.

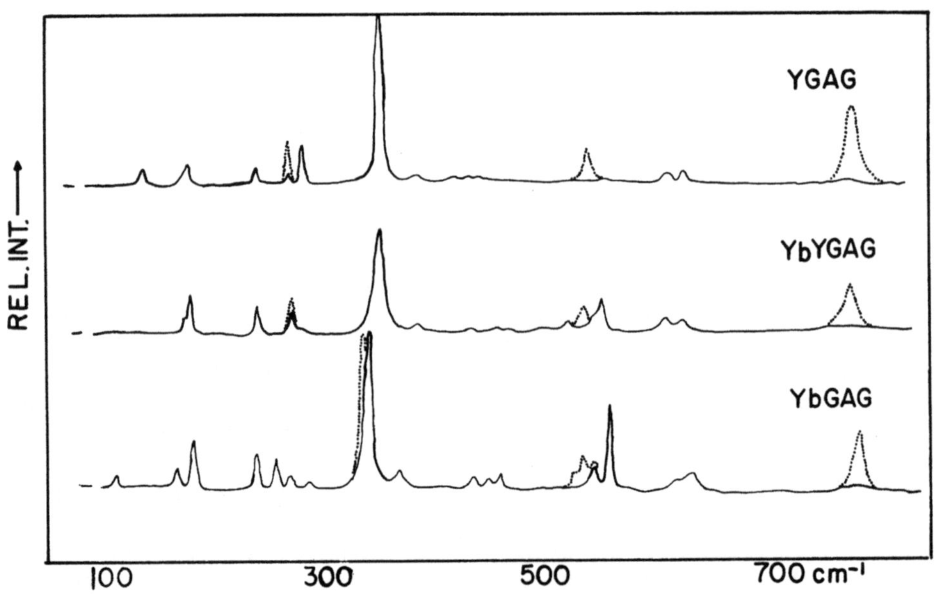

Fig. 1. The low-temperature laser-excited Raman spectra of some Gallium garnet single crystals containing the trivalent ytterbium ion. Electronic Raman transitions occur between 520 cm^{-1} and 560 cm^{-1}.

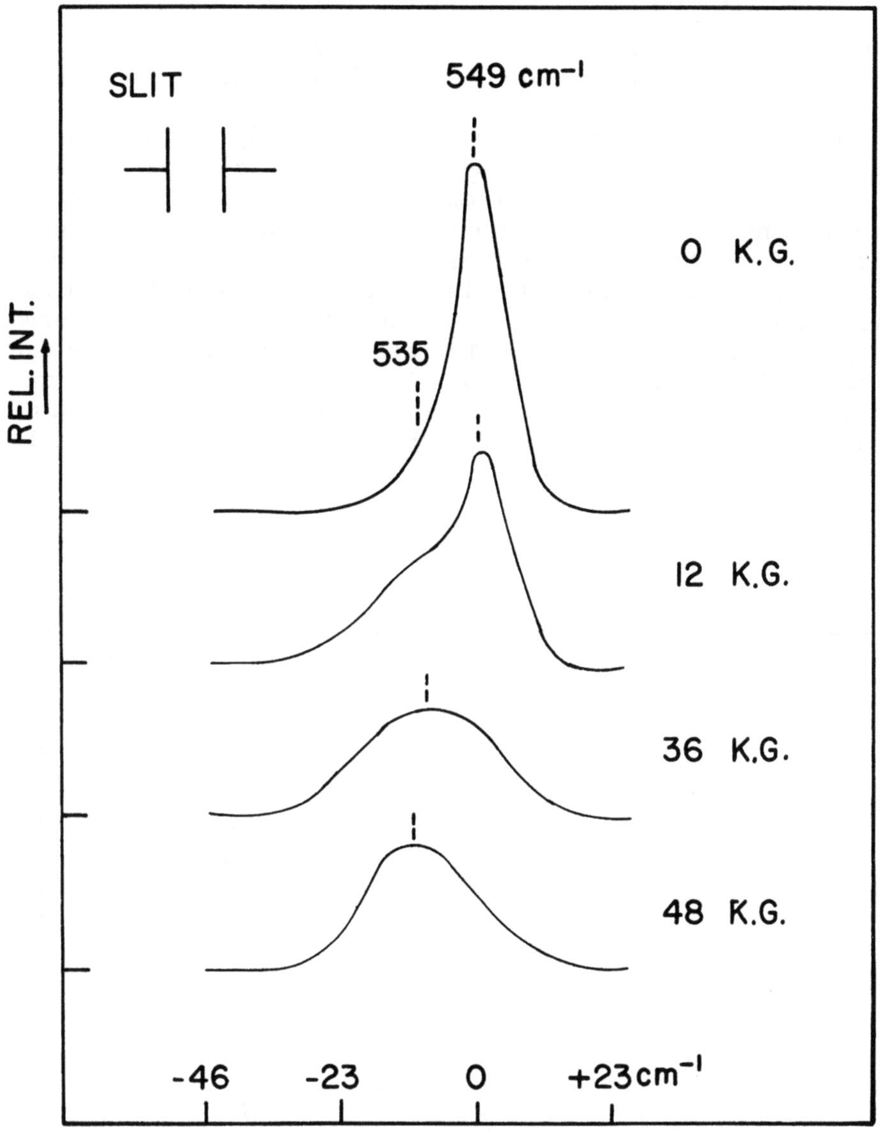

Fig.2. The Zeeman Raman spectrum of the electronic Raman transitions of Yb:YGaG. The ground state is Kramers degenerate and this degeneracy is lifted if the ion experiences the effect of the magnetic field. The crystal, however, is at $4.2^{o}K$ and the population of the first excited state decreases rapidly if the field (and thus the energy difference) increases. The excited state at 550 cm^{-1} splits also in the magnetic field and the position of the high field transition corresponds to the Raman effect of the ground state and the lowest lying level of the Kramers pair at 550 cm^{-1}.

d. Dysprosium Ion

The ground state of this ion is $^6H_{15/2}$ and the first excited manifold $^6H_{13/2}$ is situated some 3500 cm^{-1} above the ground state.

Electronic Raman scattering of the compound DyAlG has been observed for many levels of the ground state manifold[14] and to presumably all seven crystal field components of the $^6H_{13/2}$ manifold[15]. The transitions to the excited manifold are shown in Fig. 3. Here again it is found that electronic Raman spectroscopy is of importance because the position of the levels of this manifold could not be pinned down from either the low temperature absorption or fluorescence spectra of this compound. The full Raman selection rules for these transitions can only be derived if the composition of the wavefunctions of the various states are known and the interpretation of the polarization measurements awaits the results of a crystal field computation of the eigenvectors of the states.

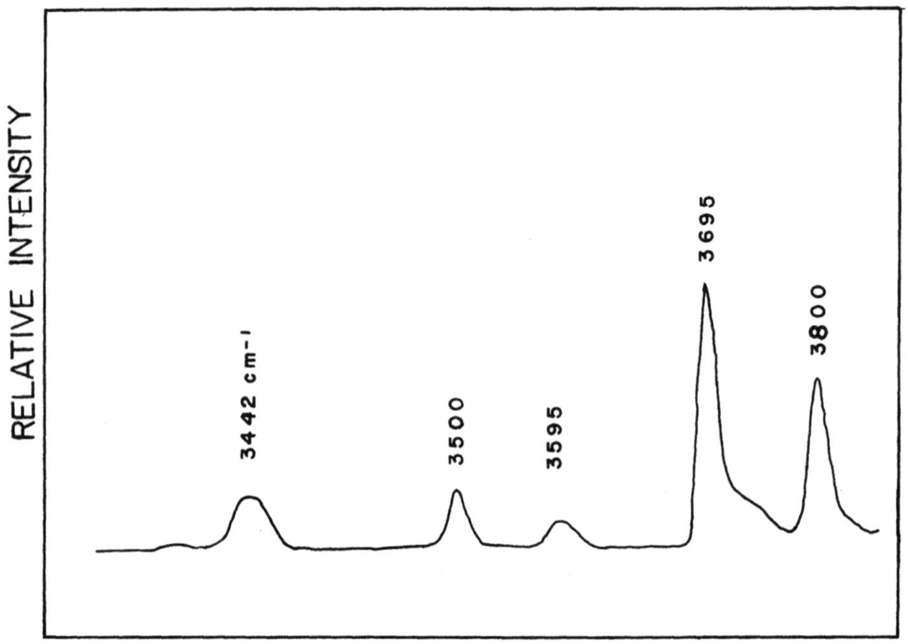

Fig.3. Electronic Raman transition of DyAlG to the $^6H_{13/2}$ manyfolds. Temperature of the sample 80°Km spectrum excited with 4880 Angstrom radiation from an A ion laser.

e. Terbium ion

Many electronic Raman transitions of the compound TbAlG have been observed[16] and their polarization properties measured. No evidence has been obtained of the occurrence of anti-symmetry in the scattering tensor of these transitions. Particularly strong scattering has been observed to electronic levels situated at some 72 cm^{-1} and 82 cm^{-1} and the polarization properties of these transitions have been investigated by rotating the electric field vector of the incident light and leaving the crystal located in the same position. The experimental results are shown in Fig. 4. The electronic transition is of symmetry T_{2g} but also the antisymmetric part of the tensor should show up in this transition. The behavior of the electronic transition is exactly identical to the behavior of all phonon lines of symmetry T_{2g} indicating that the amount of asymmetry is small. This is in good agreement with an apriori calculation of the scattering tensor[4].

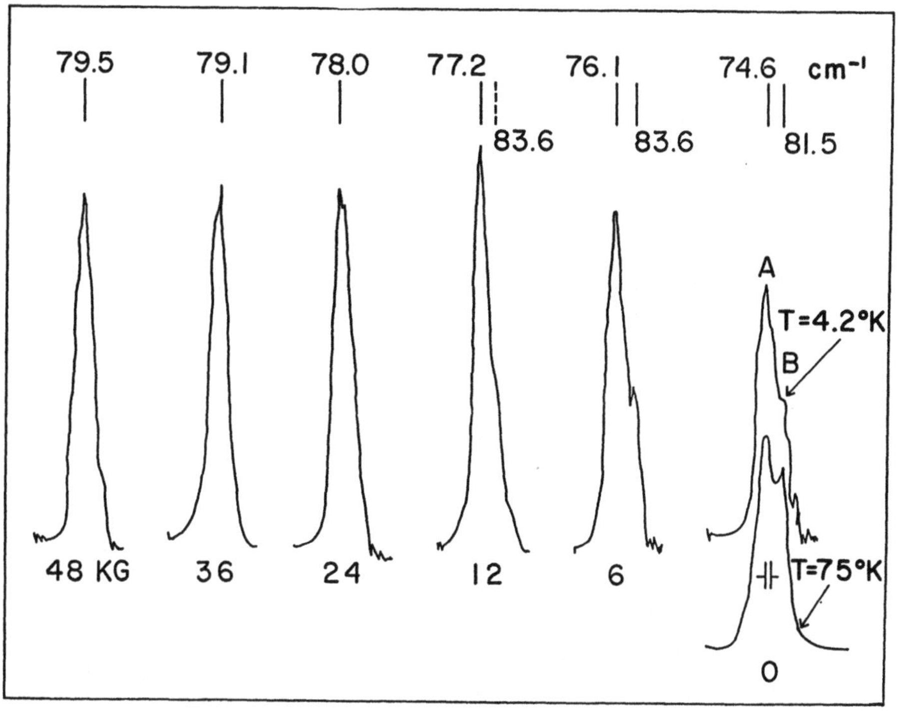

Fig.4. Zeeman Raman effect of electronic Raman transitions of TBALG. The transition at 81.5 cm^{-1} above the ground state. The energy difference of ground and excited state increases if the field becomes stronger and all intensity is found in the transition A.

The electronic transitions of the terbium ion are stronger than the effects of the Dy or Yb ions. Evidence of an electronic transition of symmetry of A_{1g} has been found of TbAlG and this transition is very strong indeed.

f. Europium ion

Electronic transitions to the ^7F states of europium in the YVO_4, YAlG and YGaG crystals have been observed[17-19]. The transitions are very weak and the polarization properties of the transitions have not yet been determined. The transitions to the $^7F_1$ levels are associated with a pure antisymmetric tensor (see section III).

g. Neodymium ion

Evidence of electronic scattering to levels of the ground manifold of NdYGaG has been observed[8] but the data remain to be investigated with regard to the polarization properties of the line.

h. Praesodymium ion

Electronic transitions of $PrCl_3$ were observed employing the 2536 Å radiation of a mercury arc[20]. The transitions take place within the $^3H_4$ ground manifold and from the ground state to crystal field levels of the $^3H_5$ and $^3F_2$ manifolds. Polarization measurements of the transition were not performed in detail. From a theoretical calculation, it follows[24] that some of the electronic transitions should be associated with highly anti-symmetric scattering tensors. The amount of asymmetry is strongly dependent of the frequency of the incident radiation.

i. Cerium ion

Transitions to the crystal field components of the $^2F_{5/2}$ and $^2F_{7/2}$ manifolds have been excited employing Argon ion excitation. Some of the scattering tensors are again highly anti-symmetric and, at least in one case, has there been perfect agreement between the computed[22] and observed[23] amount of asymmetry of the tensors.

j. The NO molecule

Employing mercury excitation electronic transition $^2\pi_{1/2} \leftrightarrow {}^2\pi_{3/2}$ of this molecule was observed[24] already in 1932. This transition should also be associated with a tensor containing some asymmetry.

SUMMARY OF PRESENT KNOWLEDGE ABOUT THE ELECTRONIC RAMAN EFFECT

Many electronic Raman transitions have been observed, specially of the trivalent Lanthanides in the garnet crystals. The Raman selection rules are strongly dependent on the wavefunctions of the states involved and although certain elements of the scattering tensor can transform and thus contribute to the Raman transition between two electronic states, only from a detailed knowledge of the wavefunctions of the states can be determined which of the possible tensor elements occurring in the product or the representations $\Gamma_{\psi_k} \times \Gamma_{\psi_n}$ (where ψ_k and ψ_n are the wavefunctions of the two states) will indeed contribute.

The problem of computation of absolute cross sections and the Raman to Rayleigh ratio of electronic Raman transitions of the Lanthanides is quite different from the case of the vibrational Raman effect. It thus seems that our theoretical knowledge of the electronic effects is in a more advanced state than in vibrational Raman spectroscopy, but many more experimental data have to be obtained before the electronic Raman spectra of the said ions are fully understood.

Many of the electronic transitions of the rare earth ions have been observed in the author's laboratory and also the theoretical background, especially with regard to the anti-symmetric tensor has been developed. At present, it can be said that the general area in which this kind of spectroscopy will be most successful is that of crystal field theory, because of the rather intricate relations between the polarization properties of the electronic Raman transitions and the symmetry and strength of the crystal field[25], which is normally expressed in terms of the crystal field parameters. The elucidation of the spectra of the trivalent rare earth ions in crystals has yielded considerable information on the said crystal field parameters as was already mentioned in the introduction of this paper.

REFERENCES

1. G. H. Dieke and H. M. Crosswhite, Appl. Opt. 2, 675 (1963).

2. G. Placzek, Mars Handbuch der Radiology, 2nd. ed. Vol. 6, 205 (1934).

3. O. S. Mortensen and J. A. Koningstein, Chem. Phys. Letters, 1, 409 (1968).

4. J. A. Koningstein and O. S. Mortensen, Phys. Rev. 168, 75 (1968).

5. O. S. Mortensen and J. A. Koningstein, J. Chem. Phys. 48, 3971 (1968).

6. J. A. Koningstein, Appl. Spectroscopy, 22, 438 (1968).

7. J. A. Koningstein (to be published).

8. J. A. Koningstein, J. Chem. Phys. 46, 2811 (1967).

9. J. A. Koningstein, Phys. Rev. 174, 477 (1968).

10. J. A. Koningstein , Chem. Phys. Letters 2, 469 (1968).

11. J. A. Koningstein and G. Mace, Chem. Phys. Letters 3, 343(1969).

12. J. A. Koningstein and Toanig Ng, Can. J. Chem. 47, 1395 (1969).

13. J. A. Koningstein (to be published).

14. J. A. Koningstein, Solid State Communication, 7, 351 (1968).

15. J. A. Koningstein (to be published).

16. J. A. Koningstein, Chem. Phys. Letters, 2, 469 (1968).

17. J. A. Koningstein and O. S. Mortensen, Phys. Rev. Letters, 18, 831 (1967).

18. J. A. Koningstein and O. S. Mortensen, Nature, 217, 445 (1968).

19. J. A. Koningstein, J. Opt. Soc. Am. 56, 1405 (1966).

20. J. T. Hougen and S. Singh, Phys. Rev. Letters, 20, 406 (1963).

21. J. A. Koningstein and O. S. Mortensen, Chem. Phys. Letters, 2, 693 (1968).

22. J. A. Koningstein and O. S. Mortensen, Chem. Phys. Letters 1, 409 (1967).

23. J. A. Koningstein (to be published).

24. F.Rassetti, Z. Phys., 66, 626 (1930).

25. J. A. Koningstein and O. S. Mortensen, J. Opt. Soc. Am., 58, 1208 (1968).

FAR INFRARED ELECTRONIC TRANSITIONS IN RARE EARTH IONS

Armand Hadni

Faculty of Sciences, University of Nancy, France

We have seen in the previous chapter that all crystals become completely transparent in the very far infrared when they are cooled. Then we can introduce metal ions in the lattice either intrinsically or as impurities, and look at their electronic transitions. This is not possible with glasses, the transmission of which is lower and does not increase by cooling.

A. Splitting of Energy Levels, Selection Rules

In this lecture we shall especially look at rare earth ions (Fig. 1). They have a partially filled 4f subshell ($n=4$, $\ell=3$). They are generated primarily by filling the 4f shell. Lanthanum, the first of the series has no 4f electrons, and lutetium the last of the series has fourteen 4f electrons.

Rare earth ions are mostly trivalent. The atom generally loses its most external two $6s^2$ electrons and often one of the 4f electrons. It is generally accepted that the ground states of the trivalent rare earth ions are of the form $4f^x$ where $x = 1$ for Ce^{3+}; $x=2$ for Pr^{3+}; $x = 3$ for Nd^{3+}; $x = 5$ for Sm^{3+}, etc...$x = 13$ for Yb^{3+}.

I. The Free Ion

To get the energy levels of the free ion in the 4f subshell, we first take into account each individual 4f electron and add their energy. In the case of Pr^{3+} which contains two 4f electrons we get a level drawn on the left of the Figure 2.

Now from Pauli principle, no two electrons can be in the same

Atomic Numbers	X-ray levels	K	L	M	N	O	P	Q	Term symbols
	Princ. quant. number n =	1	2 2	3 3 3	4 4 4 4	5 5 5 5	6 6 6	7	
	Subs. quant. number l =	0	0 1	0 1 2	0 1 2 3	0 1 2 3	0 1 2	0	
	Symbols for the energy levels	s	s p	s p d	s p d f	s p d f	s p d	s	
57	Lanthanum ..	2	2 6	2 6 10	2 6 10	2 6 1	2		$^2D_{3/2}$
58	Cerium	2	2 6	2 6 10	2 6 10 (1)	2 6 (1)	(2)		($^3H_4$)
59	Praseodymium	2	2 6	2 6 10	2 6 10 (3)	2 6	(2)		—
60	Neodymium .	2	2 6	2 6 10	2 6 10 4	2 6	2		$^5I_4$
61	Promethium..	2	2 6	2 6 10	2 6 10 (5)	2 6	(2)		—
62	Samarium ...	2	2 6	2 6 10	2 6 10 6	2 6	2		$^7F_0$
63	Europium....	2	2 6	2 6 10	2 6 10 7	2 6	2		$^8S_{7/2}$
64	Gadolinium ..	2	2 6	2 6 10	2 6 10 7	2 6 1	2		^9D
65	Terbium	2	2 6	2 6 10	2 6 10 8	2 6 1	2		—
66	Dysprosium ..	2	2 6	2 6 10	2 6 10 (9)	2 6 (1)	(2)		—
67	Holmium	2	2 6	2 6 10	2 6 10 (10)	2 6 (1)	(2)		—
68	Erbium	2	2 6	2 6 10	2 6 10 (11)	2 6 (1)	(2)		—
69	Thulium	2	2 6	2 6 10	2 6 10 13	2 6	2		$^2F_{7/2}$
70	Ytterbium ...	2	2 6	2 6 10	2 6 10 14	2 6	2		$^1S_0$
71	Lutetium	2	2 6	2 6 10	2 6 10 14	2 6 1	2		$^2D_{3/2}$
72	Hafnium.....	2	2 6	2 6 10	2 6 10 14	2 6 2	2		$^3F_2$
73	Tantalum ...	2	2 6	2 6 10	2 6 10 14	2 6 3	2		$^4F_{3/2}$
74	Tungsten	2	2 6	2 6 10	2 6 10 14	2 6 4	2		$^5D_0$
75	Rhenium	2	2 6	2 6 10	2 6 10 14	2 6 5	2		$^6S_{5/2}$
76	Osmium	2	2 6	2 6 10	2 6 10 14	2 6 6	2		$^5D_4$
77	Iridium......	2	2 6	2 6 10	2 6 10 14	2 6 7	2		$^4F_{9/2}$
78	Platinum	2	2 6	2 6 10	2 6 10 14	2 6 9	1		$^3D_3$
79	Gold	2	2 6	2 6 10	2 6 10 14	2 6 10	1		$^2S_{1/2}$
80	Mercury	2	2 6	2 6 10	2 6 10 14	2 6 10	2		$^1S_0$
81	Thallium	2	2 6	2 6 10	2 6 10 14	2 6 10	2 1		$^2P_{1/2}$
82	Lead	2	2 6	2 6 10	2 6 10 14	2 6 10	2 2		$^3P_0$
83	Bismuth	2	2 6	2 6 10	2 6 10 14	2 6 10	2 3		$^4S_{3/2}$
84	Polonium	2	2 6	2 6 10	2 6 10 14	2 6 10	2 4		$^3P_2$
85	Astatine	2	2 6	2 6 10	2 6 10 14	2 6 10	2 5		$^2P_{3/2}$
86	Radon	2	2 6	2 6 10	2 6 10 14	2 6 10	2 6		$^1S_0$
87	Francium ...	2	2 6	2 6 10	2 6 10 14	2 6 10	2 6	1	$^2S_{1/2}$
88	Radium	2	2 6	2 6 10	2 6 10 14	2 6 10	2 6	2	$^1S_0$
89	Actinium	2	2 6	2 6 10	2 6 10 14	2 6 10	2 6 1	2	($^2D_{3/2}$)
90	Thorium.....	2	2 6	2 6 10	2 6 10 14	2 6 10	2 6 (2)	(2)	($^3F_2$)
91	Protactinium .	2	2 6	2 6 10	2 6 10 14	2 6 10	2 6 (3)	(2)	($^4F_{3/2}$)
92	Uranium	2	2 6	2 6 10	2 6 10 14	2 6 10	2 6 (4)	(2)	($^5D_0$)
93	Neptunium ..	2	2 6	2 6 10	2 6 10 14	2 6 10 (5)	2 6	(2)	—
94	Plutonium ...	2	2 6	2 6 10	2 6 10 14	2 6 10 (5)	2 6 (1)	(2)	
95	Americium ..	2	2 6	2 6 10	2 6 10 14	2 6 10 (6)	2 6 (1)	(2)	—
96	Curium......	2	2 6	2 6 10	2 6 10 14	2 6 10 (7)	2 6 (1)	(2)	—
97	Berkelium....	2	2 6	2 6 10	2 6 10 14	2 6 10 (8)	2 6 (1)	(2)	..
98	Californium .	2	2 6	2 6 10	2 6 10 14	2 6 10 (9)	2 6 (1)	(2)	...
99	Einsteinium ..	2	2 6	2 6 10	2 6 10 14	2 6 10 (10)	2 6 (1)	(2)	--
100	Fermium	2	2 6	2 6 10	2 6 10 14	2 6 10 (11)	2 6 (1)	(2)	
101	Mendelevium.	2	2 6	2 6 10	2 6 10 14	2 6 10 (12)	2 6 (1)	(2)	--

Figure 1. Electron configuration of the rare earth elements (reference 8)

state. We have to introduce quantum numbers m_l and m_s associated with the orbital momentum l, and the spin momentums of each electron. We have $m_\ell = \ell$; $(\ell-1)$, $(\ell-2)\ldots0\ldots$; $-\ell$; and $m_s = \pm 1/2$. Hence $2(2\ell+1)$ possibilities for and electron with quantum number ℓ. Now in the 4f subshell, $\ell = 3$, hence a total of 14 possibilities. This is also the maximum number of electrons which can occupy the 4f shell.

1) Electronic repulsions

Then we consider the mutual repulsion between the electrons, due to their electrical charge. From wave mechanics an "exchange energy" appears with a strong coupling of electron spins, and an energy of interaction $W = -2J \cdot s_i\ s_j$, when $s_i s_j$ are the spin vectors. We shall admit that in the ground state all spins are parallel, and the total spin S is maximum. This is the first Hund's rule. The second rule tells us that orbital momenta are also coupled

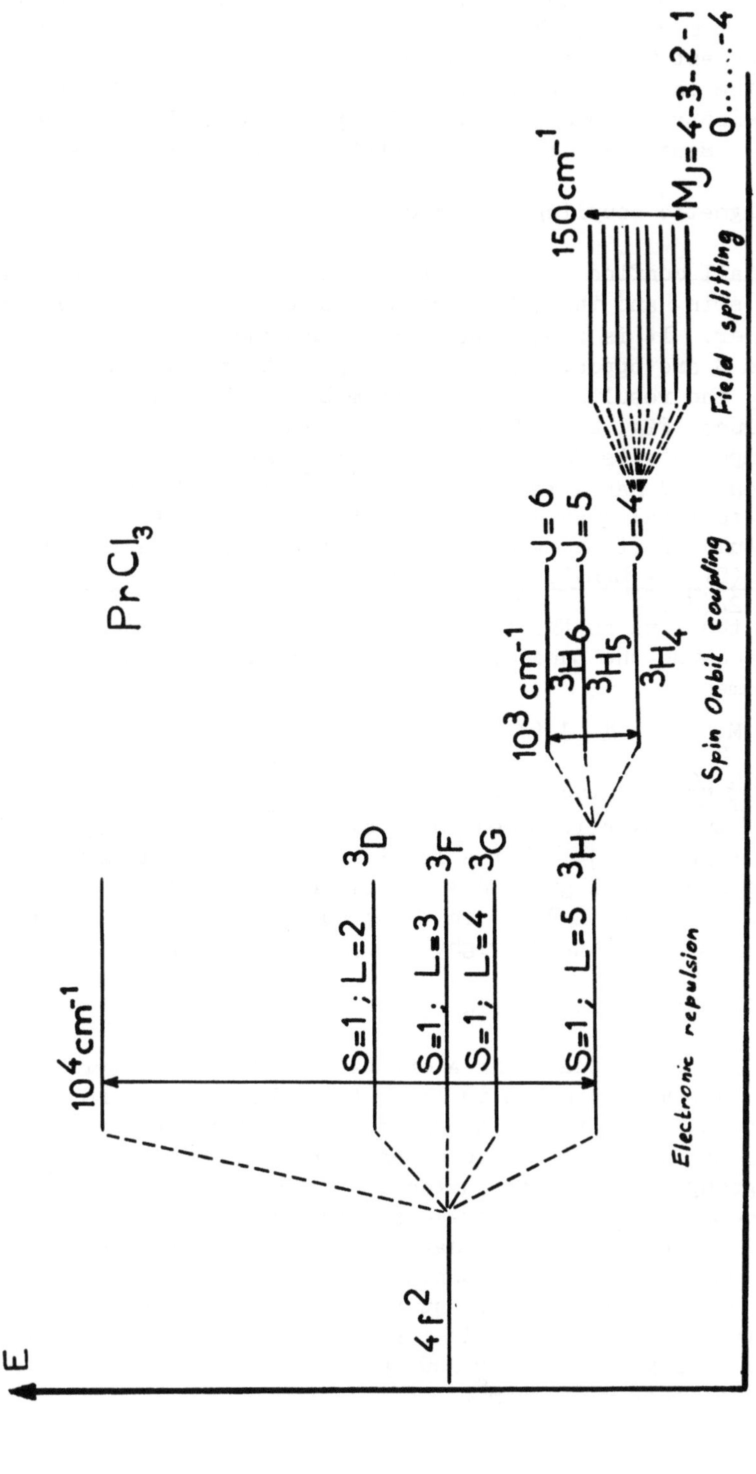

Figure 2. Splitting of the $4f^2$ energy level on Pr^{3+} in $PrCl_3$

to form a resultant with a quantum number L, the collective azimuthal quantum number. For the ground state it is shown that it has the largest value consistent with the exclusion principle. Hence in the case of a praseodymium ion in the ground state, $m_s=\frac{1}{2}$ for each electron and S = 1; m_ℓ = 3; 2; respectively for the two electrons. Hence L=5: the ground state is an H state (Fig. 2).

2) Magnetic coupling of S and L

We shall confine ourselves to Russel-Saunders coupling where the total spin S and the total orbital angular momentum L are coupled together. It is shown that for the ground state of rare earth ions L and S are oppositely directed up to Europium and in the vector model, the resultant J varies from L-S to L+S by integrally spaced values (Fig. 2). From this, it results that the total number of components is 2S + 1. Let us also recall that the total angular momentum J has a magnitude $\sqrt{J(J+1)}\,\frac{h}{2\pi}$. Table I gives the ground state total quantum numbers S (total spin), L (total orbital momentum), J (total angular momentum). The Landé factor $g = \frac{3}{2} + \frac{S(S+1) - L(L+1)}{2J(J+1)}$ gives the projection M_J of the magnetic moment of the ions on the total angular momentum J (they are not colinear because the Landé factor of the spin is twice that of the orbital moment):

$$M_J = - g\,\beta\,[J\,(J+1)]^{\frac{1}{2}}$$

where $\beta\,(= \frac{e}{m}\frac{h}{4\pi c})$ is the Bohr magneton; $p = \frac{M_J}{\beta}$ represents the projection of the magnetic moment on J expressed in Bohr magneton units: $p = g[J(J+1)]^{\frac{1}{2}}$. For a level of total angular momentum J, a magnetic field will give 2J + 1 sublevels, with M_J=J; J-1; J-2; ...-J. In the case of Pr^{3+} in the ground sublevel, J=4, hence 9 Zeeman components (Fig. 2, right side) are expected.

II. The Crystalline Field

When an ion is placed in a crystal the local electric field introduces a symmetry which is lower than the complete rotational symmetry of the free ion. It results that the degree of residual degeneracy is lower and can be determined exactly from the symmetry by group theory. For instance, for J = 2, a cubic field will give two components and a tetragonal field will give as many components as a magnetic field, that is five components. For J=3/2, they will only give two components. In fact, for half integral J values, all levels are at least two fold degenerate for any symmetry of the electric field (Kramer's theorem). Table II gives more detailed information for J=0 up to J=15/2, and Figure 3 gives an example of energy diagram[9].

Table I

Ground state of rare earth ions. All s_i corresponding to every electron are parallel up to Gd, hence S = 0, 1/2, 1, 3/2, 2, 5/2, 3; L has the maximum value allowed by Pauli principle, hence L = 0, 3, 5, 6, 6, 5, 3, 0 up to Gd, then L takes symmetric values. L and S are antiparallel up to Gd. They are parallel for heavier rare earth ions.

Z	No. of electrons in 4 f shell	Ion	Ground State	S	L	J	g	$p^2=g^2 J(J+1)$	p^2 (exp.)
57	0	La^{3+}	$^1S_0$	0	0	0	-	0	0
58	1	Ce^{3+}	$^2F_{5/2}$	1/2	3	5/2	6/7	6.43	6
59	2	Pr^{3+}	$^3H_4$	1	5	4	4/5	12.8	12
60	3	Nd^{3+}	$^4I_{9/2}$	3/2	6	9/2	8/11	13.1	12
61	4	Pm^{3+}	$^5I_4$	2	6	4	3/5	7.2	-
62	5	Sm^{3+}	$^6H_{5/2}$	5/2	5	5/2	2/7	0.7	2.4
63	6	Eu^{3+}	$^7F_0$	3	3	0	-	0	-
64	7	Gd^{3+}	$^8S_{7/2}$	7/2	0	7/2	2	63	63
65	8	Tb^{3+}	$^7F_6$	3	3	6	3/2	94	92
66	9	Dy^{3+}	$^6H_{15/2}$	5/2	5	15/2	4/3	113	110
67	10	Ho^{3+}	$^5I_8$	2	6	8	5/4	112	110
68	11	Er^{3+}	$^4I_{15/2}$	3/2	6	15/2	6/5	92	90
69	12	Tm^{3+}	$^3H_6$	1	5	6	7/6	57	52
70	13	Yb^{3+}	$^2F_{7/2}$	1/2	3	7/2	8/7	21	19
71	14	Lu^{3+}	$^1S_0$	0	0	0	-	0	0

Table II

Splitting of the ground state of rear eath ions by different fields

J Values	FIELD CHARACTER					
	H uniform	E uniform		Cubic	Octahedral and Hexagonal	Tetragonal and lower symmetry (Rhomb.)
	(2J+1)	(J+1)	(J+½)			
0	1	1		1	1	1
1/2	2		1	1	1	1 (K)
1	3	2		1	2	3
3/2	4		2	1	2	2 (K)
2	5	3		2	3	5
5/2	6		3	2	3	3 (K)
3	7	4		3	5	7
7/2	8		4	3	4	4 (K)
4	9	5		4	6	9
9/2	10		5	3	5	5 (K)
5	11	6		4	7	11
11/2	12		6	4	6	6 (K)
6	13	7		6	9	13
13/2	14		7			
7	15	8				
15/2	16		8	5	8	

Number of sublevels

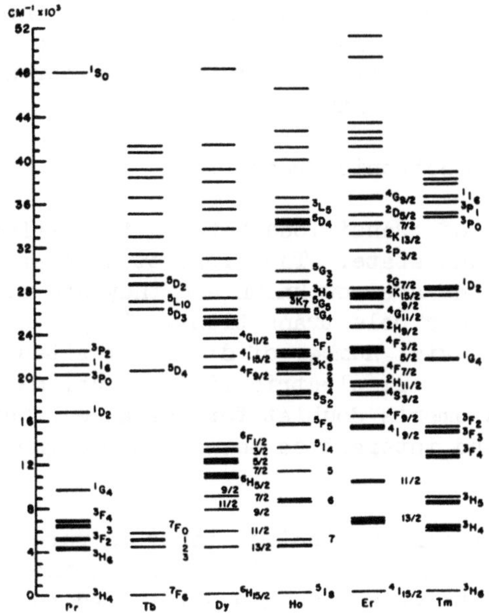

Figure 3. Energy levels of rare earth atoms.

III. Intensities and Selection Rules

1) Free ion

The intensities of spectral lines in rare earth ions have been widely considered for many years because they cannot be explained as simple first-order electric dipole transitions as was first pointed out by Van Vleck. In the transition from state i to state f in the same $4f^x$ configuration, the wave functions ψ_i and ψ_f have the same parity, while er (the operator associated with the electric dipole moment) has odd parity, so that the matrix element of electric dipole transitions $\int \psi_i^* \, (er) \, \psi_f \, d\tau$ must be zero (average dipole moment of the mixed wave functions of the two states). Hence in the free ion, the infrared active transitions must be either magnetic dipole, or electric quadrupole.

2) Crystal

We have seen that in a crystal the site symmetry is lower than in free space. If the inversion symmetry is removed, then parity is no longer conserved. The wave function will now contain both even and odd parity components. The levels of the ion are no longer true 4f levels. However, they are mostly 4f and include some admixture with the nearest configuration (probably 5d). The transitions take place between the odd parity part of one state, to the even parity part of another, and it is possible from symmetry considerations at the site of the ion, to infer selection rules for what is called "electric dipole forced transitions".

B. Far Infrared Electronic Transitions

It is a question of transitions between the different Stark components of the ground state. The first studies are from 1962. Several such transitions are now known and they are either electric dipole or magnetic dipole transitions. They can give several important information, viz. number of different sites for the ion, symmetry of each site, local electric field, local magnetic field (which can split, a Kramer's doublet for instance), interaction with lattice phonon and interaction between paramagnetic rare earth ions.

These transitions can be studied either in pure materials like $PrC\ell_3$, or in doped crystals like $LaCl_3:Pr^{3+}$. In the last case, possible interactions between ions are eliminated and sharper lines could be expected. This has not been checked up to now in the far infrared but are fairly well understood in Raman effect as explained in Dr. Koningstein's lecture. Furthermore, the introduction of rare earth ions as impurities may cause additional complexities in the spectrum due to gap or resonant band modes.

Anhydrous Rare Earth Chlorides

1. Introduction

The crystal structures of rare earth chlorides are known with an unusual degree of certainty and the symmetry of the site occupied by the rare earth ion is generally high enough $[C_{3h}]$ to make easier the calculations. Figure 4 gives the absorption of a pure $LaC\ell_3$ plate cut perpendicular to the axis at room and liquid hydrogen temperatures. The absorption has sufficiently decreased to allow studies up to 120 cm^{-1}.

The best spectroscopy is made with oriented crystals either perpendicular or parallel to the axis (Fig. 5). In Figure 5a we get the axial spectrum. In Figure 5b and 5c, we respectively get the σ and π spectra.

From elementary considerations we can deduce the character either $DE_{//}$, $DE_{\perp}$ or $DM_{//}$, $DM_{\perp}$ of the transition (Table III).

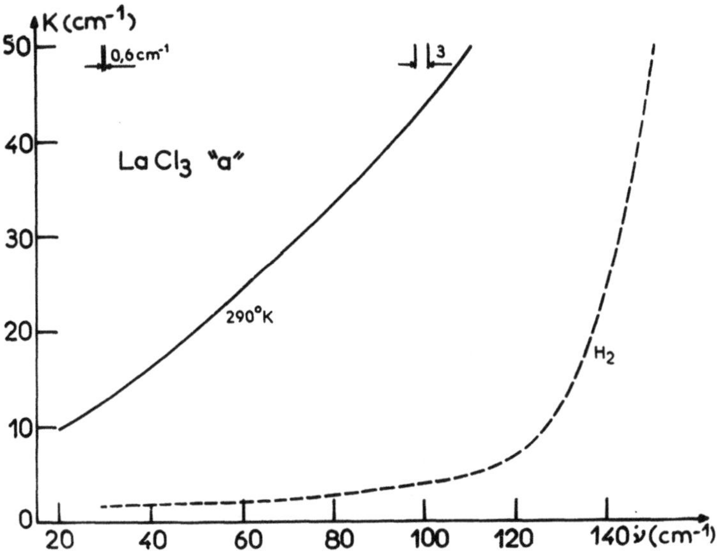

Figure 4. Absorption of a pure $LaCl_3$ plate cut perpendicular to the axis at two different temperatures.

Table III

a	+	+	-	-
σ	+	-	+	-
π	-	+	-	+
Character	$DE_{\perp}$	$DM_{\perp}$	$DM_{//}$	$DE_{//}$

$PrCl_3$

I.2 .1 Theory

There are two 4f electrons hence the ground state is denoted $^3H_4$ with S = 1, L = 5, J = 4. In a magnetic field, the ground level would be split into 2J + 1 = 9 components. In the field of C_{3h} symmetry, in the site of the ion there will only be 6 sublevels characterized by μ, Hellwege's crystal quantum numbers 0; ± 1; ± 2; ± 3. Those levels with a (±) prefix are doubly degenerate (symmetry E' or E"), the remainder of the symmetry A' or A" are non-degenerate.

The sublevels of the ground state $^3H_4$ were indirectly known from ultraviolet and visible spectroscopy at room temperature and

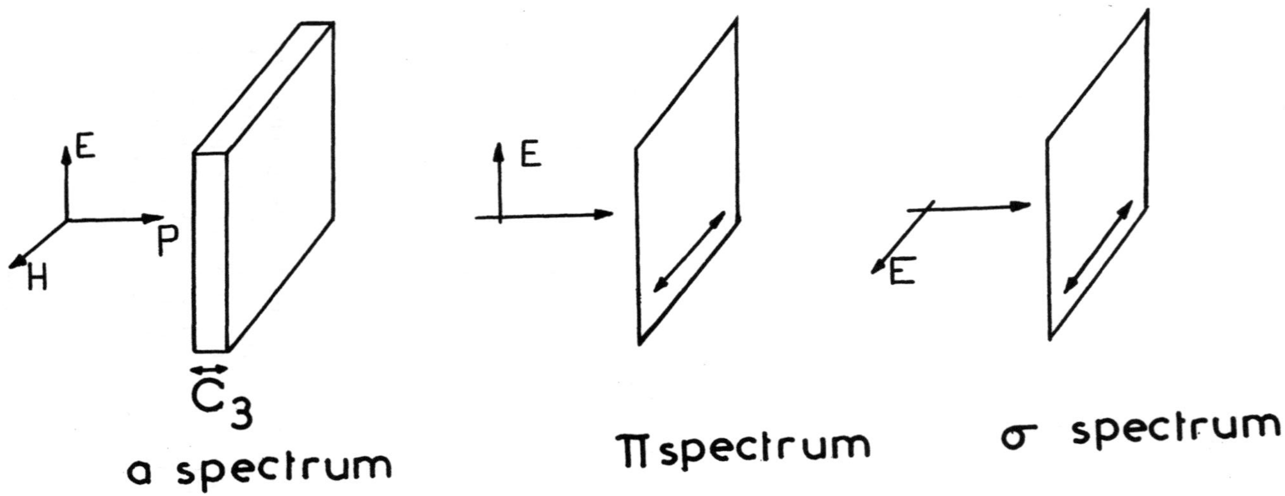

Figure 5. Definition of the a, π and σ spectra

liquid helium temperature with $LaCl_3$ doped with praseodymium[1]. Recently they were also found in Raman spectra of pure $PrCl_3$ (Fig. 6)

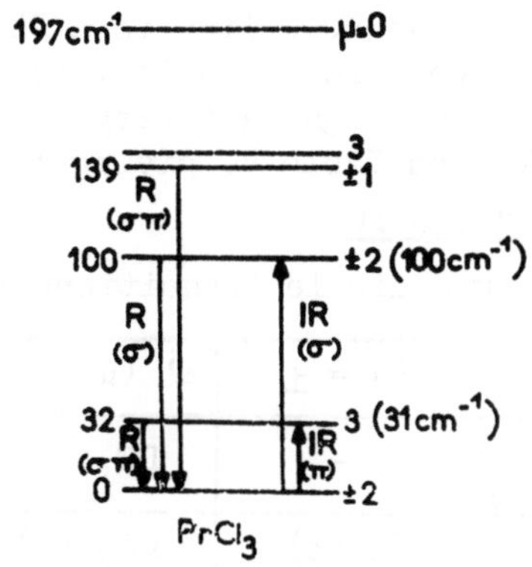

Figure 6. Stark splitting of the $^3H_4$ ground state of Pr^{3+} in pure $PrCl_3^3$ (from ref. 7)

and the authors[2] hoped to get narrower lines with a doped crystal and a brighter source.

I.2.2 Experimental

Figure 7 gives the σ, π and axial spectra of $PrCl_3$. The axial

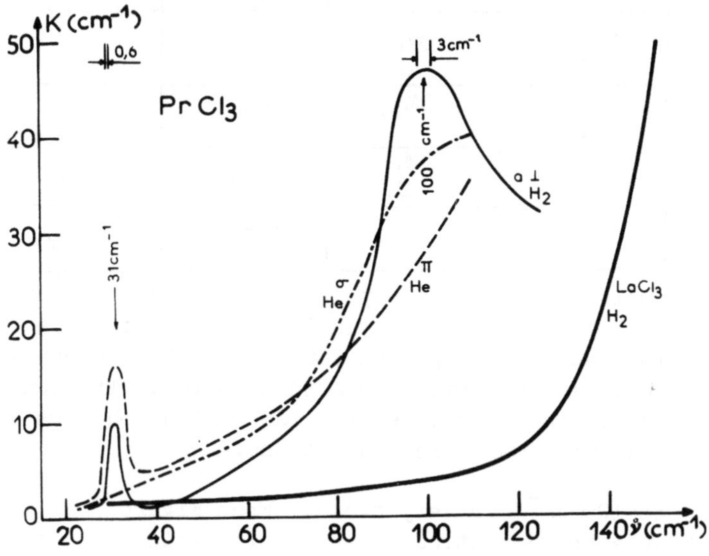

Figure 7. Absorption of $PrCl_3$ and $LaCl_3$

spectrum of a pure $LaCl_3$ plate is given for comparison. The 325μ line is of type π, the 100μ line is of type σ and both lines are found in the axial spectrum. Hence, both arise from oscillations perpendicular to the axis. The 100μ line is a $DE_\perp$ line, the 325μ line is a $DM_\perp$ line. Their positions are in good accordance with the scheme of Figure 6 and the right activity of the lines could be predicted from Table IV and Table V, giving the selection rules

Table IV

Selection rules for electric dipole transitions in $PrCl_3$

	A' (μ=0)	E" (μ = ± 1)	E' (μ = ± 2)	A" (μ=3)
A' (μ = 0)	-	-	$DE_\perp$ (σ)	$DE_{//}$ (π)
E" (μ= ±1)	-	$DE_\perp$ (σ)	$DE_{//}$ (π)	$DE_\perp$ (σ)
E' (μ = ± 2)	$DE_\perp$ (σ)	$DE_{//}$ (π)	$DE_\perp$ (σ)	-
A" (μ = 3)	$DE_{//}$ (π)	$DE_\perp$ (σ)	-	-

Table V

Selection rules for magnetic dipole transitions in $PrCl_3$

	μ = 0	μ = ± 1	μ = ± 2	μ = 3
μ = 0	$DM_{//}$	$DM_\perp$	-	-
μ = ± 1	$DM_\perp$	$DM_{//}$	$DM_\perp$	-
μ = ± 2	-	$DM_\perp$	$DM_{//}$	$DM_\perp$
μ = 3	-	-	$DM_\perp$	$DM_{//}$

for electric dipole and magnetic dipole transitions. All transitions found in the visible for Pr^{3+} in $LaCl_3$ are forced electric dipole transition, and the 325μ line found in $PrCl_3$ is probably the first magnetic dipole transition found in this ion. It shows a much smaller width. This might be an evidence that phonons do not interact very much with magnetic transitions.

The line at 100 cm^{-1} is very broad. From Figure 7 it is clear that $PrCl_3$ shows a broad continuous absorption which is not seen in $LaCl_3$. To make the things clearer, we have plotted in Figure 8 the additional absorption ΔK introduced in $PrCl_3$, vs. frequency. It

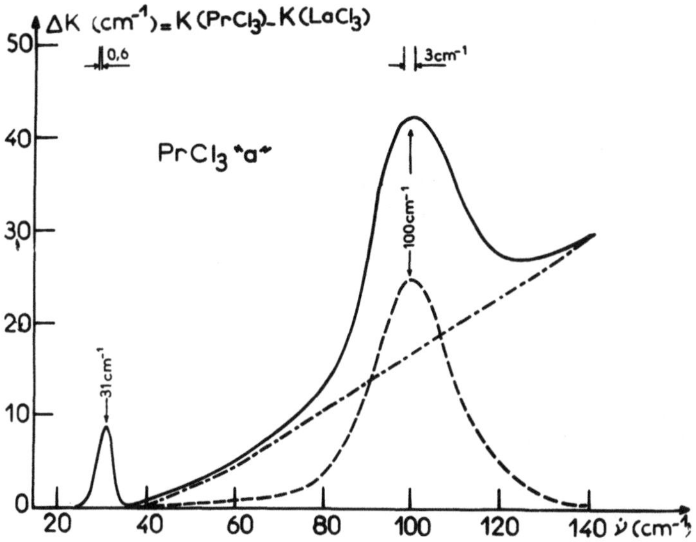

Figure 8. Additional absorption of $PrCl_3$ "a"

may be considered as the superposition of a broad absorption and the two bands at 31 and 100 cm^{-1}.

1.3 Sa^{3+} : $LaCl_3$

1.3.1 Theory

The anhydrous samarium chloride either pure or diluted in $LaCl_3$ seems also to approach the model of the isolated ion in an electrostatic field most closely. The Sa^{3+} ion possesses five 4f electrons with $\ell = 3$. The ground state is $^6H_{5/2}$ which means S=5/2, L = 5 and J = 5/2. Hence L and S are antiparallel which is the rule for the first rare earth ions. In a uniform magnetic field we should get 2J + 1 = 6 sublevels corresponding to the magnetic quantum number $M_J = \pm 5/2$; $\pm 3/2$; $\pm 1/2$. In the crystal field of symmetry C_{3h} in the site of Sa^{3+} ion, there will be a splitting into 3 Kramers doublets characterized by the crystal quantum numbers $\mu = \pm 5/2$; $\pm 3/2$; $\pm 1/2$ corresponding to the three irreducible representations of the ion in a field of C_{3h} symmetry. Here $\mu = M_J$. The positions of the Kramer's doublets are known from absorption and fluorescence spectra (Figure 9).

1.3.2 Experimental

Figure 10 gives the absorption of both $LaCl_3 : Sm^{3+}$ and pure $LaCl_3$. From it the additional absorption ΔK corresponding to Sm^{3+} ions is calculated (Figure 11). In Figure 11 we see two lines of

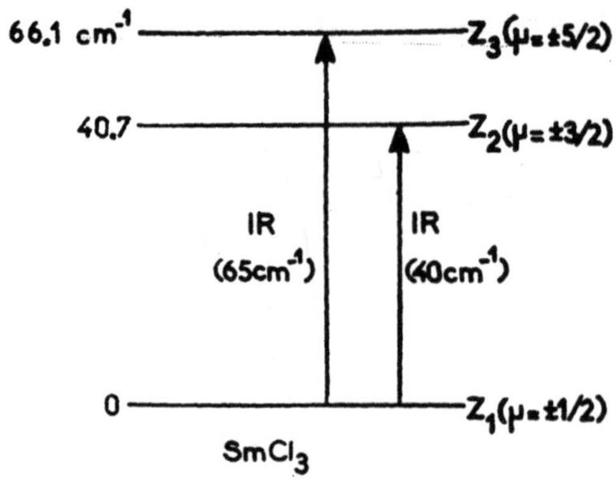

Figure 9. Stark splitting of the $^6H_{5/2}$ state of the Sa^{3+} ion in $LaCl_3$ (reference 7)

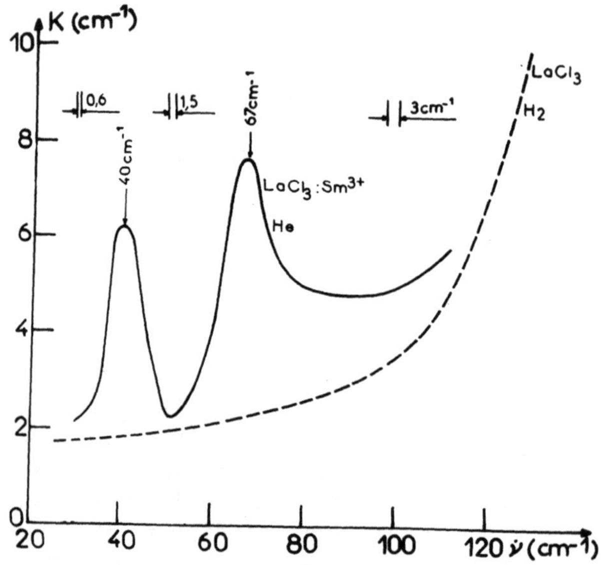

Figure 10. Absorption of $LaCl_3:Sm^{3+}$ at liquid helium temperature and pure $LaCl_3$ at liquid hydrogen temperature

nearly the same strength (dashed line) and a background absorption due probably to an induced absorption peaking near 80cm^{-1}. The two lines correspond to the transitions from the ground sublevel, to the two highest levels, with $\Delta M_J = 1$ and $\Delta M_J = 2$.

1.4 $ErCl_3$

There are 11 electrons in the 4f shell, hence 3 holes. Then

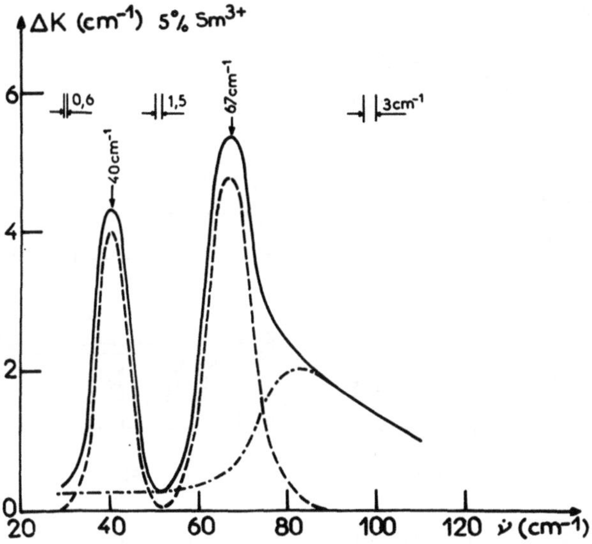

Figure 11. Additional absorption induced by samarium in $LaCl_3$

S = 3/2, L = 6 as in the case of Nd^{3+}. But here J = 15/2 in the ground state $^4I_{15/2}$. The crystal structure is monoclinic, the site symmetry is low (C_2) and 8 Kramer's doublets are expected. However, the environment of the ions is such that the true symmetry is nearly cubic. For perfect cubic symmetry only 5 components are expected. In order to produce five groups of lines some components must go very close together. This can be observed (Figure 12) and also the big change when the site gets a different symmetry ($LaCl_3$ and YCl_3).

1.5 $DyCl_3$

There are 9 electrons, hence 5 holes, Then S = 5/2; L = 5, as in the case of Sm^{3+}. But here J = 15/2 in the ground state $^6H_{15/2}$. Here again the crystal structure is monoclinic and 8 Kramer's doublets are expected. Only the 37 cm^{-1} transition is observed as a very broad band (Figure 13).

II. Garnets

The first garnets were studied by Tinkham in 1962 and he discovered antiferromagnetic transitions (Figure 14) which were strongly temperature dependent (Figure 15). We shall only look at the pure electronic transitions at 22.4 and 26.4 cm^{-1} which are independent of temperature. In the case of Yb^{3+}, there are 13 electrons in the 4f shell. There is only one hole, hence S = 1/2;

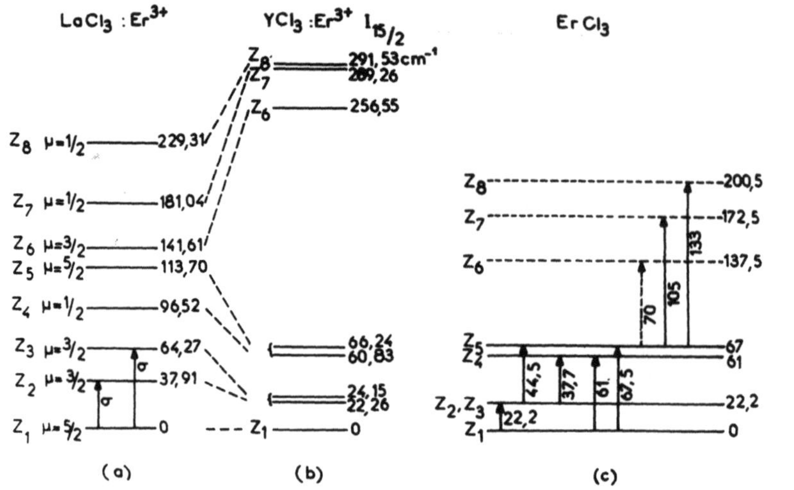

Figure 12. Crystal field splitting in the ground state $I_{15/2}$ of Er^{3+} in three different host lattices

L = 3, and the ground state is $^2F_{7/2}$: 4 Kramer's doublets are expected and they cannot be split by the Stark effect of the electric crystal field. The nice thing here is a splitting ω_2 of the lowest Kramer's doublet by the exchange magnetic field of the iron ions:

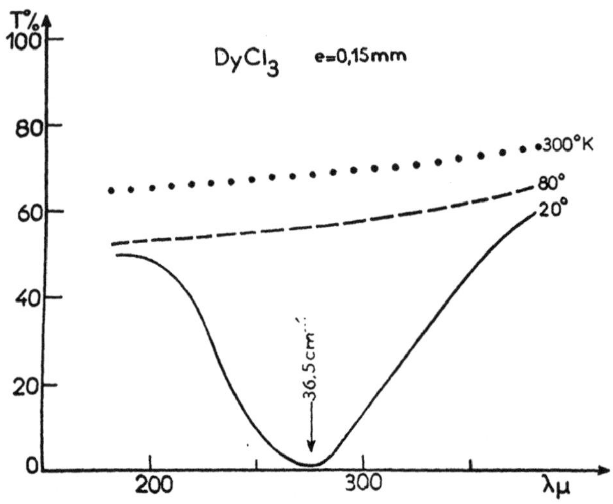

Figure 13. Transmission of a $DyCl_3$ plate (ref. 10).

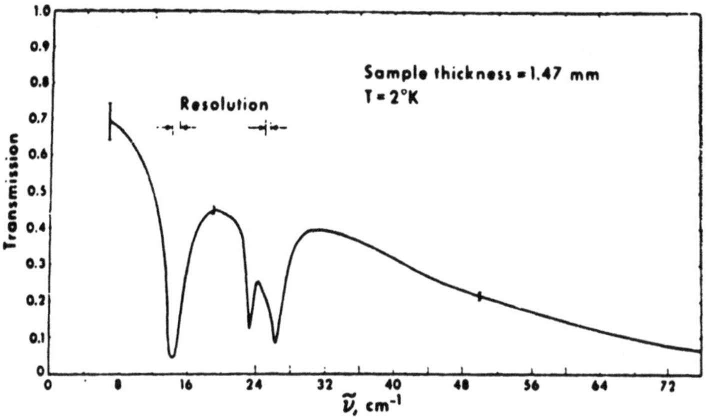

Figure 14. Transmission through a pressed powder sample of ytterbium iron garnet, 1.47 mm thick, at a temperature of 2K (ref. 11).

$$\omega_a = \gamma_a \quad \lambda \ M_1$$

M_1 represents the magnetic dipole moment of the unit cell and $\gamma_a = g\beta$ takes two values corresponding to two sites of Yb^{3+} ions. The lines at 22 and 26 cm^{-1} correspond to transitions from the ground state to the upper states of the ground Kramer's doublet.

III. Rare Earth Fluorides

They represent a very complicated case where a number of

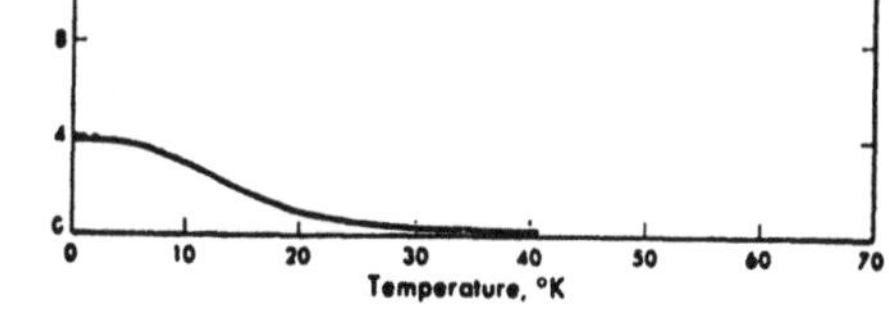

Figure 15. Different temperatures dependence of the electronic transitions and magnetic resonance frequencies in ytterbium iron garnet (ref. 11).

lines appear. The low C_2 symmetry does not seem sufficient to explain this large number of lines (see for instance, Figure 16 for the case of PrF_3). There are probably at least two slightly different sites for the rare earth ions in the crystal.

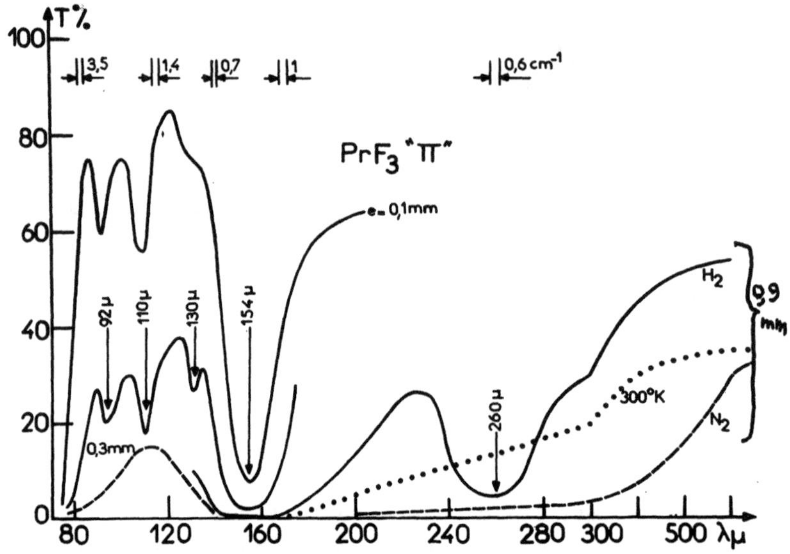

Figure 16. Transmission of PrF_3 for different thickness and different temperatures, between 80 and 600 microns

IV. Hydrated Crystals

4.1 Water in ionic crystals

We have studied several crystals of inorganic chlorides, bromides and iodides either anhydrous or deuterated. By deuteration it has been possible to differentiate between several types of absorption bands. (i) Low frequency absorption bands not sensitive to deuteration are related to motions involving mainly the heaviest atoms of the unit cell. Their frequencies, generally still smaller than those of the anhydrous crystals, are located between 40 and 90 cm^{-1}. (ii) Absorption bands mainly related to water molecules may be further subdivided into a very broad perturbed valence vibration, two angular vibrations arising from the coupling of two water molecules, several librations around the three principal molecular axes of inertia, with frequencies ranging from 400 to 600 cm^{-1}, and six translational vibrations ranging from 90 to 250 cm^{-1}.

This is the case of $BaCl_2$, $2H_2O$ (Figs. 17-18), NaI, $2H_2O$ and

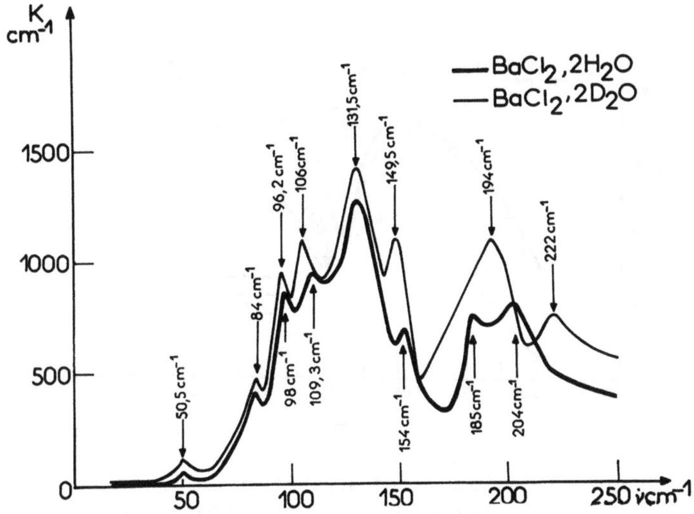

Figure 17. Absorption coefficient of $BaCl_2$, $2H_2O$ and $BaCl_2$, $2D_2O$ between 0 and 250 cm^{-1} at room temperature (ref. 12)

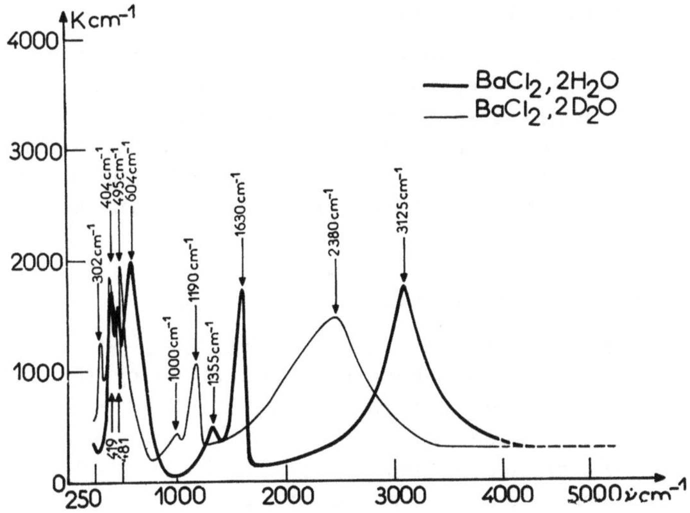

Figure 18. Absorption coefficient of $BaCl_2$, $2H_2O$ and $BaCl_2$, $2D_2O$ between 250 and 5000 cm^{-1} at room temperature (ref. 12).

also several rare earth chlorides which adhere to $6H_2O$. When hydrated they often show a band at 70 cm^{-1} which does not disappear when the temperature is raised up to the room temperature, but shifts toward lower frequencies. This is the case for $TmCl_3$, $6H_2O$ (Fig. 19) and $DyCl_3$, $6H_2O$.

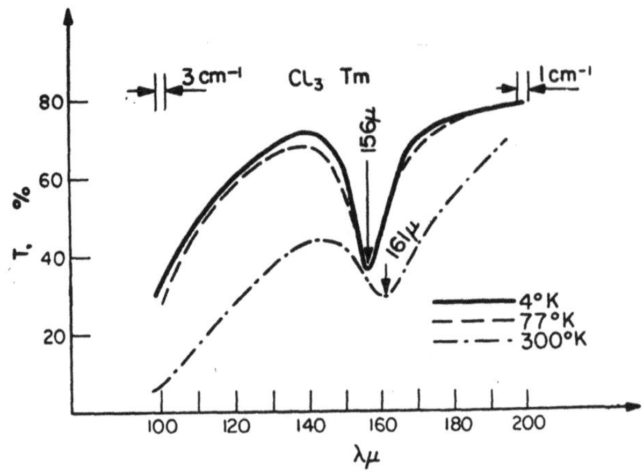

Figure 19. Apparent transmission of hydrated $TmCl_3$ in terms of wavelength for three different temperatures (ref. 13)

4.2 Rare earth nitrates

(i) Neodymium and samarium

They are easy to grow from solutions where they take $6H_2O$ to form triclinic crystals. From the work of Hellwege's school[3], an electronic transition is expected at 33 cm^{-1} for Nd^{3+} and at 41 cm^{-1} for Sm^{3+}. Both spectra are effectively similar except at 33 cm^{-1} where there is one line for Nd^{3+} and no absorption for Sm^{3+}. The line at 41 cm^{-1} is hardly visible in the Sm^{3+} spectrum.

(ii) Cerium double nitrate

The absorption spectra of $Ce_2Mg_3(NO_3)_{12}$: $24H_2O$, and the similar zinc nitrate have been described by Thornley*. A lattice vibration is seen in both spectra at 50 cm^{-1} (Fig. 20) and an electronic line is observed at $\nu_1 = 36$ cm^{-1} and $\nu_1 = 30$ cm^{-1} respectively for the magnesium and the zinc salt in good accordance with estimates from relaxation measurements. The spectra are not obtained with polarized radiation and it is difficult to establish unequivocally the character of the transition.

The electronic transition for each salt has the expected temperature behavior: its frequency is temperature independent, whereas its intensity is proportional to the population difference between the electronic states concerned. The intensity agrees within a factor of two with that expected for magnetic dipole transitions. The electric dipole matrix element is not necessarily zero as the

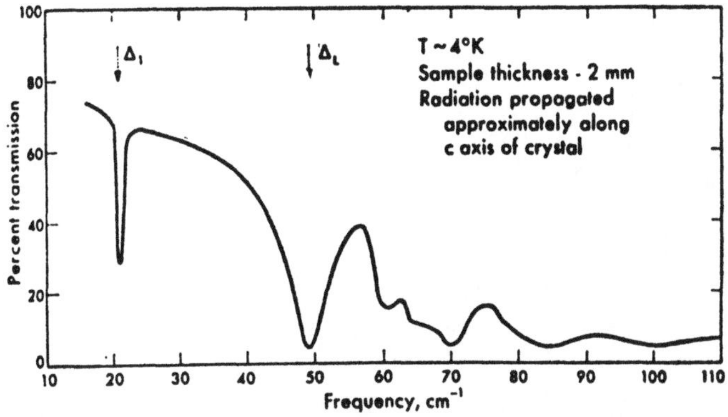

Figure 20. Transmission of cerium magnesium nitrate (ref. 4)

environment of the Ce^{3+} ion does not have inversion symmetry. It is possible to admit into the $4f^1$ ground configuration a little of the excited configuration $5d^1$ of opposite parity which has an excitation energy of about 40,000 cm^{-1}. Electric dipole matrix elements within the perturbed $4f^1$ configuration are then allowed. However, recent crystallographic studies indicate that the departure from an even parity arrangement of the twelve nearest neighbor oxygen atoms is small ($\delta \simeq 0.03$ Å for a Ce-O bond length $\simeq$ 2.65 Å). Thus the odd-parity components of the crystal field would be expected to be at least 50 times smaller than the even parity components and the transition probability for electric dipole transitions would be several orders of magnitude less than that for magnetic dipole transitions (from Thornley[4]).

4.3 Ethyl sulphates

They have a complicated formula $RE(C_2H_5SO_4)_3 \, 9H_2O$ and are not very attractive at first sight. However, they can be grown easily and the problem of infrared active phonons is again limited to frequencies higher than 45 cm^{-1}.

4.3.1 Phonon lines

It is assumed that the rare earth ions, water molecules ethyl radicals, and sulphate radicals all move as units. There are 16 particles in a molecule and since there are two molecules per unit cell, a total of 32 particles are found per unit cell. By application of the factor group analysis in which account is taken of translational modes of the crystal as a whole, it is found that there are six Γ_{5u} (Γ_{6u}) infrared active modes. The total of fifteen is consistent with the observation of fourteen, and possibly fifteen, modes which have the same frequency for La, Er, Dy and Sm (Fig. 21).

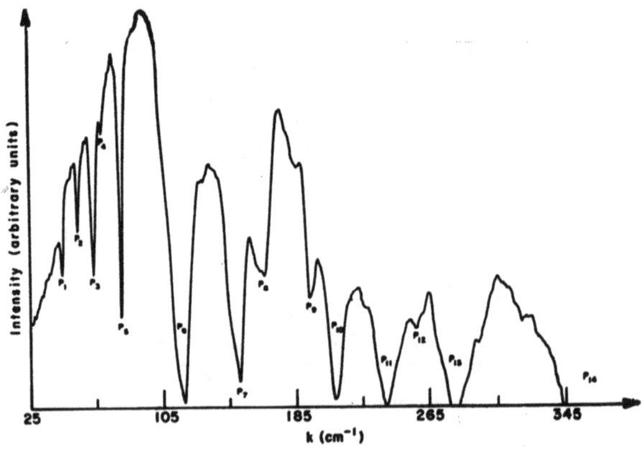

Figure 21. Lanthanum-ethyl (sulphate LaEs) spectrum with absorption lines P_1 - P_{14} produced by the excitation of phonons (ref.5)

The range of frequencies extend from 44 to 360 cm^{-1}. All these lines are certainly phonon lines and those of highest frequencies might arise from librations of water molecules. However, there are four additional bands labeled P_i^* which are only observable in Er, Dy and Sm and are tentatively ascribed to impurity modes. They do not shift at all when a magnetic field up to 80 kG is applied (from Hill et.al.[5]).

4.3.2 Crystal field lines

a) Erbium ethyl sulphate

(i) Electronic transitions

The ground configuration of erbium is $4f^{11}$, equivalent to three 4f holes, since the closed 4f shell contains 14 electrons. The ground manifold is $^4I_{15/2}$ and under the influence of the C_{3h} crystal field in the ethyl sulphate lattice, it splits into eight Kramer's doublets. For a crystal at low temperatures where only the lowest doublet will have an apprecialbe population, a maximum of seven transitions within the ground manifold can be observed at zero field. Four of these transitions were actually found (Fig.22)[5].

By using a magnetic field up to 80 kG, the electronic character of the transition is definitely demonstrated. The Zeeman effect observed is linear (Figs. 23-24)

(ii) Magnetic dipole character of the transitions

No polarization data are available to show definitely the character (electric dipole or magnetic) of the transitions. To replace these data the integrated intensity of the Δ 1.6 transition of Er^{3+} for H = 42.3 kG//C has been calculated. At H = 42.3 kG this line is fairly well separated from nearby phonons and its absorption coefficient was measured as a function of frequency. For

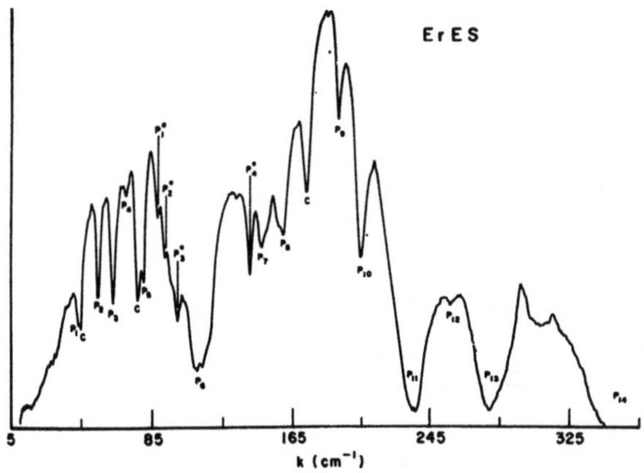

Figure 22. Erbium-ethyl sulphate (ErES) spectrum with phonon lines labeled P_1 - P_{14}. The lines labeled with a C are crystal-field lines while those labeled P_1* - P_4* are the additional phonon lines (ref. 5)

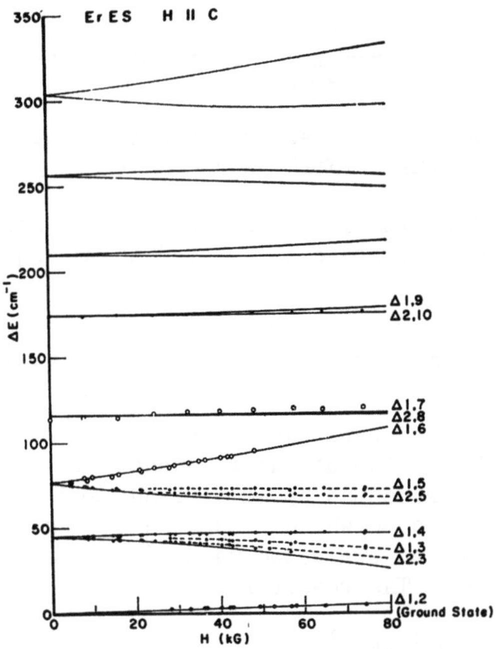

Figure 23. Calculated and observed transition energies for erbium ethyl sulphate (ErES) and H//c. The calculated values are represented by solid lines and the measured points by open circles whose size is indicative of the experimental error. The sixteen levels are numbered 1-2... 16, and Δ 1.2 for instance represents a transition between the two components of the ground Kramers doublet (ref.5)

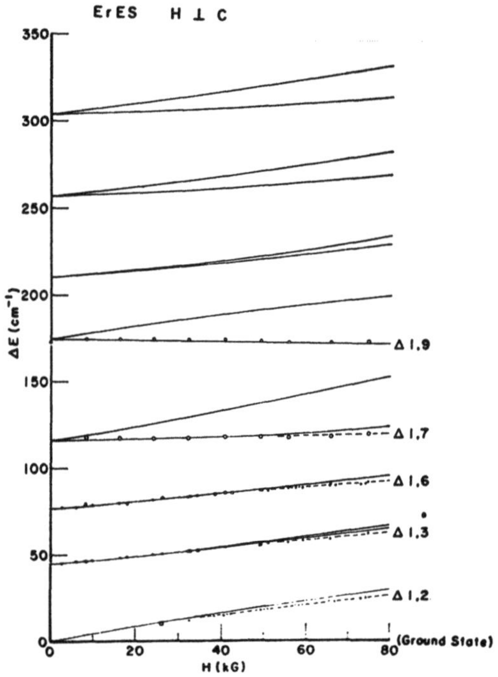

Figure 24. Calculated and observed transition energies for erbium ethyl-sulphate (ErES) and $H \perp c$ (ref. 5).

a magnetic dipole transition the integral of the absorption coefficient over all frequency space is related to the square of the matrix element of J by

$$c\int_0^\infty \alpha(k)dk = \rho\, n\, \frac{2\pi}{3}\, \frac{h\, k\, e^2\, g_L^2}{m^2\, c^2}\, |\langle\psi_i|J|\psi_f\rangle|^2$$

where ρ is the number of rare-earth ions per cm^3, n is the index of refraction of the medium at the frequency k, e and m the electronic charge and mass, c the speed of light and g_L, the Landé g factor for the manifold to which ψ_i and ψ_j belong. The right hand side of the equation was evaluated using the calculated value given by Krupke and Gruber. The result is

$$\sim 9 \times 10^{12}\ cm^{-1}\ sec^{-1}$$

The left side was approximated by integrating the measured $\alpha(k)$ to obtain $\sim 25 \times 10^{12}\ cm^{-1}\ sec^{-1}$. This remarkable agreement is taken as definitely confirming the conclusion that the observed transitions are primarily magnetic dipole transitions, a few of these probably being induced by phonon interaction.

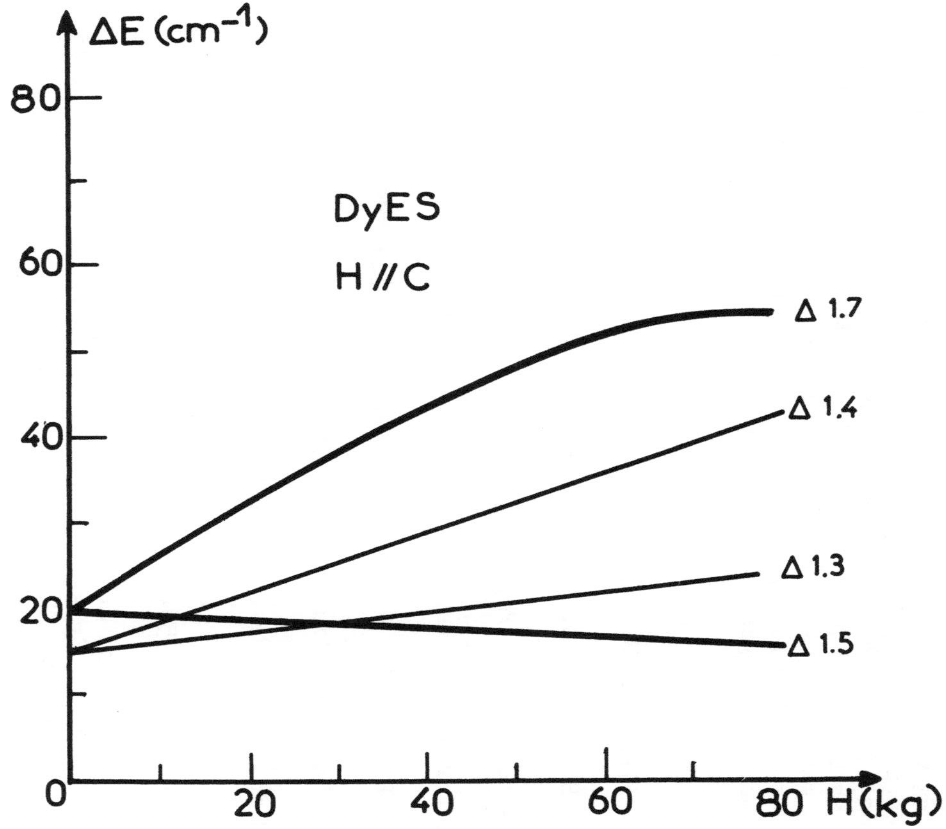

Figure 25. Calculated and observed transition energies for dysprosium ethyl sulphate (DyES) and H//c (reference 5)

b) Dysprosium and samarium sulphate

Here a few significant deviations have been found when the magnetically shifted electronic line approaches a TO phonon line. A hypothetical phonon level is believed to lie betwen 5 to 10 cm^{-1} and to couple to electronic states from which transitions are induced. Unfortunately our spectrometer is not capable of supplying data in this region. If the phonon is observed, its energy should vary as an applied magnetic field is varied. This would then be a convincing proof of the postulated electron-phonon interaction. With H parallel to C, the C_{3h} axis of the hexagonal crystal, the splitting of the three different states labeled Z_1, Z_2 and Z_3 is written as $\Delta\nu = g\beta H (M_j - M_i)$. Here $M_j = -M_i = \pm 1/2$; $\pm 3/2$; $\pm 5/2$ and $\Delta\nu = g\beta H$; $3g\beta H$ and $5g\beta H$, respectively. Figure 25 shows the measured splitting for H//C, and it is clear that the first excited doublet has the largest shift and must be the $\pm 5/2$ state.

Bibliography

1. G. H. Eieke and R. Sarup, J. Chem. Phys. 29, 741 (1958).
2. J. T. Hougen and S. Singh, Proc. Roy. Soc. A277 193 (1964).
3. H. Lammermann, Z. Phys. 150 551 (1958).
4. J.H.M. Thornley, Phys. Rev. 132 1492 (1963).
5. J. C. Hill and R. G. Wheeler, Phys. Rev. 152, 482 (1966).
6. J. P. Mathieu, L'infrarouge Lointain.
7. A. Hadni, Phys. Rev. 136, A758 (1964).
8. H. Remy, Treatise on Inorganic Chemistry, Elsevier Publishing Co., Amsterdam 1956, Vol. II, p. 773.
9. L. Esterowitz and J. Noonan, Optical Properties of Ions in Crystals, H. M. Crosswhite and H. W. Moos, editors, Interscience Publications, New York 1967, p. 487.
10. A. Hadni, G. Morlot and P. Strimer, IEEE J. Quantum Electr. QE-3, 111 (1967).
11. M. Tinkham, Science 145, 240 (1964).
12. A. Hadni, G. Morlot and F. Brehat, Spectrochim Acta 24A, 1167 (1968).
13. A. Hadni, J. Claudel and X. Gerbaux in Lattice Dynamics, R. F. Wallis, ed., Pergamon Press, New York, 1964.

LATTICE ABSORPTION IN THE VERY FAR INFRARED

Armand Hadni

Faculty of Sciences, University of Nancy, France

INTRODUCTION

We shall first consider the case of pure single crystals beyond the reststrahlen where a considerable absorption is seen at room temperature. This absorption disappears when the crystal is cooled and can be explained by phonon difference processes. Then we shall consider additional absorption induced by impurities and defects and especially the case of glasses and liquids.

I. THE ABSORPTION OF PURE SINGLE CRYSTALS BEYOND THE RESTSTRAHLEN

I.1 Experimental

It has been known for a long time that there are no transparent materials in the far infrared at room temperature beyond the reststrahlen and up to millimeter waves except crystalline quartz, silicon, diamond and polyethylene. The multiphonon processes described by Professor Mitra[1] are responsible for this considerable absorption. I would like to insist a little bit on it because of its importance in most studies when the temperature is not lower than 2°K for instance.

In Figure 1 the transmission of ZnSe as a sintered material (Irtran IV) is given as a function of wavelength at room temperature, liquid nitrogen and liquid hydrogen temperatures and shows three types of absorption encountered in the far infrared. The band at 49μ is the fundamental absorption of lowest frequency. By

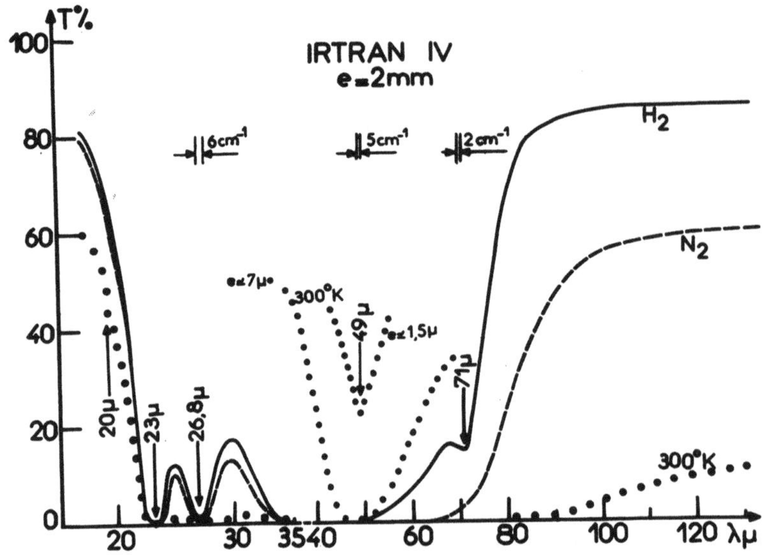

Figure 1. Transmission of zinc selenide crystalline plate (reference 2).

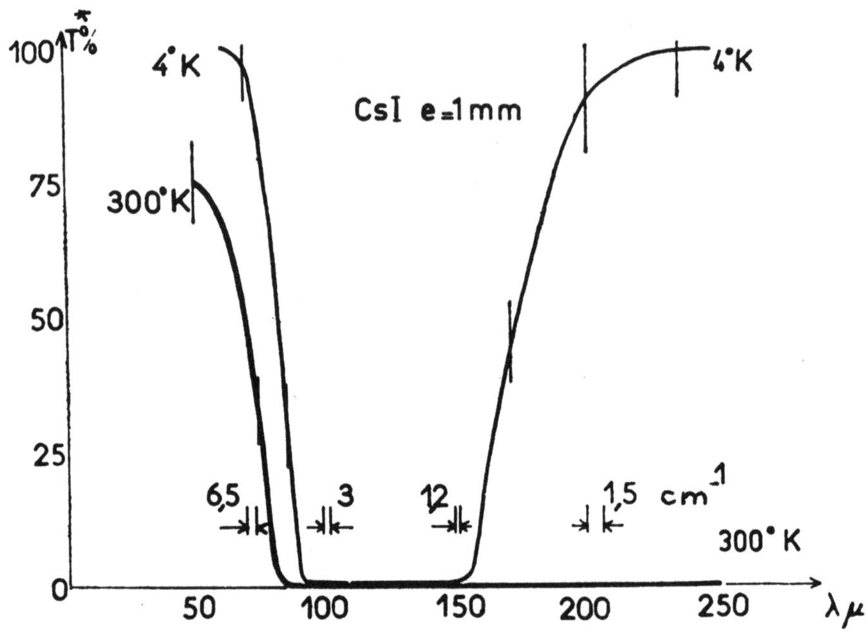

Figure 2. Transmission of cesium iodide (reference3)

fundamental we mean an absorption band where one photon disappears to give one phonon with a wave vector $q \simeq 0$. It is seen that the intensity of such a band is nearly temperature independent. At longer wavelengths, the absorption band at 71μ does not disappear by cooling and is still clearly visible at liquid hydrogen temperature. It is interpreted as a two phonon absorption band where two transverse acoustic (TA) phonons are created while the far infrared photon disappears. Such bands are not very temperature sensitive as was explained in Mitra's talk, but they rarely occur beyond the reststrahlen. Then an intense background absorption is seen at room temperature, decreases at liquid nitrogen temperature and disappears at liquid hydrogen temperature for $\lambda > 80\mu$. This absorption is very temperature sensitive and can be explained by difference processes. In the simplest two phonon process, one phonon disappears simultaneously with the photon, to give one phonon of higher energy. Such a process disappears at low temperatures because of the lack of phonons.

In Figure 2 the transmission of a 1 mm thick cesium iodide is given vs. wavelength. For $\lambda \simeq 250\ \mu$ the pure transmission T* (corrected from reflection losses) goes from zero to 100°/o when temperature is decreased down to 4°K. This perfect transparency at low temperatures ("Supertransparency") peculiar to CsI is probably due to the lack of natural isotopes for both cesium and iodine. The sharp cut-off may be used to cut the middle infrared, and a cooled plate of cesium iodide appears as a very good filter completely transparent for the far infrared.

Figure 3 gives the transmission of a KBr plate. The transmission is null at room temperature. Supertransmission is observed at liquid helium temperature for a pure crystal (curve n°1). The cut-off is not so sharp as in the case of cesium iodide. K^{39} has a 93 percent natural abundance, but BR^{79} only 50.5 percent. The intensity of the band at 107μ increases when chlorine impurities are added. When sodium impurities are introduced (curve n° 2), a broad background absorption is induced, and also sharp modes at 99-120 and 140μ appear. The case of KBr has been especially considered because phonon dispersion curves are well known in this case from neutron diffraction. Two laws have been found experimentally:

(i) The variation of the absorption coefficient is linear with temperature (Figure 4).

(ii) The variation of the absorption coefficient is proportional to the square of the frequency (Figure 5).

Under some simple assumptions both laws can be demonstrated as resulting from two phonon difference processes.

I.2 Theoretical

If we take into account both absorption and stimulated emission

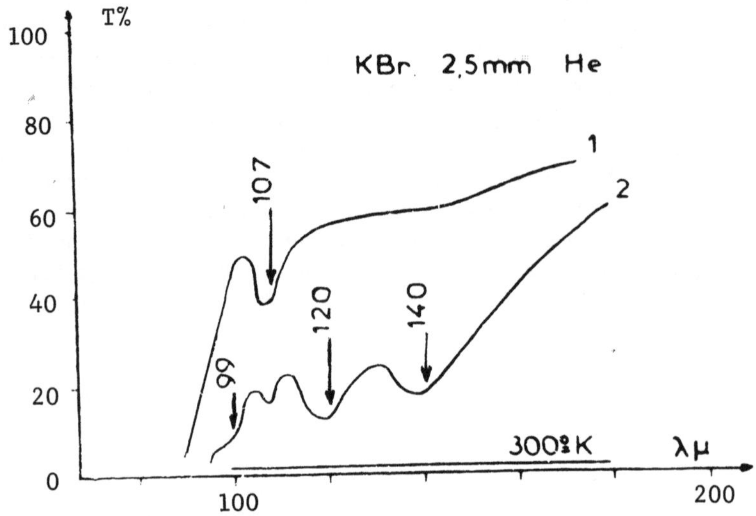

Figure 3. Transmission of pure potassium bromide (curve No.1) and doped with sodium (curve No. 2), at liquid helium temperature. The lower curve corresponds to room temperature (reference 4).

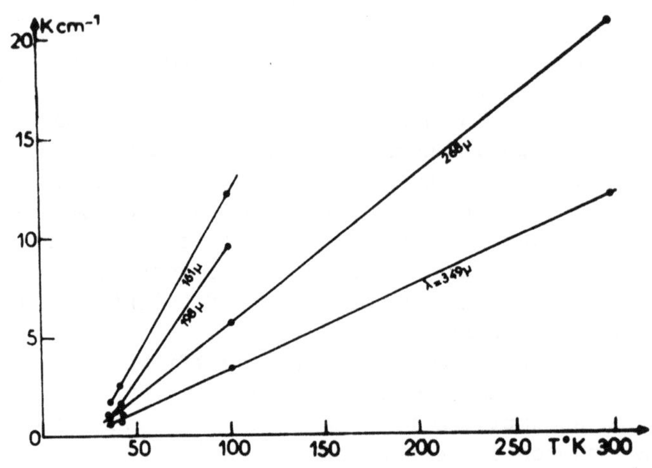

Figure 4. From 50 to 300K the absorption coefficient is a linear function of temperature, for $T \geqslant 40^{\circ}K$ and $\lambda \geqslant 161$ | (reference 5)

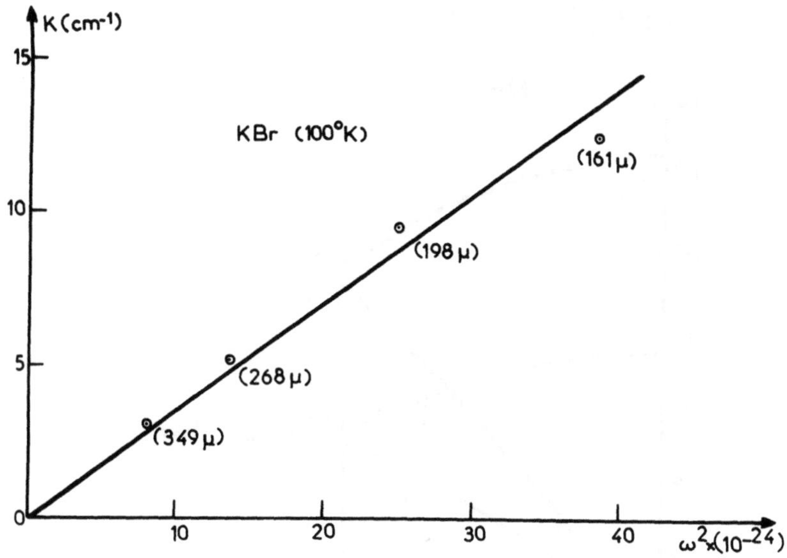

Figure 5. At 100K, the absorption coefficient is a linear function of the square of the frequency for $\lambda \gg 161\mu$.

of radiation, the number of photons which disappear in one second for a vertical transition of wave vector q_i (Figure 6) is written:

$$dn(q_i) = (N_i - N_j)B' u_\omega$$

with $N_i = \dfrac{1}{e^{h\nu_i/kT} - 1}$ and where $u_\omega = du/d\omega$ represents the specific density of energy of angular frequency $\omega = 2\pi(\nu_j - \nu_i)$. This loss of photons brings a contribution $\Delta K(q_i)$ to the absorption coefficient K_ω:

$$\Delta K(q_i) = A\ (N_i - N_j) \qquad (1)$$

with $A = \dfrac{D \hbar\ \omega}{(\omega^2 - \omega_T^2)\ \omega_i \omega_j}$ x number of vectors corresponding to $\Delta\omega = 1$ (2)

and D being a constant, characteristic of the material.

In the assumption that difference processes are due to the anharmonic mechanism described by Professor Mitra which is generally accepted for alkali halides, it is shown that:

$$D = \frac{\pi\ (\varepsilon_o - \varepsilon_\infty)}{2\ n\ c} B^2\ \omega_T^2 \qquad (3)$$

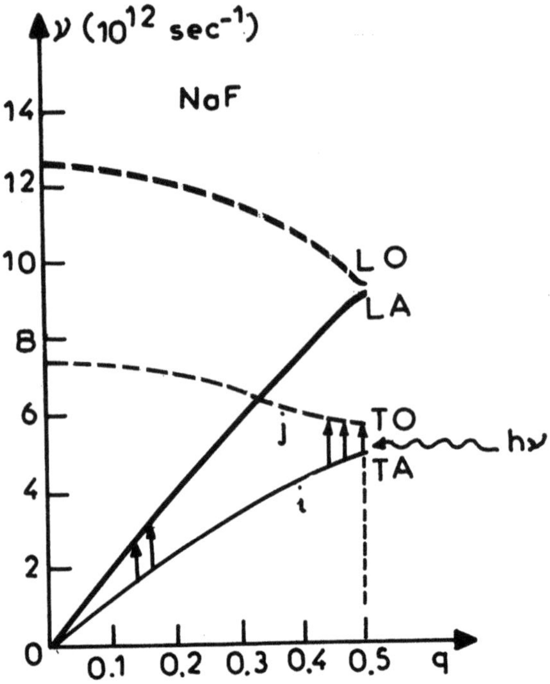

Figure 6. Dispersion curves of NaF in the <111> direction (reference 6).

From equation (2) it is seen that the absorption increases when ω approaches the lattice resonance. In fact, the amount of mechanical energy in the mixed mode increases and hence an increase of anharmonicity.

From equation (3) it is seen that for a material with no ionic charge, and no reststrahlen, $\varepsilon_0 - \varepsilon_\infty = 0$, and there would be no far infrared absorption due to the anharmonic mechanism. This is the case of germanium, silicon, and diamond which are transparent in the far infrared, but however give some absorption in the middle infrared from addition processes which can arise from second order term in the dipole moment.

(a) High temperature limit:

If $h\nu \ll kT$, from equation (1) we can write:

$$\Delta K = A\, kT \left(\frac{\nu_j - \nu_i}{\nu_i\, \nu_j}\right)$$

hence $\Delta K = \dfrac{D\, h\, \omega^3\, kT}{(\omega^2 - \omega_T^2)\,(\omega_i\, \omega_j)^2}$ x number of wave vectors corresponding to $\Delta\omega = 0$. (4).

From this relation, three conclusions can be drawn:

(i) $K = \Sigma \Delta K$ is proportional to T

(ii) if we add the condition $\omega \ll \omega_T$ (very low frequencies), ΔK, hence $K = \Sigma \Delta K$, is proportional to ω^2

(iii) transitions at the zone boundary give the contribution of a larger number of wave vectors, but the frequencies ω_i, ω_j of the phonons used for the difference are higher. The result is that the contribution of all wave vectors, and not only boundary wave vectors is important, except if selection rules forbid for instance TA → LA transitions. In that case the two phonon difference processes would be limited to a lower frequency TO - TA but this does not mean a sharp cut-off where the crystal should become transparent. Three phonon difference processes can occur and give some absorption at much lower frequencies.

I have no information about such selection rules for difference processes.

(b) Low temperatures:

The approximation $h\nu \ll kT$ does not hold for temperatures much lower than room temperature and to get ΔK, we have to make the difference between two Bose Einstein functions without any simplification (Figure 7). ΔK goes to zero with T but the curve

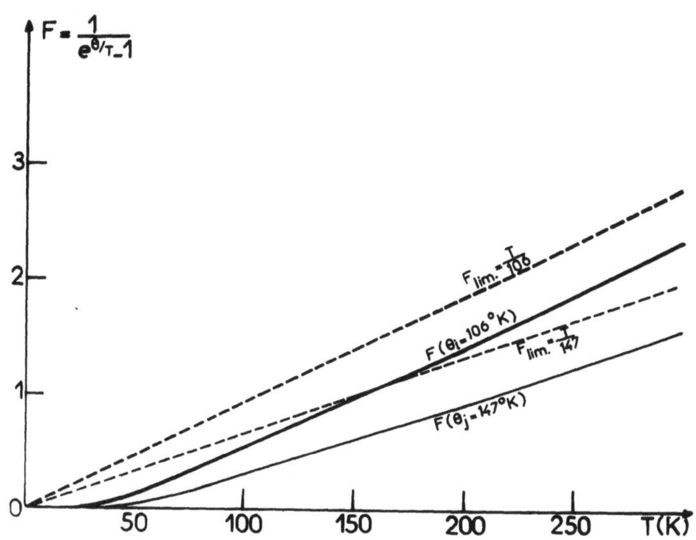

Figure 7. Variation of function $F = \frac{1}{e^{\Theta/T} - 1}$ vs temperature for $\Theta = 106^{\circ}$K and 147°K.

reaches the linear asymptot passing through zero only at high temperatures (Figure 8). When dispersion curves are known, as is the case of KBr, and if we limit ourselves to the case of the few wavelengths where boundary phonons could give a major contribution, the calculation is easy, and an accurate check is possible with laser radiations where transmission as low as 0.1 % can be measured with a one percent accuracy. The decrease with temperature seems more rapid than predicted by two phonon difference processes and might be an evidence of some contribution from three phonons processes.

I.3 Three phonon difference processes

For instance two phonons disappear and a more energetic one appears, or one phonon disappears and two others are created (Figure 9). It can be shown that the absorption coefficient should here follow a T^2 dependence fairly closely, and should vary with frequency as ω^2 again, but much more generally than for the two phonon processes.

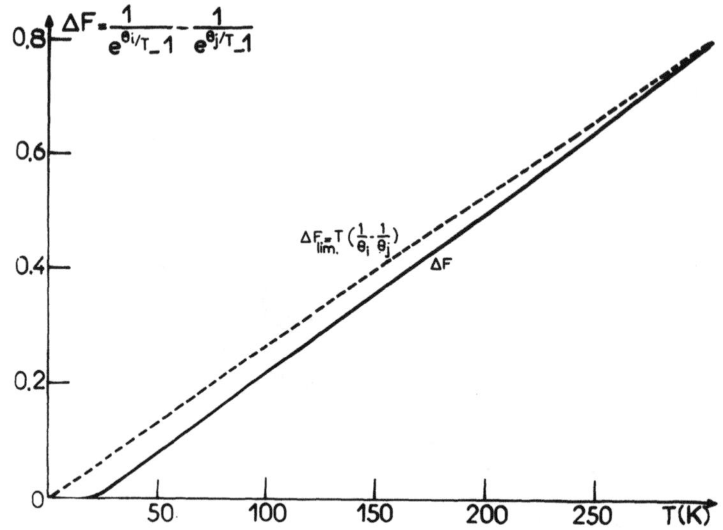

Figure 8. Variation of the absorption coefficient vs temperature at $\dot{\nu}$ u $\lambda = 349\ \mu$ for K Br.

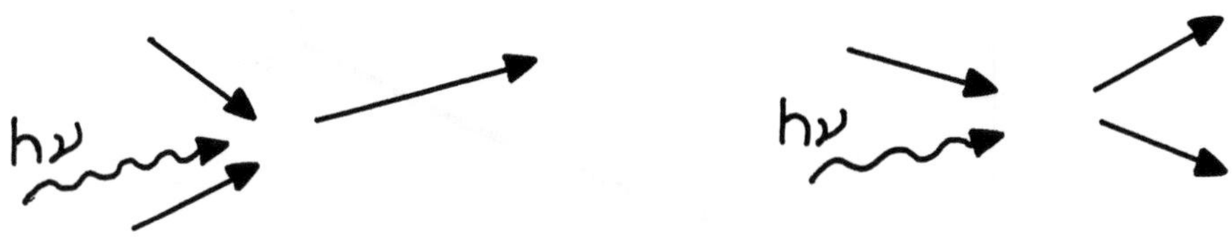

Figure 9. Scheme of three phonon difference processes.

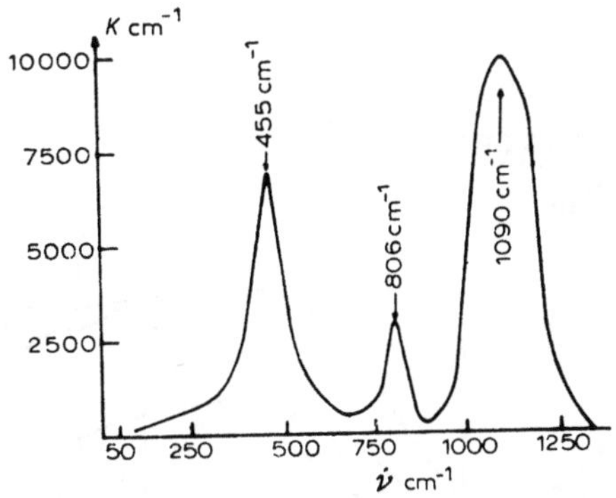

Figure 10. Absorption of fused quartz. The band at 455 cm^{-1} corresponds to the $\nu_3(F_2)$ asymmetric bending; the 806 cm^{-1} to the $\nu_1(A_1)$ symmetric stretching and the 1090 cm^{-1} to the $\nu_4(F_2)$ asymmetric stretching vibration of the SiO_4 group. (Reference 7).

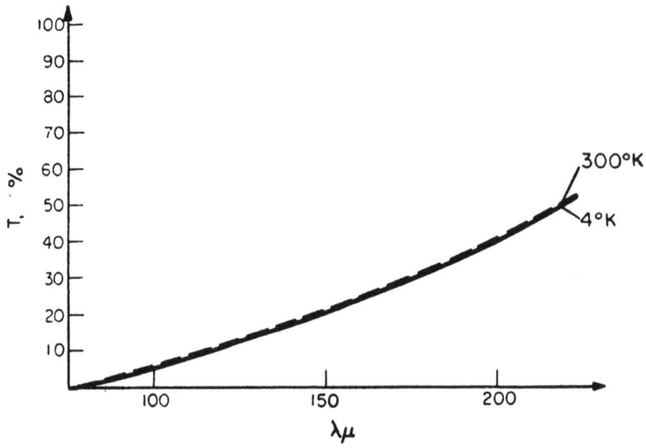

Figure 12. The transmission of a 2 mm thick plate of silica is not modified by cooling. For a soda glass e=0.2 mm, the transmission curve is nearly the same hence an absorption coefficient ten times larger. (Reference 8).

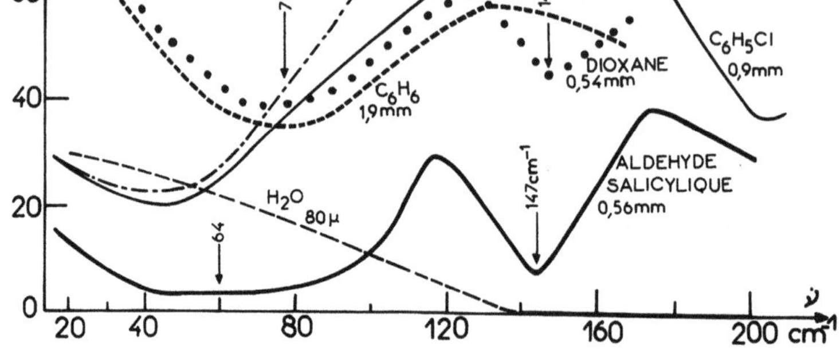

Figure 13. Polar liquids show an intense absorption band in the very far infrared.

II. THE ABSORPTION OF GLASSES AND LIQUIDS BEYOND THE RESTSTRAHLEN

II.1 Experimental

If we compare a crystal and a glass of the same composition (i.e. crystal quartz, and fused quartz), it is seen that for the glass the absorption lines in the middle infrared are much broader and mixed together to give only a few wide degenerate absorption bands (Figure 10). The band at 455 cm^{-1} corresponds to ν_4 (F_2) (asymmetric triply degenerate stretching vibration); the one at 860 cm^{-1} to ν_1 (A_1) (symmetric stretching vibration), and the one at 1090 cm^{-1} to ν_3 (F_2) (asymmetric triply degenerate bending vibration). The broadening corresponds to a coupling between identical neighbouring oscillators. The coupling gives two symmetrical frequencies $\nu = \nu_0 \left(1 \pm \frac{C}{2}\right)$, where C is the coupling coefficient. For a glass C is different from one place to another and hence the broadening. In the far infrared the glass has a much higher absorption coefficient than the corresponding crystal (Figure 11). This absorption is not temperature sensitive (Figure 12). Polar liquids show an intense absorption in the far infrared (Figure 13). Nevertheless even non polar liquids like

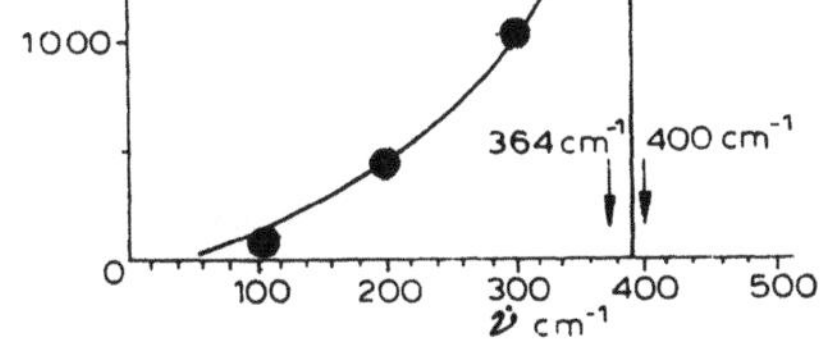

Figure 11. The difference ΔK between the absorption coefficients of fused quartz and crystal quartz is plotted vs frequency. The circled dots correspond to an approximate law in ω^2, $K = 10^{-2}\nu^2$. For frequencies higher than 300 cm^{-1}, the observed absorption is higher because of the proximity of the ν_4 band. For frequencies lower than 100 cm^{-1}, the observed absorption is also higher and other processes have probably to be considered or a law in ω^m, with $m>2$. (Reference 7).

CCl_4 or cyclohexane (probably the most transparent liquid in the far infrared) show some broad absorption in the far infrared (Figure 14).

II.2 Theoretical

II.2.1 The case of liquids

II.2.1.1 Polar molcules

As we have seen it, all liquids, specially polar liquids, give a broad and intense absorption in the far infrared.

1°) Debye absorption

For polar liquids the most important part of the absorption arises from Debye absorption which for a long time could only be studied from dielectric measurements at radio frequencies.

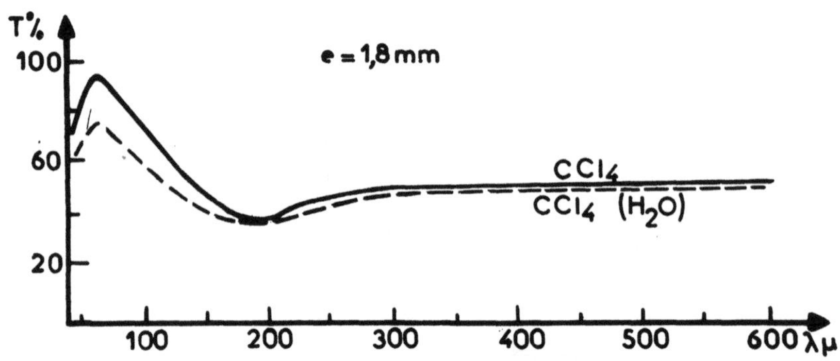

Figure 14. Transmission of carbon tetrachloride in the far infrared. (Reference 9).

Figure 15 recalls the principal features of Debye relaxation where:

$$\varepsilon' = \varepsilon_{RI} + \frac{\varepsilon_{RS} - \varepsilon_{RI}}{1 + \omega^2 \tau^2} \qquad (5)$$

$$\varepsilon'' = \frac{\varepsilon_{RS} - \varepsilon_{RI}}{1 + \omega^2 \tau^2} \qquad (6)$$

$\varepsilon = \varepsilon' - j\varepsilon''$ represents the dielectric constant; ε_{RS} and ε_{RI} are respectively the limits for null and high frequencies, τ is the relaxation time.

Similar curves can be found for several solid ferroelectrics like triglycine sulphate (TGS), then the origin of the high dielectric constant for $\omega = 0$ is not a soft mode but arises from Debye relaxation processes (Figure 16). They are also found in "plastic crystals" such as $(CH_3)_3CCl$ at temperatures which are not much below the melting point (Figure 17).

Figure 15 corresponds to one relaxation time $\tau = \frac{4\pi \eta a^3}{kT}$, for a spherical molecule of radius a, in a non-binding solvent of viscosity η.

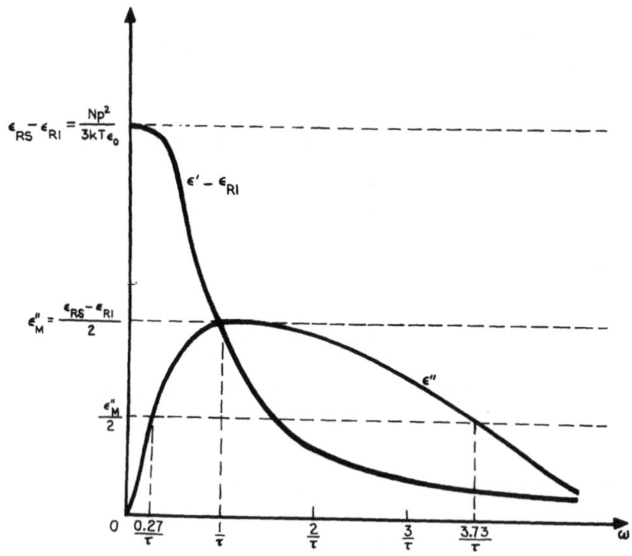

Figure 15. Contribution to the dielectric constant at temperature T of a one relaxation time Debye process of molecules (the electric dipole moment is p).

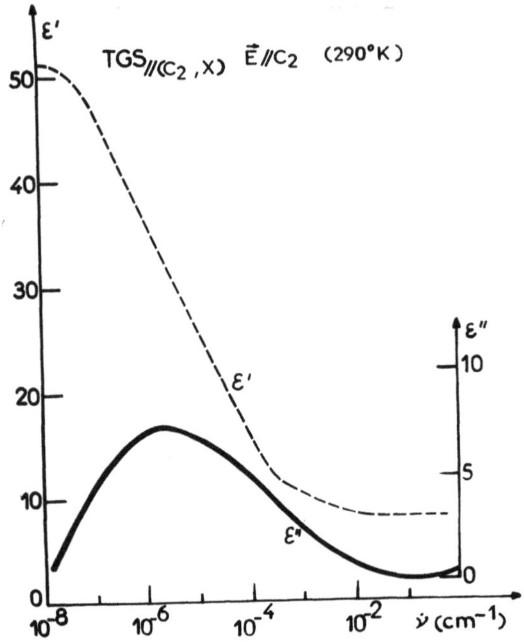

Figure 16. Dielectric constant of TGS vs frequency for radio waves.(Reference 10).

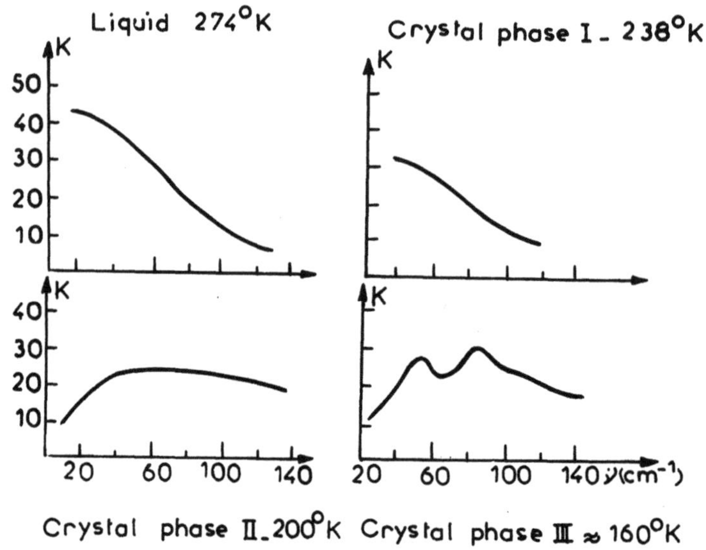

Figure 17. Submillimeter wave spectra of liquid and crystalline tributyl chloride. (Reference 11).

For high frequencies, the second Debye equation gives:

$$\mathcal{E}'' = \frac{\omega\,\tau}{1 + \omega^2\,\tau^2}\,(\mathcal{E}_{RS} - \mathcal{E}_{RI});\ \mathcal{E}'' \simeq \frac{\mathcal{E}_{RS} - \mathcal{E}_{RI}}{\omega\,\tau},$$

the imaginary part of the dielectric constant goes to zero when ω increases, but only as $1/\omega$. Then $2nk = \mathcal{E}''$, hence $k \simeq \dfrac{\mathcal{E}_{RS} - \mathcal{E}_{RI}}{2\,n\,\omega\,\tau}$; $K = \dfrac{\mathcal{E}_{RS} - \mathcal{E}_{RI}}{c\,n\,\tau}$ = constant, where c is the velocity of light in vacuum and n the refractive index for the low frequency considered. This non vanishing value of the absorption coefficient is not confirmed by experience: most polar liquids are transparent in the visible range.

The difficulty can be solved as was shown by Rocard (1933) and more recently by Leroy and Constant (1967) by taking into account the moment of inertia I of the molecule (still supposed to be spherical), and a second relaxation time $\tau_2 = \dfrac{I}{8\,\pi\,\eta\,a^3}$ is introduced. Then

$$K = \frac{\mathcal{E}_{RS} - \mathcal{E}_{RI}}{c\,n}\quad \frac{\omega^2\,\tau_1}{(1 + \omega^2\,\tau_1^2)\,(1 + \omega^2\,\tau_2^2)}$$

goes to zero when ω increases to infinity (Figure 18). This simple theory is perfectly checked by experiment for some polar liquids in solution. For instance 20°/o trichlorethane in normal hexane.

2°) Additional absorption

For most polar liquids diluted in a non polar solvent there is an additional absorption (Figure 19). There are different interpretations according to different cases. In the case of acid solutions there appears a broad absorption band, centered around 150 cm^{-1} for HCl (Figure 20). This looks like the envelope of a pure rotation spectrum. Thus there is the possibility of both resonant contributions and non resonant ones. In the case of acid solutions the resonant contribution is mainly limited to rotational transitions. But even in this case there might be some coupling between translation and rotation. The problem of HCl in a CCl_4 solution has been treated by Galatry and collaborators. They consider a cage model where one HCl molecule has twelve CCl_4 neighbours at the same distance of the center of the case. The displacement of the HCl molecule to the center of the cage and the orientation of the HCl molecules are considered as variables but also the displacements of the twelve carbon tetrachloride molecules from their equilibrium position. It is shown that the potential energy is minimum for a significant distance of the HCl molecule from the center of the cage. The average electric field acting on HCl is very strong and the Stark effect gives a number of components for

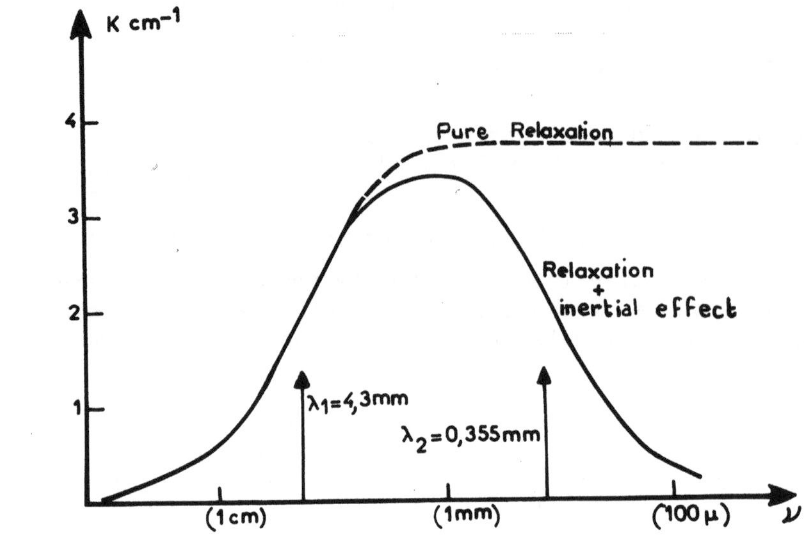

Figure 18. The absorption in the far infrared of a 20°/o trichlorethane solution in hexane is entirely explained when account is taken of relaxation ($\lambda_1 = 4.3$ mm) and inertial effect ($\lambda_2 = 0.355$ mm). (Reference 12).

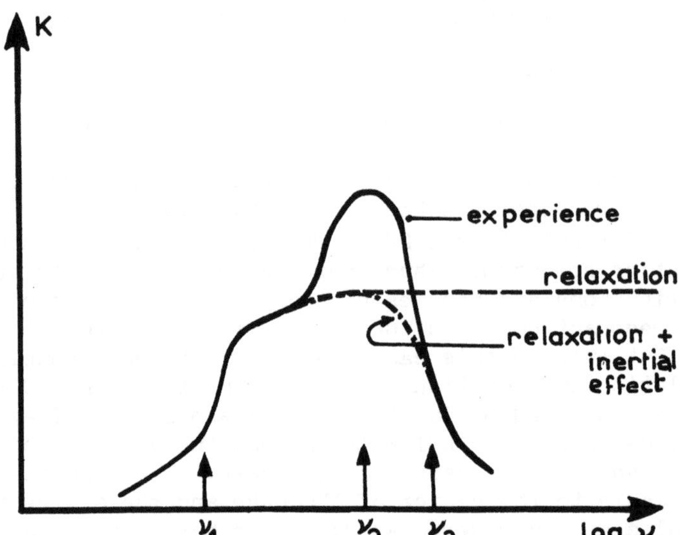

Figure 19. The absorption of bromoform in the far infrared is higher than that predicted from dipolar relaxation theory, even corrected for inertial effects. Here $\lambda_1 = 35.7$ mm, $\lambda_2 = 83$ μm, $\lambda_3 = 280$ μm (Ref. 13).

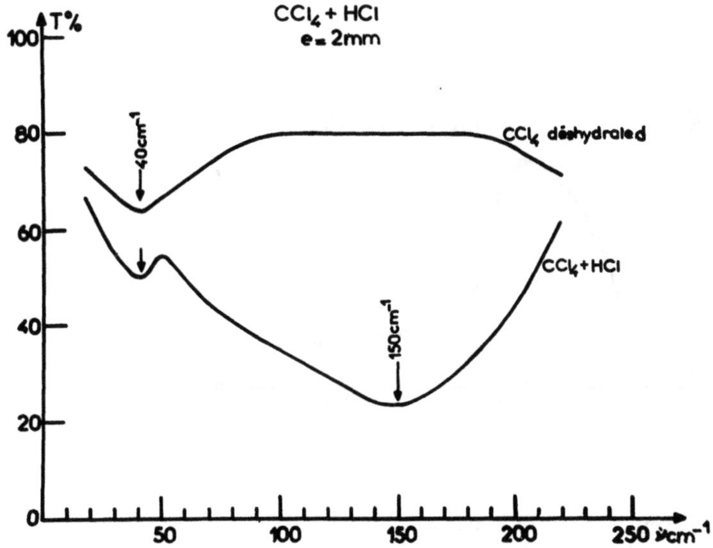

Figure 20. Introduction of HCl to carbon tetrachloride increases the band at 40 cm^{-1} peculiar to CCl_4 and lets appear the broad envelope of HCl rotation spectrum.

each J level as soon as $J \neq 0$. It is then shown that the $\Delta J = 1$ transitions give the major contribution to the far infrared absorption, and the envelope of the spectrum is calculated. The maximum absorption is located at 150 cm^{-1} for room temperature, and the non resonant absorption (Debye relaxation) is negligible. From the cage model the effect of temperature could be important especially close to the critical point where fluctuations are considerable.

b) Liquid crystal vibrations

For most cases there exists some local type of order where the polar molecule and the non polar solvent are associated. This is the case of C_6H_6 and chloroform for instance.

Strong associations are found in the case of pure polar liquids and Debye theory is not strictly applicable. Figure 21 shows the absorption of pure liquid hydrochloric acid and Figure 22 the absorption of liquid water. In that case a structure has been seen by X-ray diffraction where a given water molecule is surrounded by four others at the apexes of a tetrahedral (Figure 23).

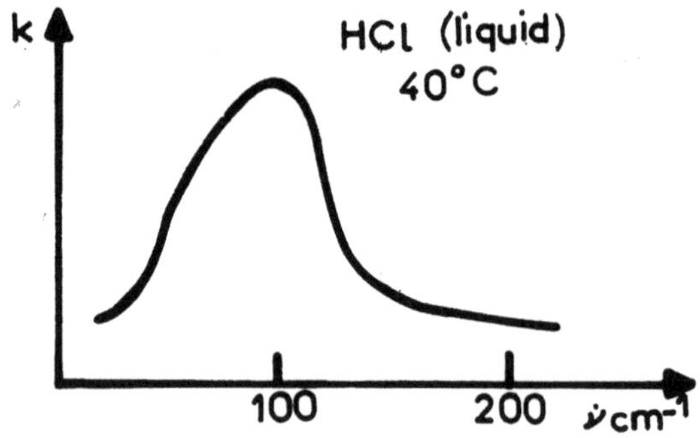

Figure 21. Absorption spectrum of liquid hydrochloric acid. The absorption index $k = c/2\omega$ is proportional to the transition probability of the molecule. (Reference 14).

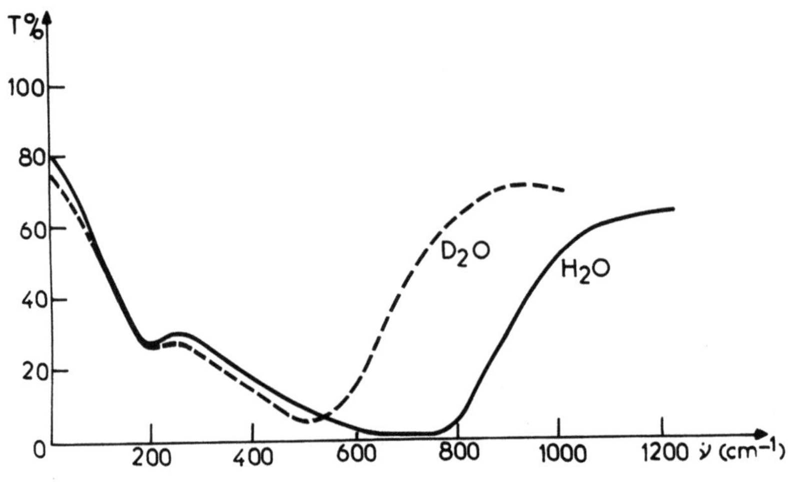

Figure 22. Spectral transmittance of water (10 micron thick) at 30° C as a function of frequency in the region from 32 to 1250 cm^{-1} (full line) and spectral transmittance of a 10 micron layer of liquid D_2O_2 at 30°C in the region from 32 to 1000 cm^{-1} (doted line). (Reference 15).

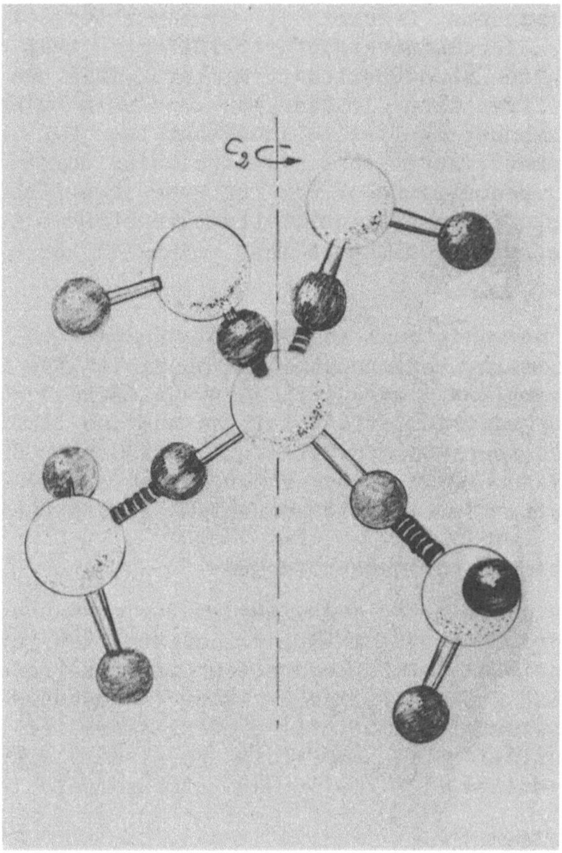

Figure 23. H_2O molecules hydrogen bended. Structure of C_2 symmetry. Small spheres, H atoms; large spheres, O atoms; disk refer to hydrogen bonds (Reference 16).

The order is only good at short distance and such a structure has a limited lifetime of the order of $\tau \simeq 10^{-11}$ sec., the relaxation time. The bands around 700 and 500 cm^{-1} are explained by libration of the water molecules around axis u and w since a change of orientation of the dipole moment is observed during these librations. The band around 170 cm^{-1} occurs from translational vibrations of water molecules where the length of the hydrogen bond is modified. The band at 60 cm^{-1} arises from a bending of the hydrogen bond. Both bands at 170 cm^{-1} and 60 cm^{-1} are not significantly affected by deuteration.

Similar absorption bands ascribed to translational vibrations or librations are found in hydrated salts. They are much narrower and show several components (Figure 24). Concentrated solutions of these salts have a localized structure similar to that of the hydrated crystals which are deposited from them. This has been shown from X-rays diffraction. It has also been seen by Raman spectra which can demonstrate for instance that the NO_3^- radical has no center of symmetry as in the nitrate calcium crystal for instance. Recently measurements of far infrared absorption spectra of aqueous solutions of strong electrolytes have shown a striking resemblance to the spectrum of the solvent, even for the highest concentrations.

In conclusion the additional absorption of polar liquids seem to be explained by models where rotation is possible like in the clathrate (see farther), or models with a crystal structure order at short distances where translational vibrations and librations are seen as in the corresponding crystal. A broad absorption at smaller frequencies can be explained either as Debye absorption or one phonon processes in the acoustic continuum (see farther).

c) Effect of temperature and pressure:

The band can be shifted to lower frequencies when the temperature is raised (chlorobenzene). This may indicate that pure rotation is not responsible for the absorption (for pure rotation, the direction of the shift would be towards higher frequencies), but librations could be possible. The effect of pressure for chlorobenzene has been considered by Chamberlain and also a large shift towards higher frequencies (100°/o for 4400 atmospheres) is indicated.

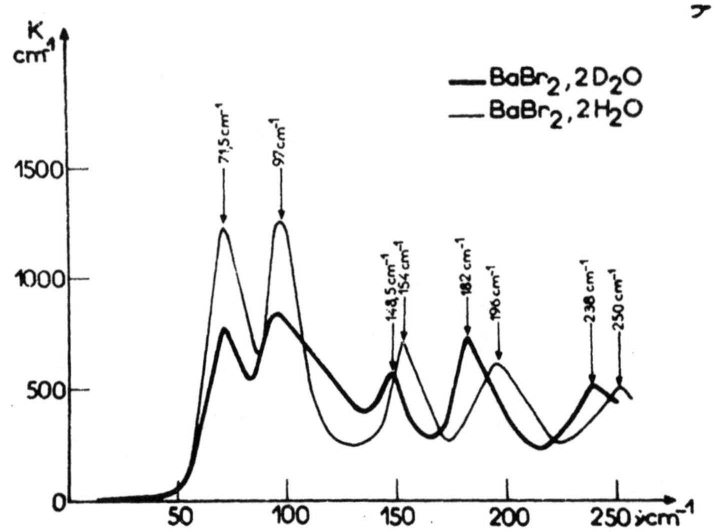

Figure 24. Absorption coefficient K of $BaBr_2$, $2H_2O$ and $BaBr_2$, $2D_2O$ from 20 to 250 cm^{-1} at 290°K. (Reference 17).

II.2.1.2 Non polar molecules

Liquids made of non polar molecules show a broad absorption band in the far infrared. For carbon tetrachloride this absorption has the same order of magnitude as the additional absorption of polar liquids (Figure 14). It is noteworthly that it had been predicted by Poley (1955) from the fact that $\varepsilon_{RO} - n^2_{IR}$ was slightly but significantly different from zero. Several explanations have been considered.

a) Pure rotation

Strauss, because of the qualitative similarity of a pure rotation spectrum, tried to analyse the absorption as if the CCl_4 molecule possessed a fixed electric dipole moment. The order of magnitude is 0.2 D which corresponds to bending one C-Cl bond by about 10°. However the corresponding moment of inertia would be 24 times smaller than the moment of inertia of CCl_4.

b) Liquid crystal vibrations

A more reliable analysis can be made from the fact that the absorption of liquid CCl_4 occurs at the same place where the crystal exhibits several lines (Figure 25). However, it is much

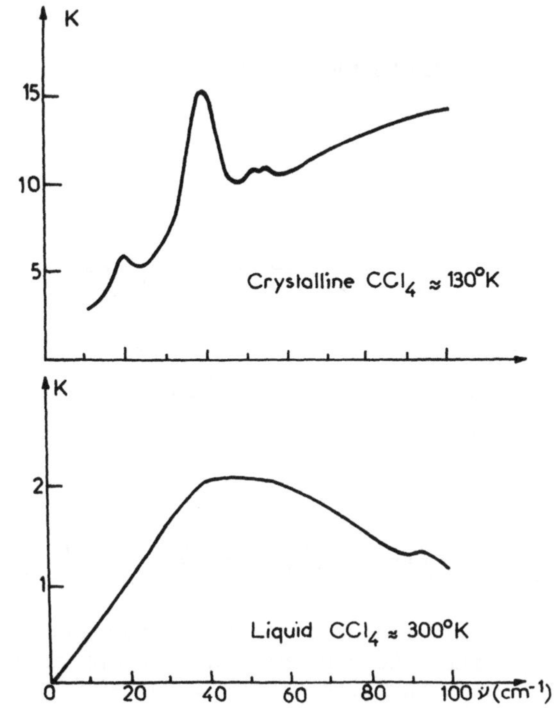

Figure 25. Submillimeter wave spectra of non polar liquid CCl_4 and crystalline CCl_4. (Reference 18).

broader and the integrated absorption is about twice (140 cm^{-2} instead of 80 cm^{-2} for the solid). In microwaves there is still some absorption. The absorption coefficient is proportional to the square of the frequency. For carbone disulphide and benzene, the same broadening of the external vibration bands of the crystal can be considered, and the integrated absorption is still higher for the liquid (400 against 230 cm^{-2} for CS_2; 450 against 350 cm^{-2} for C_6H_6). Cyclohexane is very transparent, liquid nitrogen is more transparent, and liquid oxygen still more. The infrared activity of librations and translations in a non polar molecule arises probably from the dipole induced by the quadrupoles (or higher terms) of the neighbouring molecules. Its value may be very low and the liquid is then very transparent in the infrared if the polarizability is also very small (cyclohexane for instance). For several molecules the electric quadrupole moment is known: CS_2 (3.1 x 10^{-26} esu), CO_2 (3.1); C_2H_4 (2.3); N_2 (1.5); C_2H_6 (<1.3); O_2 (<0.5). Addition of impurities can increase the far infrared absorption of non polar molecules (Figure 26). The mechanism may

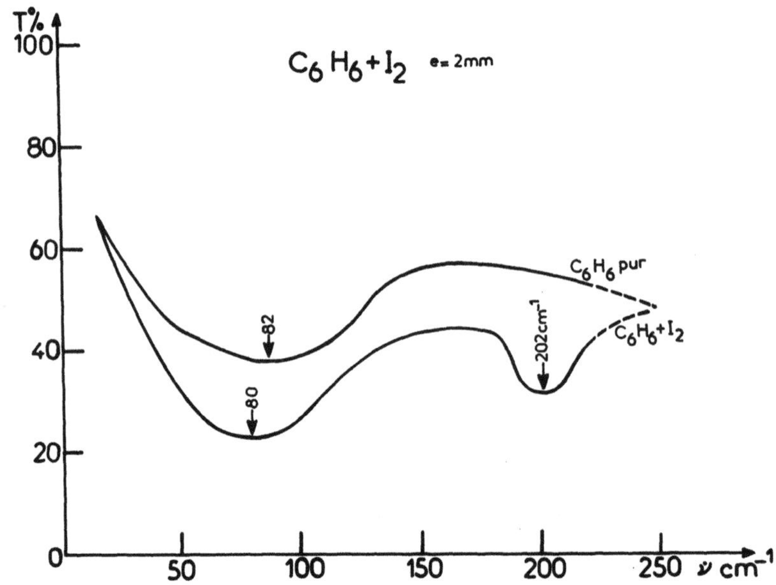

Figure 26. The addition of iodine to benzene increases the band at 82 cm^{-1} peculiar to benzene and lets appear the band at 202 cm^{-1} characteristic of the I_2 stratching vibration. (Reference 7).

be similar to that described to explain the enhancement of near infrared absorption from the H-H vibration in compressed hydrogen (92 amagats) when 6 amagats of HCl are added: the effect is ten times higher than for 6 amagats of N_2.

The HCl electric dipole moment is responsible for the polarization either of a couple of hydrogen atoms or of a couple of carbon tetrachloride molecules.

The effect of increasing the temperature is generally to increase the transmission. This can be seen in Figure 27 corresponding to cyclohexane. The effect of pressure is a shift of the maximum absorption towards high frequencies in the case of CS_2 and an increase of the integrated absorption.

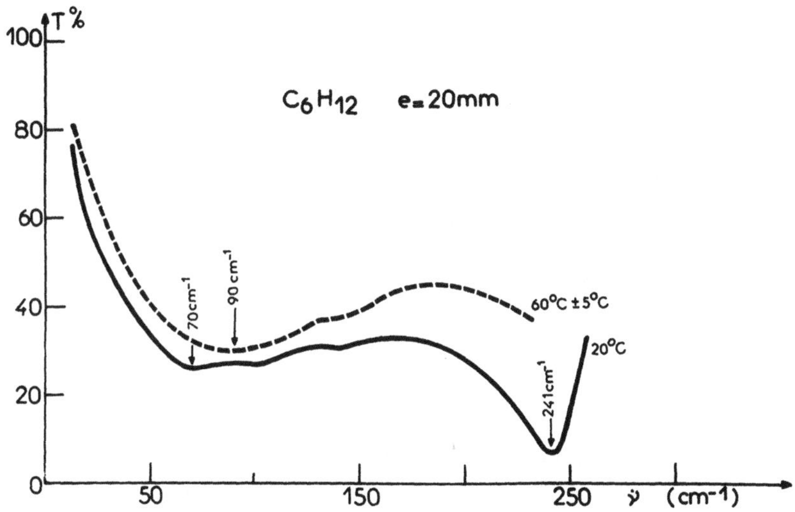

Figure 27. The far infrared transmission of cyclohexane increases when the temperature is raised.

II.2.2 The case of crystals with disordered charge distribution

A theory of dielectric losses to be expected in non-conducting, ionic crystals containing disordered charge distribution is described by Vinogradov, Schlössman, etc. In this theory the disorder in the distribution of mass and force constant which accompanies the disordered charge distribution is disregarded. Detailed results are derived for crystals with the inverse spinel structure. It is generally believed that in such crystals two- and three-valent ions are distributed over equivalent sites in a random fashion. It is shown that with a random distribution of charges, acoustic vibrations become optically active and lead to far infrared absorption:

$$K = \frac{2}{3} \frac{N_i}{c\sqrt{\varepsilon'}\,\rho c_T^3} \langle q^2 \rangle \cdot \omega^2.$$

where c_T is the transverse velocity of sound, c the speed of light, ρ the density, ε' the dielectric constant, N_i the number of randomly distributed charges, $\langle q^2 \rangle$ the average of the square of the deviations from the average charge. As an example we shall take a cube as a primitive unit cell with + and - sites (Figure 28). The

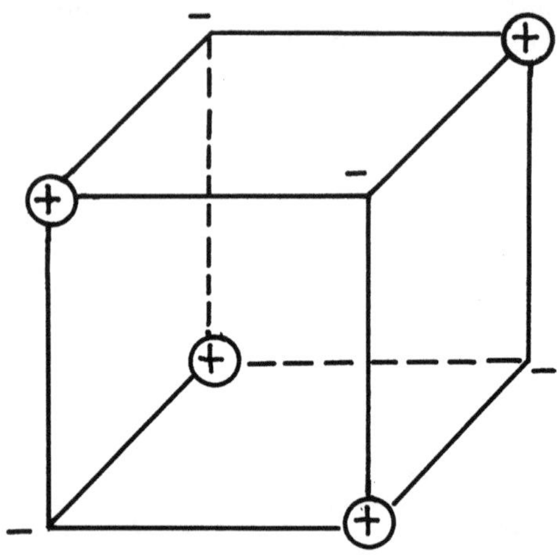

Figure 28. The charge defects in a spinel are indicated by + and -. They are randomly distributed at the corners of the cube but their sum is zero.

positive sites can be occupied either by ion 'a' with a charge $(1+\delta)e$, or by ion 'b' with a charge $(1-\delta)e$. However, the charge deviations are strongly correlated in such a way that charge neutrality is maintained within each unit cell of the crystal. Hence the unit cell will contain two 'a' ions, and two 'b' ions. Then $\langle q^2 \rangle = \frac{4\delta^2}{8} e^2$. If the charge deviations were substantially uncorrelated, K would be proportional to the frequency.

It is tempting to apply such a treatment to glasses where in fact K seems to follow a law in ω^2. It would be of great interest to study crystals with the inverse spinel structure.

Another approach has been made by Whalley and Bertie[19] for translational lattice modes in a class of crystals in which the molecules are arranged on regular positions in space but are irregularly oriented. The irregular orientation of the molecules frequently perturbs only slightly the translational lattice vibrations of the crystal. For a first approximation, then the crystal can be considered to have mechanically regular vibrations. The irregular orientation of the molecules will cause the dipole-moment change due to a displacement of a particular molecule to depend in magnitude and direction upon its own orientation and upon the orientation of its neighbors. The crystal is thus in a sense electrically irregular and mechanically regular for translational vibrations. The theory shows that the intensity of absorption or of Raman scattering by any particular normal vibration of the crystal that is active only because of the irregularity of the orientations is proportional to the square of its frequency. Then for long enough wavelengths both the density of states in the acoustical branches and the intensity of absorption are proportional to the square of frequency, so the total intensity should be proportional to the fourth power of the frequency. This is not the case for glasses in the far infrared and Schlössman's theory appears more realistic for this case.

II.2.3. The case of clathrates and solid inert matrix

In some cases, molecules trapped in a host lattice retain, to a large extent, their gas-phase character and enjoy a certain degree of rotational freedom. Two such systems are the clathrate inclusion compound and solid inert matrix. The molecules H_2O, HCl, CH_4 , NH, NH_2 either exhibit a large frequency interval between rotational lines or a low rotational barrier and will give nearly pure rotation lines either in Argon or in some clathrates.

For instance, HCl in β-Quinol clathrate at 1.2°K show one rotation line at 19 cm^{-1} (instead of 21 cm^{-1} for the free gas) and one translational line at 52 cm^{-1} . For higher concentrations the line is shifted to 28 cm^{-1} and its width increases up to 26 cm^{-1}. When temperature is raised up to 18°K, and for the higher concentrations, ν decreases down to 24 cm^{-1}, but the width increases up to 38 cm^{-1}. On the other hand, molecules like O_2, N_2 NO, CO could not rotate because of a high rotational barrier.

The clatrates are a good illustration of the cage model. But the molecules on the walls can be considered as fixed, which is not the case for liquids, and give an additional term in the expression of the potential energy.

III. CONCLUSIONS

The absorption of liquids in the very far infrared may be

explained by different schemes according to different cases. We have seen the possibility of nearly pure Debye absorption (solution of dichlorethane in carbon tetrachloride), the case of nearly pure rotation spectrum (solution of HCl in CCl_4), the case of liquid lattice absorption (CCl_4, CS_2, C_6H_6, etc.). In that case the absorption bands of the crystal lattice are broadened, mixed together and the integrated absorption seems higher by a factor of two for the liquid than for the crystal, probably because of some relaxation of the selection rules forbidding the excitation of acoustical lattice vibrations. The far infrared absorption of liquids is a question of specific cases and it was known for a long time from thermodynamic considerations (entropy of mixing), that some liquids are much ordered and others are less. For instance, liquid bromine shows some structure at short distance and only librations can be considered. On the other hand, free rotations have to be considered when bromine molecules are diluted in carbon tetrachloride.

Far infrared spectroscopy thus appears to be a very instructive tool to look at the degree of order of a liquid.

REFERENCES

1. S. S. Mitra in Optical Properties of Solids (S. Nudelman and S.S. Mitra, editors) Plenum Press, New York, 1969, p. 333-451.

2. A. Hadni, J. Claudel and P. Strimer, Phys. Stat. Sol. 26, 241 (1968).

3. A. Hadni, J. Claudel, E. Decamps, X. Gerbaux and P. Strimer, Comptes Rendus 255, 1595 (1962).

4. A. Hadni, J. Claudel, D. Chanal, P. Strimer and P. Vergnat, Phys. Rev. 163 836, (1967).

5. A. Hadni, P. Henry, J. P. Lambert, G. Morlot, P. Strimer and C. Chanal, J. Phys. 28, C1-118 (1967).

6. W. J. L. Buyers, Phys. Rev. 153, 92 (1967).

7. J. M. Munier and A. Hadni, J. Mol. Struc. 1, 249 (1967/68).

8. A. Hadni, J. Claudel and X. Gerbaux, in Lattice Dynamics (ed. R. F. Wallis), Pergamon Press, New York (1964) p. 381.

9. A. Hadni, J. Phys. 28, 978 (1967).

10. A. Hadni, D. Grandjean, F. Brehat, J. Claudel, X. Gerbaux, P. Strimer, R. Thomas, F. Vermillard and Sir R. Thomas, J. Phys. (in press) 1969.

11. C. Brot, B. Lassier, G. W. Chantry and H. A. Gebbie, Spectrochim. Acta 24A, 295 (1968).

12. E. Constant, Y. Leroy, J. L. Barois and P. Desplanques, Comptes Rendus (1966).

13. Y. Leroy and E. Constant, Comptes Rendus (1966).

14. P. Marteau, J. Bonnemain and H. Vu, Comptes Rendus, 260, 3583 (1965).

15. D. A. Draegert, N.W.B. Stone, B. Curnutte and D. Williams, J. Opt. Soc. Amer. 56, 64 (1966).

16. G. E. Walrafen, J. Chem. Phys. 40, 3249 (1964).

17. A. Hadni, G. Morlot and F. Brehat, Spectrochim. Acta 24A, 1167 (1968).

18. G. W. Chantry, H. A. Gebbie, M. Lassier and G. Wyllie, to be published.

19. E. Whalley and J. E. Bertie, J. Chem. Phys. 46, 1264 (1967).

AUTHOR INDEX

SUBJECT INDEX

MIX
Papier aus verantwortungsvollen Quellen
Paper from responsible sources
FSC® C105338

If you have any concerns about our products,
you can contact us on
ProductSafety@springernature.com

In case Publisher is established outside the EU,
the EU authorized representative is:
Springer Nature Customer Service Center GmbH
Europaplatz 3, 69115 Heidelberg, Germany

Printed by Libri Plureos GmbH
in Hamburg, Germany